中英双语版

（英）罗伯特·亚当　著
（英）德里克·布伦特纳尔　绘图
刘艳红　王文婷　徐培文　梁甜甜　王福刚　译

Written by Robert Adam
Illustrations by Derek Brentnall

古典建筑完全手册

CLASSICAL ARCHITECTURE

A Complete Handbook

辽宁科学技术出版社
·沈阳·

图书在版编目（CIP）数据

古典建筑完全手册 ：中英双语版 /（英）罗伯特·亚当（Robert Adam）著；刘艳红等译. —沈阳：辽宁科学技术出版社，2017.8（2018.10重印）
ISBN 978-7-5591-0282-9

Ⅰ.①古… Ⅱ.①罗… ②刘… Ⅲ.①古典建筑—手册—汉、英 Ⅳ.①TU-091

中国版本图书馆CIP数据核字（2017）第130735号

出版发行：辽宁科学技术出版社
（地址：沈阳市和平区十一纬路25号 邮编：110003）
印 刷 者：辽宁鼎籍数码科技有限公司
经 销 者：各地新华书店
幅面尺寸：240mm×300mm
印　　张：41
插　　页：4
字　　数：480 千字
出版时间：2017 年 8 月第 1 版
印刷时间：2018 年 10 月第 2 次印刷
专业审读：田英莹
特约编辑：苏珊·威尔逊（Susan Wilson） 雪 晴
责任编辑：闻 通
封面设计：周 周
责任校对：李 霞

书　　号：ISBN 978-7-5591-0282-9
定　　价：158.00元

联系编辑：024-23284740
邮购热线：024-23284502
投稿信箱：605807453@qq.com
http://www.lnkj.com.cn

For
Sarah, Jamie, and Charlotte

献给
莎拉、杰米以及夏洛特

介　绍

罗伯特·亚当（中文名“叶栋”）是英国的一位建筑师、城市设计师、学者和作家。他创建了ADAM（亚当）建筑事务所，这是世界上专长于古典建筑的最大的建筑师事务所。他和另外五位建筑师董事一起领导着这个事务所的工作，他们是：纳吉·安德森、保罗·汉威、休·派特、乔治·苏马兹·史密斯和罗伯特·卡。这些董事在各个年龄层次，并且为了保持公司的活力，将会有更加年轻的建筑师加盟。

ADAM（亚当）建筑事务所以设计大型别墅、商业建筑、公共建筑、小型住宅开发以及新城和住区规划而闻名。事务所董事将他们传统和古典建筑设计的专长运用于所有的项目。他们的城市规划用先驱方式分析地方脉络，并将结果应用于新建区域设计。

一座6000平方米的别墅（*a*），位于英格兰南部的一个私家公园中。它由法国进口石材建造，按照英国乡村别墅的传统设计。丰富的石材细部均由计算机控制切割。平面布局将建筑的一侧朝向花园景观，在另一侧建造了入口庭院。大型的古典柱式有着特殊设计的柱头，其设计受到意大利16世纪伟大建筑师安德烈·帕拉迪奥（1508—1580年）的启示。小型的古典柱式设计注重人体尺度。入口门廊留有一个大型的浮雕创作。

一座新建办公楼（*b*），地处伦敦皮卡迪利——伦敦西区最主要的历史大道之一。该建筑采用多种传统古典形式，力求呼应每个立面所对应的不同城市文脉，并充分利用其独特的位置条件。面向主街的立面非常壮观，运用了不同尺度的古典柱式，柱头出自英国著名的传统雕塑家之手。在建筑转向次路的转角，使用简化过的小柱子，以产生较为谦逊的感觉。

在英格兰南部的三幢新建的公寓楼项目（*c*），重现了20世纪初源于美国、后来几十年中成为一种全球风格的古典高层建筑。这种风格的原则是将塔楼比作柱式，底部柱础的细节与街道一致，上有简单的垂直线条模拟柱身，顶部楼层富有表现力，轮廓远观之下就像柱头。

位于英格兰西部的一个新建村庄（*d*），遵循了典型英国村庄的不规则街道格局，并包含了一系列大小不同、设计各异的传统石屋。这些元素将此地打造得独特而迷人，曲折狭窄的街道、一幢大型公共建筑和地方商店，无不反映出当地的风格特色。

INTRODUCTION

Robert Adam is a British architect, urban designer, academic, and author. He founded ADAM Architecture, the largest architectural firm in the world specialising in Classical architecture, and works with five other directors who are also leading practitioners: Nigel Anderson, Paul Hanvey, Hugh Petter, George Saumarez Smith, and Robert Kerr. The directors are of varying ages and, to maintain the vitality of the firm, younger practitioners will join the practice.

ADAM Architecture is well-known for the design of large houses, commercial architecture, public buildings, developments of smaller houses, and the planning of new towns and neighbourhoods. The directors bring their specialist knowledge of traditional and Classical design to all their projects. Their urban planning has pioneered the analysis of local context and its application to the design of new areas.

A 6,000 square metre house (*a*) in southern England is set in a private park. It is built from imported French stone and designed in the English country-house tradition. The rich Classical stone detailing has been cut by computer. The plan gives garden views on one side and creates an entrance court on the other. Large Classical columns have specially designed capitals inspired by the great Italian sixteenth-century architect Andrea Palladio (1508-80). Smaller Classical columns give a human scale. The entrance portico includes a large sculptural composition.

A new office building (*b*) in London is located on Piccadilly, one of the most important historic thoroughfares in the West End of the city. The building uses traditional Classical forms in varied ways to relate to the different urban context on each façade and to take advantage of its unique location. Using Classical columns of different scales, the façade to the main street is imposing, with column capitals sculpted by one of Britain's leading traditional sculptors. As the building turns the corner into minor streets, the smaller columns are simplified to create a more modest appearance.

A project for three new residential towers (*c*) in a southern English city revives the tradition of the Classical skyscraper that originated in the USA in the early twentieth century and became a global style in the following decades. This tradition established the principle of a tower designed as a column, with a base to give detail to the street, a simple vertical arrangement above acting as a column shaft, and expressive upper floors as a column capital seen from a distance and in silhouette.

A new village (*d*) in the west of England follows the informal street pattern of typical English villages and incorporates a series of traditional stone houses of different sizes and different designs. Together they create a distinctive and attractive place with twisting narrow streets, a major public building, and local shop that together reflect the character of the area.

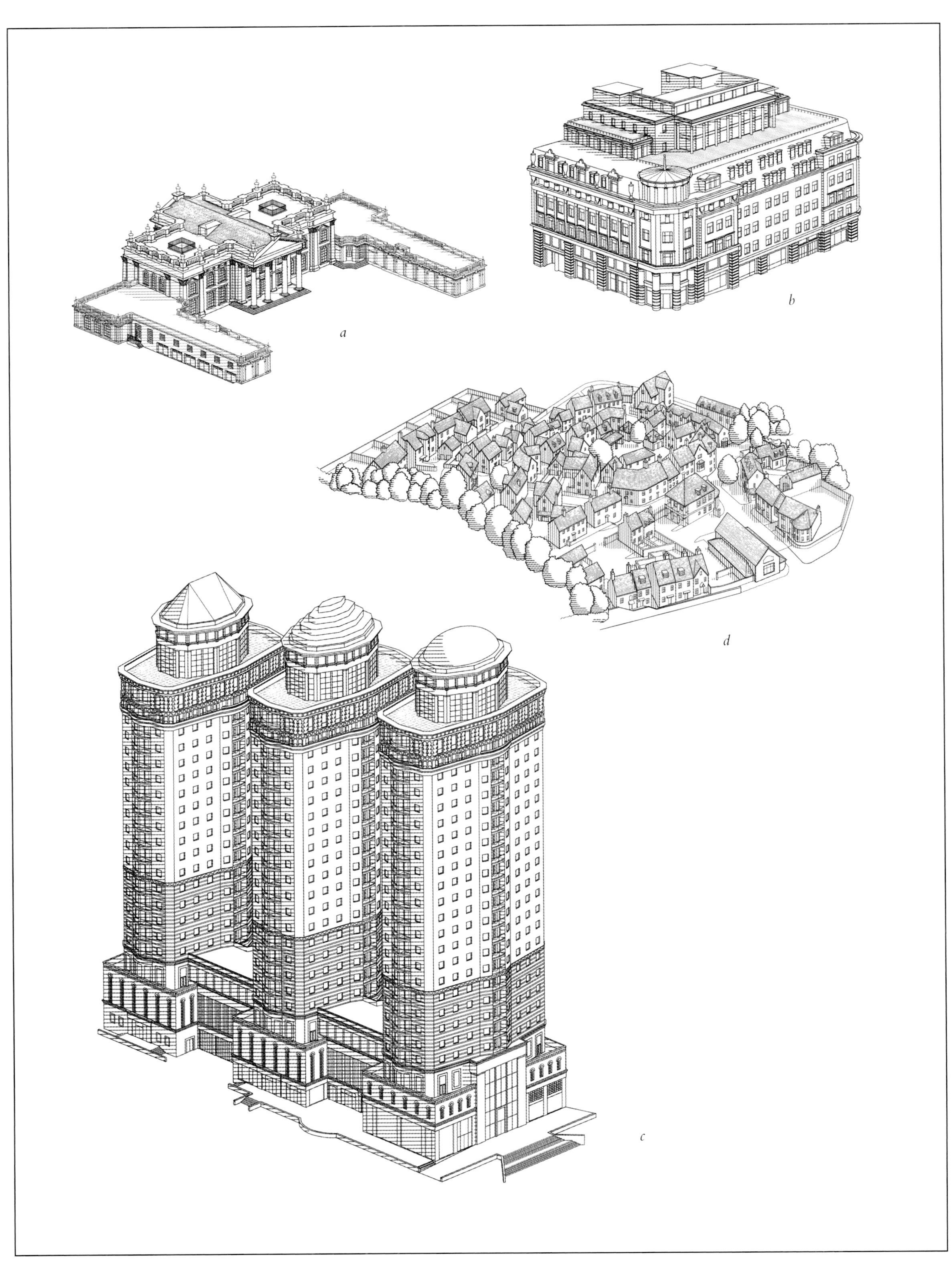
a
b
d
c

致　谢

我必须感谢客户、学生以及使得古典建筑保持活力的同行专家，更重要的是，比起二十五年前本书首次出版时，他们使得本书与当代建筑更加息息相关。我特别感谢我在ADAM（亚当）建筑事务所的董事们，作为同事与推动古典建筑先驱的纳吉·安德森、保罗·汉威、休·派特、乔治·苏马兹·史密斯和罗伯特·卡。我还必须感谢辽宁科学技术出版社将本书介绍给中国读者，苏珊·威尔逊完成修订版的编辑工作，杰里米·德雷克为本书增加图片，田英莹为本书翻译提供意见，以及雪晴对编辑中文翻译提供的帮助。

罗伯特·亚当（叶栋）2017
温彻斯特，英格兰，英国

ACKNOWLEDGEMENTS

I must thank the clients, students, and fellow professionals who have kept Classical architecture alive and, if anything, have made this book more relevant to contemporary architecture than it was when the book was first published more than twenty-five years ago. I am particularly grateful to my fellow directors at ADAM Architecture, Nigel Anderson, Paul Hanvey, Hugh Petter, George Saumarez Smith, and Robert Kerr, as both colleagues and leaders in the advancement of Classical design. I must also thank the Liaoning Science & Technology Publishing House for offering this work to Chinese readers, Susan Wilson who has led the editing of the revised edition, Jeremy Drake for additional graphics, Yingying Tian for her advice on the translation, and Qing Xue for her assistance with editing the Chinese translation.

Robert Adam (Ye Dong) 2017
Winchester, England, United Kingdom

目录 CONTENTS

本书的使用

古典建筑是我们日常生活中很熟悉的一部分。两千年以来的古典建筑传统已经成为西方文明的必要组成部分，这一漫长的联系使我们对建筑语汇有了更深刻的了解。

正如文学作品一样，经典的设计可以采用多种形式并且以无数不同的手法呈现出来。像一部伟大的小说，一座宏伟的古典建筑给每个普通人传递的是简单的信息，然而给有学识的观察家提供的则是一层层的深刻内涵。本书将引导读者进入我们居住艺术领域的创造层面。

古典设计从它的历史中汲取力量和变化，这一点与语言是相通的。从历史当中衍生出的神话和符号，给最简单的设计增添色彩。每一个建造手法和微小的细部设计都蕴含着自己的故事。本书将着眼这部分历史，介绍所有的元素，追溯几个世纪以来它们的演变过程并为其应用提供实际的建议。

本书将首先审视古典建筑伟大传统的丰富性和多样性与其历史内涵和使用之间的关系。再进一步描述隐藏在所有古典建筑背后不同的正规设计体系，即柱式。然后逐一解读建筑的组成部分和设计风格，生成古典建筑所有组成部分的全景图。

本书的每一页都是一篇完整的文章，其相对的页面上则是图示，通常是按照比例绘制的，其目的是为文章做图解。这种版式设计使读者既能够逐章学习，又能够挑选感兴趣的页面和章节来阅读。

由于是专门学科，古典建筑包括大量术语。将本书划分为一系列相关联的章节，意味着读者可能不知道或没有读到过某些晦涩却必不可少的术语。当这些名词术语也没有在同一页进行说明时，在术语表中会有简短的解释。目录中指明该条目在哪一页进行更深入的讨论。

读者很快会意识到古典建筑的研究是如此庞大与复杂。本书只能为其作简要介绍，其中好多页面内容可以成为几本书的题材，因此选材是十分严格的。在选择上力求客观，对于非重点内容进行必要的删减。所选实例体现古典建筑的多样性、深度和活力，来反映随着其所服务的社会传统而发生的演变。

USING THE BOOK

Classical architecture is a familiar part of our everyday lives. For two thousand years the tradition of Classical building has been an essential part of western civilisation and this long association has given us all a deep-rooted understanding of its language.

Like literature, a Classical design can take many forms and can be expressed in countless different ways. Like a great novel, a great Classical building will have a simple message for everyone while offering the knowledgeable spectator layer upon layer of more profound meaning. This handbook will help to guide the reader into the creative depths of the works of art that we inhabit.

In common with the spoken language, Classical design draws its strength and variety from its history. From this history have grown myths and symbols which colour even the simplest designs. Methods of construction and small details each have their own story to tell. *Classical Architecture* will look at this history and introduce all of these elements, tracing their evolution through the centuries and offering practical advice for their use.

The richness and diversity of the great heritage of Classical buildings are first examined in relation to their historical context and use. The book goes on to describe the different formal systems of design, or Orders, which lie behind all Classical architecture. Parts of buildings and types of decoration are then examined one by one to create a comprehensive picture of the Classical building in all its aspects.

Each page of text is a self-contained essay and is placed opposite a page of drawings, usually to scale, which have been prepared specially to illustrate the text. This format allows the reader to study the book chapter by chapter or to select pages or sections of specific interest.

As it is a specialised subject, Classical architecture has a large number of technical words. The division of the book into a series of linked essays means that the reader may not know or have read previously the definition of some of the more obscure but unavoidable expressions. Where these are not defined on the same page a short definition can be found in the glossary. The index gives directions to pages where the subject is discussed in more depth.

The reader will soon realise just how vast and complex is the study of Classical architecture. This handbook can only be an introduction. When many of these pages can be the subject of several books, the choice of material has to be highly selective. The selection is intended to be objective and there is no significance in the inevitable list of omissions. Examples have been chosen to show the variety, depth, and dynamism of Classical architecture and to trace the evolution of the tradition as it changes with the society it serves.

1. 古典传承

起源

从远古时代人类学会耕种土地和蓄养牲畜起，一些部族便定居在地中海东部沿岸土地肥沃的地区。

这些史前人类的生活和传说成为西方文明的基石。在未开化的古老村落里，他们为神修建与自己同样的原始民居建筑，这些房屋与庙宇就是古典建筑的起源。

当这些部族走出无知的晦暗，便将祖先的神圣传统精心地保存下来，并且记录了已经很古老的口头流传下来的神和英雄的传说，这就是西方文学的源头。他们的房屋和庙宇体现逐渐复杂的生活方式，他们将祖先简单的建筑构造编织在新的建筑设计与装饰中。

埃及的石头神庙和坟墓在形状、雕刻和绘画装饰上常常模仿尼罗河下游早期农民用芦苇建造的房屋。时至今日，在美索不达米亚南部河畔的乡村仍然能找到相似的民居（*b*）。

公元前9世纪，希腊科林斯附近的陶土模制建筑是献祭给女神赫拉的。图例中复原的建筑物（*a*）可能是这类神庙的模板。公元前7世纪，拉丁部族定居于现在罗马所在的山丘上，将死者的骨灰安置在黏土制的小型棚屋中。图（*c*）是基于这些容身之处的遗迹的复原图。

在这片古典建筑诞生的土地上，这些用乡间天然材料建造的房屋中，仍然居住着迁徙的牧民。至今，这些建筑仍然是世界上贫穷地区人们的固定住所。在古代，不仅普通的乡间民居能够时刻提醒古典石头神庙的存在，而且到了1世纪，在雅典和罗马，人们还虔诚地保存着茅草屋顶的神庙。

在古典建筑漫长的历史中，这些遗留下来的传统逐渐被湮没，有时似乎已被遗忘，有时浮出表面并得以短暂复兴，但从未彻底消失过。每一种古典建筑都是这些原始棚屋的后裔，使用古典手法是现代与西方文明起源之间鲜活的纽带。

1. CLASSICAL HERITAGE

ORIGINS

From the earliest days, when men learnt to till the soil and herd their animals, tribes settled the fertile regions around the eastern perimeter of the Mediterranean Sea.

The life and legends of these prehistoric peoples became the foundations of western civilisation. In rough villages they housed their gods in the same primitive dwellings they built for themselves, and in these houses and temples lie the roots of Classical architecture.

As these tribes emerged from illiterate obscurity they jealously preserved their sacred ancestral traditions and recorded their already ancient spoken tales of gods and heroes to create the fountainhead of western literature. As their houses and temples grew to reflect the increasing sophistication of their lives, memories of the simple structures of their forebears were woven into the design and decoration of the new buildings.

The stone temples and tombs of Egypt often imitate in their shape, carvings, and painted decorations the reed houses of the first farmers of the Lower Nile. Even today, similar dwellings (*b*) can be found in riverside villages in southern Mesopotamia.

In the nineth century BC, pottery models of buildings were offered to the goddess Hera, near Corinth in Greece. The reconstruction illustrated (*a*) is probably a model of such a temple. In the seventh century BC, the Latin tribes that settled the hills that were to become Rome buried the ashes of their dead in miniature huts of clay and illustration (*c*) is a reconstruction of a hut based on the remains of these vessels.

Made from the natural materials of the countryside, buildings such as these still house migrant shepherds in the lands where Classical architecture was born. To this day they can serve as permanent homes in the poorer regions of the world. In antiquity not only would everyday rural dwellings have served as a constant reminder of the ancestry of stone temples, but up until the first century AD, ancient thatched temples were reverently preserved in both Athens and Rome.

Throughout the long history of Classical architecture the influence of these inherited origins has lain below the surface, sometimes half-forgotten, sometimes emerging to the surface to revitalize the strain, but never totally absent. Every Classical building is a descendant of these primitive huts and the use of Classical architecture is a living bond with the cultural origins of western civilization.

a
b
c

埃及

西方纪念碑式建筑的基础源于埃及。当希腊首次出现文明的迹象时，埃及建筑师已经有一千五百年修建大型建筑的历史了。

尼罗河带来极其肥沃的泥土，这一自然资源创造了富庶的文明，由被奉为神明的国王即法老统治。繁复的宗教信仰需要消耗大量劳动力，用以修建大型建筑物，献祭给半兽身的神和法老的来世。

执着地追求永生并创造永恒的思想主宰了古埃及的建筑理念。皇家陵墓成为膜拜已逝法老的神庙。到公元前2800年，开始修建巨大的、人造的、石料砌面的山丘，即金字塔，并建造在墓室上，用以安放精心涂抹防腐香料的法老遗体和他所喜爱的世间财物。金字塔是最早的西方世界用切割石料修建的纪念碑。

在古代，金字塔随处可见，但是由于其作为坟墓的单一功能和它们巨大的体量，对其北部出现的希腊文明并无特殊影响。另一方面，巨大的封闭石头神庙有时是金字塔建筑群的附属品，有时独立存在，对希腊的商业和贸易产生了更大的影响。

卡纳克的月亮神庙（*a*）修建于公元前1198年，虽然是个比较小的例子（附近的阿蒙神庙是它的6倍），但是比较典型。它矗立在一个由围墙包围（图中没有显示）的大面积区域内，其中包括服务楼和圣水湖。神庙向内朝向供奉有神舟室的圣所，通过位于礼仪轴线的引道，经过两座塔楼之间的大门、一个开放式庭院和列柱厅可以到达圣所。圣所周围环绕着长廊通往礼拜堂。

我们应该看到，埃及建筑的单个特征对古典风格的发展产生的重大影响，以及细部设计是如何被罗马建筑采纳的。然而，虽然埃及的建筑风格维持了一千五百年而未发生改变，在境外却没有留下埃及风格的古迹。它好像对于周边贫困地区的文化来说过于特殊与保守。公元前1300年源自古希腊克里特文明和迈锡尼文明的阿特柔斯宝库（*b*）和继希腊民族之后六百年的克里特民族在普林尼亚斯修建神庙（*c*），显示了埃及发达富庶的文明与其北部邻国的落后文明之间的差异。我们应当看到，比起重要建筑物上精美的石料装饰，埃及建筑对主流古典主义最重大的贡献在于神庙中梁柱式结构的使用，以及对修建宏伟壮观建筑的热爱。

EGYPT

The foundations of western monumental architecture were laid in Egypt. By the time the first signs of civilisation began to emerge in Greece, Egyptian architects had been building massive structures for fifteen hundred years.

Soil of unusual richness was carried down the River Nile and the exploitation of this resource produced a wealthy civilisation ruled by god-kings, or pharaohs. A complex religion consumed the surplus labour force in the erection of huge buildings dedicated to half-animal gods and the afterlife of the pharaohs.

A preoccupation with eternity and the creation of permanence dominated the architecture of ancient Egypt. Royal tombs were temples for the worship of the deceased pharaoh. By about 2800 BC huge, artificial, stone-faced mountains, or pyramids, were built over rooms containing the elaborately embalmed body of the pharaoh and his valued worldly possessions. The pyramids were the first major monuments in the western world to be constructed with cut stone.

The pyramids were famed throughout the ancient world, but due to their solely sepulchral function and huge size they had no practical influence on the emerging Greek civilisation to the north. The great enclosed stone temples on the other hand, that sometimes accompanied and sometimes stood apart from the pyramid complexes, must have made a greater impression on Greek traders and mercenaries.

The Temple of Khons at Karnak (*a*) was built in 1198 BC and is quite a small example (the nearby Temple of Amon is six times the size), but typical. It stood inside a large walled enclosure (not shown) which would have contained service buildings and a sacred lake. The temple faces inwards to the sanctuary containing the sacred barge, which is approached on an axial processional route by way of a large door between two tall pylons, an open court, and a colonnaded room, or hypostyle hall. The sanctuary is surrounded by a corridor giving access to chapels.

We shall see how individual features of Egyptian architecture had a major influence on the development of the Classical style, and how details came to be incorporated in Roman architecture. Yet, although it remained virtually unchanged for fifteen hundred years, no Egyptian style became established outside Egypt in antiquity. It was, it seems, too specific and too conservative for adaptation to the needs of the poorer surrounding cultures. The Treasury of Atreus (*b*) of about 1300 BC, from the Greek Minoan and Mycenaean civilisation, and the Temple at Prinias (*c*), erected in Crete by the emerging Greek peoples six hundred years later, illustrate the contrast between the advancement and wealth of Egypt and the comparative backwardness of its northern neighbours. We must look rather to the adoption of finely dressed stone for buildings of importance, the use of the column and beam system (trabeation) for temples, and an appetite for large and impressive structures as the most significant contributions of Egyptian architecture to the mainstream of Classicism.

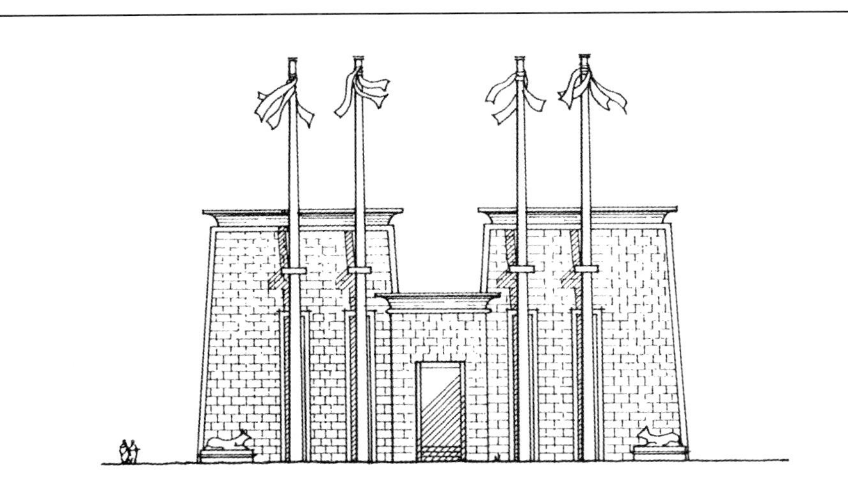

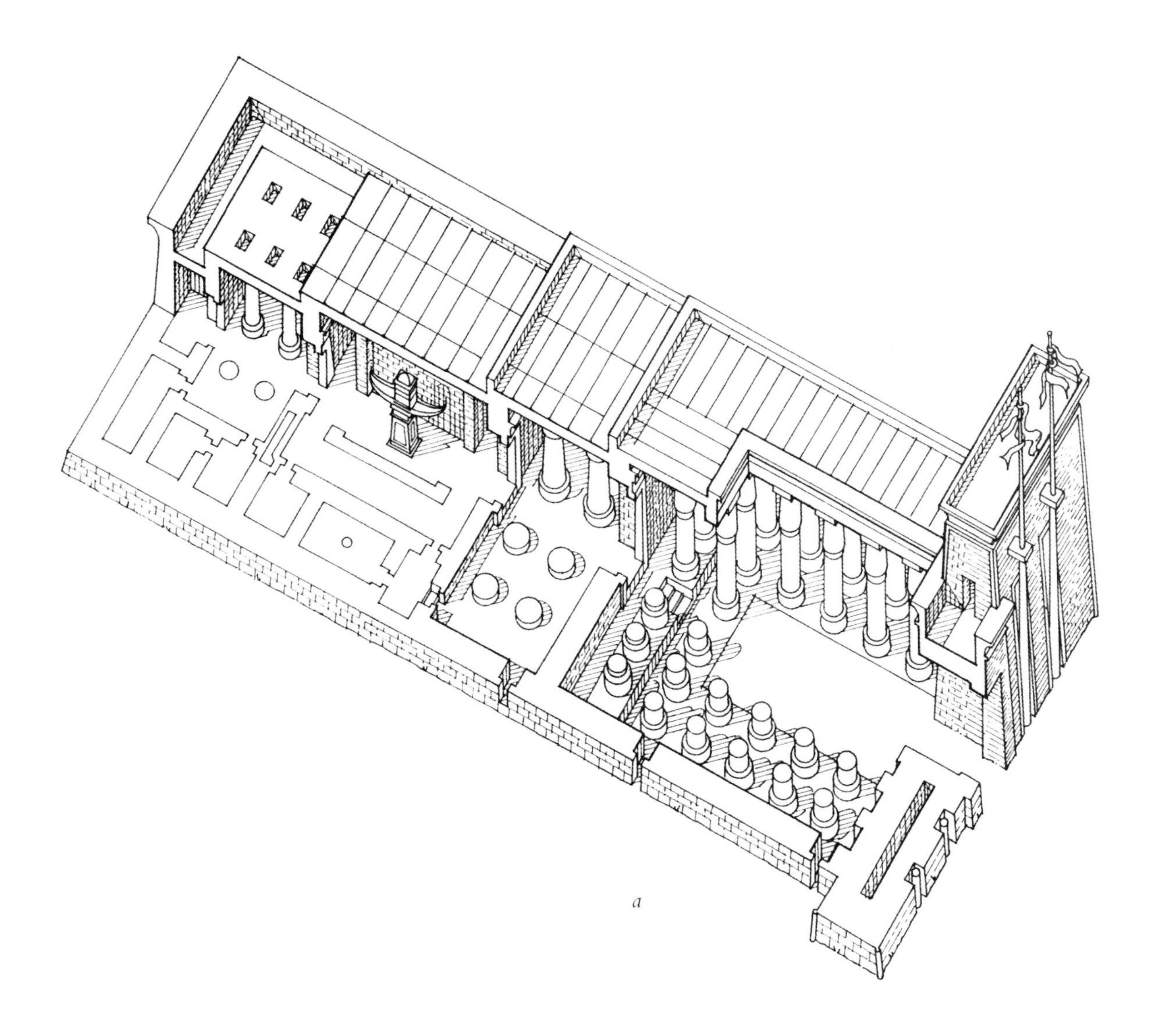

a

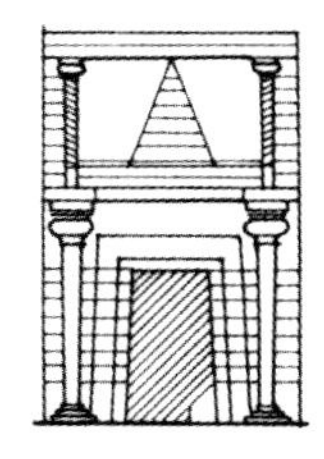

b

m 10 *ft* 30

c

希腊神庙

克里特文明和迈锡尼文明在公元前1100年被毁，希腊文明也在随后的四个世纪中进入到灰暗的时代。到公元前800年，希腊分裂，城邦开始在希腊、爱琴海诸岛和土耳其西部沿岸出现。随后的两个世纪中，这些城邦轮流对西西里、意大利南部、法国南部和黑海地区进行殖民统治。

希腊城邦因其占据地中海发达的通商路线而获利，并且直到公元前700年在文化上是统一的，在政治上则不然。这些分散的部族创造出一种文明，将自由思想与内部斗争结合起来，带来了知识与艺术的革新。

希腊文明的兴起将西方世界带入到政治、知识和艺术的洪流中，并将其推动了两千五百年。然而，当希腊走出灰暗时代后，理性思想的到来并没有驱除野蛮的神，为他们修建的庙宇成为西方建筑的试金石。

以泥砖和茅草屋顶建筑为基础，如P3图中所示，已经演变出一种新的神庙设计。最初的神庙有着泥质或茅草材质的平屋顶，但是到了公元前8世纪，发明了大片的黏土屋瓦，使得带有山墙，又称三角墙和平缓屋顶的简单长方形平面成为现实，山墙体现了古典建筑特征。在更加重要的神庙中使用环绕在神庙四周的开放式柱廊来支撑屋顶。

在公元前630年修建的希腊色蒙的阿波罗神庙（*a*）中可以找到这些新的特征。该神庙采用木质支柱和屋顶。这一基本模式在随后几个世纪的重要神庙建筑中都有保留，但不到四十年，建造者已经将暴露的木质特征元素改为石头，例如科孚岛的阿耳忒弥斯神庙（*b*）。

其后的一百五十年主要是细化与扩建。尝试细微的改动，随着把握的不断增大，开始采用更加修长的比例。但正如公元前460年在意大利帕埃斯图姆修建的赫拉神庙（*c*）所展示的，主要的神庙设计从未脱离基本形制。 封闭的圣所中供奉神明；如果需要的话，内部数排柱子用以帮助支撑屋顶；通常由门廊进入，后部也是如此，整个建筑物被柱子包围。当建筑物过于庞大，柱子不能支撑其屋顶时，如西西里的阿格里真托的奥林匹亚姆（*d*），需要修建承重墙并在其上装饰仿造的柱子。

古希腊的建筑师花费两个多世纪来完善这一种建筑形制，创造出的建筑如此不朽，以至于它作为世界通用的典范，存在了两千多年。

THE GREEK TEMPLE

The Minoan and Mycenaean civilisations were destroyed in 1100 BC and Greece descended into four centuries of obscurity. By 800 BC the Greek peoples had dispersed and city-states were created in Greece, the Aegean islands, and on the west coast of Turkey. In the next two centuries these cities in turn colonised Sicily, southern Italy, the south of France, and the Black Sea.

The Greek states came to profit from their position on the rich trade routes of the Mediterranean and by about 700 BC were unified in culture but not in government. These diverse communities were to create a civilisation where freedom of thought combined with internal rivalry to bring about an intellectual and artistic revolution.

The rise of Greek civilisation launched the western world into a political, intellectual, and artistic current which has carried it forward for two and a half thousand years. The dawn of rational thought did not, however, dispel the savage gods that followed the Greeks out of their dark past, and the houses that were built for them became the touchstone of western architecture.

From the mud-brick and thatched buildings illustrated on page 3 a new temple design evolved. The first temples had flat mud roofs or thatch, but in the eighth century BC, the invention of large clay roof-tiles made it more practical to build a simple rectangular plan with the shallow roof and gable-end, or pediment, which characterise Classical buildings. On more important temples this roof came to be carried on an open colonnade which surrounded the temple.

All of these new features can be found in the Temple of Apollo at Thermon in Greece (*a*), built in about 630 BC. This temple was constructed with wooden columns and a timber roof. Its basic form remained the pattern for all major temples for centuries to come, but within forty years the builders of temples such as that of Artemis in Corfu (*b*) had translated all the exposed timber features into stone.

There followed one hundred and fifty years of refinement and enlargement. Small changes were tried out and, as confidence increased, more slender proportions were used, but major temples never strayed far from the basic type illustrated by the Temple of Hera at Paestum in Italy (*c*) of 460 BC. An enclosed sanctuary housed the image of the god; if necessary, rows of inner columns helped to support the roof; it was entered through a porch, which was often repeated on the back, and the whole structure was surrounded by columns. When a building was of such a huge size that the columns could not support the roof, such as the Olympieum at Agrigento in Sicily (*d*), an imitation colonnade was added as decoration to the supporting walls.

Ancient Greek architects spent more than two centuries trying to perfect just one building type and created an architecture of such enduring excellence that it has survived for more than two thousand years as a universal ideal.

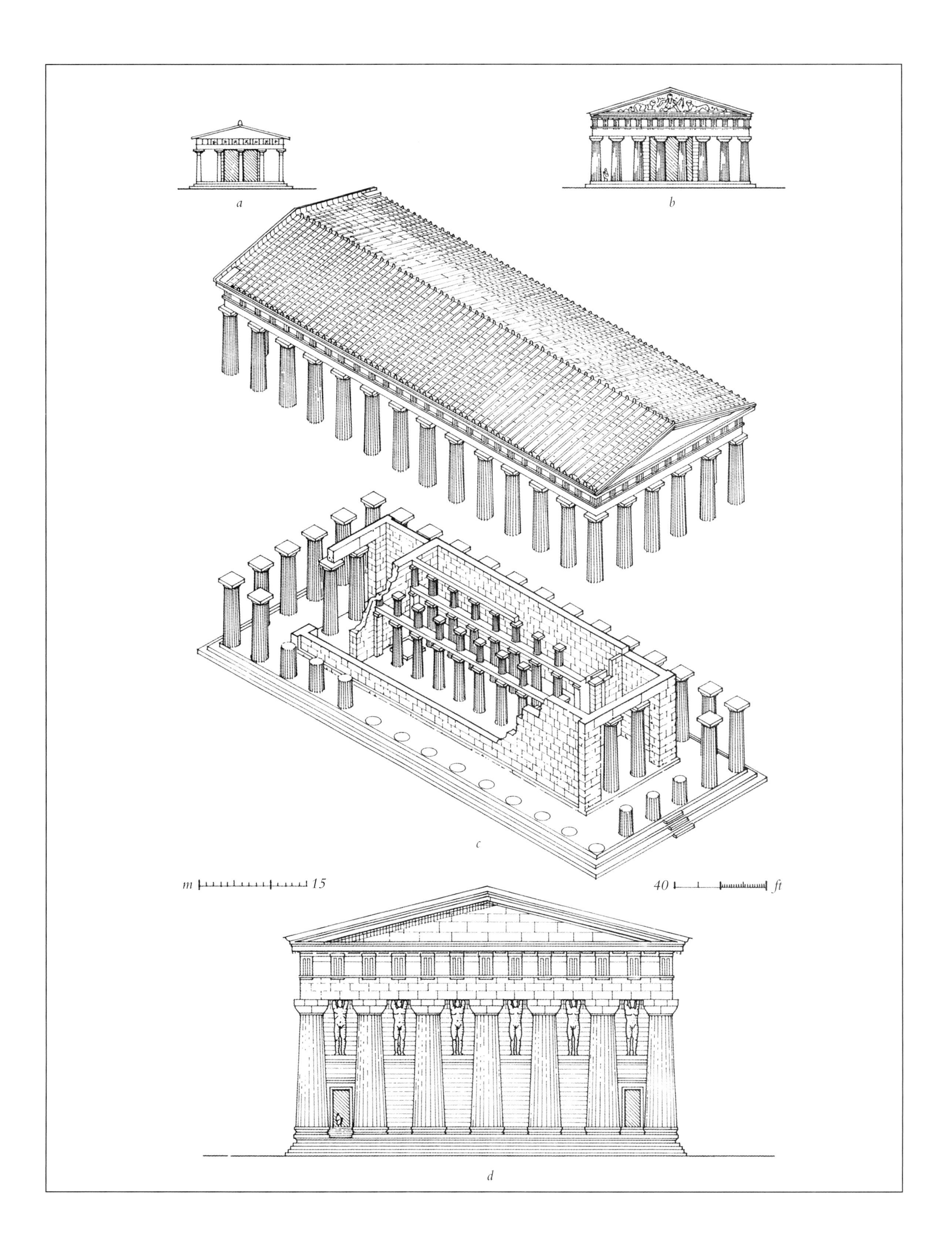
a
b
c
m 15
40 ft
d

希腊完美主义

希腊人于公元前5世纪中叶推翻波斯帝国后，他们独特文明的智慧积淀得以实现，拥有了一个世纪的财富与权力。文学、艺术和哲学繁荣发展，新的建筑极其精美，其细部设计成为范本，对后世西方建筑的发展产生影响。

精湛的领导力和强大的海上实力使得雅典人能够带领盟军打败波斯人，随后进入和平时代。由伯里克利（公元前495年—公元前429年）——雅典蛊惑民心的政客，所统帅的海军联盟调拨经费在雅典修建政权基地，制订了公共建筑的修订计划，创造出了前所未有的、高级复杂的建筑形式。

这些建筑中最伟大的就是新的雅典娜神庙——帕提农神庙（*a*），通常被认为是传统希腊神庙和随其演变的多立克柱式建筑体系的顶峰之作。这座新的神庙，修建于公元前438年，用雕像装饰得富丽堂皇，出自两位建筑师——卡里克拉特（逝于公元前420年）与伊克蒂诺斯（活跃于大约公元前5世纪）之手。

它的设计融合了一些视觉手法，通常是肉眼无法分辨的，但可能是为了校正长方形设计造成的视错觉效果。这些手法包括柱身逐渐收窄，阶梯四周向中间轻微拱起，地面也同样拱起，柱身稍向内倾斜，柱体间距变化，角柱略粗以及建筑物上部向外倾斜。

17年后，在附近兴建了厄里希翁神殿。这座建筑有着不同寻常的复杂平面，可能是因为建造在不平整场地上和有着多种用途功能。神庙有3个门廊，其中一个如图所示（*b*），其余两个有不同的装饰体系即柱式。这种柱式称为爱奥尼亚式，以其柱身上柱头的螺旋装饰细部，即卷涡装饰，以及与多立克柱式相比更加优美的比例著称，细部设计更加丰富。

公元前334年，雅典人吕西克拉特为纪念其唱诗班在比赛中获得胜利而修建了一座纪念碑。这座精巧的建筑物（*c*）是现存最早将另一种建筑柱式应用于外部的实例，这种柱式发明于80年前，应用于多立克柱式神庙内部。与爱奥尼亚柱式相似，这种柱式就是科林斯柱式。其柱身特点是更加修长，将柱头卷涡装饰替换成精心雕琢的长形叶片纹案。

当希腊城邦享受短暂的财富高峰时，他们为后代创造了我们今天所说的古典建筑。三种主要柱式得以确定，尽管在未来的一千年里发生了巨大的改变，古典建筑仍从这一纯净的源头不断汲取精神力量。

THE PERFECTION OF GREECE

When the Greek peoples turned back the Persian Empire in the middle of the fifth century BC, their unique civilisation realised its accumulated wisdom in a century of wealth and power. Literature, art, and philosophy flourished, and new buildings displayed such a level of refinement that their details became prototypes that would influence all future development in western architecture.

Skilful leadership and naval power gave the city of Athens dominance over the alliance that defeated the Persians, and the peace that followed. The appropriation of allied naval subscriptions by Pericles (495–429 BC), the Athenian demagogue, to cultivate a popular power base in Athens generated a public building programme that produced architecture of unprecedented sophistication.

The greatest of these buildings was the new temple to the goddess Athena, the Parthenon (*a*), often considered to be the peak of the development of the traditional Greek temple and of the architectural system, the Doric Order, that had evolved with it. The new temple, dedicated in 438 BC, was richly decorated with sculpture and was the work of two architects, Callicrates (*d*.420 BC) and Ictinus (*fl*.C5 BC).

The design incorporates several optical devices, often not in themselves visible to the human eye, but probably intended to correct undesirable visual effects created by rectangular forms. These devices include a subtle tapering of the column, a minute rise to the centre in the surrounding steps, an equally gentle rise in the floor, an inward slant to the columns, variations in the column spacing, an increased thickness of the corner columns, and an outward slant in the upper part of the building.

Seventeen years later, work began on a nearby temple called the Erechtheion. This building had an unusually complicated plan, perhaps suggested by the uneven ground on which it was constructed and its multiple dedications. It had three different porches, one of which is illustrated (*b*), and two of them have a quite different decorative system, or Order. This Order, the Ionic, is most notable for the spiral details, or volutes, on the capitals over the columns, for its more elegant proportions, compared with the Doric, and for its richer details.

In 334 BC an Athenian called Lysicrates won a prize for his choir and decided to build a monument to celebrate his victory. This exquisite little building (*c*) is the earliest surviving exterior use of another architectural Order which had been invented for the interiors of Doric temples about eighty years before. Similar to the Ionic, this is the Corinthian Order. It is characterised by columns that are made more slender by replacing the spiral volutes with a tall decoration of finely carved leaves.

While the Greek city-states enjoyed the brief peak of their fortunes, they created for posterity the architecture we call Classical. The three principal Orders were established and, although great changes would be made in the following millennia, Classical architecture would always draw its spiritual energy from this pure spring.

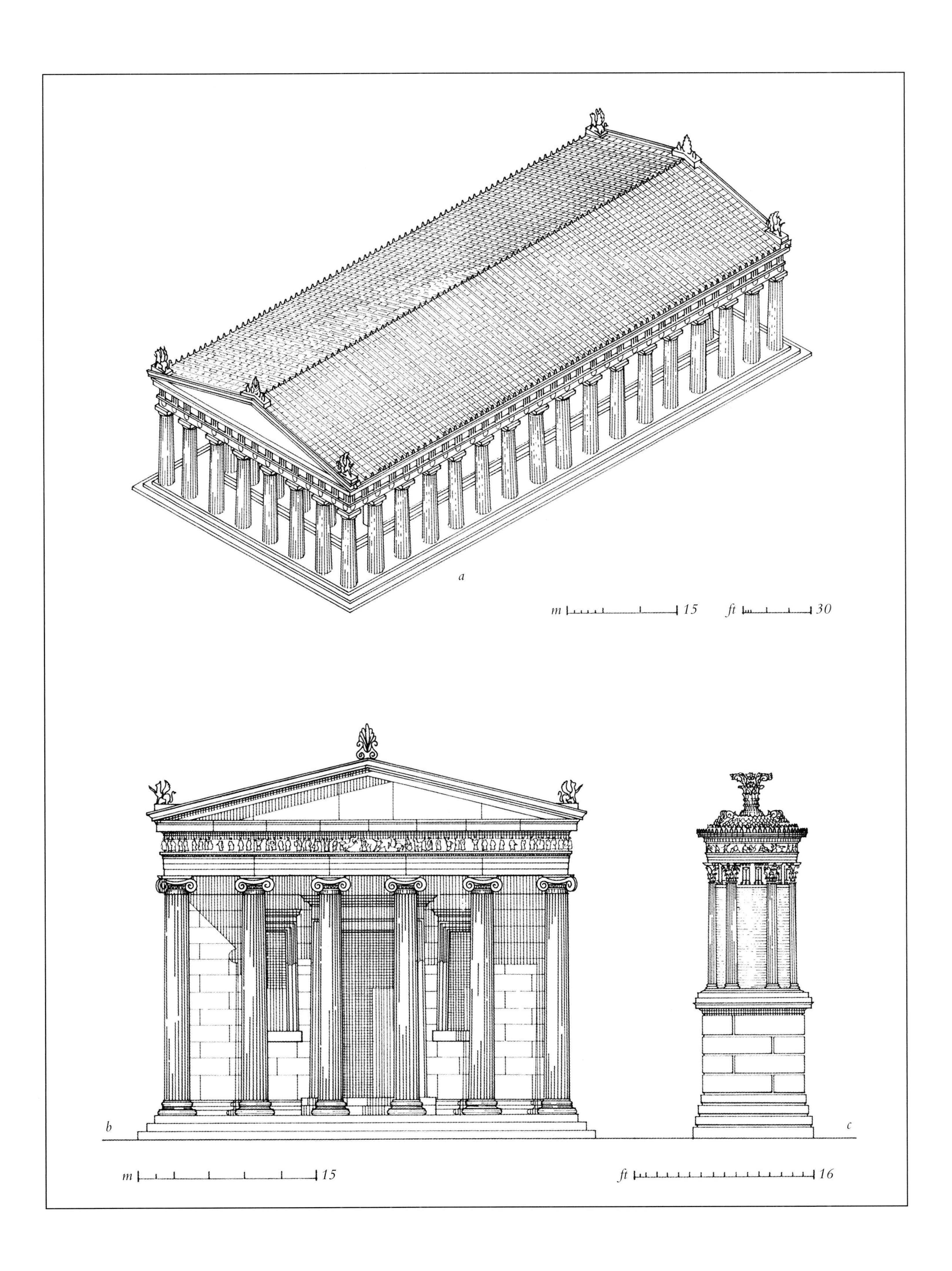
a
m 15
ft 30
b
c
m 15
ft 16

希腊风格

希腊城邦的独立是他们毁灭的根源。对抗带来血腥的战争，直到公元前338年，希腊大陆沦落到由马其顿国王菲利普（公元前382年—公元前336年）统帅的北方半野蛮军队的控制之下。尽管君主厌恶希腊的共和思想，但这些马其顿的征服者们还是被神秘的希腊文化所吸引。菲利普在征服希腊两年后遇刺身亡，他年轻的儿子亚历山大继承了政权和对希腊文明的崇敬。在其短暂且辉煌的十三年执政期间里，亚历山大三世（史称亚历山大大帝）（公元前356年—公元前323年）沿袭希腊文化并带领他的军队向东远征到印度。在地中海地区，他创建了帝国，确保在他父亲征服的小共和国内所宣传的文化革新得以延续。

亚历山大传奇的统治开启了我们所说的长达一百五十年的希腊风格。亚历山大逝世后，帝国被他的将军们分割，发展出新的希腊文明以适应超越了希腊城邦之上的、更加多元化的世界。

希腊建筑基于一种建筑形式的发展——神庙。这种建筑形式难以适应逐渐成形的、复杂社会的不同功能需求。神庙建筑早已被改变成简单的会议厅、大门和坟墓。但是这一时期采用更加重要的改良，确立了神庙建筑体系作为所有建筑形式恰当的装饰形式的地位。这一改良成为后期古典建筑发展的基础。

米利都的议会厅（*a*），修建于公元前170年，由两座神庙式的建筑物分别作为大门和封闭的大厅，由柱廊庭院连接。帕加马圣坛（*b*）是在约同一时期修建的，极大地偏离了希腊传统。圣坛原来是神庙外部狭长的平台，但这一希腊圣坛在宏伟和炫耀方面建立了新的标准。帕加马圣坛主要是由平台和四周包围的柱廊构成。

到大约公元前200年，罗马的平民军队和他们的意大利盟军最终打败了迦太基人，推动衰落的亚历山大帝国残存的部分继续前进。起初只是希腊世界的军事前哨，随着罗马帝国的扩张而成为文明与享乐的中心。不久，罗马就摒弃了他们朴素版本的希腊多立克式建筑，罗马建筑师们将神庙改为更加贴近他们希腊邻国的都市风格。罗马的波图努斯神庙（*c*），修建于公元前2世纪末，保留了传统的罗马神庙入口处的阶梯和坚固的边墙，但是采用了希腊爱奥尼亚柱式神庙的外观。当古典建筑的进程传递到罗马，激励其发展的不是希腊共和国的历史纪念碑，而是当时希腊王国的建筑。

HELLENISM

The independence of the Greek states was their undoing. Rivalry led to bloody wars until in 338 BC the Greek mainland fell to the army of a half-barbarian king from the north, Philip of Macedon (382-336 BC). Although monarchy was abhorrent to the republican mentality of the Greeks, their Macedonian conquerors were absorbed into the mystique of Greek culture, and when Philip was assassinated only two years after his conquest, both his dominance and veneration of Greece were inherited by his young son, Alexander. In a brief and brilliant thirteen-year reign, Alexander Ⅲ (known as Alexander the Great) (356-323 BC) took his armies and his adopted Greek culture to the east as far as India. In the Mediterranean he created an empire that would ensure the continuation of the cultural revolution instigated by the small republics his father had defeated.

With Alexander's legendary reign begins the one-hundred-and-fifty-year epoch we call Hellenistic. On Alexander's death his empire was divided among his generals and this new Greek civilisation developed to accommodate the more diverse world beyond the introspective states of Greece.

Greek architecture had been based on the development of one building, the temple. This building type could not be used for all the different functional requirements of the more complex society that was now emerging. Temple architecture had already been adopted for simple meeting houses, gates, and tombs, but now a much more significant adaptation took place that would establish the temple architectural system as an appropriate decorative form for all building types. This adaptation lies at the root of all subsequent developments in Classical architecture.

The Council House at Miletus (*a*), built in 170 BC, is two temple forms adapted to act as a gate and a large covered hall, linked by a colonnaded courtyard. The altar enclosure at Pergamon (*b*) was built at about the same time and is a dramatic departure from Greek tradition. Altars were originally long narrow platforms outside temples, but Hellenistic altars set new standards of grandeur and ostentation. At Pergamon the altar is dominated by its platform and surrounding colonnade.

By about 200 BC the citizen armies of Rome and its Italian allies had finally defeated the Carthaginians and were making advances into the weakening remnants of Alexander's empire. At first an outpost of the Hellenistic world, Rome's expansion brought it into the centre of civilisation and luxury. Before long Rome had abandoned its provincial version of Greek Doric architecture and Roman architects started to adapt their temples to the more cosmopolitan style of their Hellenistic Greek neighbours. The Temple of Portunus in Rome (*c*), built in the late second century BC, retains the entrance steps and solid side-walls of the traditional Roman temple, but adopts the appearance of a Greek Ionic temple. When the progress of Classical architecture passed to Rome, the inspiration was not by then the historic monuments of the Greek republics, but the living architecture of the Hellenistic kingdoms.

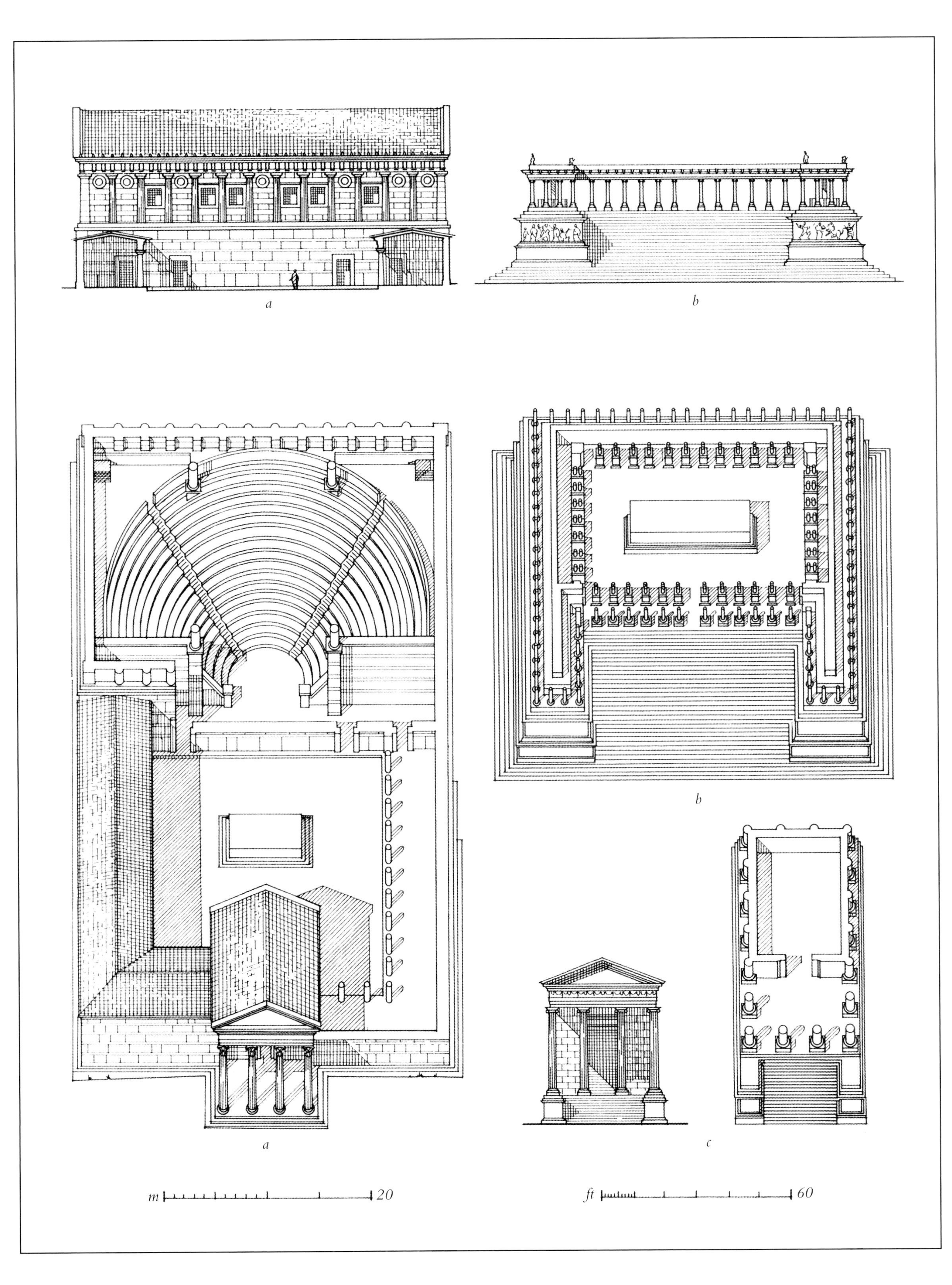

a
b
b
a
c
m 20
ft 60

罗马帝国

在长达两个世纪的时间内，罗马帝国不断推进对地中海地区的统治和本国的城邦民主，并努力寻求两者之间的平衡。强大权力和丰厚财富的诱惑导致内战和新帝国的出现，这些都是煞费苦心奉行民主伪装的产物。新的君主即皇帝，在5个世纪的时间内，巩固他们对一片广袤复杂国土的政权及统治，这片疆域从大西洋延伸至波斯湾，从苏格兰到尼罗河上游地区。

在帝国的创生期，1世纪末，使用火山沙土与石灰的混合材料给建筑带来技术革新并且改变了古典建筑形式。砖块覆面内部浇注天然水泥，凝固后创造出坚固持久的体量，这种形式不适用于希腊建筑的梁柱系统，但最适宜塑造拱形结构和穹顶。罗马建筑师们沿用希腊流传下来的建筑体系，并采用新的材料。

帝王们继承民主的责任，在罗马城维护建筑物及其声望。帝国的财富使得后来的君主得以修建大面积的建筑物，供大众娱乐并美化城市。

其中最著名的恐怕要数古罗马圆形大剧场（*b*）,于公元前70年，由国王维斯帕先（公元前79年—公元前9年）开始修建。这座巨大的建筑物可容纳5万名观众观赏在露天圆形竞技场中进行残忍的娱乐活动。其功能与结构问题的解决方案直到今天也令人备感崇敬，其外部装饰着层叠的柱子和拱形结构为古典建筑增加了新的维度（见P128）。

罗马人对我们所称的土耳其浴室情有独钟，这是他们通过辉煌建筑体现帝国慷慨的又一途径。他们修建了一系列的巨大浴场，其中包括图书馆、讲堂、健身房甚至是室内公园。卡拉卡拉皇帝（188—217年）浴场（*c*）修建于216年，是典型的代表建筑，可容纳1600人洗浴。广阔的公共建筑需要宽敞的覆有屋顶的大厅，结构和建筑体系包括跨度达27米的无支撑水泥拱顶。

万神殿（*a*），由哈德良皇帝（76—138年）于120年修建，给我们以最完整的帝王建筑概念。传统的神庙式门廊矗立在巨大的水泥镶砖鼓形座前，上面覆盖40米宽的穹顶，创造出引人注目的天窗空间，并装饰有彩色大理石柱和拱形结构。这种新的对内部空间的重视和对希腊及希腊风格由传统向水泥结构新机遇的改良，改变了古典传统，而且在帝国威望的支持下，将对其进一步发展产生影响。

IMPERIAL ROME

For two centuries the city of Rome struggled to reconcile its progressive domination of the Mediterranean world with its own city-state democracy. The temptations of great power and the enormous wealth of conquest led to civil war and the emergence of a new kind of monarchy, born out of an elaborate pretence of democratic conformity. The new monarchs, or emperors, consolidated their power to rule for five centuries over a vast and complex empire which stretched from the Atlantic to the Persian Gulf and from Scotland to the Upper Nile.

At about the time of the creation of the imperial monarchy, at the end of the first century BC, the use of a volcanic sand mixed with lime revolutionised building construction and transformed Classical architecture. This natural cement was laid inside a brick skin and hardened to create a solid durable mass; it was not suited to the beam and post system of Greek architecture, but worked best when moulded into arches and domes. Roman architects extended the range of the architectural system that had come down to them from the Greeks to accommodate the new material.

The emperors inherited a democratic obligation to maintain the buildings and their own popularity in the city of Rome. The wealth of the empire allowed successive emperors to erect vast buildings for the entertainment of the people and the adornment of the city.

Perhaps the most famous of these is the Colosseum (*b*), begun in AD 70 by the Emperor Vespasian (79-9 BC). This huge structure seated some 50,000 spectators for the brutal entertainments of the amphitheatre. The solution of the functional and structural problems inspires respect to this day and the tiers of columns and arches decorating the exterior added a new dimension to Classical architecture (see page 128).

The Roman fondness for what we would call Turkish baths provided another outlet for imperial munificence in architectural splendour. A series of ever larger baths was built and came to include libraries, lecture halls, gymnasia, and even enclosed parks. The Baths of the Emperor Caracalla (AD 188-217) (*c*), built in AD 216, are typical and could accommodate 1600 bathers. These vast public buildings required large, roofed halls and a structural and architectural system evolved for unsupported spans of up to 27 metres in vaulted concrete.

The Temple to the Pantheon of the Gods (*a*), built by the Emperor Hadrian (AD 76-138), in AD 120, gives us the most complete impression of the great buildings of the emperors. Its conventional temple porch sits in front of a huge brick-clad concrete drum covered with a 40-metre-wide dome creating a dramatic top-lit space decorated with rows of coloured marble columns and arches. This new concentration on interior spaces, and the adaptation of Greek and Hellenistic tradition to the new opportunities of concrete construction, transformed the Classical tradition and, supported by the prestige of the empire, would continue to influence its further development.

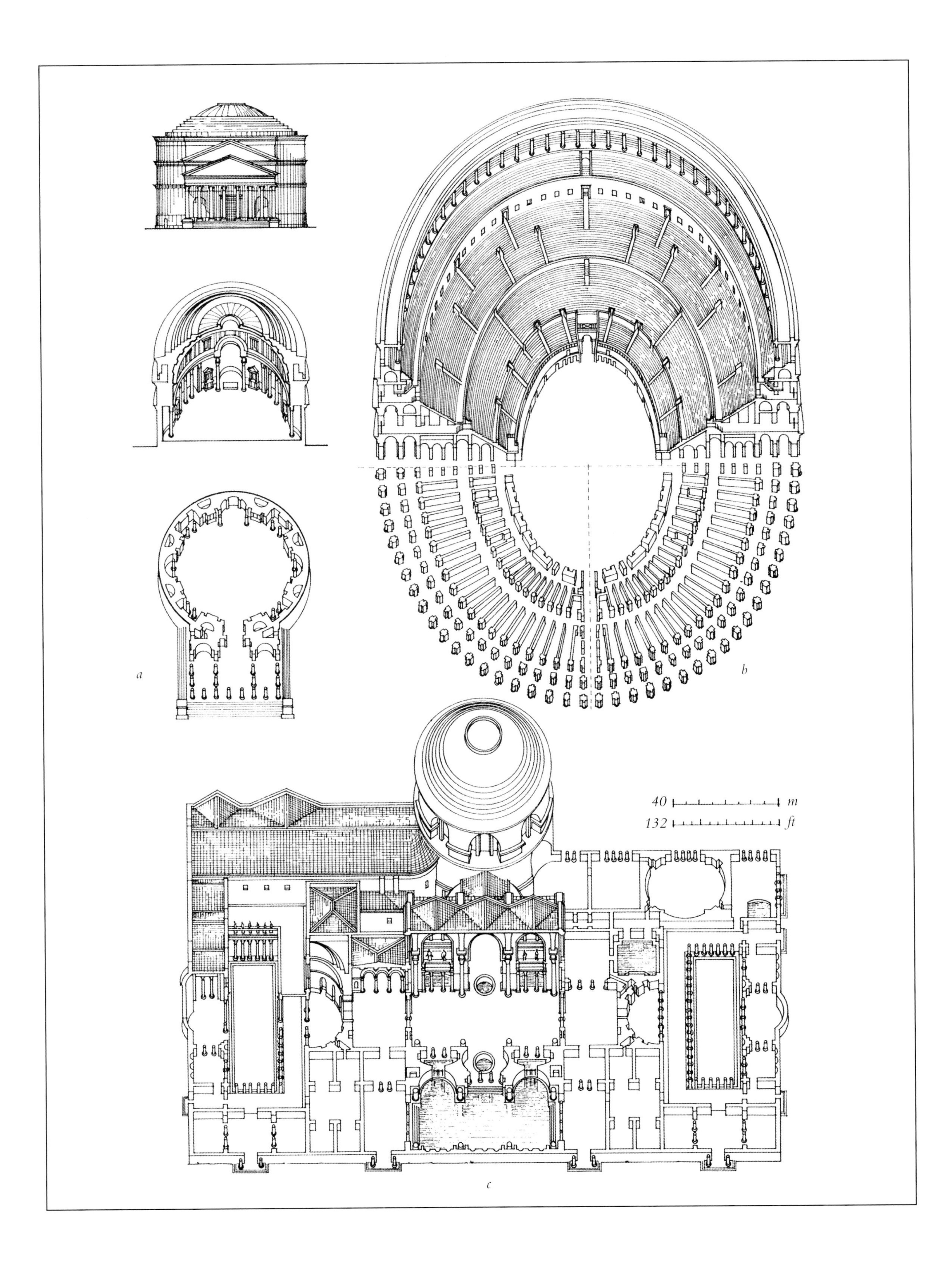
a
b
40 m
132 ft
c

罗马帝国后期

罗马帝国在一百年内已将其统治范围扩大到意大利沿岸地区。对于建国民族来说，帝国的疆域太广阔而难以治理。3世纪，其北部和东部边境的民族压力带来的战争和瘟疫最终侵入罗马帝国腹地。军队成倍地增长，受野蛮的雇佣兵控制，他们从自己的阶层当中自立皇帝，不断地进行内战，其破坏力不亚于侵略战争。通货膨胀和赋税破坏了古代世界文明的象征——城邦文化。古老的宗教变成近东地区的神秘邪教，他们抓住了民众惴惴不安的心。

3世纪末期，戴克里先皇帝（244—313年）倡导区划、改革来拯救帝国。他的继承人康斯坦丁大帝（272—337年）将区划正规化，并建立康斯坦丁堡（伊斯坦布尔），确立新的宗教——基督教为国教。

重新确立皇权之后，古典文化就不再掌握在古老的地中海民族手中。希腊的传统成为遥远的回忆，手工艺人逐渐消失，在军营里受过严格训练的士兵们逐渐掌握政权。然而，帝国精神所代表的力量及其表现方式却通过雄伟的建筑得以保存。

戴克里先皇帝的改革预见到了他本人的退位，300年，他为自己在原南斯拉夫的家乡建立了一座宫殿。宫殿规模宏大如一座城镇，并模仿兵营形式修建。住所入门（*b*）处的庭院展现出叙利亚行省的古典风范，内部的建筑却模仿早期帝国时代的古迹而建。

康斯坦丁大帝于312年在罗马城外打败马克森提乌斯皇帝（278—312年），登基称帝。马克森提乌斯皇帝在古罗马广场未完成的广厅即巴西利卡（*a*），被修改成为康斯坦丁堡的一部分。这一宏伟的开放式大厅，外部构造简单，内部则采用大胆的几何设计。与传统的巴西利卡廊台不同，其形式来源于古罗马浴场的中央大厅设计。

康斯坦丁大帝的女儿康斯坦丁娜（出生于307年之后，317—354年之前）之后称为圣康斯坦斯，葬于罗马。她的基督教陵墓（*c*）称为圣康斯坦娜教堂，同样将简单的外部设计与丰富的几何形制结合起来，装饰着精美的传统上为异教所使用的马赛克。

后期罗马的建筑灵感并非来自遥远的希腊，而是早期罗马帝国的建筑和对后来统治罗马的行省文化的不完全解读。熟练地使用混凝土拱顶和拱门，构造各种可能的几何形制，采用具有区域特色的改良设计，例如在柱子上直接构建拱顶，增加古典建筑传统并且创造最初的基督建筑形式。随后几千年的西方基督教国家对这一时期的建筑风格情有独钟。

LATE ROME

The Roman Empire became too large to be dominated by its founding race and, within a century, power had passed beyond the shores of Italy. In the third century AD, pressure from peoples beyond the northern and eastern borders brought warfare, plague and, finally, invasion to the heart of the empire. As the armies grew in numbers they came to be dominated by barbarian mercenaries who created emperors out of their ranks and fought civil wars as damaging as invasions. Inflation and taxation began to destroy the city-based culture that had characterised the civilisations of the ancient world. The old religions were reduced to ceremony as mystery cults from the Near East gripped the insecure population.

As the third century drew to a close, the Emperor Diocletian (244-313 AD) instituted reforms to divide, transform and save the empire. His successor Constantine the Great (272-337 AD) formalised the division by founding Constantinople (Istanbul) and established one of the new cults, Christianity, as the official religion of the state.

Once the empire was restored, Classical culture was no longer in the hands of the old Mediterranean races. The Greek tradition had become a distant memory, craftsmen had died out and power lay with dour soldiers educated in the austere discipline of the camps. Yet the strength of the imperial idea and its expression through great buildings survived.

Diocletian's reforms foresaw his own retirement and in the year 300 he built himself a palace in his homeland in Yugoslavia. As large as a town, the palace took the form of a fortified military camp. The entrance court to his apartments (*b*) displays provincial Classicism characteristic of the Syrian provinces, but leads to buildings that deliberately echo the monuments of an earlier, imperial age.

Constantine defeated the Emperor Maxentius (AD *c*.278-312) outside Rome in AD 312 and assumed power. Maxentius' unfinished public hall, or basilica (*a*), in the Forum was modified and completed as part of Constantine's building programme for the city. This large open hall, with plain exterior and bold geometric interior, was unlike traditional galleried basilicas and owed its origins to the central halls of baths.

Constantine's daughter, Constantina (*b*.after AD 307/before AD 317-354) later known as Saint Constance, was buried in Rome. Her Christian mausoleum (known as Church of Santa Costanza) (*c*) combines the same exterior simplicity with a geometrically complex interior, decorated with fine, traditionally pagan, mosaics.

Late Roman buildings draw their inspiration not from a distant Greek source but from earlier imperial buildings and the impure interpretations of the provinces which now gave Rome her rulers. Complete familiarity with the geometric possibilities of the concrete vault and arch and the adoption of regional innovations, such as the arch rising directly off the column, added to the Classical tradition and produced the first specifically Christian buildings. It was this period that the following thousand years of western Christendom would view with particular devotion.

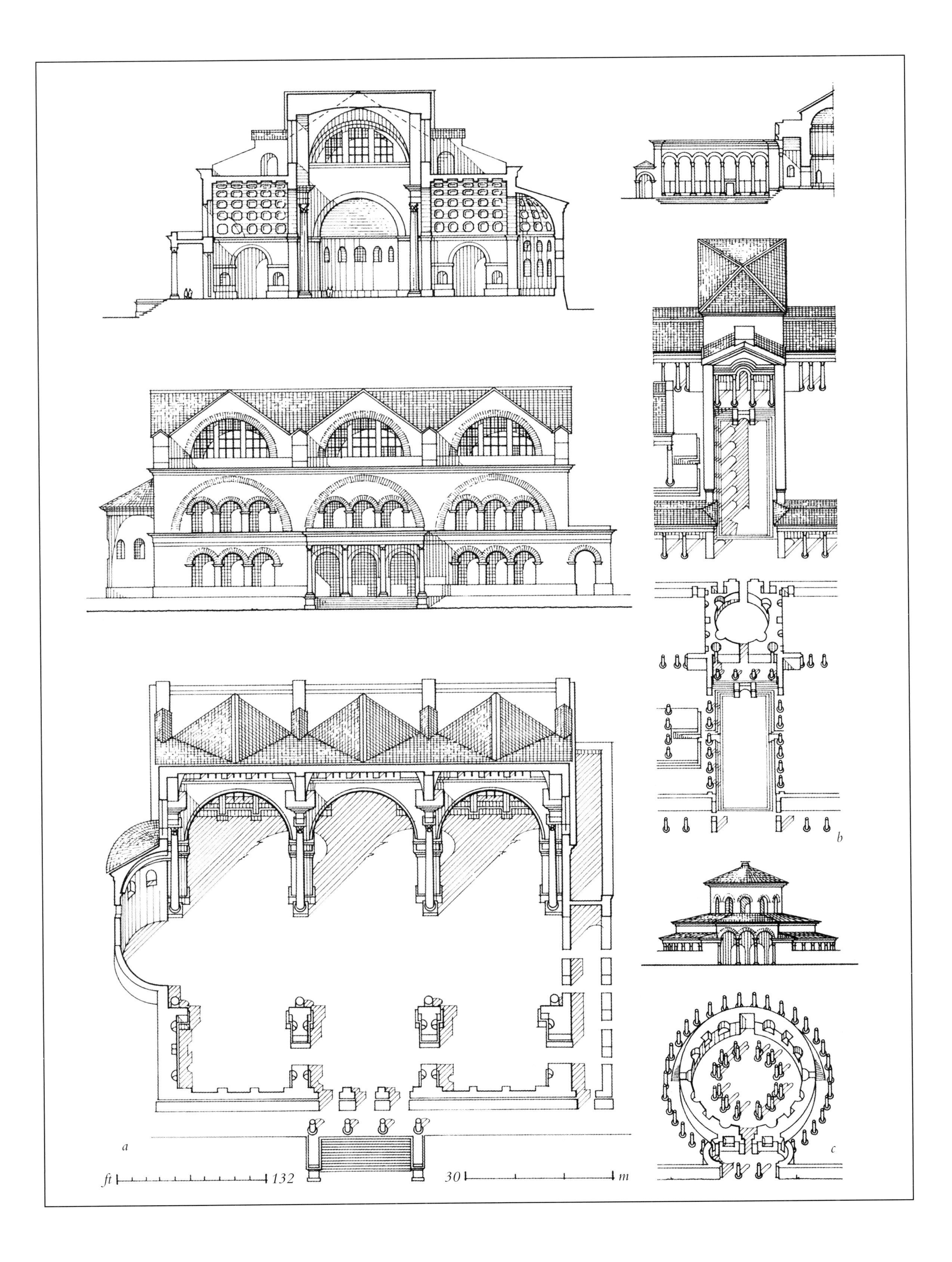

a
b
c
ft 132
30 m

拜占庭

西部的罗马帝国最终于476年向侵略者妥协，帝国执政权力的徽章被移回东部的君士坦丁堡，或称拜占庭。拜占庭人认为自己是罗马第一位皇帝的继承者。如果仅从地理条件来说，拜占庭建筑是古典主义的延续。

早期的拜占庭建筑，例如帖撒罗尼迦的圣德米特里厄斯教堂（*e*），建于大约475年，与罗马帝国后期其他建筑实例大同小异。早在一个世纪之前，罗马建造的大型教堂或巴西利卡模式基本相同。但是在内部细部设计上逐渐表现出明显的东方风格，使用色彩斑斓的石头和金色马赛克拼成的抽象图案，在柱子上装饰切割成蕾丝网状的叶子纹案。

由于战争，东方的君主无法在东部边境重新统一帝国，但是532年，查士丁尼皇帝（483—565年）与波斯人签订条约，使其军队能够再次征服（尽管十分短暂）在非洲失去的领土和罗马地区。

东方帝国最辉煌的时期，矗立起最伟大的建筑，君士坦丁堡的圣索菲亚大教堂（*d*）。这座拥有宏伟穹顶的大教堂，采用罗马后期的内部几何形制，使这种设计手法达到顶峰。建筑师安提缪斯（420—472年）和伊西多尔同时也是数学家，这绝非偶然。55米高的穹顶坐落在方形平面上，由拱形结构支撑，两侧有两个半穹隆圆顶，每个圆顶又连接三个更小的半穹隆圆顶。大胆的建筑构造使得建筑师能够在穹顶上高高地排列着窗户，并没有支撑穹顶的拱形结构，因此看来，穹顶仿佛飘浮在空中。几何构造使得改良的古典细部设计相形见绌。成排的廊柱和拱门是早期罗马时代有柱式的古典内部设计的模糊记忆。不再使用罗马的水泥，巨大的穹顶由坚固的砖石砌成。

拜占庭式对古典建筑独特的解读已崭露头角并且由新的传统柱头设计诠释。图例（*a*）中的柱头雕刻，采用古老的向上叶片纹案，将其变成好似被风扫过的自然形态，而（*b*）所示的柱头中传统的细部设计已消失，取而代之的是抽象纹案和基督教的象征纹案。

西部野蛮的侵略和东部伊斯兰教的兴起使得胜利的喜悦迅速变为长期的隐退。但在随后的9个世纪中，伟大的查士丁尼时代设计手法仍见于小型建筑中，这是一些建筑师执着追求并不断复杂化的几何结构，例如图例（*c*）帖撒罗尼迦建于1315年的圣使徒教堂中的拱顶砖石结构。

1453年，当最后的罗马皇帝离开君士坦丁堡，拜占庭古典时代结束了，但它的传奇在欧洲东部和土耳其的征服者手中得以延续。

BYZANTIUM

When the Roman Empire in the west finally succumbed to invaders in AD 476, the imperial badges of office were sent back to the eastern emperor in Constantinople, or Byzantium. The Byzantines considered themselves to be the successors of the first emperors in Rome and Byzantine architecture is a direct, if geographically limited, continuation of the Classical tradition.

Early Byzantine buildings, such as the church of St Demetrius in Thessalonica (*e*), built in about AD 475, differ little from other examples in the late Rome Empire. The large churches or basilicas built in Rome a century earlier were designed to an almost identical pattern. But in the details of the interior, distinctly eastern features have started to appear, with abstract patterns in multicolored stone and golden mosaics, and leaf decoration over the columns cut like lacework.

The eastern emperors were prevented from reunifying the empire by war on their eastern borders, but in AD 532 the Emperor Justinian (AD 483-565) signed a treaty with the Persians which freed his army to reconquer – albeit briefly – the lost provinces of Africa and even Rome itself.

In this its most glorious period, the greatest building of the Eastern Empire was erected, Hagia Sofia in Constantinople (*d*). This huge domed structure takes the interior geometry of late Rome to its conclusion. It is no accident that the architects, Anthemius (AD 420-472) and Isidorus, were known as mathematicians. The 55-metre-high dome sits on a square plan supported by huge arches and flanked by two half-domes each containing three more half-domes. The audacious structure allows windows to be placed high up in the dome and its supporting arches so that the dome appeared to contemporaries to float in the air. The geometry overshadows the transformed Classical details. Rows of columns and arches are faint memories of the ordered Classical interiors of earlier Rome. Roman cement had by now fallen out of use and the great domes are made of solid brick.

A distinct Byzantine interpretation of Classicism had now emerged and is illustrated by novel versions of the traditional column capitals. The carving in (*a*) takes the old upright leaf pattern and sweeps it round as if blown by the wind, and in (*b*) all the traditional details have gone, to be replaced by abstract pattern and Christian symbols.

Barbarian invasion in the west and the rise of Islam in the east soon turned victory into a long retreat. But for the following nine centuries the great age of Justinian lived on in small buildings whose architects continued the pursuit of increasingly complex geometry, such as in the domed brick structure of the church of the Holy Apostles in Thessalonica (*c*) of 1315.

When the last Roman emperor left Constantinople in 1453, Byzantine Classicism ended, but its legacy endured in the buildings of eastern Europe and even in those of its Turkish conquerors.

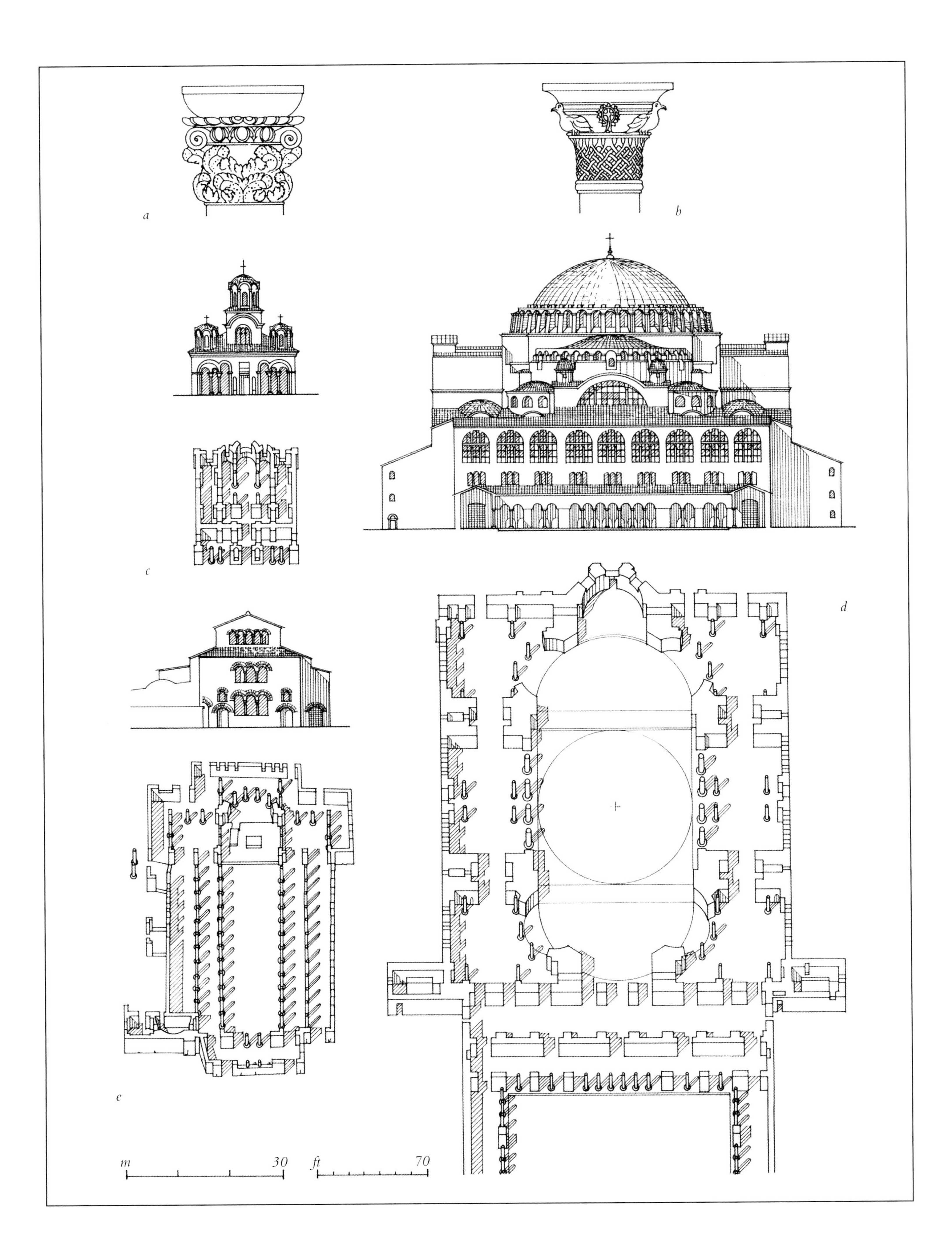

a
b
c
d
e
m
30
ft
70

走出黑暗时代

当西方帝国疲惫地进入末世时期，罗马公民已经与不断增加的野蛮居民共同生活了两个世纪。在法国、意大利、西班牙，这些不安分的邻居处在当地部族的统治之下，西方帝国的破败在最后的年代成为不争的事实。

帝国的政治梦想仍在继续，建筑成为遗失的技艺和有组织的市民安定生活的回忆。野蛮人国王渴望帝王头衔所赋予的身份和对国土的统治权，并试图在自己粗糙的房屋中模仿现存的伟大建筑。

在战争带来的移民骚乱之中，法兰克部族的国王，查理曼大帝（逝于814年）获得了对法国、德国和意大利大部分地区的统治权。800年，他在罗马由教皇利奥三世（750—816年）加冕，成为西方的皇帝，即神圣罗马帝国的皇帝。查理曼大帝开启了罗马文明在文学和绘画方面的伟大复兴。他所修建的建筑认真而不成熟地模仿了罗马伟大的建筑。例如，洛尔施修道院（*c*）的大门，吃力地试图复兴宏伟的凯旋门。

神圣罗马帝国在内部分裂以及来自斯堪的纳维亚、斯拉夫和阿拉伯等部族的外部压力下，很快陷入混乱。罗马教堂和教皇、罗马大主教，成为分裂的欧洲中唯一稳定的组织。随着人们对僧侣生活的热情不断高涨，每一个新涌入的野蛮民族都转化为基督教信徒，加上教皇对神圣罗马皇帝强有力的控制权，西方教堂成为最伟大的欧洲建筑。

在罗马的废墟之上，查理曼短暂的统治期间，在罗马后期建筑与东部帝国并不完善的建筑知识的启发之下，出现了欧洲野蛮古典主义。为了向罗马建筑的启迪致敬，这种建筑称为罗马式，在英国则称为诺曼式建筑。

蜂拥而至的朝圣者们寻找先圣遗骸并对其神秘魔力着迷，拥有这些令人质疑骸骨的人们因此得到了巨大的财富。宏伟的教堂，如1080年修建的西班牙圣地亚哥教堂（*a*），为朝圣者们而建，并提升掌管这些大教堂主教们的地位。建造技术得以改进，石造建筑试图再现周围不断腐朽的罗马式建筑内部的壮丽景象。

巨大的石造结构遵循遗失的水泥构造技艺而设计，其装饰则努力模仿野蛮人从古希腊继承的几何主题。北至英格兰的达拉谟（*b*），南到西西里，建筑走出了黑暗时代。消失的传统在幼稚的权力中得以复兴并成就伟大，将欧洲建筑带入中世纪。

OUT OF THE DARK AGES

When the Western Empire came to its weary end, Roman citizens had been living alongside increasing numbers of barbarian settlers for two centuries. In France, Italy, and Spain these uneasy neighbours came together under local tribal rulers, and the fragmented nature of the Western Empire in its final years became a recognised fact.

The political ideal of the empire lived on and the buildings served as a constant reminder of the lost skills and settled life of organised civilisation. Barbarian kings longed for the status and territorial control implied by the imperial title and in their crude buildings tried to imitate the great architecture they had come to inhabit.

Amid the turmoil of migration and war, the king of the Frankish tribes, Charlemagne (*d.*AD 814), gained control of France, Germany, and much of Italy. In Rome in AD 800 he was crowned by Pope Leo Ⅲ (AD 750-816) as the emperor of the west, the Holy Roman Emperor. Charlemagne set in motion a remarkable revival of Roman culture in literature and painting. His buildings were a sincere if rudimentary attempt to copy the greatness of Rome; the abbey gateway at Lorsch (*c*), for example, struggles to revive the splendour of the triumphal arch.

The Holy Roman Empire fell quickly into disorder with internal division and external pressure from Scandinavians, Slavs, and Arabs. The Roman Church and the Pope, the Bishop of Rome, became the only stable organisation in a divided Europe. Strengthened by a growing passion for monastic life, by the conversion to Christianity of each new barbarian influx, and by papal control over the still potent title of Holy Roman Emperor, the western Church became Europe's greatest builder.

Out of the ruins of Rome, and Charlemagne's brief rule, inspired by late Roman buildings and incomplete knowledge of the great architecture of the Eastern Empire, a European barbarian Classicism emerged. In recognition of its Roman inspiration, this new architecture is called Romanesque. In Britain it is called Norman.

An obsession with the magical powers of the mortal remains of saints brought the possessors of these doubtful relics great wealth from the hordes of pilgrims who sought them out. Great churches such as Santiago de Compostela in Spain (*a*) of 1080 were built to house these pilgrims and to raise the status of the powerful bishops who controlled them. Building skills improved and stonemasons attempted to recapture the great interiors of the decaying Roman buildings that surrounded them.

Massive stone structures were designed in such a way as to follow the forms of the lost skills of concrete construction and, in their decoration, tried to imitate with barbarian geometry the themes inherited from ancient Greece. As far north as Durham in England (*b*), and as far south as Sicily, architecture emerged from a dark age. The lost traditions were revived with a naïve power that achieved a kind of greatness and launched European architecture into the Middle Ages.

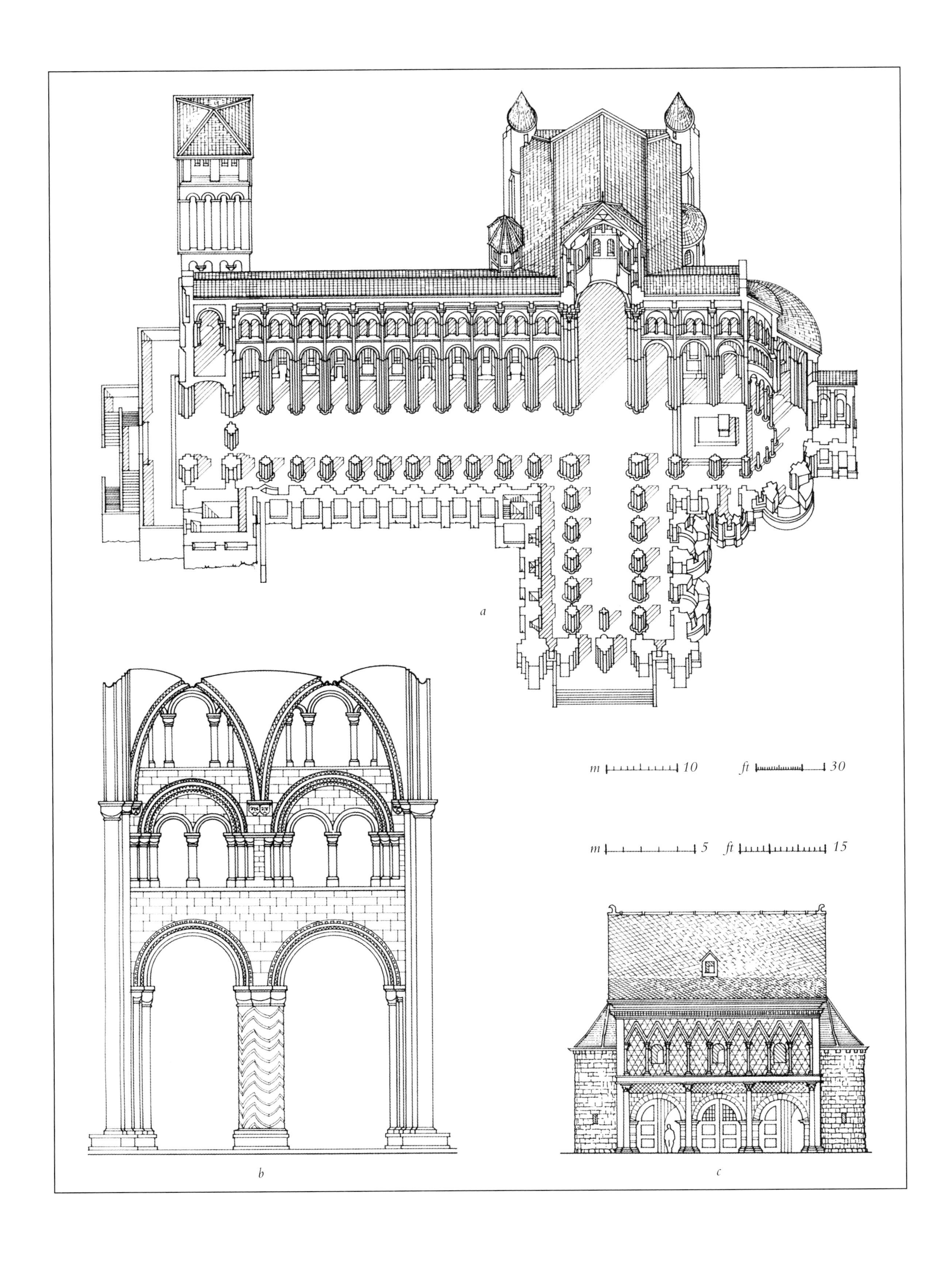

a
m 10
ft 30
m 5
ft 15
b
c

哥特式的诞生

经历了几个世纪的侵略、战争和破坏，西方世界终于在12世纪安定下来，进入相对稳定的和平时期，天主教城市发展成为文化中心。十字军阻止了阿拉伯扩张，使得通往欧洲南部和朝圣地的旅途更加安全，人们转向关注保留在修道院图书馆中的文化成就。拉丁语仍然是现用语言，在古代文学的复兴过程中变得更加纯粹。古代作家们在科学与哲学方面的成就，成为教育的基石。富有且见多识广的牧师，参观在圣地修建的罗马古典教堂与康斯坦丁纪念碑后，重新审视自己的教堂，认为这些带有巨大支柱、粗糙雕刻和低矮内部空间的教堂十分粗俗。

中间出现断层的几个世纪并没有毁掉希腊的魅力，但是文明的连贯统一却消失了。历史的进程感不复存在，艺术家们幼稚地用现代的外衣解读古典纹案。迷信和基督教神秘仪式，将人们的思想转向注重谜一般的符号和神话世界，获得了与有形世界同样的真实感。当大教堂的建造者们寻找再现古代建筑的方法时，他们并不打算建造完全相同的复制品，而是仿造建筑特征中具有含义的尺度与数字。

1089年，在法国东部，克吕尼教堂（*a*），即克吕尼修会教士权力总部得以重建，力求复兴罗马圣彼得大教堂。新的建筑富于古典装饰，但也包括早期从阿拉伯、西西里、西班牙和圣地建筑中模仿而来的早期尖拱。

这些拱券与罗马圆拱的区别并不明显，但尖拱逐渐呈流行趋势。这一时期法国北部富有的克吕尼修会主教们，要求建筑师将教堂的墙体打开，安装大面积的彩色玻璃，在内部排列优美的古典柱体上投射出神秘的光影。为实现这一效果，建筑师们减少了石拱的重量，将柱子之间的拱进行十字交叉。这些拱又叫拱肋，将其所承受重量向上托起至斜倚在外墙上的半拱，即飞扶壁处。

建造技术的巨大改进与尖拱的结构优势共同发展，使得建筑师能够修建更高的建筑和更大面积的窗户。从诺扬大教堂（*b*）到拉昂大教堂（*c*），随着结构的不断完善，不间断排列的柱子设计终于得以实现，正如在后期罗马教堂中所见。

到本世纪结束，拉丁复兴结束了。大教堂结构的革新有了新的机遇，在夏尔特大教堂（*d*）的设计中，古典理念淹没在高度与玻璃中。中世纪建筑已经拥有自己的风格灵感，但这一风格是在对罗马朦胧的观察中诞生的。

THE BIRTH OF GOTHIC

After centuries of invasion, war and destruction, in the twelfth century the West settled down to a period of relative peace and stability and centres of learning developed in cathedral cities. The crusades halted Arab expansion and allowed safer travel in southern Europe and the Holy Land. Attention turned again to the intellectual achievements of Rome, preserved in monastic libraries. Latin, still a living language, was purified by a revival of the literature of antiquity, and the science and philosophy of the ancient authors became the cornerstone of education. Wealthy and sophisticated clerics, visiting the Classical churches of Rome and Constantine's monuments in the Holy Land, saw their own churches, with their massive piers, crude decoration, and squat interiors, as barbaric.

The intervening centuries had not destroyed the spell of Rome, but the unbroken continuity of civilisation had been lost. The sense of the progress of history had gone and artists innocently depicted ancient figures in modern dress. Superstition and a mystic form of Christianity turned minds to a cryptic world of symbol and myth which gained a reality equal to the physical world. When the builders of cathedrals sought to recreate ancient buildings they had no interest in exact reproduction, but duplicated symbolic dimensions and numbers of architectural features.

In eastern France in 1089, Cluny (*a*), the headquarters of the powerful order of Cluniac monks, was rebuilt to rival St Peter's in Rome. The new building was rich in Classical decoration but included an early example of the pointed arch, copied from Arab buildings seen in Sicily, Spain, or the Holy Land.

The difference between these arches and round Roman arches was not regarded as significant and the fashion for pointed arches spread. At this time the rich Cluniac bishops of northern France were asking their architects to open up the walls of their churches for large areas of stained glass to cast a mystic light on elegant internal rows of Classical columns. To achieve this architects reduced the weight of stone vaults by crossing arches diagonally between columns. These arches, or ribs, then pushed the weight they carried outwards to half-arches, or flying buttresses, leaning against the outside walls.

Remarkable structure improvements were developed which, together with the constructional advantages discovered in pointed arches enabled architects to create higher and higher buildings with ever larger windows. As the structure improved from Noyon Cathedral (*b*) to Laon (*c*) the unbroken row of columns, such as those found in late Roman churches, was finally achieved.

By the end of the century, the Latin revival was over. In the cathedrals the structural innovations had opened up new opportunities. At Chartres Cathedral (*d*) the Classical ideal was overwhelmed by height and glass. Medieval architecture now had its own style for inspiration, but it was a style born of a clouded vision of Rome.

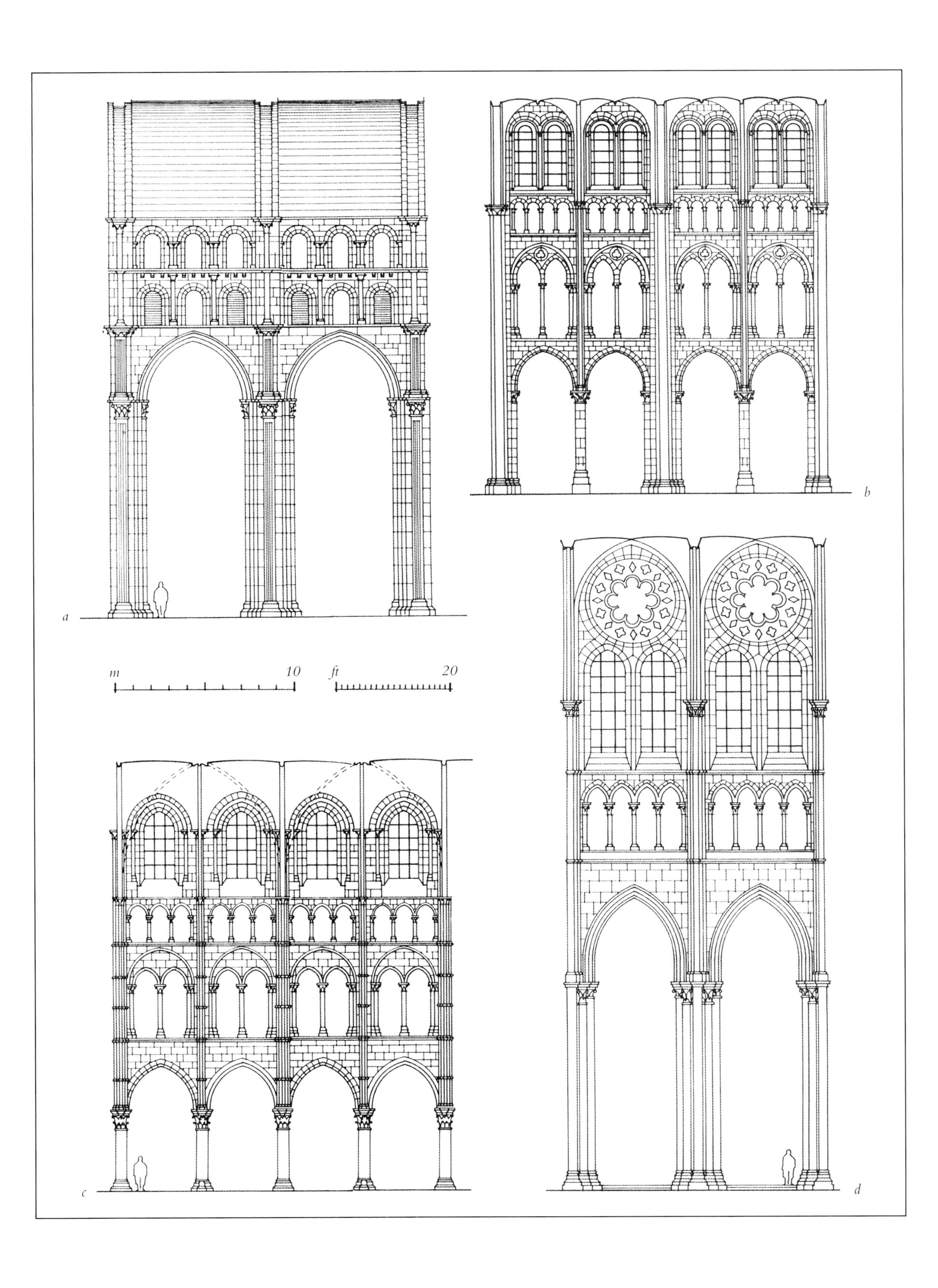
a
b
c
d
m
10
ft
20

文艺复兴

意大利城市生活从黑暗时代和中世纪的毁坏中幸存，一些城邦成为自治的共和国，不再遭受拥有土地的公爵与国王的奴役。自由促进了城市文明和制造业的迅速发展。伫立在欧洲、伊斯兰和拜占庭的交会处，这些城邦中的贵族商人成为欧洲的银行家，聚集了巨额财富与强大权力，仅仅受到明显存在于城邦内部的政治压力的局限和小的城邦在贸易、土地和贵族头衔方面激烈竞争的影响。

高哥特式建筑的消瘦与高贵在意大利并无用武之地，与拜占庭艺术结合激发了13世纪现实主义的革新，此时的艺术由符号与传统主宰。14世纪对古罗马文学的迷恋挽救了修道院图书馆中被遗忘的手稿，创造了新的哲学、人文主义，并且调和了基督教与异教在思想上的矛盾。人文主义的第一故乡在佛罗伦萨，这里重要家族都是满腔热情的赞助人。

这场伟大的艺术革命史称文艺复兴，从15世纪早期富饶的土壤中滋生出来。古罗马文化激发出的自我意识，排斥哥特文化，文艺复兴的艺术家们迸发出来的创造力改变着社会。建筑师们研究古罗马遗址，创造建筑，满足从中世纪封建社会形式中逐渐发展起来的社会需求。

美第奇家族府邸（*a*）由米开罗佐（1396—1472年）于1440年设计，具备家族居住、商业经营和城堡等功用。饱受野蛮仇家困扰的意大利城市，通常设计这样的巨大城市住宅来进行防卫，但是米开罗佐在传统的设计中增加了古典的柱式设计。三层墙垣逐渐趋于平整，每层由水平石头腰线分割，同时支撑着规则排列成直线的拱形窗户。整个建筑顶部覆盖巨大的凸出屋檐，即檐口，取代了传统的城垛。

伟大的学者及建筑师莱昂·巴蒂斯塔·阿尔伯蒂（1404—1472年）晚年于1470年，修建了曼图亚的圣安德烈教堂（*b*），他的灵感明显来源于康斯坦丁堡的巴西利卡（见P14）中巨大的古罗马内部拱顶结构。他遵循已有的狭长教堂平面，将侧廊改为小圣堂，用于支撑沉重的石头屋顶。在外部设计上，他使用古罗马凯旋门作为正立面，将其颂扬帝王的功用改为赞美上帝。

早期的意大利文艺复兴是一场知识的探险，推动欧洲进入现代社会。历史的觉醒中注入了原创的活力，即创造全新的、壮丽的古典建筑。

THE RENAISSANCE

Italian city life survived the devastation of the Dark Age and, in the Middle Ages, some city-states became self-governing republics, free from the obligation of servitude to landed dukes and kings. Freedom accelerated the sophistication of urban culture and the growth of manufacturing industry. Poised at the crossroads of Europe, Islam, and Byzantium, the merchant aristocracy of these city-states became the bankers of Europe and accumulated great wealth and power, restrained only by the remarkable violence of internal city politics and the fierce competition between the small states for trade, territory, and artistic prestige.

The brittle elegance of High Gothic found a home in central Italy and contact with Byzantium art stimulated a revolutionary realism in the thirteenth century, when art had been dominated by symbol and convention. In the fourteenth century a passionate interest in the literature of ancient Rome led to the rescue of forgotten manuscripts from monastic libraries and a new philosophy, humanism, reconciling Christian and pagan thought. The first home of humanism was the city-state of Florence where the leading families were enthusiastic patrons.

The great artistic revolution, known as the Renaissance, blossomed from this fertile ground in the early fifteenth century. Self-consciously inspired by ancient Rome and rejecting the Gothic, Renaissance artists transformed society in an explosion of creativity. Architects studied the forgotten ruins of Rome to create buildings that served the needs of a society only gradually emerging from the feudalism of the Middle Ages.

The palace of the Medici family (*a*) was designed by Michelozzo (1396-1472) in 1440 to act as a home, business premises, and fortress. Protection from the savage vendettas that plagued Italian cities had always set the design of these great city houses but Michelozzo added Classical Order to the traditional scheme. The walls become progressively smoother in three tiers, each divided by a horizontal band of stone supporting a regular line of arched windows. The whole composition is crowned by huge Classical overhanging eaves, the cornice, which take the place of traditional battlements.

When the great scholar and architect Leon Battista Alberti (1404-72) designed the church of S. Andrea in Mantua (*b*) in 1470 at the end of his life, he had clearer Classical examples to follow and drew his inspiration from large Roman vaulted interiors such as the Basilica of Constantine (page 14). He followed the established long church plan, but changed the side-aisles to chapels to support the heavy stone roof. On the exterior he used the design of a Roman triumphal arch for the façade, changing its role from the glorification of an empire to that of the Christian God.

The early Italian Renaissance was an intellectual adventure which launched Europe into the modern world. A fresh consciousness of history was infused with the vigour of originality to create a new Classical architecture of great beauty.

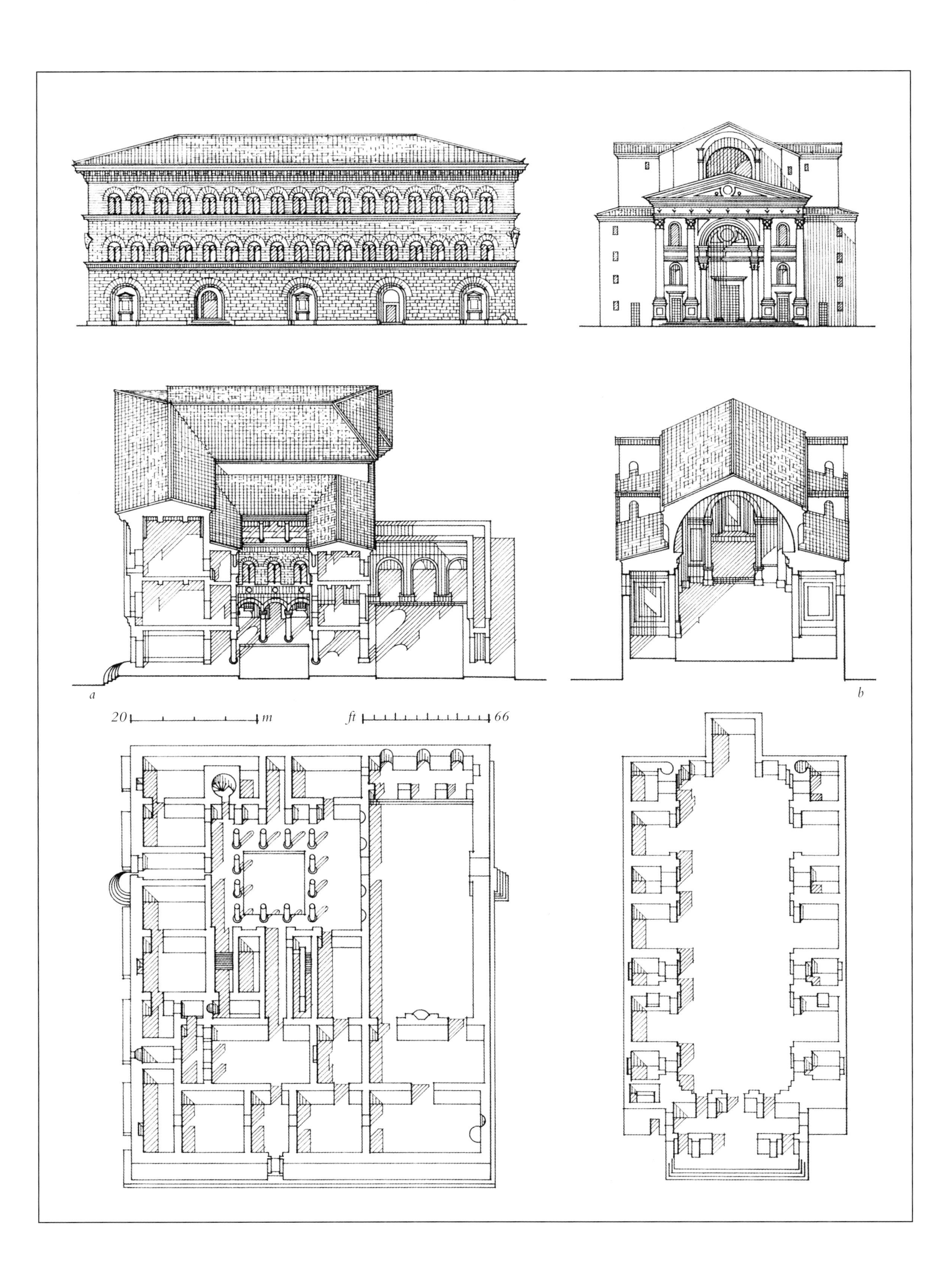
a
b
20 m
ft 66

文艺复兴全盛期

1417年，教皇马蒂诺五世（1369—1431年）登基，其余三位竞争对手被废黜，因此结束了东西教会大分裂。这一时期的大分裂耗尽了教皇享有的意大利国土资源，并败坏了罗马教廷的声誉。30年后，教皇尼古拉五世迁入梵蒂冈居住，标志着教廷财政状况复苏，进而开启了建筑业的竞争。文艺复兴的艺术家和建筑师们齐聚罗马。此时，几个世纪以来罗马教廷选举积累的宿怨与长年混战，已将罗马变为废墟中的村落。对古罗马重新燃起的兴趣，引发对众多现存纪念碑的潜心研究。不幸的是，寻找旧石建造新建筑对它们造成了破坏。

罗马成为重要的文化中心和文艺复兴艺术与建筑发展进程的核心。坎榭列利亚宫（*a*和*b*），1486年由佚名建筑师设计，引入了新的复杂工艺，建成我们如今熟知的文艺复兴时期建筑正立面。底层成为带有两层方形支柱，即壁柱的厚重底座，这些柱子不是均匀排列而是成对设计。凸出的屋檐又称檐口，仅与上层壁柱而不是整栋建筑的高度成比例。早期宫廷中的拱形窗户逐渐消失，平顶带有檐口的窗户逐渐出现。整体立面以浅面凸出的末端，即亭台结束。

1502年，建筑师多纳托・布拉曼特（1444—1514年）设计了小礼拜堂（*c*）来纪念传说中圣彼得的殉难之地。设计基于现存罗马圆形的神庙基础上，这座小教堂包括了一些现代元素如护栏、栏杆以及特殊的基督教标志。作为文艺复兴全盛期的一颗明珠，它得到了广泛的赞赏，很快获得了与古代建筑同等的声誉。

宏伟的抚慰圣母教堂（*d*），位于意大利中部小镇托迪城外，由一位鲜为人知的建筑师科拉・达・卡普拉洛拉于1508年修建。这座教堂由一系列圆形和方形设计构成，同时也是众多带有中央穹顶建筑中的一座。尽管十分明显，其灵感来自古代建筑，这些不切实际的独立式教堂带着对学术的迷恋，呈现出纯粹的几何形制，最终带来了对古罗马遗址的记录与分类。

文艺复兴的全盛期为古代世界带来了充满激情的佛罗伦萨派这样伟大的古典学派。随着时间的推移，人们完全掌握了古典设计的各种可能性，并且在不受中世纪无意识遗留的传统开放思想的阻碍下，创造出各种新事物。

HIGH RENAISSANCE

In 1417 Pope Martin V (1369-1431) was elected and three rival popes deposed, thereby ending the Great Schism which had impoverished the Pope's Italian territories and discredited the papacy. Thirty years later Pope Nicholas V moved his residence to the Vatican, signalling a recovery in papal finances and the start of an extensive building campaign. Renaissance artists and architects flocked to Rome which had been reduced to little more than a village among the ruins by centuries of vendettas and battles over papal elections. The new interest in ancient Rome led to the careful study of the many surviving monuments yet, tragically, the search for stone for the new buildings led to their destruction.

Rome became a major cultural focus and the centre of progress in Renaissance architecture and art. The Cancelleria, (*a*) and (*b*), designed in 1486 by an unknown architect, introduced a new level of sophistication to the by now familiar Renaissance palace façade. The lower floor has become a heavy base for two layers of flat columns - pilasters - which are not evenly spaced but grouped in pairs. The projecting eaves, or cornice, are in proportion to the top row of columns only, rather than to the total building height. The windows are starting to lose the arches of earlier palaces and flat-topped windows with their own cornice have been cautiously introduced. The whole elevation is completed with very shallow projecting end-wings, or pavilions.

In 1502 the architect Donato Bramante (1444-1514) designed a small chapel (*c*) to mark the place where St Peter was traditionally supposed to have been martyred. Closely based on the remains of surviving circular Roman temples, this tiny building also includes modern features such as railings, or balustrades, and specific Christian symbolism. A jewel of the High Renaissance, it came to be so widely admired that it was soon given the unique privilege of being placed on a level with the buildings of the ancients.

The huge church of S. Maria della Consolazione (*d*), just outside the small town of Todi in central Italy, was designed by a little-known architect called Cola da Capraola in 1508. The plan is made up from a series of circles and squares and is one of several similar buildings with one central domed space. Although obviously inspired by ancient buildings, the architects of these rather impractical free-standing churches played with pure geometric shapes with the same sort of academic fascination that led to the recording and classification of the ruins of ancient Rome.

The High Renaissance brought a greater knowledge of antiquity to the Florentine passion for the ancient world. The passage of time brought complete familiarity with the possibilities of Classical design and opened up creative possibilities unhindered by the unconscious legacy of the Middle Ages.

a

b

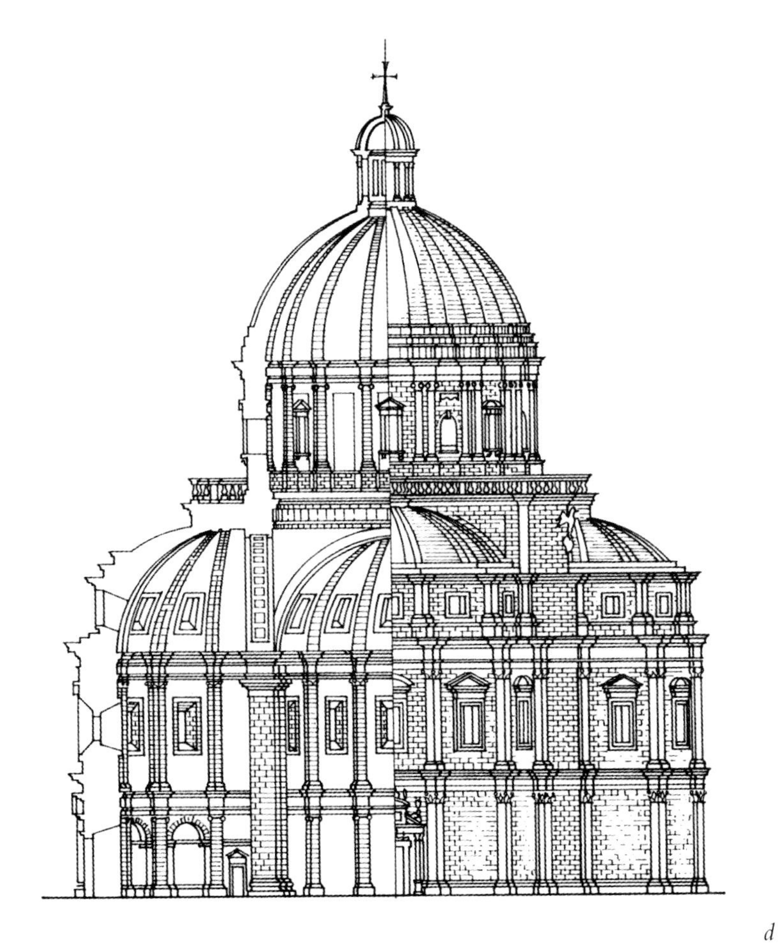

d

c

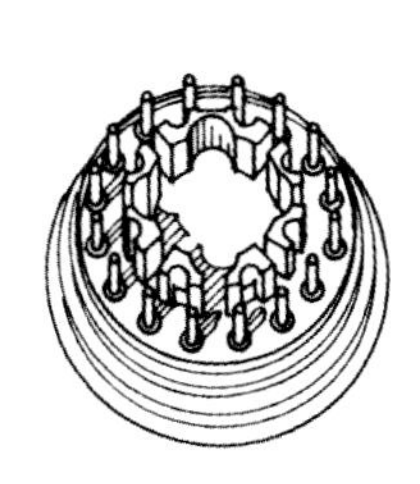

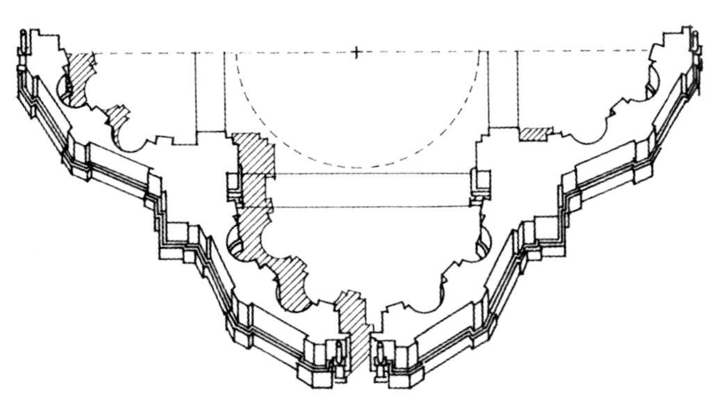

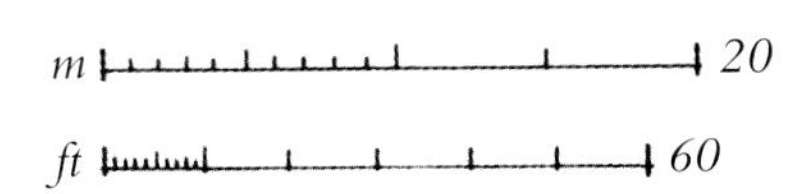

矫饰主义

1527年，历史上一直肩负保护教皇任务的神圣罗马帝国军队洗劫了罗马城，教皇沦为阶下囚。之前的短短几年，刚刚发生了路德与教会决裂的重大历史事件，这场发生在意大利的战争结束了自由散漫、腐化堕落的文艺复兴时期教皇制度。西班牙皇帝的胜利大军为意大利带来了新的灵性。在艺术领域，古罗马浪漫气息与人文主义思想中孕育的世俗理性激发出非凡的创造力和学识，然而，这些早已过时。疲惫的熟悉感和古典艺术的可能性与其他同时期的世俗气息结合，引导艺术家与建筑师寻找新的潜在灵感。扭曲的文艺复兴艺术思想觉醒了，矫饰主义便在这样自省的氛围中诞生。

佛罗伦萨的门廊（*d*）由贝尔纳多·布翁塔伦蒂（1531—1608年）于1574年设计，扭曲了几乎所有环绕大门的传统细部设计。支柱缩减为薄薄的轮廓，包围着一个独特的边框，又称额枋。其上，普通的弧形顶部即山花被切成两半，将每一半翻转后，成为较大半身像的小底座，实际目的是隐藏后面不协调的半圆形窗户。

朱利奥·罗马诺（1499—1546年）修建德泰府邸（*a*和*b*）的工程始于西哥特人洗劫罗马的前一年。建筑的每一面都不同，并各具特色，甚至两边相同的庭院都有各自独特的奇趣。北侧（*a*）有三个稍稍偏离中心的拱门，每个拱门两侧有悖常理地设计了不规则排列的支柱。西侧（*b*）大门立于正中，支柱规则排列，但支柱中间的石材却不一致，其细部设计十分生动。天花板之下，支柱之间的一块石头呈垂落之势，其下部，每扇窗户上方的拱顶石与之相接，将上面的石头撑起。

将古典设计从刻板的一致性中解放出来，伟大的艺术家米开朗琪罗（1475—1564年）在这方面的成就要远远超过其他人。他运用非凡的建筑艺术手法，重新启用古代建造原理和细部设计。他最后的作品，1561年在罗马修建的教皇皮乌斯四世之门（*c*） 只是一个单薄的立面，然而门与门层层相叠，这一结构以全新手法使用熟知的古典细部设计，赋予其无限的想象空间。

矫饰主义依赖观察者对古典建筑细部的熟识，创造出惊讶与戏剧性。没有文艺复兴的成功，矫饰主义便不会存在，尽管一些细部设计过于奇特而无法复制，但是许多这一时期的创造，在古典传统中永远占有一席之地。

MANNERISM

In 1527 the city of Rome was sacked by the army of the Pope's historic protector, the Holy Roman Emperor, and the Pope became his prisoner. Coming only a few years after Luther's momentous break with the Church, the war in Italy brought the carefree, corrupt, Renaissance papacy to an end. A new wave of spiritualism entered Italy with the Spanish emperor's conquering army. In the arts a century of frantic creativity and scholarship, inspired by the romance of ancient Rome and the worldly reasonableness of humanist thinking, had run out of steam. Weary familiarity with the possibilities of Classical art combined with the other-worldly atmosphere of the period to lead architects and artists to seek new inspiration in hidden meaning. A self-conscious, artistic distortion of Renaissance thought, Mannerism, grew out of this introspective climate.

On a doorway in Florence (*d*), designed by Bernardo Buontalenti (1531-1608) in 1574, almost all the traditional details that surround the door have been distorted. The columns are reduce to a thin profile on the outside of an unusual surround, or architrave; above them, the normal curved top, or pediment, has been cut in half and each half has been turned round, forming an undersized base for an oversized bust which in turn virtually hides an incongruous arched window.

The construction of Palazzo del Tè, (*a*) and (*b*), by Giulio Romano (1499-1546), began a year before the Sack of Rome. Each side of the building was different and had its own peculiar character. Even two sides of the same courtyard had individual quirks. On the north side (*a*) the three arches are just off-centre and on each side of them the columns are spaced with perverse irregularity. On the west side (*b*) the door is central and the columns have a regular pattern, but the stones between the columns are out of line and the details themselves seems to come alive. Just below the roof in the spaces between columns a stone appears to be falling out, and above each of the windows underneath, the central stone rises up to meet it, pushing apart the stones overhead.

The great artist Michelangelo di Lodovico Buonarroti Simoni (1475-1564) (known as Michelangelo) more than any other artist freed Classical design from rigid conformity. His buildings treated the principles and details so recently reclaimed from the ancient world with an extraordinary artistic licence. His last work, the Gate of Pope Pius IV in Rome (*c*) of 1561, is little more than a thin façade and yet one door is layered over another and the composition is given a restless drama by using familiar Classical details in unfamiliar ways.

Mannerism relied on the knowledgeable familiarity of the onlooker with traditional Classical details to create shock and drama. It could not have occurred without the success of the Renaissance and, although some details were too idiosyncratic to be repeated, many of the inventions of the period found a permanent place in the Classical tradition.

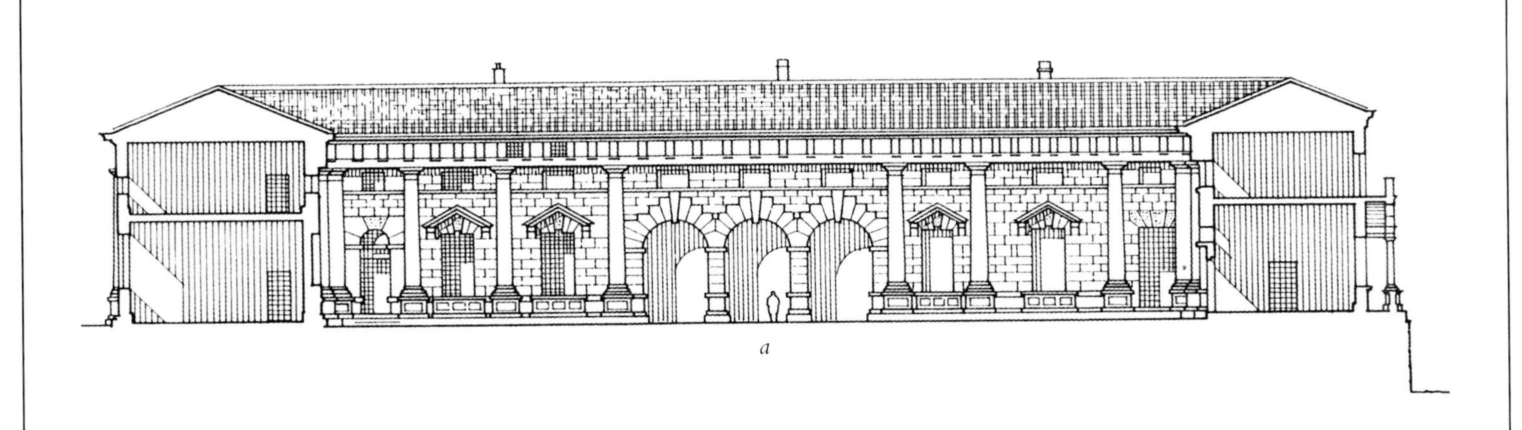

a

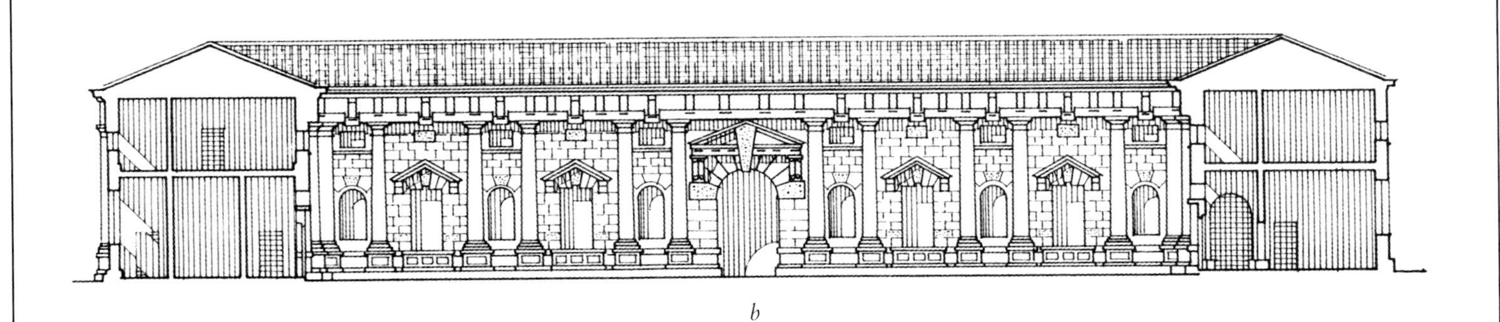

b

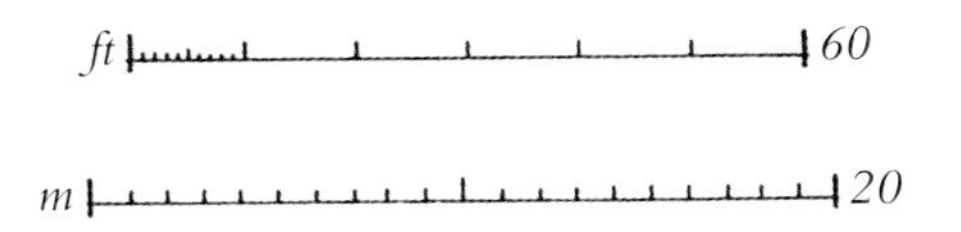

c

d

北方的文艺复兴

长达上百年的时间，基于古罗马纪念碑研究之上而修建的建筑一直保留着意大利风格。教会宣传古典文学中的人文精神，来自北方的游历画师带回文艺复兴的现实主义，但在这些国家中，哥特式建筑风格根深蒂固，他们的建筑依然坚定地保持着这一传统。

1494年，意大利入侵法国，将其卷入意大利政治体系，并使其成为首先受到新建筑风格影响的国家。意大利城邦与侵略者通过联盟寻求利益，他们的建筑师也来到了法国。但是，没有坚实的意大利文明基础，新式风格仅仅为年代久远的哥特式建筑技艺增添了时髦的装饰而已。

壮丽的尚博尔城堡采用中世纪城堡式设计，带有护城河，以环绕的建筑物作为城墙，并建有宏伟的城堡主塔，又称城堡主楼（*b*）。但这座城堡是一位意大利建筑师，多米尼克·达·科尔托纳，于1519年设计。引人注目的十字形平面与传统的城堡平面勉强匹配，外部设计生动别致，精美的意大利式细部设计别出心裁地叠加于庞杂的哥特式屋顶与角楼上。不久，本土建筑师开始探索新风格的意大利源头，但是法国的风俗得以保存，产生了独具特色的混合风格，并演变成法国对古典建筑的特殊贡献。

新教脱离罗马教廷使得许多北部国家与其天主教邻国分离。大西洋地区的贸易机遇为北部国家带来新的财富。到了这一时期，意大利建筑师摒弃了古代世界单纯的传统，用矫饰主义将其取代，这一风格可见于北部印刷出版并配以图例的图书中。在荷兰，这一特性被荒诞地夸大，其奢华的装饰形式传播到信仰新教的邻国。

1561年，科内利斯·弗洛里斯·德·弗里恩特（1514—1575年）修建了安特卫普市政厅（*c*），为低地国家富裕且思想独立的城市树立新的古典风格城市建筑模板。尽管弗洛里斯在意大利进行了研究，屋顶陡斜、窗户面积较大、极富装饰效果的垂直中央山墙，这些哥特式风格元素都来自他的祖国。1572年，罗伯特·史密森（1535—1614年）为富庶的朝臣在英国乡村朗利特（*a*）修建庄园。他增加了修长对称的古典支柱装饰和英国哥特式建筑中稀疏的方形细部设计，并带有凸出的大窗和平屋顶。

令人惊讶的是，在交流不发达并且欧洲出现分化的时代，古典建筑如此吸引人，以至于颠覆了长达几个世纪，存在于意大利的强大哥特式传统。这种风格在最初出现时，受到区域传统影响，发生了很多改变，这一点并不奇怪。

THE RENAISSANCE IN THE NORTH

Architecture founded on the study of the monuments of ancient Rome remained an Italian style for a hundred years. The Church had spread the humanist passion for ancient literature and travelling painters from the north had brought Renaissance realism home with them, but architecture remained firmly Gothic in those countries where its roots ran deep.

France was the first to come under the influence of the new architecture when the invasion of Italy in 1494 drew her into the web of Italian politics. Italian city-states sought advantage in alliance with the invader and their architects visited France. But, without its strong Italian foundations, the new style could be little more than fashionable decoration on long-established Gothic building techniques.

The great Chateau de Chambord was planned just like a medieval castle with a moat, a ring of buildings for walls, and a huge central donjon, or keep (*b*), but it was designed by an Italian, Domenico da Cortona, in 1519. A dramatic cross-shaped floor plan fits awkwardly into a traditional castle plan and the exterior is a picturesque combination of fine Italian details inventively superimposed on a Gothic jumble of roofs and turrets. It was not long before native architects sought out the Italian origins of the new style, but French custom survived and a distinctive hybrid was developed to become a uniquely French contribution to the Classical tradition.

The Protestant break with Rome isolated many of the northern states from their Catholic neighbours. New wealth had been created in the northern countries by the opening up of trading opportunities in the Atlantic. By this time Italian architecture had left behind the purity of the ancient world for the distortions of Mannerism and it was this style that found its way north in printed and illustrated books. In Holland these curiosities were grotesquely magnified and this extravagant form of decoration was exported to her Protestant neighbours.

When Antwerp Town Hall (*c*) was built by Cornelis Floris de Vriendt (1514-75) in 1561 it established the new Classical style as the model for civic buildings in the rich and independently minded cities of the Low Countries. Although Floris had studied in Italy, the steep pitch of the roof, the large areas of window and the richly decorated vertical central gable all come from the Gothic traditions of his native city. So too, when in 1572 Robert Smythson (1535-1614) designed a new house for a wealthy courtier at Longleat (*a*) in the English countryside, he added a thin symmetrical decoration of Classical columns and details to the sparse square shape of a late English Gothic building, with its distinctive large windows and flat roof.

In an age of poor communication and emerging divisions in Europe it is in some ways remarkable that Classical architecture should have been so universally attractive as to overturn centuries of a much stronger Gothic tradition than had existed in Italy. It is less surprising that it should first appear heavily modified by regional traditions.

a
40 ft
m 10
b
c

国际巴洛克风格

从16世纪中叶开始，欧洲卷入了长达百年的战争，新教徒和天主教徒为信仰而战，国家为独立而战，贸易阶级为自由而战。罗马教廷，在新教徒革命的冲击之下，经受了野蛮的西班牙军队以及旨在复兴新教的反宗教改革的残酷洗礼。到大约1625年，反宗教改革者相信，新教即使没有被打败，也受到了重创，于是放弃了严酷的信条，改为奉行繁盛与华丽。艺术与建筑风格开始迎合这一纵欲的新精神。这一风格就是横扫欧洲的巴洛克风格，每一个地区都以这样或者那样的形式从战争的伤害中复兴。到这一时期，古典艺术无处不在，巴洛克风格是整个欧洲大陆自13世纪以来首次共同采纳的一种建筑风格。

巴洛克发展时期最重要的人物是罗马雕刻家和建筑师吉安·洛伦索·贝尔尼尼（1598—1680年）。1658年，他在罗马设计的奎利那雷圣·安德烈教堂（*a*），其平面并非采用有序的方式将各种不同的古典元素和设计分割开来，而是把它们融合在一起，将整个建筑塑造成一座巨大的雕塑，扩充了熟悉的主题并将它们用一系列美丽的曲线连接起来，令观看者为之震撼，并提升了建筑给人们带来戏剧般的兴奋之情。

最早出现在法国的建筑，丢掉哥特式传统残余，使用意大利矫饰主义手法创造出了巴洛克风格。早期引入这一风格，在法国孕育出更加克制的传统，因而避免了后期出现于意大利和西班牙的极端风格。雅各·勒梅希亚（1585—1654年）设计的索邦神学院的教堂（*b*）始建于1635年，是法国最早采用这种意大利风格的建筑。华丽且具有研究价值的古典风格装饰，集中于中央大门的一排支柱间距各不相同，其上都冠以引人注目的穹顶，与60年前罗马的建筑相仿。

17世纪法国的政治权力将这种新建筑风格传入北部邻国。当英国走出内战，皇室也结束了在法国的流亡生活，巴洛克建筑便应用于充满活力与激情的重建项目中，创造出伟大的建筑，如建筑师约翰·范布勒（1664—1726年）和尼古拉斯·霍克斯莫尔（1661—1736年）设计的布伦海姆宫（*c*）。

巴洛克风格凝重的戏剧效果为古典建筑传统做出了重大并且长久的贡献。尽管巴洛克风格的自由与创新源于打破文艺复兴的一致性与矫饰主义的矫揉造作，但是在很多方面，这一建筑风格都与古代罗马的古典自由风格相近。

INTERNATIONAL BAROQUE

From the middle of the sixteenth century Europe was submerged in a hundred years of virtually continuous warfare, as Protestants and Catholics fought for their beliefs, nations fought for their independence and the trading classes for their freedom. The Roman Church, shocked by the Protestant revolution, purged itself in a savage, Spanish-inspired, militant austerity, the Counter-Reformation, which was designed to turn back the tide of Protestantism. By about 1625 the counter-reformers, confident that Protestantism was checked, if not defeated, abandoned austerity in favour of exuberant and glamorous propaganda. An artistic and architectural style grew up to serve this new spirit of self-indulgence. This, the Baroque style, swept across Europe, in one form or another, as each region recovered from the deprivations of war. By this time Classical architecture had made itself felt everywhere and with the universal adoption of the Baroque the whole continent shared one architecture for the first time since the thirteenth century.

One of the most important figures in the development of the Baroque was the Roman sculptor and architect Gian Lorenzo Bernini (1598-1680). In his church of S. Andrea al Quirinale in Rome (*a*) of 1658 the various Classical elements and the plan are not separated out in an orderly way, but merged together and moulded to make the building into a vast sculpture which, by enlarging familiar themes and uniting them with a series of sensuous curves, overwhelms the onlooker and allows the building to lift the emotions in a surge of theatrical exhilaration.

The first buildings in France to cast off the remnants of the Gothic tradition used the Italian Mannerist style that had given birth to the Baroque. This early introduction created a more restrained tradition in France which avoided the temptation to indulge in the later extremes of Italy and Spain. The church of the Sorbonne (*b*) by Jacques Lemercier (1585-1654), begun in 1635, was one of the first examples of this Italian style in France. The rich but scholarly Classical decoration, the varied spacing of the screen of columns concentrating on the central door, all crowned by a dramatic dome, resembled buildings in Rome of sixty years before.

The political power of France in the seventeenth century helped to introduce the new style to her northern neighbours. When England emerged from her civil war and the monarchy returned from French exile, Baroque architecture was adopted for a vigorous and enthusiastic rebuilding programme which saw the creation of such great buildings as Blenheim Palace (*c*) by the architects John Vanbrugh (1664-1726) and Nicholas Hawksmoor (1661-1736).

The heavy drama of the Baroque made a major and lasting contribution to the Classical tradition. Although it relied for its freedom and inventiveness on the break with Renaissance conformity and the deliberate perversity of Mannerism, in many respects Baroque architecture was closer to the Classical freedom of ancient Rome.

a
15 m
ft 50
b
c

帕拉迪奥及其传承

强大的海上城邦威尼斯巧妙地逃脱了分别于1527年和1530年毁灭罗马和佛罗伦萨的可怕战火蹂躏。伟大帝国庞大的军队将意大利半岛变为荒地，还试图说服威尼斯共和国巩固其领土内的财富，并放弃危险的扩张进程。随着威尼斯财富的累积，被放逐的艺术家与学者开始在这片中立的土地上寻求庇护。战争创造出的闭塞思想，农业革新促进乡村的发展，激励富裕的家庭修建房屋，并且管理庄园。

伟大的建筑师安德烈亚·帕拉迪奥（1508—1580年）在这一进步的环境中获得成功。他的建筑设计别出心裁地将矫饰主义的自由灵活与文艺复兴全盛期的成果结合起来。帕拉迪奥个人对罗马遗址与当时艺术准则的研究开阔了他的眼界，使其看到了未被发现的古代建筑的自由风格。维琴察梵蒂冈镇的基耶里凯蒂宫（*b*）结合了两层支柱的传统应用与新颖的由内翻向外侧的正立面设计，即开放式庭院的露台朝向前面。这种不那么威严的宫殿设计显示了威尼斯共和国稳定的政治环境。同时也为城镇中的乡村别墅设计带来了开放式风格。帕拉迪奥式乡村别墅为古典建筑开启了新的篇章。

1570年，帕拉迪奥出版了《建筑四书》，成为古典设计的范本。书中除了记录他对古典建筑的研究之外，还收录了他的教堂、宫殿和别墅设计。他为威尼斯贵族设计的用于管理与娱乐的乡村庄园别墅，自古以来，首次将建筑与周围的景观结合起来。

在英国，历时简短却屡被近现代所采用的帕拉迪奥建筑风格，淹没在内战当中。巴洛克风格于1715年以巴洛克风格矫正手法的形式再度出现。英国作为新教的主要信仰国家的兴起，赋予帕拉迪奥风格以国家认同与国际地位。总之，传统的英国乡村庄园带有大面积房舍与农场，采用帕拉迪奥设计，便具有了简单华丽以及令人难以抗拒的魅力。威廉·肯特设计的霍克曼大厅（*c*）、帕拉迪奥乡村别墅等建筑样式已成为英国的传统。

帕拉迪奥《建筑四书》的翻译及其经久不衰的盛名，加之大英帝国作为工业革命之父，其国际地位的提升，为帕拉迪奥风格带来了非凡的生命力和广泛的关注。这一风格受到大众的推崇，直至1796年，刚刚从英国殖民地获得自由的革命者之一托马斯·杰斐逊（1743—1826年），在设计其美国的蒙蒂塞洛宅邸（*a*）时，就借用了他的自然建筑风格。

PALLADIO AND HIS LEGACY

The powerful maritime city-state of Venice was skilful enough to escape the ravages of the terrible war that had brought down Rome in 1527 and Florence in 1530. The great armies of the huge kingdoms that had laid waste most of the Italian peninsula convinced the Venetian republic that it should consolidate its wealth within its borders and abandon the perilous course of expansion. There followed a rise in Venetian fortunes when artistic and intellectual exiles sought safety in the neutral republic. The war created a siege mentality and agricultural reforms improved the countryside, encouraging wealthy families to build houses to manage their estates.

The great architect Andrea Palladio (1508-80) flourished in this stimulating atmosphere. His buildings are a unique combination of the freedom of Mannerism and the scholarship of the High Renaissance. Personal study of the ruins of Rome and the artistic licence of the time opened his eyes to the undiscovered liberty of ancient architecture. His Palazzo Chiericati in the Venetian town of Vicenza (*b*) brings together a traditional application of two layers of columns and a novel inside-out façade where the open courtyard terraces are brought out to the front. This less threatening form of palace perhaps shows the political stability of the Venetian republic, but it also brings the open character of countryside villa design into the town and it is Palladio's country villas that start a new chapter in Classical architecture.

In 1570 Palladio published his *Four Books of Architecture*, which became a standard textbook on Classical design. Alongside his studies of ancient buildings were his own designs for churches, palaces, and villas. His villas for the Venetian noblemen's management and enjoyment of their improved country estates combined, for the first time since the ancient world, architecture and the total surrounding landscape.

A brief and virtually contemporary adoption of Palladian architecture in England was submerged by civil war and the Baroque, only to re-emerge in 1715 as a sober antidote to the Baroque. The rise of Britain as a major Protestant power gave the new Palladianism a national identity and an enduring international status. Above all, the traditional British country estate with its large house and working farms found Palladio's creation of a simple grandeur in a rural setting irresistible and, in buildings such as Holkham Hall (*c*) by William Kent (1685-1748), Palladio's Venetian villas became a British tradition.

The translation and lasting fame of Palladio's *Four Books* and Britain's rising position in the world as an imperial power and father of the Industrial Revolution gave Palladianism a remarkably long life and widespread popularity. The style became so generally accepted that, as late as 1796, one of the first colonial revolutionaries to gain independence from Britain, Thomas Jefferson (1743-1826), could regard Palladianism as a natural style for his own house, Monticello (*a*), in the United States of America.

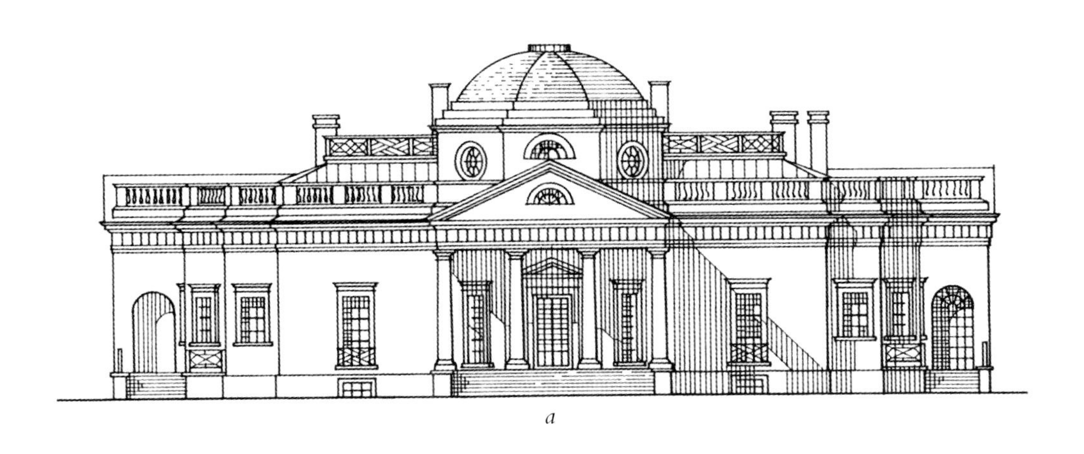

a

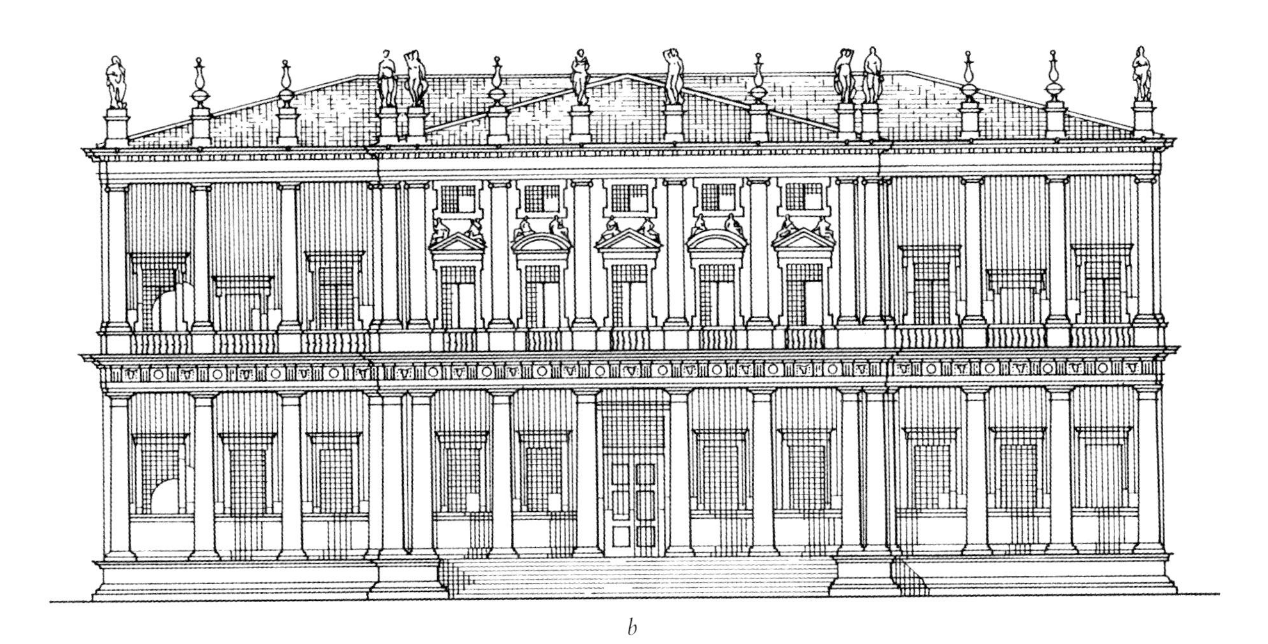

b

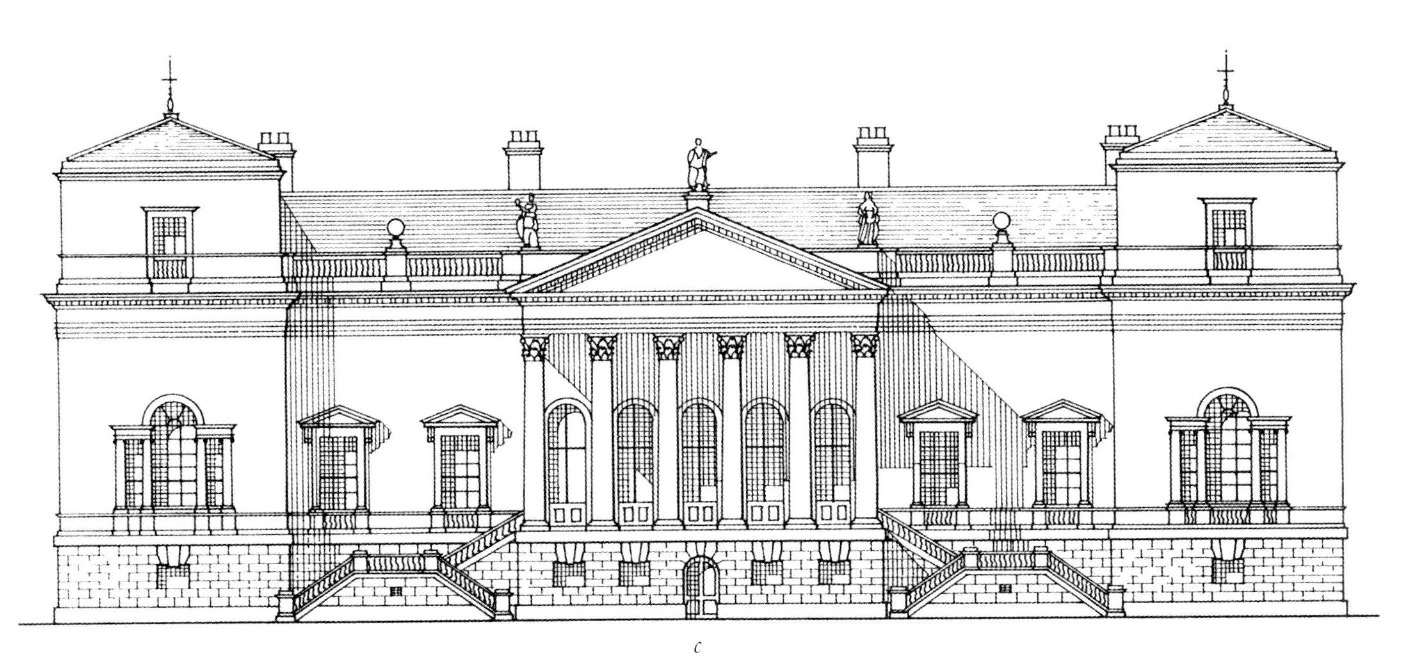

c

30 ft   m 10

洛可可风格

当英国成为主要的新教国家并摒弃了生机勃勃的巴洛克风格时，罗马天主教国家却将这一风格设计推进到不可超越的浮夸与奢华的水平。18世纪上半叶，在西班牙、法国、意大利以及欧洲中部的国家，巴洛克风格笨拙的效果被精妙地分解为坚固结实的单个元素，汇集成极其丰富的表面装饰与视觉幻象。这一逃避主义的伟大迸发称为洛可可风格。

洛可可风格似乎是从西班牙自然演变而来的。在这里，古典建筑单纯的形式一直只是表面的印象。八百年后，穆斯林的统治结束于1494年，但阿拉伯复杂的装饰艺术传统为建筑保留了奢华感。17世纪修道士推崇的严谨风格被抛弃，18世纪自然巴洛克风格再度出现，首次在外来的指引下得以强化。马德里省立孤儿院大门（*d*）由佩德罗·德·里韦拉（1681—1742年）在1722年设计，整个建筑浸润在矫饰古典风格的背景中，石刻外墙上迸发出象征着花朵、珠宝、水果的纹案，还有瀑布般悬垂的布帘和无数变形的传统古典细部装饰设计。这种几乎与古典风格无关且疯狂的艺术形式被战胜了印第安文明的美洲中部和南部的人们所接受，他们的本土艺术在本质上与这种精神十分接近。

像德国这样的小国逐渐忘却了30年战争的恐惧，奥地利也终于松了一口气，于1683年将土耳其人从维也纳打出国门，中部欧洲出现了辉煌灿烂的社会，皇帝、主教、王公贵族们开始竞相追逐艺术声誉。本国的建筑师取代了意大利移民建筑师，创造出极其精美的本国洛可可风格，应用在教堂和宫殿中。这种新风格与教堂、宫殿的气氛十分协调，因此也毫不费力地融入日常生活的艺术中。德累斯顿的茨温格宫，举行仪式的场所建有宏伟的大门（*c*），由马特乌斯·丹尼尔·培尔曼（1662—1736年）于1711年设计，比西班牙皇宫建筑更加精致优美。建筑师们不是压抑古典形式，而是将传统细部设计拆分后，以令人叹为观止的方式重组，构成曲线与双弧线、阴影与凸出装饰。

洛可可基本上是一种与其所应用的建筑完美结合的装饰艺术。它可以应用于内部装饰、家具和同种类的纺织品上，通过法国进入了以更加朴素的建筑为设计背景的英国。正如图中两种柱头（*a*）和（*b*）所示，这一风格并不顾及古典建筑理想，其卓越的自由与偶尔的不对称性在古典设计中独树一帜。

ROCOCO

While Britain adopted the role of the leading Protestant power and turned its back on the exuberance of the Baroque, the Roman Catholic countries took Baroque design to unsurpassed levels of illusion and extravagance. During the first half of the eighteenth century in Spain, France, Italy, and the states of central Europe, the ponderous drama of the Baroque was transformed by a fragile disintegration of everything that appeared solid and substantial into a riot of surface decoration and visual make-believe. This remarkable outburst of escapism is called Rococo.

The natural home of Rococo seemed to be Spain, where Classical architecture in its purer forms had never made anything more than a superficial impression. Eight hundred years of Muslim rule finally ended in 1494, but the legacy of the sophisticated decoration of the Arabs had, however, preserved a taste for extravagant architecture. A severe style, introduced by monkish monarchs in the seventeenth century, was abandoned in the eighteenth century and the natural Baroque reappeared, reinforced for the first time by a common direction abroad. The door of the Hospicio Provincial in Madrid (*d*), designed by Pedro de Ribera (1681-1742) in 1722, almost submerges the Mannerist Classical background in a crust of stone carved in an explosion of decoration representing flowers, jewels, fruit, cascades of drapery, and innumerable distortions of traditional Classical details. This barely Classical, frenzied art was eagerly absorbed by the conquered Indian civilisations in Central and South America who had a native art surprisingly similar in spirit.

As the small states of Germany emerged from the horrors of the Thirty Years War, and Austria breathed a sigh of relief having turned back the Turks from the gates of Vienna in 1683, a brilliant society emerged in central Europe where emperors, bishop princes, and archdukes competed for artistic prestige. Local architects replaced migrant Italians and a highly sophisticated regional Rococo developed around the church and the palace. This new style was so attuned to the sentiments of the population that it was effortlessly incorporated into everyday art. Great buildings such as the Zwinger in Dresden, a ceremonial enclosure with lavish gates (*c*), by Matthäus Daniel Pöppelmann (1662-1736) in 1711, are more fragile and elegant than their Spanish counterparts. Rather than overwhelm the Classical forms, the architect has taken the traditional details apart and reconstructed them in a breathtaking composition of curves and double curves, shadows, and projecting decoration.

The Rococo was essentially a type of decoration which at its best united with the architecture to which it was applied. It could be used for interior details, furniture, and fabrics and in this form, principally by way of France, it found its way into the more austere architectural background of Britain. As two column capitals (*a*) and (*b*) show, in its remarkable freedom and occasional asymmetry it is virtually unique in Classical design in paying no regard to the ideals of ancient architecture.

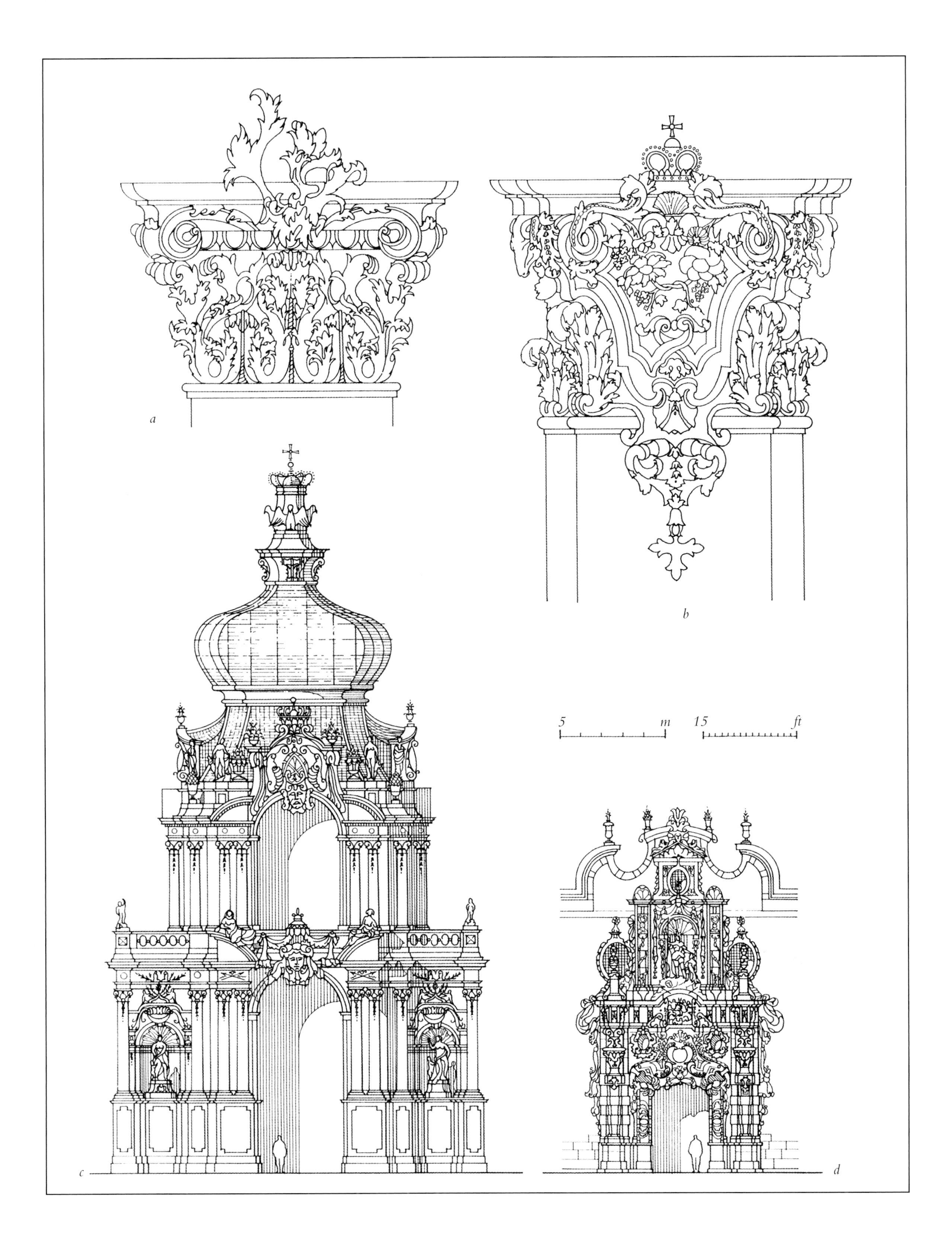
a
b
5 m
15 ft
c
d

新古典主义

18世纪，欧洲开始摆脱长达几个世纪政府体系。新的哲学、更好的教育和贸易带来的财富增长以及工业革命使得新兴阶级对国王与皇后世袭的专政感到不安。18世纪中叶，一些发展最终带来了美国与法国的共和革命并且削弱了大英帝国。

当人们摆脱了长久以来的文化历史羁绊，新的思维方式出现了，人们开始追溯简单纯粹的理想之源，并试图使其得以复兴。在寻求真理基础的过程中，哲学家审视脱离了社会习俗的自然人的生活与特性。对历史本身的研究与考古学的出现使得艺术家与建筑师通过认真重建过去来寻求灵感。

领主们模仿废墟修建建筑物，英国绅士旅行参观真正的希腊罗马废墟。哥特式风格开始复兴，欧洲和美洲进入与历史浪漫关系时期，一直延续到今天。1740年，罗马法国学院的建筑师基于他们对该城市古代纪念碑的解读，创造出新的设计，开启了新古典主义方向。普鲁士考古学家约翰·约阿希姆·温克尔曼（1717—1768年）于1754年再度将希腊艺术引入西方。1762年，《古希腊建筑图解》出版，由英国建筑师詹姆斯·斯图尔特（1713—1788年）和尼古拉斯·里维特（1720—1804年）编著，展示出的简单设计令现代人震撼。同时，苏格兰建筑师罗伯特·亚当（1728—1792年）以其出版的著作而闻名，他的书中主要撰写对古罗马更加复杂建筑的研究以及对英国帕拉迪奥式建筑变体进行更加精妙的解读，例如1772年修建的皇家艺术协会（*a*）。

改革领袖在世纪末将古希腊与古罗马共和人士看作精神上的祖先，新的建筑给他们的理想以生动的表达。希腊和罗马文化在美国北部参与革命的各州中同样受到热烈欢迎。托马斯·杰斐逊，新共和国第三任总统，亲自设计了弗吉尼亚州夏洛茨维尔的新大学，1817年修建的图书馆（*b*）就是以罗马万神殿为模板建造的。

法国的政治和视觉革命进行得更加激烈。建筑师们，比如艾蒂安·路易斯·布勒（1728—1799年）和克劳德·尼古拉斯·勒杜（1736—1806年），希望政治革命带来类似古希腊和古罗马宏伟纪念碑般极简的原始建筑设计。他们的质朴、有创造力的项目包含了理想城镇规划和纪念碑，表达了政治理念并象征理想。设计如勒杜府邸（*c*）表达了新古典主义后期，人类在即将进入19世纪重组世界时的威严与刻板。

NEO-CLASSICISM

In the eighteenth century, Europe began to cast off its centuries-old systems of government. New philosophies, better education, and increased wealth from trade and the Industrial Revolution made the growing middle classes uneasy with the inherited dictatorship of kings and queens. By the middle of the century, a process was under way that would lead to republican revolution in America and France and a weakened monarchy in Britain.

A new way of thinking emerged as people turned from the burden of their accumulated cultural history and sought rejuvenation in the imagined purity and simplicity of their primitive origins. In the search for fundamental truth, philosophers looked to the life and character of the natural man, free from the conventions of society. The study of history for its own sake and the invention of archaeology led artists and architects to seek inspiration in a carefully reconstructed past.

Landowners built mock ruins and English gentlemen travelled to the real ruins of Greece and Rome. The Gothic Revival began and Europe and American entered into a romantic relationship with the past that has lasted to the present day. In 1740, architects at the French Academy at Rome produced designs based on their interpretations of the ancient monuments of the city which were the first examples of a new Classical direction, Neo-Classicism. A Prussian archaeologist Johann Joachim Winckelmann (1717-68) reintroduced Greek art to the west in 1754. *Illustrations of the architecture of ancient Greece,* published in 1762 by the British architects James Stuart (1713-1788) and Nicholas Revett (1720-1804), revealed a simplicity that shocked contemporaries. At the same time, the Scottish architect Robert Adam (1728-92) made his name from books on the more complex buildings of ancient Rome and transformed the Palladian architecture of Britain with elegant interpretations of his studies, such as the Royal Society of Arts (*a*) of 1772.

The revolutionary leaders at the end of the century saw their spiritual forefathers as the republicans of ancient Greece and Rome, and the architecture gave eloquent expression to their aspirations. In the revolutionary states of North America, Greece, and Rome were embraced with equal enthusiasm and Thomas Jefferson, third president of the new republic, personally designed a new university in Charlottesville, Virginia, in 1817 with a library (*b*) closely modelled on the Pantheon in Rome.

In France the political and visual revolutions were more violent. Architects such as Étienne-Louis Boullée (1728-99) and Claude Nicholas Ledoux (1736-1806) anticipated the political revolution with designs of a brutal simplicity inspired by the primitive buildings of ancient Greece and the vast monuments of Rome. Their stark, inventive projects included imaginary schemes for towns and monuments which expressed political theories and symbolic ideals. Designs such as the house illustrated (*c*) by Ledoux were early expressions of the imposing severity of the last phase of Neo-Classicism as it entered the reshaped world of the nineteenth century.

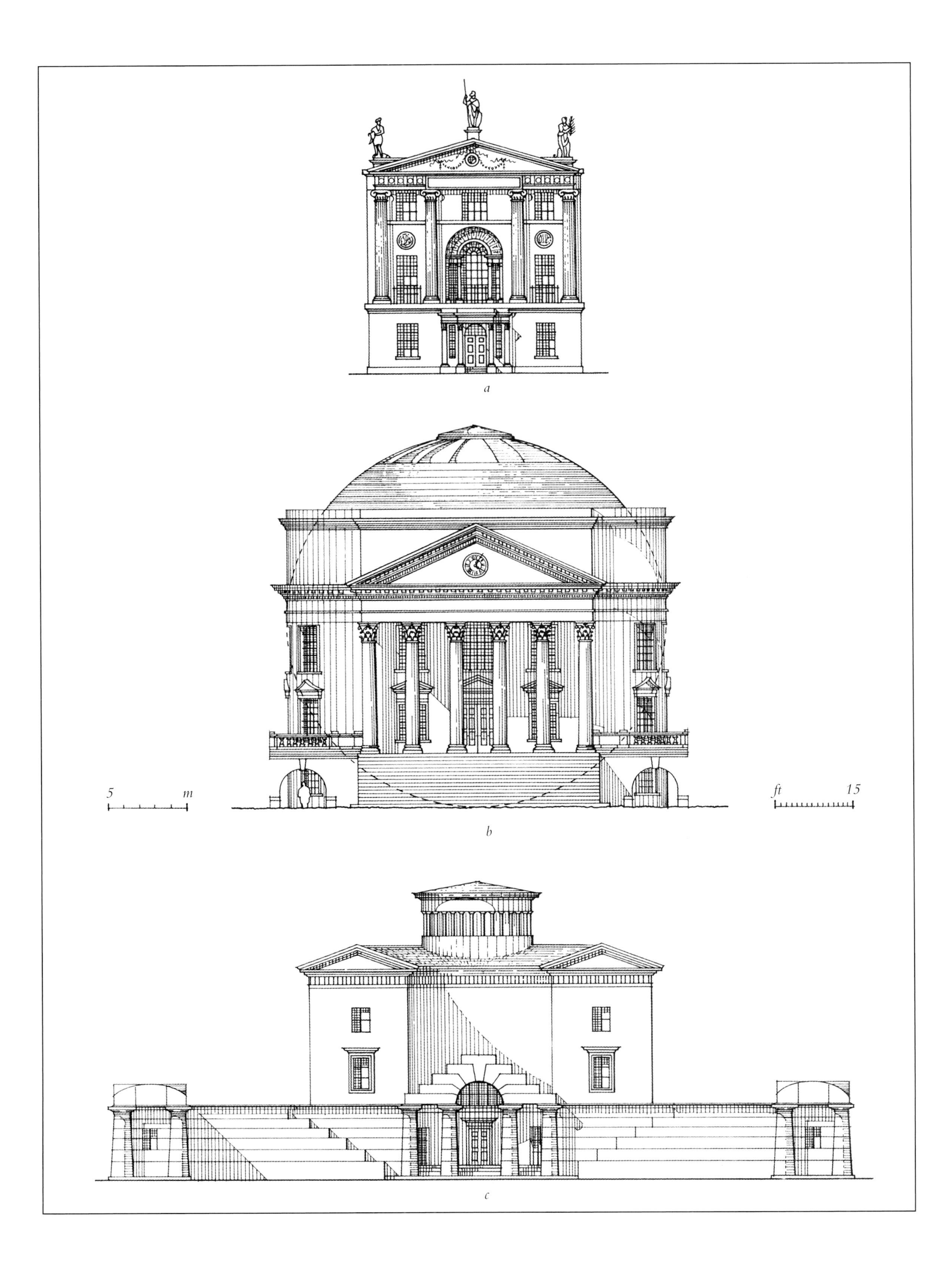

a
5
m
ft
15
b
c

新文艺复兴建筑

到19世纪中叶，奥地利、德国、瑞士、比利时、意大利和英国，大范围的反抗运动表明法国大革命的影响不仅局限于本国。工业革命带来的财富及其造成的痛苦，冲破英国本土，加之政治动乱后的民主，创造出一个在需求上比以往更加复杂与多样化的社会。

人们最初传统意义上对潜在理性秩序的信仰已经遗失，对历史荒谬的解读却已成为主导。宗教信仰与文学的复兴建立在激情与恐惧的基础上。比起讨好雇主，艺术家更热衷于自我表达，在阁楼上忍饥挨饿，建筑师也厌倦了新古典主义建筑的条条框框。哥特复兴主义者将自己与宗教联系起来，并坚决地蔑视与古典建筑轻浮地结合。致力于典雅装饰的哥特式，以及其设计中为满足更加复杂的工业社会所融入的自由元素，与这一世纪的品位与需求相一致。

新古典主义的建筑师在这一世纪中继续他们的设计，但武断地与古代世界的形制结合造成了居住功能上的困难。新文艺复兴解决了新时代的问题：其优雅的比例以及装饰与旧式新古典建筑形成对比。多种多样的建筑形式与自由创造更多建筑类型的可能性，满足了复杂平面的需求。早期文艺复兴式建筑与哥特式建筑之间模糊的关系，以及文艺复兴时期多样的发展，创造了自由的混合风格与细部设计，成为19世纪建筑的特点。

文艺复兴的影响在本世纪的最初10年首先出现于巴黎。1828年，伟大的普鲁士新古典建筑师卡尔·弗里德里希·申克尔（1781—1841年）为一位陶土制造商设计了一幢柏林住宅（*a*），使其雇主能够用简单文艺复兴正立面展示其装饰产品。在随后的10年中，查尔斯·巴里（1795—1860年）于1837年设计的改革俱乐部（*b*）确立了复兴地位，成为英国的先驱。巴里简单的文艺复兴全盛期建筑外观设计，功能强大的内部构造，巧妙结合绅士俱乐部中不同的居住需求，周围环绕意大利式庭院并带有玻璃屋顶用以抵御英国的恶劣天气。由查尔斯·弗伦·麦金（1847—1909年）、威廉·卢瑟福·米德（1846—1928年）、斯坦福·怀特（1853—1906年）在纽约设计的亨利·G·维拉德宅邸（*c*），将六幢住宅连接起来围绕着庭院，其中一面开放面向街道。 这一卓越的城市住宅设计修建于1882年。它出现较晚，见证了19世纪新文艺复兴在将功能与视觉需求结合起来方面取得的伟大成就。

RENAISSANCE REVIVAL

By the middle of the nineteenth century, popular revolt in Austria, Germany, Switzerland, Belgium, Italy, and Britain demonstrated that the explosion of the French Revolution could not be contained within national boundaries. Both the wealth and the misery of the Industrial Revolution also broke out of their British home and, together with the democratic consequences of political unrest, created a society with more complicated and varied needs than ever before.

Belief in an underlying rational order in our primitive origins was lost and an irrational interpretation of the past came to the fore. There was a revival of religious faith, and literature dwelt on passion and horror. Artists, more interested in self-expression than patrons, starved in their garrets and architects tired of the bare orderliness of Neo-Classical buildings. The Gothic Revivalists allied themselves with religion and decisively spurned their frivolous association with Classical architecture. The busy decorative elegance of Gothic, and the freedom it allowed in designing for the more complex demands of an industrial society, were in tune with the taste and needs of the century.

Neo-Classical architects continued to design throughout the century, but by dogmatically adhering to the forms of the ancient world made the accommodation of new functions difficult. The revival of the Renaissance solved the problems of the new age: its elegant proportions and use of decoration contrasted with old-fashioned Neo-Classical buildings. The large range of building types, and the freedom to create more, answered the need for complicated plans. The ambiguous relationship between early Renaissance and Gothic architecture, and the varied development of the Renaissance, were behind a free mixing of styles and details that became characteristic of the nineteenth century.

The influence of the Renaissance first appears in Paris in the first decade of the century. In 1828 the remarkable Prussian Neo-Classical architect Karl Friedrich Schinkel (1781-1841) designed a house in Berlin (*a*), for a manufacturer of terracotta, which displays his client's decorative products on a simple Renaissance façade. In the following decade the revival became established and with his Reform Club (*b*) of 1837 Charles Barry (1795-1860) became a pioneer in Britain. Barry's simple High Renaissance exterior has a functional interior which skillfully organises the varied accommodation required for a gentleman's club around an Italian courtyard with a glass roof to keep out the English weather. The Henry G.Villard houses (*c*) by Charles Follen McKim (1847-1909), William Rutherford Mead (1846-1928), Stanford White (1853-1906) in New York City, bring together six houses around a courtyard, opening on one side to the street. This distinguished urban design was built in 1882. Its late date is a witness to the great success of the Renaissance revival in meeting the functional and visual needs of the nineteenth century.

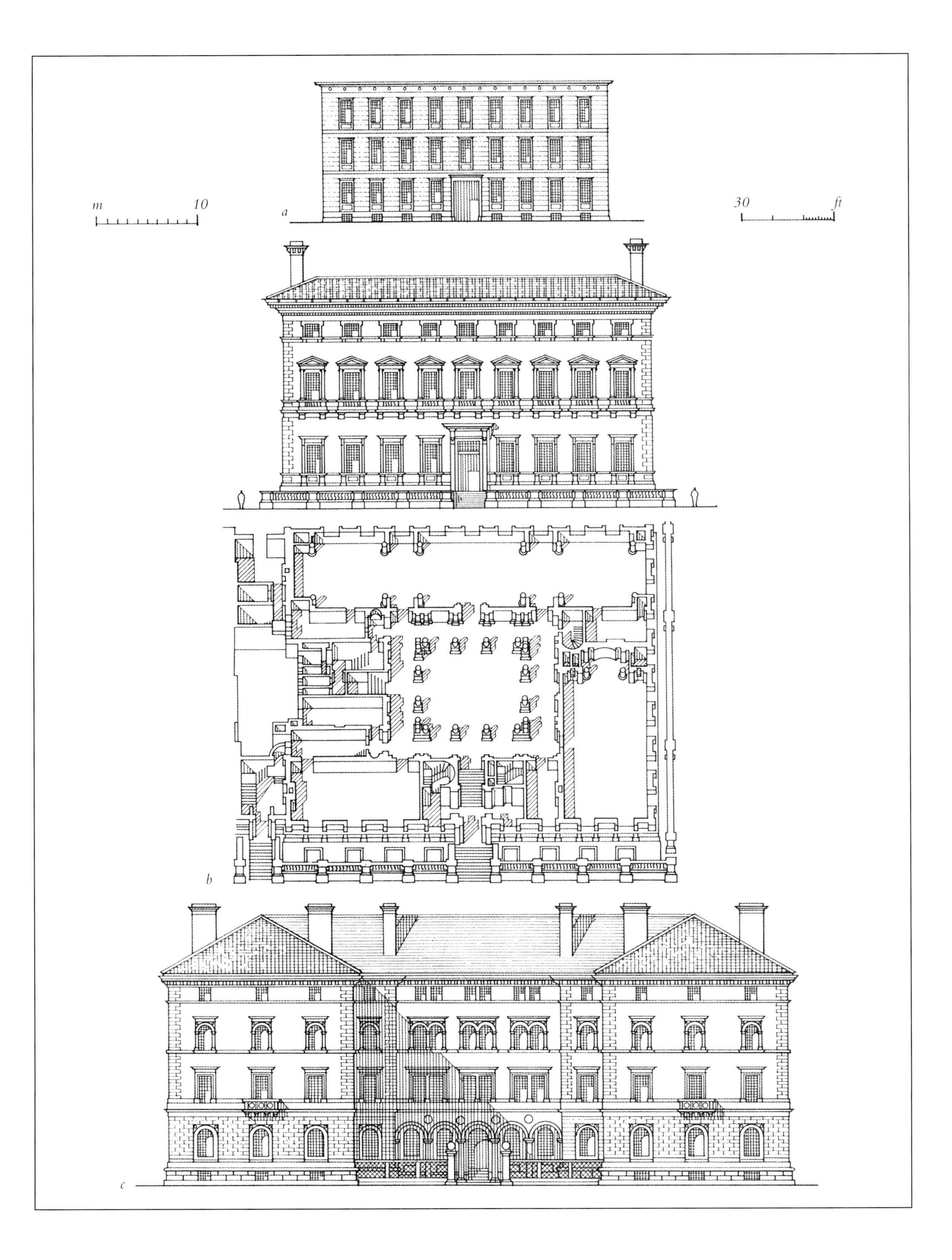

m
10
a
30
ft
b
c

工艺美术运动

工业革命为英国带来最初的权力与财富，但是也是肮脏和剥削的根源。制造业导致恶劣的居住环境，改革制造过程被慈善家们认作解决上述问题的方法。同时，哥特式复兴不仅是基督教与英国的特色，中世纪被理想化地认为是道德的时代，不知名但却内心满足的工人们能够手工制造出极其精美的产品。

19世纪60年代约翰·罗斯金（1819—1900年）的作品以及威廉·莫里斯（1834—1896年）的文章、产品和影响，和其他人一起带来了社会和艺术的融合，开启了工艺美术运动。运动的先驱者们制造高品质的家具、陶器、纺织品和金属制品。这些作品在实用特征的基础上，装饰朴素、符合道德标准。尽管这一运动并未带来工业改革，其产品却受到有鉴别力的卖家们的推崇。

在建筑中，早期工艺美术风格的建筑物很少。这些大都是独立的房屋，并尝试创造出本土哥特式建筑风格。这一运动起初并未对建筑产生重大影响，直到1884年艺术工作者行会的成立。

作为奠基人的建筑师们都来自诺曼·肖（1831—1912年）工作室，正如其1875年在伦敦修建的天鹅宅邸（*a*）所展示的，这一时期进步的建筑师开始逐渐脱离哥特式建筑风格而形成自由独特的英国风格，这一风格在精神上是古典主义，同时受到18世纪早期安妮女王时期本土建筑的影响。

这一风格自然且迅速地复制并传播到美国，例如，1879年约翰·阿普尔顿·威尔逊（1868—1920年）为麦克金夫人修建的巴尔的摩府邸（*b*）。这一古典风格导向并不是工艺美术运动的主流，但是自由古典元素常与哥特式平面以及这一运动的显著特征混合出现。一些建筑师继续研究莫里斯严谨的哥特式理念，但这种纯粹的理想通常是短命的。到世纪交替的时候，独具特色的古典建筑，例如由欧内斯特·牛顿（1856—1922年）在沃金厄姆的拉科蕾（*c*）修建的建筑物就是由许多工艺美术运动的老建筑师设计的。这些建筑大多数遵守了这一运动的基本原则：支持手工艺、就地取材并且创造独特的地域风格。

工艺美术运动影响巨大，但相对而言创造出的产品却不多。这一运动既影响了在现代化运动中寻求终结古典运动的人们，又通过建筑师影响了20世纪的古典建筑，在这一点上不免有些矛盾。

THE ARTS AND CRAFTS MOVEMENT

The Industrial Revolution brought Britain early power and wealth, but was also a source of squalor and exploitation. As manufacturing industry was the cause of the bad living conditions, a reform of the manufacturing process seemed to some philanthropists to be the solution to the problem. At the same time the Gothic Revival was not only considered to be uniquely Christian and English, but the Middle Ages were idealised as a moral age where anonymous but contented workers hand-made products of great beauty.

In the 1860s the writings of John Ruskin (1819-1900) and the writings, products, and influence of William Morris (1834-1896), amongst others, brought socialism and art together, creating the Arts and Crafts Movement. Pioneers of the movement made furniture, pottery, textiles, and metal-work of the highest quality. The practical features of these pieces were embellished with restrained decoration applied according to moral rules. Although the movement failed to reform industry, its products were much admired by discerning buyers.

In architecture early Arts and Crafts buildings were few. These were mostly individual houses which tried to create a native Gothic domestic character. The movement did not really start to have a major architectural impact until the foundation of the Art Workers Guild in 1884.

All the founding architects were from the office of Norman Shaw (1831-1912) and, as his Swan House in London (*a*) of 1875 shows, by this time progressive architects were starting to move away from the Gothic to a similarly free and distinctly British style, Classical in spirit and inspired by early-eighteenth-century domestic architecture in the reign of Queen Anne.

This style was readily copied and quickly spread to the United States in buildings such as the Baltimore house (*b*) by John Appleton Wilson (1868-1920) in 1879 for Mrs. McKim. This Classical direction did not dominate Arts and Crafts architecture, but free Classical elements are often to be found mixed with Gothic planning and some of the distinctive features of the movement. Some of the architects continued to observe the strict Gothic ideals of Morris, but this purity was generally short-lived. By the turn of the century, distinctly classical buildings such as Luckley at Wokingham (*c*) by Ernest Newton (1856-1922) were being designed by most of the old Arts and Crafts architects. Most of these buildings remained true to the principles of the movement in the support of the craftsmanship, the use of local materials and the creation of a distinctly regional character.

The Art and Crafts Movement had an influence out of proportion to the small quantity of its products. The movement was paradoxically to influence both those who sought the end of the Classical tradition in the Modern Movement and, through its own architects, the course of twentieth-century Classical architecture.

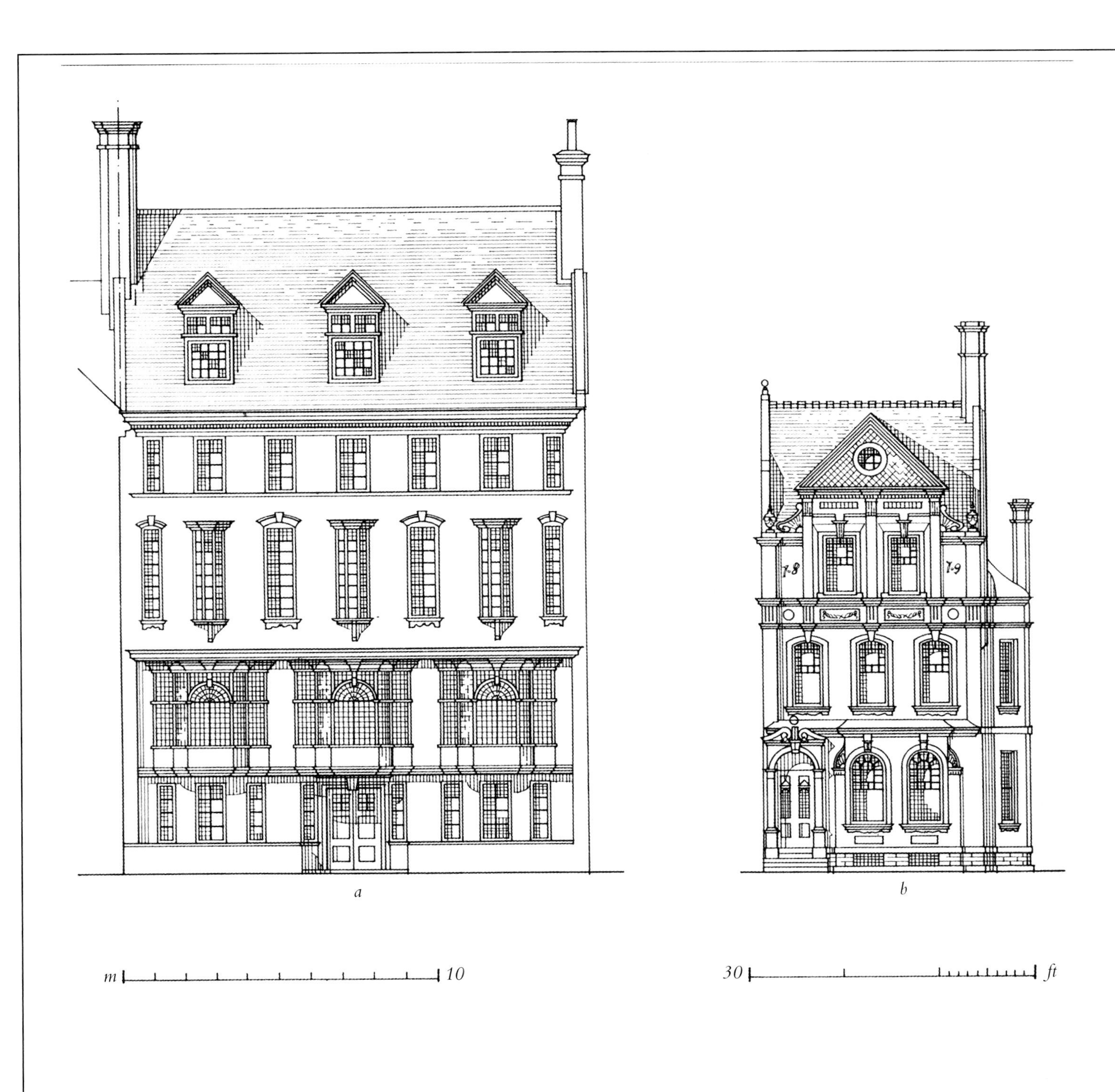

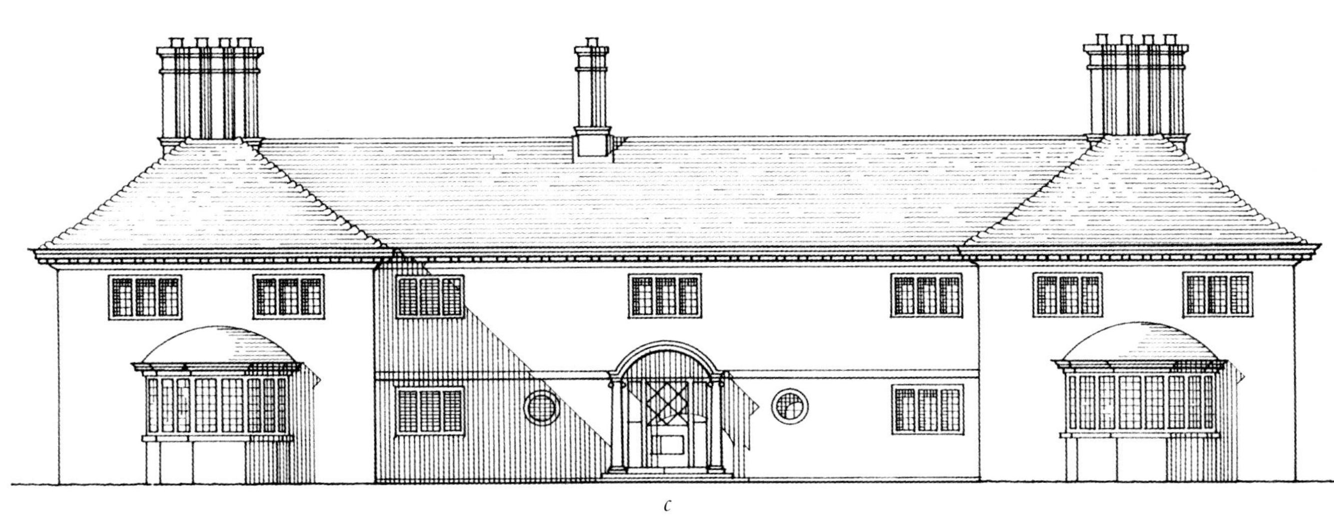

c

巴洛克风格复兴与学院派

1871年欧洲两大势力，意大利与德国联合起来。随后进入了激烈竞争时期，新旧国家争夺产品市场，后来者试图在海外划分版图，军国主义势力也盛行起来。美国从国内战争中逐渐复原，随后进入了帝国主义时代。东方的日本也蠢蠢欲动。19世纪中期，宽松的自由贸易时代宣告结束，世界某些国家开始为自己的帝国进行宣传造势。

根植于罗马荣耀之上的帝国理想在建造艺术与文学的推动下不断膨胀。整个西方世界，通过经济扩张获得繁荣，禁锢的帝国市场在富丽堂皇的建筑工程上挥金如土。英国与法国都求助于他们17世纪的建筑师，这些建筑师将巴洛克式炫耀的创造力与本国特色结合在一起，使得新建筑物既富于创新又享有盛名，且不受考古学条条框框的限制。

在法国，古典传统通过一种独特的官方建筑教育体系，在整个19世纪都保持着繁荣与活力，这种教育体系培养建筑师设计富丽堂皇的公共建筑物，还包括派往罗马的法国学院进修。到19世纪末期，这一风格被称为学院派，因巴黎建筑学院而得名。类似查尔斯·吉罗（1851—1932年）设计的小皇宫这样的建筑（*b*和*c*），为1900年巴黎世界博览会而修建，展现出风格独特、烦琐复杂、恢宏壮丽的中轴平面，以及丰富、壮观的立面。这一风格在美国产生了强烈的反响，在这里自信的新兴富人阶级修建大量的公用建筑与商业建筑，例如建于1911年的纽约公共图书馆（*a*），设计师为约翰·卡雷拉（1858—1911年）和托马斯·黑斯廷斯（1860—1929年）。

在英国，巴洛克复兴被工艺美术运动所取代，这些建筑师们将自己的独创与本国特征灌输其中。工艺美术运动与哥特式复兴都在生机勃勃、美轮美奂的古典风格建筑物面前相形见绌。例如，位于兰卡斯特市由约翰·贝尔彻（1841—1913年）与约翰·乔亚斯（1868—1952年）在1907—1909年设计的阿什顿纪念碑（*d*）。

帝国主义国家思想是第一次世界大战的关键导火索，也是国际社会主义革命兴起的重要原因。新的奢华古典风格，以及理性与庄重的哥特式复兴的终结推动了现代化运动，一场艺术革新正在酝酿之中，这场革新摒弃所有旧的象征手法，寻求创造新的社会主义世界。艺术在即将到来的世纪中，将采用新的技术服务于社会。

BAROQUE REVIVAL AND THE BEAUX ARTS

Eighteen hundred and seventy-one saw the unification of two major European powers, Italy and Germany. A period of fierce competition ensued, new and old nations struggled to maintain markets for their products, the late starters tried to carve out overseas territories for themselves and an atmosphere of militant nationalism prevailed. As the United States recovered from the Civil War, it too entered the imperial arena while Japan started to flex its muscles in the east. The relaxed, free-market days of the middle of the nineteenth century were over and the major nations of the world set in motion a propaganda campaign for their empires.

Establishment art and literature inflated imperial ideals and dwelt on the glories of Imperial Rome. Throughout the western world the prosperity gained from expanding economies and captive imperial markets was lavished on grandiose building projects. Both England and France turned to their seventeenth-century architects, who combined ostentatious inventiveness of the Baroque with national individuality, allowing new buildings to be creative and prestigious without the restrictions of archaeological correctness.

In France the Classical tradition had been vigorously maintained throughout the nineteenth century by a unique official architectural education system training architects in the design of grand public buildings and included study at the French Academy in Rome. By the end of the century the style came to be called Beaux Arts, after the architectural school in Paris. Buildings such as Charles Girault's (1851-1932) Petit Palais, (*b*) and (*c*), built for the Paris Exhibition of 1900, show the characteristic, complex, majestic axial plan and the rich and imposing elevations of the style. This was particularly influential in the United States where a new self-confident wealth was stimulating the construction of a large number of impressive public and commercial buildings such as the New York Public Library (*a*) of 1911, by John Carrére (1858-1911) and Thomas Hastings (1860-1929).

In Britain the Baroque Revival was adopted by architects of the Arts and Crafts Movement, imbuing it with their own brand of originality and nationalism. Both the Arts and Crafts Movement and the Gothic Revival declined in the face of the exuberant, ostentatious Classical architecture of buildings such as the huge Ashton Memorial (*d*) by John Belcher (1841-1913) and John Joass (1868-1952), built (1907-09) in memory of Lord Ashton's second wife, in the town of Lancaster.

Imperial nationalism was one of the contributing causes of First World War and of the rise of international revolutionary socialism. The new sumptuous Classical style and the demise of the rational and moral Gothic Revival helped to fuel the Modern Movement, a revolutionary artistic undercurrent which rejected all that the past represented and sought to create a new socialist world served by an art that looked to the new technologies of the coming century.

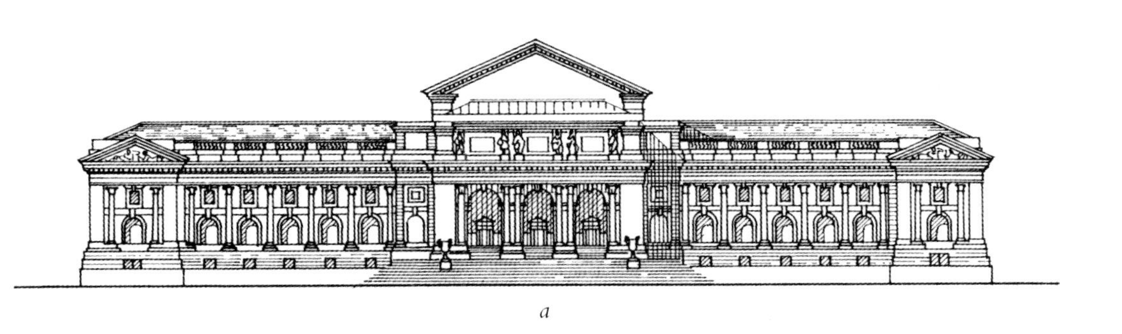

a

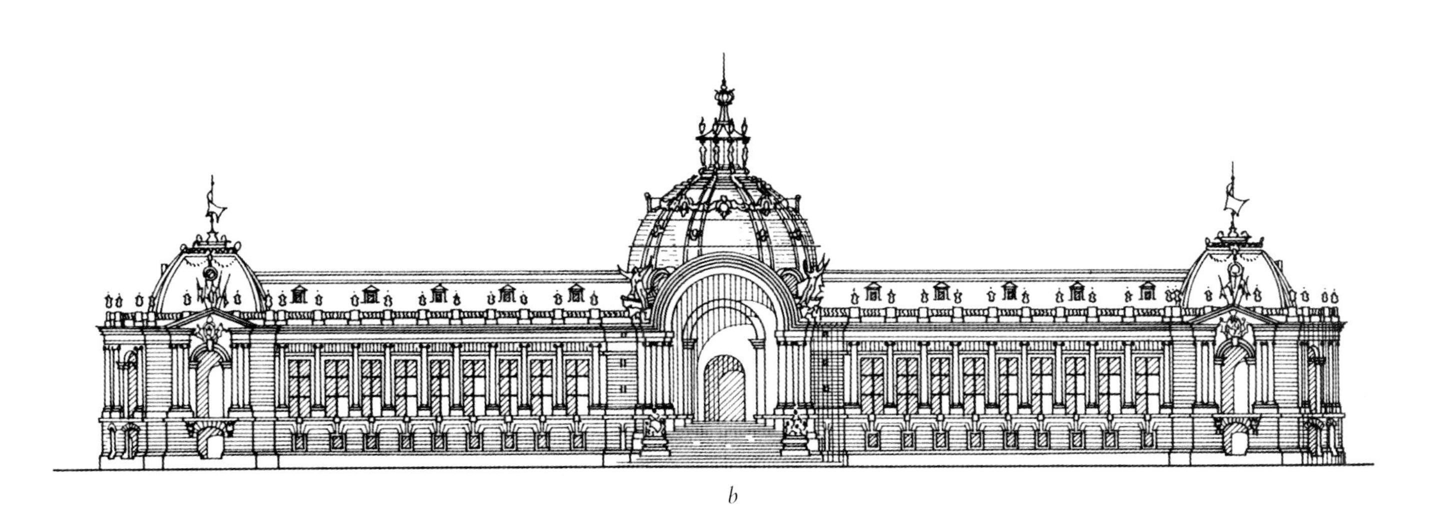

b

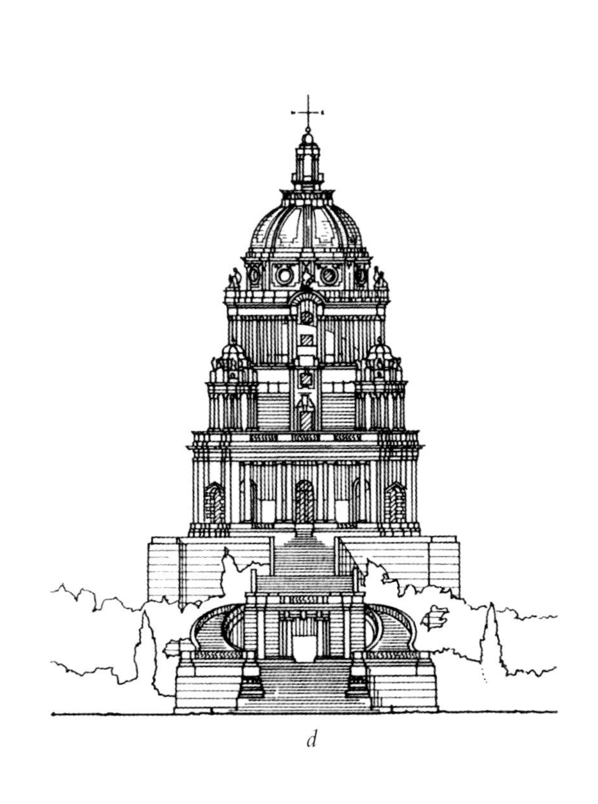

d

m 20 ft 66

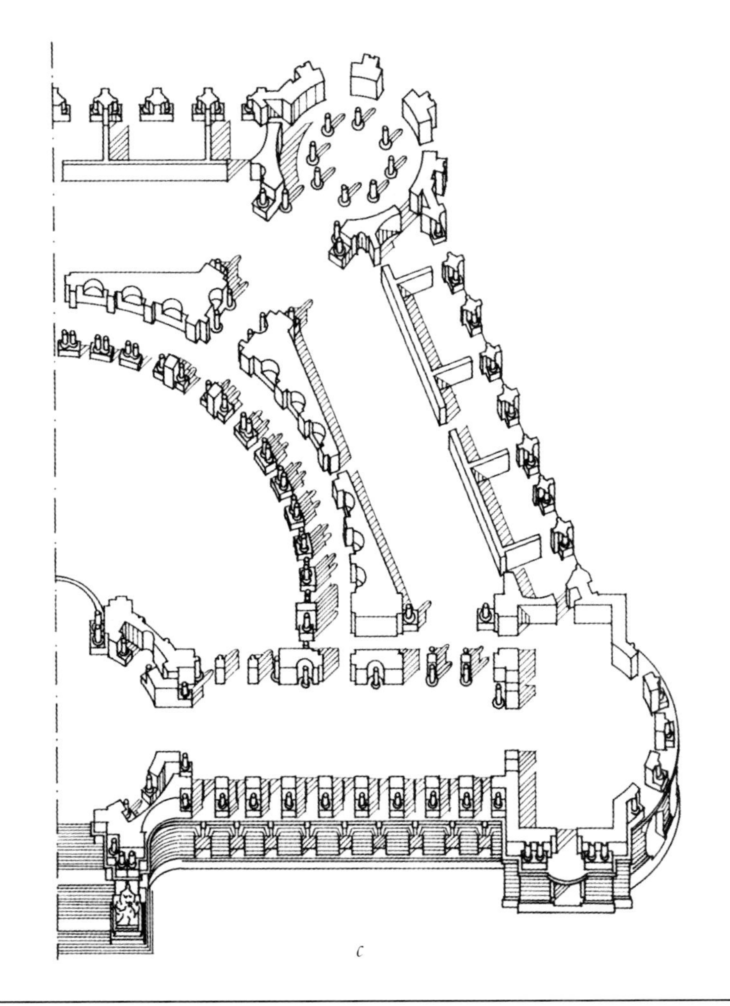

c

两次世界大战期间

第一次世界大战的余波，将西方国家卷入数十年的混乱与革命中，在此期间滋生出了为利益而进行更加血腥冲突的各种因素。战前，一些民主程度各不相同的帝国主义国家为贸易与影响力而进行争夺。战后，西方的政治局面发生了巨大的变化，怀揣着不同甚至是相悖的政治理想的国家政府之间，关系日趋紧张。

在战胜国中，美国、英国和法国保持了他们历史遗留下来的民主，尽管法国政府的不稳定已是臭名昭著，欧洲的帝国开始分裂。意大利，作为战胜国之一，滋生出法西斯主义，作为操控战后经济与社会动荡局面的独裁手段。法西斯主义受到战败国、德国和奥地利，以及内战胜利后的西班牙这些国家的狂热推崇。“俄国”国内进行了共产主义革命，造成了当时的政局不稳，国际社会主义貌似对其他国家的稳定产生了“威胁”。中立国，如瑞士和瑞典，未受这些动荡的影响，仍维持其民主体系。

第一次世界大战后的“俄国”与德国，尤其是德国，现代化运动的艺术家与建筑师们试图创造新的艺术潮流，摒弃旧的传统。这在两个国家中由于毫无怜悯之心的独裁者的出现而戛然而止，他们坚持保留古典风格，这种西方世界主要的建筑形式。现代化运动在民主世界里成为少数派风格。

在这20年中，古典建筑代表所有不同的西方政治势力。在德国，阿道夫·希特勒（1889—1945年）对建筑的兴趣，既源于个人因素又把其看成是维系权力的宣传媒介。他的建筑师贝托尔德·康拉德·赫尔曼·阿尔伯特·斯佩尔（1905—1981年）受到普鲁士新古典主义建筑师申克尔的影响，为美化政权设计了纪念碑，修建了一系列的政府办公大楼，例如柏林的总理府（*a*）。在英国，帝国摇摇欲坠，国家没有任何政治连贯性可言。建筑师们，如埃德温·勒琴斯爵士（1869—1944年）继续建造更加内敛的战前巴洛克式建筑，例如伦敦的英伊石油公司的新办公大楼（*c*）。一股简单克制的风潮浮出水面。在英国，称这一风格为新乔治亚式，但它对首创社会福利的斯堪的纳维亚国家影响更大。瑞典建筑师贡纳·阿斯普隆德（1885—1940年）是北欧古典主义的主要拥护者，这一风格可见于斯德哥尔摩市图书馆（*b*）。这种古典主义的简单风格，使得现代化运动在斯堪的纳维亚国家发展出与古典建筑并不矛盾的关系。

BETWEEN THE WARS

The aftermath of the First World War threw the western nations into a decade of confusion and revolution out of which emerged the ingredients for a further bloody collision of interests. Before the war a series of imperial monarchies with different levels of democracy were competing for trade and influence. After the war the political complexion of the west changed dramatically and national governments of quite different and opposing political ideals viewed each other with increasing unease.

Among the victors, the United States, Britain, and France retained their historic democracies, although the French government was notoriously unstable and the European empires began to break up. Italy, one of the victors, invented fascism as an authoritarian way out of the financial and social chaos caused by the war. Fascism was taken up with enthusiasm by the losers, Germany and Austria, and emerged triumphant out of a civil war in Spain. The communist revolution that had taken Russia out of the war was responsible for much of the political instability of the period, as international socialism seemed to threaten the stability of other countries. Neutral nations such as Switzerland and Sweden, unaffected by these major upheavals, maintained their own democratic systems.

In the years after the First World War in Russia and, in particular, in Germany, artistic and architectural innovators of the Modern Movement tried to create a new artistic direction by abandoning tradition. This was abruptly halted in both countries by unsympathetic dictators who insisted on retaining the Classical tradition that was the principal form of architecture in the other western nations. The Modern Movement continued as a minority style in the democracies.

Classical architecture in these two decades represented all the differing political systems in the west. In Germany, Adolf Hitler (1889-1945) had an interest in architecture both personally and as a propaganda medium that sustained his power. His architect Berthold Konrad Hermann Albert Speer (1905-81) designed monumental schemes, influenced by the Prussian Neo-Classical architect Schinkel, to glorify the regime, and a number of official buildings, such as the Chancellery buildings in Berlin (*a*), were constructed. In Britain, the empire teetered on, and the country experienced, no political discontinuity. Architects such as Sir Edwin Lutyens (1869-1944) continued to build in a more subdued version of the pre-war Baroque, such as the new offices of the Anglo-Iranian Oil Company, London (*c*). A new air of restrained simplicity was emerging. In Britain it was called Neo-Georgian, but it was most influential in the pioneering welfare states of Scandinavia. The Swedish architect Gunnar Asplund (1885-1940) was one of the principal exponents of this Nordic Classicism, seen in buildings such as his Stockholm City Library (*b*). The simplicity of this style of Classicism allowed the Modern Movement to develop a less contradictory relationship with Classical architecture in the Scandinavian countries.

a
b
c
m
20
70
ft

新的开端

第二次世界大战的结束带来了“俄国”影响之外的民主的联合，也开创了西欧中间偏左的政府选举。在重建的同时，人们还致力于创造更加美好的社会。战争的压力带来技术的革新，新工业为战争造成的住宅问题提供解决办法，并注入了巨大的信心。胜利的狂喜使人们相信，通过社会工程来创造更好的社会是可行的。社会的动荡和顽固不化的欧洲等级社会制度的腐朽，使得欧洲人民为进行翻天覆地的变革做好了准备。

这对于现代主义运动来说是绝佳的发展环境，拥护者们坚决反对保留传统，并且乐观地相信这一运动在艺术方面的优越性，从其在全球被采纳的程度来看就完全得以证明。现代派的建筑师长久以来一直为社会工程研发设计方案，其设计基于新技术的采用。这一运动因被希特勒禁止而获得了心理上的优势。其结果就是完全占领了艺术与建筑学派，同时抑制了古典传统的残余影响。

在现代化运动完全占据主导地位的25年中，其社会功用不断受到质疑，人们也没能学会欣赏这种艺术，并且逐渐演变成强烈的厌恶之情。1974年石油危机，动摇了人们对现代科技的信心，一些艺术家和建筑师开始寻求更能引发人们共鸣的表达方式，这带来了古典传统的复兴。

现代主义运动对新古典建筑影响最广泛的是后现代主义运动。巴黎城外的大规模建筑群，除了西班牙建筑师里卡多·博菲尔（生于1939年）设计的阿伯拉克萨斯建筑群（*a*），外形上明显是古典主义，却有着对新材料本身的展示，也表现出典型的现代主义运动风格中令人瞠目结舌、别具一格的扭曲形态。另一个发展方向就是否认，不仅否认现代主义运动，同时还否认大部分19世纪对完整历史风格的精心再现，例如，建筑师艾伦·格林伯格（生于1938年）在康奈狄克州修建的新农舍（*b*）。在这两个极端之间是类似汉普郡的新办公楼（*c*），设计师探索古典主义传统的延续，并不否认新科技的存在或者古典主义传统发展的可能性。

少数不断增长的建筑师群体回归古典主义传统，但是现代主义运动仍然支配着艺术与建筑的确立。三千年的流行艺术传统，拥有足够的能力突破五十年的文化垄断。

A NEW BEGINNING

The end of the Second World War brought democratic unity to the nations outside the influence of Russia, and the election of left-of-centre governments in western Europe. The task of rebuilding was accompanied by a determination to create a better society. The technical advances made under the pressure of war instilled a great confidence in the possibilities of new industrial solutions to inadequate housing and war damage. The exhilaration of victory inspired a belief in the feasibility of making a better society through social engineering. The upheavals of war and an erosion of the rigid European social hierarchy prepared the population of Europe for the introduction of radical change.

This was the perfect climate for the Modern Movement whose exponents were radically opposed to the retention of tradition and were optimistic enough to believe that the movement's artistic superiority was so self-evident as to merit universal adoption. Modernist architects had long been developing schemes for social engineering and their designs were based on the exploitation of all that was new in technology. The movement also enjoyed the psychological advantage of having been banned by Hitler. The result was the complete takeover of all artistic and architectural institutions and the suppression of all remnants of the Classical tradition.

After twenty-five years of the total dominance of the Modern Movement, its social assumptions were increasingly called into question and the population, far from learning to appreciate it, had developed a strong distaste for its aesthetic. When the oil crisis of 1974 undermined confidence in technology, some artists and architects began to look for more sympathetic means of expression. This led to a rebirth of the Classical tradition.

The influence of the Modern Movement remains in the most widespread form of the new Classical architecture, the Post-Modern. The huge housing complex outside Paris, Espaces d' Abraxas (*a*) by the Spanish architect Ricardo Bofill (*b.*1939-), while obviously Classical in form, relies on a display of new materials for their own sake and a shocking novelty of distortion that is typical of the Modern Movement. Another direction has been the denial of not only the Modern Movement but much of the nineteenth century in careful recreations of a complete historic style such as the United States architect Allan Greenberg's (*b.*1938-) new farmhouse in Connecticut (*b*). Between these two extremes are buildings such as the new office building in Hampshire (*c*) by the author which seek to continue the Classical tradition without denying the existence of new technology or the possibilities of progress within the Classical tradition.

A small but growing number of architects have returned to the Classical tradition, but the Modern Movement still dominates the artistic and architectural establishment. The weight of three thousand years of popular artistic tradition has, however, sufficient energy to overcome fifty years of cultural dictatorship.

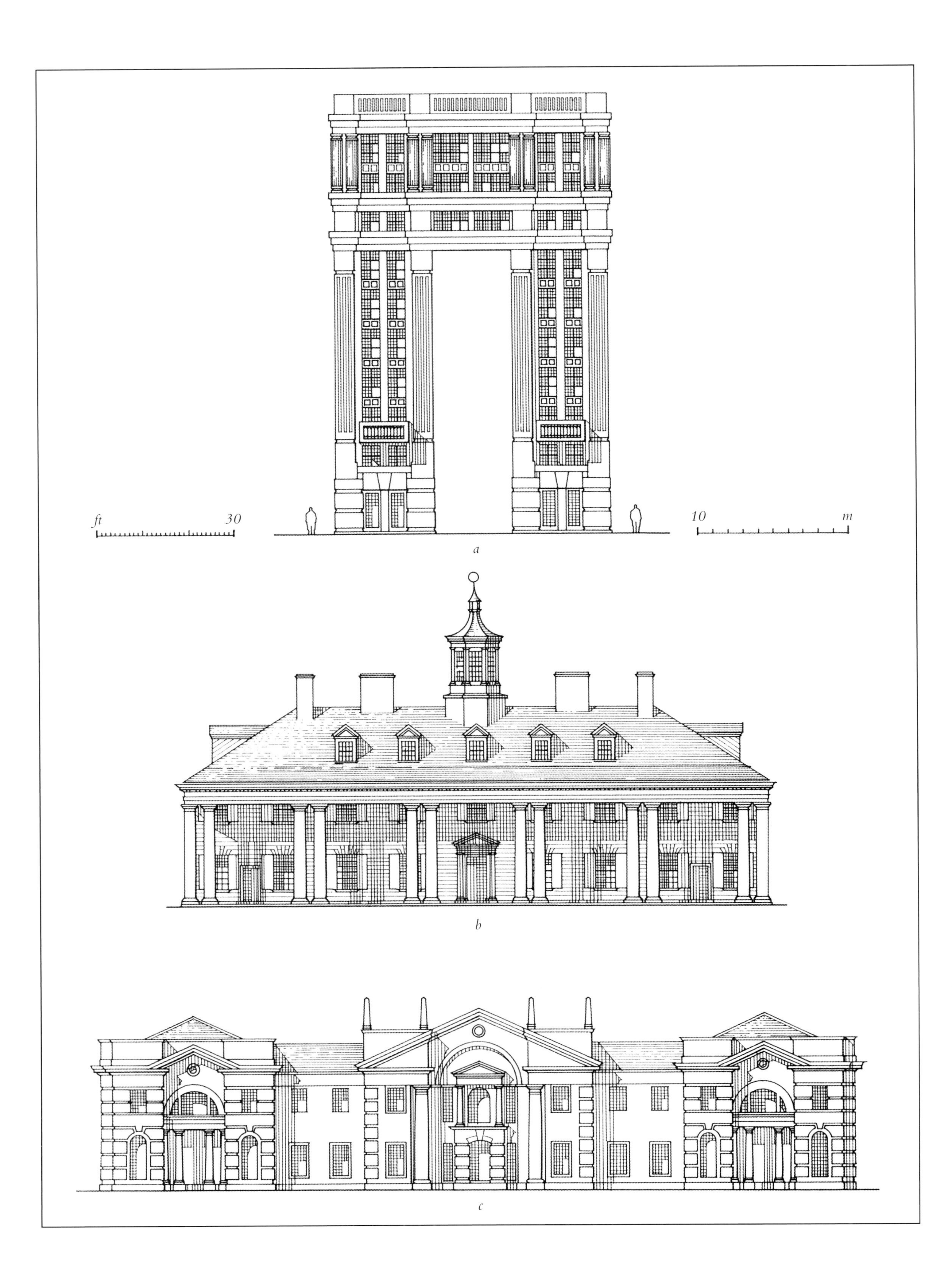
ft
30
10
m
a
b
c

2. 古典主义建筑

神庙

在过去的几个世纪中，几种类型的古典建筑在几代建筑师的努力下不断进化演变。有时，依据实际应用的原因而选择建筑类型，有时则依据建筑物的关联意义。通常是综合上述原因而做出选择，建筑类型的发展历史通常会激励建筑师在新建筑与其先祖之间做出创新的应用。

最能够代表古典建筑传统的就是神庙。在P6～P8中，描述了其发展为古典建筑进程设定模式的过程。这些最终使得许多建筑师开始设计神庙式的建筑。

由于这一建筑形式的最初意图是为了膜拜神，因此神庙式设计通常应用于教堂、犹太礼拜堂以及象征着国家荣耀的建筑物上。巴黎的玛德莲教堂（*c*）由皮埃尔·亚历山大·维尼翁（1763—1828年）于1807年设计，为纪念拿破仑·波拿巴（1769—1821年）军队的不朽荣耀而建，在外观上超越了宏伟的罗马神庙。18年后，雅克·玛丽·胡维（1783—1852年）将其内部重新改造成为教堂，以罗马浴场的设计风格为原型。

英格兰汉普郡庄园（*a*）由威廉·威尔金斯（1778—1839年）在1809年从一座老旧建筑改造为早期希腊神庙式建筑。空心柱与深门廊的戏剧化效果，置于高高的山脊上俯瞰河流，这些往往比居住者的方便更加重要。这种不妥协的设计并不常见。

建筑师海因里希·斯特拉克（1805—1880年）和弗里德里希·奥古斯特·施蒂勒（1800—1865年）巧妙地设计出柏林国家美术馆（*d*），融入罗马神庙风格。建筑物的高度、给人留下深刻印象的台阶和皇帝雕像展示了设计中的民族自豪感，这些都是十分重要的。

华盛顿特区的林肯纪念堂（*b*），作为民族英雄的纪念碑，成功地创造了与古代神庙相似的功能。这是由亨利·培根（1866—1924年）于1911年为安放前总统的巨大雕像而修建的，象征着国家与总统的权力。它基本上与雅典的帕提农神庙相近，但是在平面上做了改变，入口设在长边上。

2. THE CLASSICAL BUILDING

THE TEMPLE

Over the centuries certain types of Classical building have evolved through the contribution of successive generations of architects. Sometimes building types are chosen for practical reasons, sometimes for their associations. Often, a combination of these reasons lies behind the choice. The history of a building type can always inspire the architect to make creative use of the relationship between a new building and its predecessors.

Above all others the building type that has come to symbolise the Classical tradition is the temple. On page 6, 7, and 8 we have seen how its development set the pattern for the progression of Classical architecture. This has led many architects to design temple-like buildings.

As a building originally intended for worship, the temple, has often been adopted as the pattern for churches, synagogues, and buildings that glorify the state. La Madeleine in Paris (*c*), designed by Alexandre Pierre Vignon (1763-1828) in 1807 to celebrate the secular glory of Napoleon Bonaparte's (1769-1821) army, has the outward appearance of a large Roman temple. Eighteen years later the interior was redesigned as a church by Jacques-Marie Huvé (1783-1852), in a style derived from Roman baths.

The Grange in Hampshire, England (*a*) was remodelled by William Wilkins (1778-1839) in 1809 from an older house to resemble an early Greek temple. The drama of the open columns and the deep porch, positioned high on a ridge overlooking a river, was more important than the convenience of the occupants. Uncompromising designs of this kind were not often repeated.

The architects Heinrich Strack (1805-1880) and Friedrich August Stüler (1800-1865) skillfully designed the National Gallery in Berlin (*d*) to fit into a temple of a Roman type. The height of the building, the impressive flight of steps and statue of the Kaiser indicate that the expression of national pride in the design was of great significance.

The Lincoln Memorial in Washington D.C. (*b*) is a monument to a national hero which successfully fulfils a function similar to that of an ancient temple. It was designed by Henry Bacon (1866-1924) in 1911 to house a huge statue of the former president and includes symbolic references to the state and presidency. It is loosely based on the Parthenon in Athens but the plan is modified to provide an entrance on the long side.

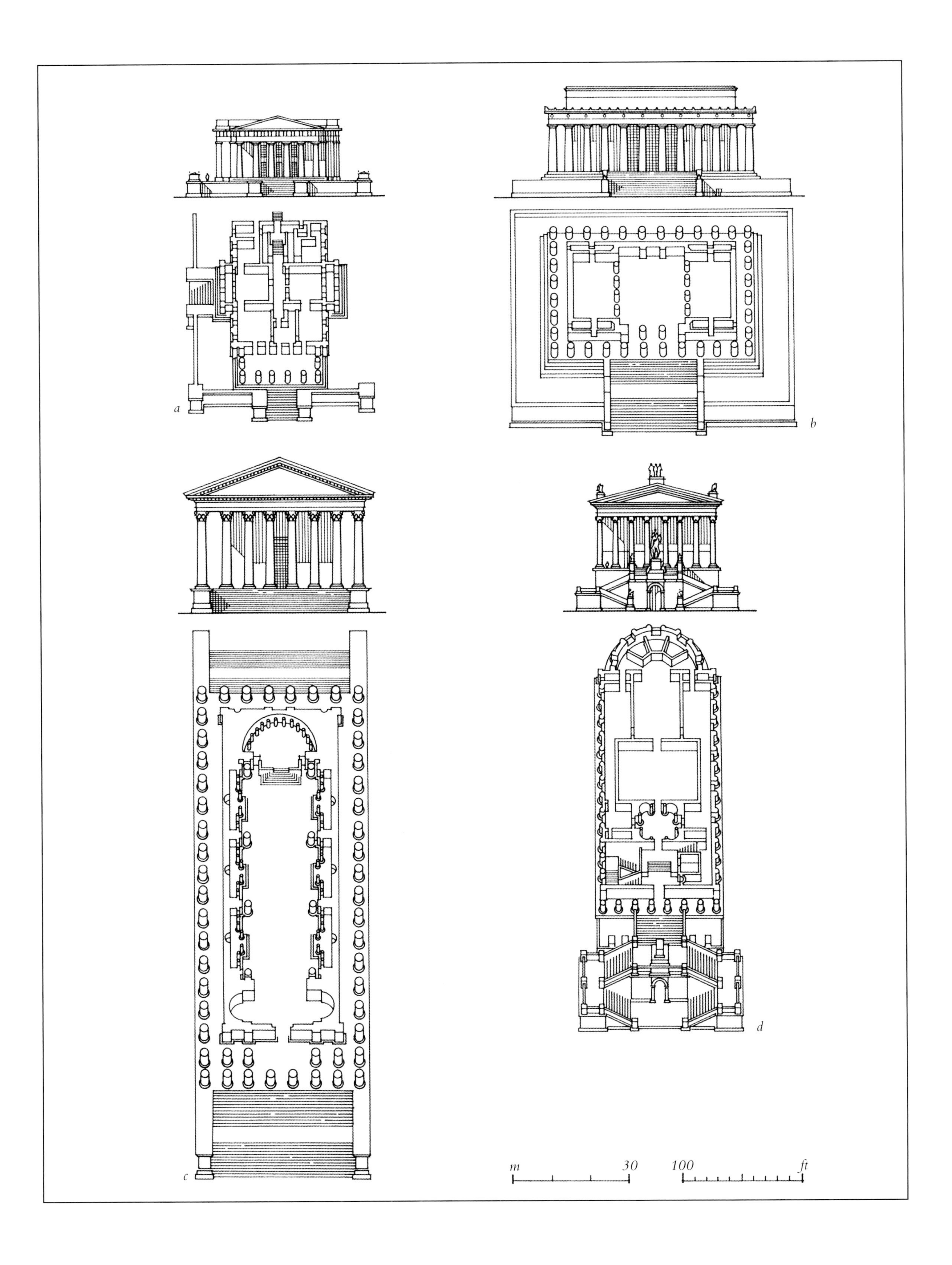
a
b
c
d
m
30
100
ft

巴西利卡与圆形神庙

修建宏伟的封闭大厅，或者称为巴西利卡，满足了罗马城市的法制和商业功能，也成为贯穿整个罗马帝国时期的建筑特征。尽管这一名称的意思是希腊国王之厅，罗马人却将巴西利卡演变成为一种普及的建筑形式。其内通常建有法庭和公共大厅。法庭、裁判所位于建筑物的一侧尽头或两侧均有，通常呈半圆形平面，包括一个升高的法官席。裁判所的前面有一个公共大厅，可用于集会和商贸活动。这些大厅大小各异，但是当时的屋顶结构使得建筑物的宽度有限，因此需将中央大厅即教堂中殿向两边扩建，并且只能通过用支柱，又称支墩来支撑主墙，并且环绕边界增建狭长的过道、走廊。这一设计能够将天花板中央抬高，置于边廊天花板之上，因此光线就可以通过建筑物中央高层的天窗透入。

巴西利卡是古罗马文明最伟大的建筑成就。极富声望的宏伟建筑，例如乌尔比亚巴西利卡（*a*），大约公元100年由罗马皇帝图拉真（53—117年）修建于新的罗马广场之上。乌尔比亚巴西利卡有两排走廊，一个内部及外部的上层画廊，按照惯例还建有两个法庭。

巴西利卡的设计通常应用于小型会议厅或餐厅，当神秘的近东宗教开始在帝国盛行时，他们开始在这些小的厅堂内举行一些秘密仪式。教会的神像或神坛安放在位于一端的半圆形教堂后殿。基督教兴起，并最终成为帝国后期正式认定的宗教，其建筑延续了这种与巴西利卡的早期关联。雄伟的巴西利卡例如罗马城外的圣保禄大殿（*d*）矗立在神圣之地，其规模与设计既努力尝试将古老帝国的伟大建筑与新兴宗教联系起来，又行使公共集会建筑的功能。这些早期的基督教建筑时至今日仍为教堂设计提供样板。

罗马圆形神庙同样也流传至基督教时期。圆形神庙，例如罗马城外蒂沃利的希贝尔神庙（*b*）修建于公元前1世纪，可能从原始时期的意大利圆形棚屋演变而来。这类简单的圆形建筑通常用于坟墓的设计。早期基督教洗礼堂，例如修建于430年的罗马康斯坦丁洗礼堂（*c*），通常呈圆形或者近似圆形平面。采用圆形神庙设计的建筑与死亡相关，将洗礼当作从死亡中得以解脱的象征意义，这些建筑物的设计师以新基督教的传统保留了古代罗马建筑模式。

BASILICAS AND ROUND TEMPLES

The legal and commercial functions of Roman cities were answered by the construction of large covered halls, or basilicas, which became a characteristic feature of Roman communities throughout the empire. Although its name suggests the hall of a Greek king, the basilica was developed to a universal form by the Romans. It generally housed both a court and a public hall. The court, or tribunal, sat in one or both ends of the building in a space, often identified by a semicircular plan; that contained a raised platform for a judge. In front of the tribunal was a public hall which could be used for assembly or trade. The sizes of these halls varied, but the limited width allowed by the roof structures of the time meant that an enlargement to the sides of the open central hall, the nave, could only be achieved by supporting the main walls on columns, or piers, and adding narrower passages, aisles, around the perimeter. This arrangement made it possible to elevate the central roof above the roofs of the side-aisle and so to admit light to the centre of the building through high-level, clerestory windows.

The basilica became one of the great architectural achievements of the ancient Roman civilisation. Huge and prestigious examples such as the Basilica Ulpia (*a*), built by the Emperor Marcus Ulpius Traianus (AD 53-117) in his new Forum in Rome in about AD 100, were famous throughout the empire. The Basilica Ulpia had two rows of aisles, an internal and external upper gallery and, unusually, two tribunals.

The basilican layout was also used for small meeting or dining halls and, when near-eastern mystery religions became popular in the empire, some of their secret rites were practised in these small halls with the cult figure or altar in a semicircular apse at one end. When Christianity rose from among these cults to become the official religion of the late empire, its architects developed this early association with the basilica. Great basilicas such as S. Paolo fuori le Mura in Rome (*d*) were erected on sacred sites to a size and design that was not only intended to associate the new religion with the greatest buildings of the old empire but also perform a similar function of public assembly. These early Christian buildings have set the pattern for the design of churches to this day.

The Roman circular temple also found its way into the architecture of Christianity. Circular temples, such as the Temple of the Sibyl at Tivoli outside Rome (*b*) from the first century BC, were probably derived from primitive Italian circular huts. Simple round buildings of this type were often used for tombs. Early examples did not have an outer ring of columns. Early Christian baptistries, such as that of Constantine in Rome (*c*) of A.D. 430, were often of a circular or near-circular plan. By adopting the architecture of the circular temple, associated as it was with death, and introducing the symbolism of baptism as a release from death, the designers of these buildings preserved an ancient Roman form in a new Christian tradition.

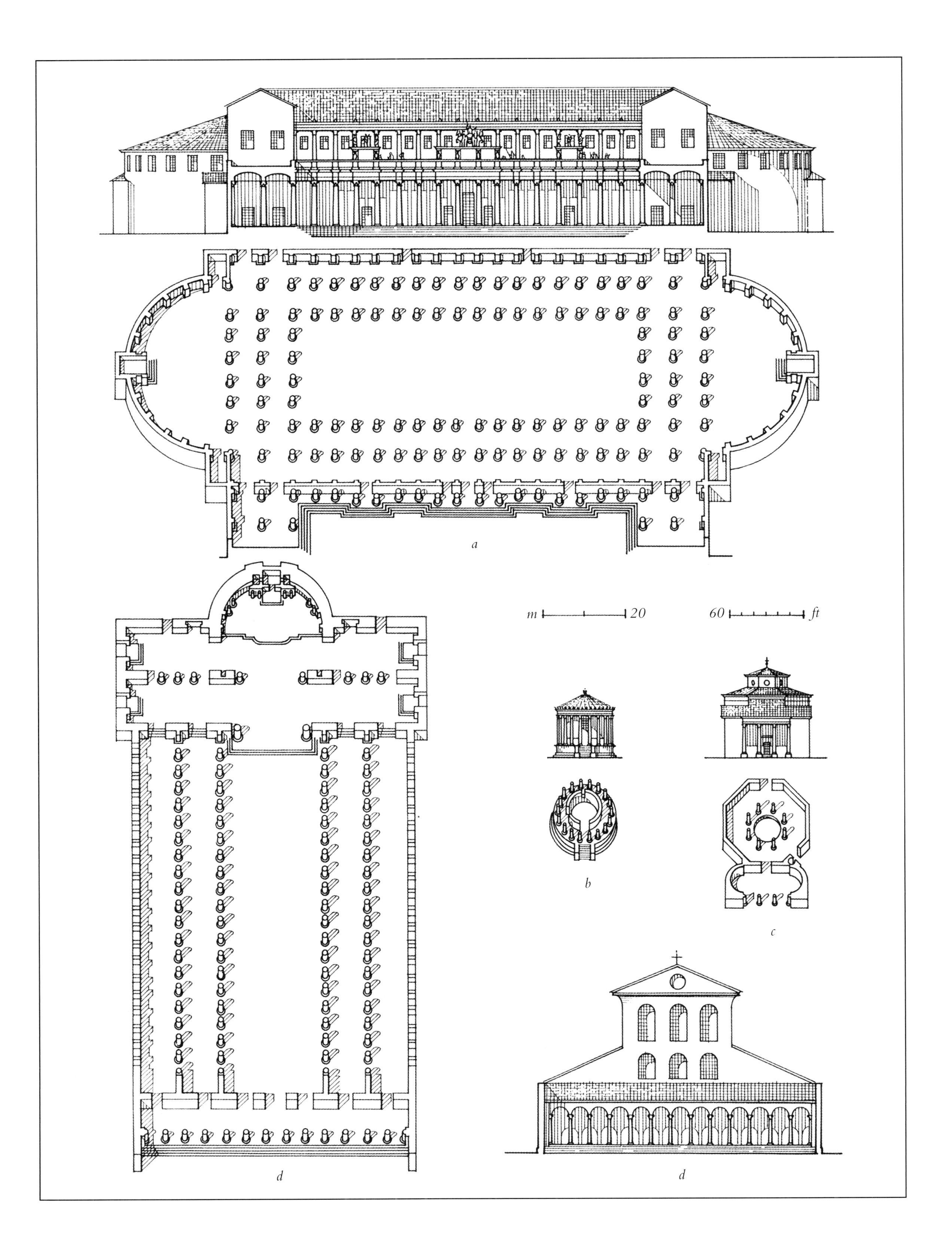
a
m 20
60 ft
b
c
d
d

文艺复兴和巴洛克教堂

西方的基督教延续了在教堂中礼拜的传统，因而保留了巴西利卡的形制。人们习惯性地选择在做礼拜的地方举行宗教仪式，并演变为一种习俗，巴西利卡的平面形式成为教堂建筑设计的固定传统。文艺复兴的建筑师开始改变哥特式教堂建筑的传统，他们首先合理地考量中世纪教堂的巴西利卡平面，同时改变其装饰语汇，使他们新的兴趣与古代世界的建筑协调一致。

佛罗伦萨的圣·洛伦佐教堂（*a*），由菲利波·布鲁内列斯基（1377—1446年）于1419年设计，是自古代以来第一座完整的古典教堂设计。这一教堂平面与当地的哥特式教堂十分相似，但也展示了对纯粹几何学设计的兴趣。教堂中殿的宽度是两边侧廊的两倍，走廊的每个分区都是正方形。这些分区，又称开间，环绕神坛，以每十个小礼拜堂为单位重复。教堂中殿的天花板是平的，装饰有小的方形嵌板，即花格镶板。走廊中每一个分区以及每一个礼拜堂的天花板都是同样的方形中带有穹顶的设计。巨大的穹顶是它们的四倍，位于交叉点的中央，这里是教堂的中殿和十字形翼部相交的地方。这一精确的几何设计与古典的细部设计，赋予内部以规则与理性的氛围，与早期哥特教堂差异很大。

对于理想几何形式完美表现的追求，引导文艺复兴时期建筑师审视古代圆形建筑以及拜占庭时期希腊人的穹顶与十字形平面（见P16）。这些平面似乎通过其理性上的纯粹来反映上帝的纯洁。蒙特普齐亚诺的麦当娜圣比亚（*b*）修建于1518年，由老安东尼奥·达·桑迦洛（约1453—1534年）设计，是数量众多的带有巨大中央穹顶的意大利教堂之一，平面为圆形、十字形或者方形内含十字形。希腊十字的等臂削弱了传统的巴西利卡中殿设计，缺乏连续的路线来强调神坛的重要性。中央空间支配着建筑的内部和外部设计，巨大的穹顶坐落于高高的鼓形座窗户之上，更倾向于拜占庭风格而不是古典风格。

然而，这些理想的平面仍然应用于纪念礼拜堂和其他特殊的教堂，对于教区教堂和大教堂的日常需求来说并不实用。由于巴洛克建筑去除了考究的文艺复兴气氛，回归更加贴近哥特式风格的激情，将穹顶下中央空间的戏剧化效果与传统的从中殿向神坛推进的设计方式相结合。巴洛克式雕刻承重墙的设计使得建筑物，例如克里斯多弗·雷恩（1632—1723年）于1675年设计的伦敦圣保罗大教堂（*c*）能够毫不费力地结合了文艺复兴时期的希腊十字与传统的教堂中殿以及英国哥特式的唱诗班设计风格。

RENAISSANCE AND BAROQUE CHURCHES

Western Christians continued to worship in churches that retained the form of the basilica. The basilican plan became a firm tradition in church design as the rituals of religious service became established through customary use of the building as a place of worship. When Renaissance architects began to change the Gothic inheritance of church architecture, at first they only rationalised the basilican plan of the medieval church and changed the decorative vocabulary to accord with their new interest in the architecture of the ancient world.

The church of S. Lorenzo in Florence (*a*) by Filippo Brunelleschi (1377-1446), designed in 1419, was the first complete Classical church since antiquity. The plan of the building is very similar to local Gothic churches but displays a new interest in geometric purity. The central nave is exactly twice the width of each of the side-aisles and each division in the aisles is a square. These square divisions, or bays, are repeated for each of the ten chapels around the altar. The ceiling of the nave is flat and is decorated with small square panels, or coffers. The ceiling of each division in the aisles and of each chapel has an identical dome in a square. A large dome, four times their size, sits in the centre of the crossing, where the nave and the transept cross. This precise geometry combines with Classical details to give to the interior an ordered and rational atmosphere quite unlike that of earlier Gothic churches.

The search for perfection in ideal geometric forms led Renaissance architects to look at ancient circular buildings and at the domed, cross-shaped plans of the Byzantine Greeks (see page 16). These plans by their intellectual purity seemed to reflect the purity of God. The Madonna di S. Biago at Montepulciano (*b*) of 1518, by Antonio da Sangallo the Elder (*c.*1453-1534), was one of a large number of churches designed in Italy that centred on large domes and had plans that were circular, cross-shaped, or cross-shaped inside a square. The equal arms of the Greek cross eliminate the traditional nave of the basilica and the lack of a processional route emphasises the importance of the altar. The central space dominates the building both inside and out while the huge dome, sitting on the windows of its tall drum, owes more to Byzantium than to antiquity.

While these ideal plans continued to be used for memorial chapels and other specialised churches, they were impractical for the everyday requirements of parish churches and cathedrals. As Baroque architecture moved away from the rarefied atmosphere of the Renaissance back to a passion more akin to the Gothic, the drama of the domed central space was unified with the traditional progressive approach down the nave to the altar. The Baroque sculptural treatment of the supporting walls allowed buildings such as St Paul's Cathedral in London (*c*) of 1675, by Christopher Wren (1632-1723), effortlessly to combine the Greek cross of the Renaissance with the traditional nave and choir layout of the English Gothic cathedral.

a
b
m
40
ft
100
c

古代住宅

在古代，人们居住在各种不同类型的房屋中。农民住在棚屋中，罗马帝国时期大部分的人居住在巨大的公寓式建筑中。然而古典文明以其大面积城市社区为特征，由当地有权势的精英掌控。大量的联排住宅被认为是典型的古代建筑。

这些房子内部朝向各自的封闭庭院。很少有窗户朝向外面，因此光线、通风都通过这些中央开放空间获取。这种庭院的构建，以及房间环绕庭院的设计，使得房屋在不同时期与不同地区都有自己的风格。

原始希腊人的房屋包括独立的长方形房间，入口门廊处有柱子支撑。建筑物、中央大厅与P3展示的庙宇同样简单。到公元前5世纪，希腊的房屋是环绕着庭院修建的。在欧洲的希腊社区，这些庭院设计得没有条理，有时是两层，带有开放式门廊由两面、三面或者四面的柱子支撑。在希腊，小亚细亚的少数民族房屋中，中央大厅作为独特的高级元素建于集会用的房间中，包括重要的居住功能。这些中央大厅由设计各异的庭院环绕（*b*）。到公元前1世纪，庭院遍布希腊世界各个角落，演变成为规范设计的列柱廊。

罗马的房屋从其伊特鲁里亚的先祖起就是围绕一个独特的庭院——中庭修建的，这就是房屋的中心。中庭的中央开放空间由横梁或者柱子支撑，使得雨水能够落入中央长方形水池。从街上可通过一个小厅进入，厅内陈设重要家具，如祖先神龛和保险箱等。早期的罗马房屋两侧安置卧室，最里面才是餐厅，并且重要的会客室朝向中庭，但是可以用折叠门封闭起来。

公元前1世纪，由于希腊人在意大利南部奢侈的生活方式而将列柱廊引入罗马。这种柱阵式的庭院包括中央花园，并且带有单独的会客室。经过一段时间后，如图所示（*a*），大部分的会客室都设计在庭院四周，与中庭直接相通。

罗马人没有把全部时间放在他们的联排住宅上，而是享受海景与自然风光。富有的罗马人通常都有乡村宅邸，即别墅。在众多现存的庞贝壁画中就有沿那不勒斯海岸排列的海滨别墅的图画。图示（*c*）便是其中一幅，展示的是临海的弧形正立面，以及建在水边的中央观景塔与防波堤。

HOUSES OF ANTIQUITY

In antiquity people lived in many different types of house. Peasants lived in huts and in Imperial Rome much of the population lived in large apartment buildings. Classical civilisation was, however, characterised by its widespread urban communities, dominated by powerful local élites. Their substantial town houses have been regarded as the typical houses of antiquity.

These houses looked inward to their own enclosed courts. These were few windows to the outside, so light and ventilation were gained from these central openings. The construction of the courts, and the way the rooms were organised around them, gave the houses of each period and locality there distinctive characteristics.

The primitive Greek house consisted of a single rectangular room with an entrance porch supported on columns. This building, the megaron, is the same as the simple temple illustrated on page 3. By the fifth century BC, however, Greek houses were built around courtyards. In the Greek communities in Europe these courts were informally planned, sometimes on two storeys, with open porches supported on columns around two, three or all four sides. In Greek Asia Minor the megaron, survived as a distinctive higher element in the assembly of rooms, containing the principle accommodation. These megarons were surrounded by courtyards of varying design (*b*). By the first century BC the courtyard had developed throughout the Greek world into a more formal design known as a peristyle.

Roman houses from their Etruscan origins onwards were built around a special kind of court, the atrium, which was the heart of the house. The atrium had a central opening supported on beams or columns which allowed rain-water to fall into a central rectangular pool. It was entered from the street by a small lobby and contained important furnishings such as the family shrine and the strong-box. On either side, in early Roman houses, lay the bedrooms and at the far end a dining-room and a principle reception room faced the atrium, but could be closed off by folding doors.

Exposure to the luxurious lifestyle of the Greeks in southern Italy in the first century BC introduced the peristyle to the Romans. This colonnaded courtyard contained a central garden and had its own special reception room. At first the peristyle was just put behind the principal reception room, but in time, as in illustration (*a*), most of the reception rooms came to be placed around it and it was linked directly with the atrium.

The Romans did not spend all their time in their town houses but enjoyed the seaside and natural scenery. Wealthy Romans usually had a country house, or villa, and some of these villas also operated as working farms. In many of the surviving wall paintings in Pompeii are pictures of the seaside villas that spread along the bay of Naples. Illustration (*c*) is an interpretation of one of these paintings showing a curved façade facing the sea, a central viewing tower and jetties at the water's edge.

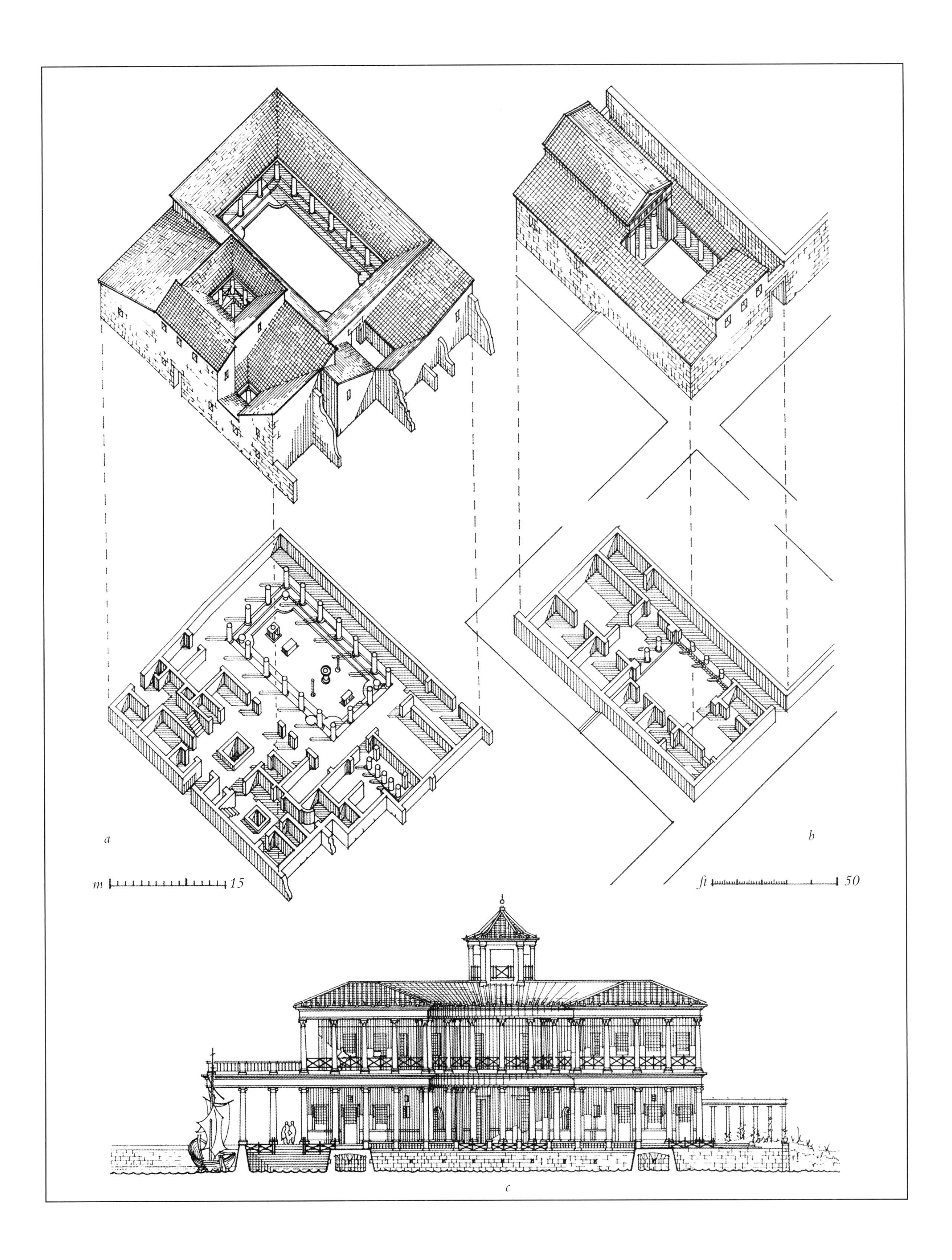
a
b
m 15
ft 50
c

文艺复兴、巴洛克宫殿与别墅

中世纪欧洲的贵族，出于安全考虑一般都居住在堡垒式建筑中。在意大利，长久以来的都市生活传统使得商业贵族开始修建城市堡垒或者宫殿，而不是孤立的城堡。向内朝向庭院，成为内部循环的重要方式。在文艺复兴之前很长一段时间，这些巨大的房屋，都是以优雅与宽敞为传统的。15世纪，新兴的对古典遗迹与比例的关注并未削减对防御的需求。文艺复兴建筑师，比如贝奈德托·达·迈亚诺（1432—1490年），于1489年设计的诗特洛奇宫（*a*），在令人生畏的正立面和隐藏的庭院设计中，关注创造精准的古典细部设计和比例，并且将整体平面结合起来形成合理对称的布局。

16世纪中更加稳定的政治生活使得贵族迁出他们的城堡，到更加舒适的住所中去。在欧洲北部，贵族阶级保留他们位于乡村领地中心的主要居住地。法国的城堡，法语称为château，仍然有护城河环绕，正面按照惯例在前面建有开阔的庭院，但是主要的建筑区域已经不再具备防御功能。建筑师现在能够集中精力关注建筑方法与内部设计的富丽堂皇，外观则注意与近处的景观相融合。自由的空间，以及当代巴洛克风格全神贯注于创造出令人印象深刻的雕塑效果，允许建筑师们，例如路易斯·勒沃（1612—1670年）于1657年，在维孔宫城堡（*c*）主建筑的设计中创造出动态布局以及建筑内部体量，这对后来的建筑设计产生了深远的影响。

在意大利，城市传统得以延续，但是在针对古代别墅研究的启发之下，贵族们为自己在城外修建娱乐用的楼阁。其中最具影响力的就是朱利亚别墅（*b*），由伽科英·巴罗兹·达·维尼奥拉（1507—1573年）、巴托洛梅奥·阿曼那提（1511—1592年）和其他建筑师于1550年为教皇朱利奥三世（1487—1555年）在罗马城外修建。建筑外观并不包含在设计之中，因为建筑物内侧朝向为其精心规划的内部景观。朴素的花园庭院四周环绕着开放式弧形正立面，装饰有彩绘天花板和精工雕刻的拱形游廊，其中有古代神的雕塑。这一布局的最末端是一个小拱门，通过这一拱门可进入带有喷泉的下沉式花园，称为休憩场所，上面建有围墙花园。

几乎同一时期，威尼斯共和国的贵族在乡村庄园修建别墅。这些通常都是朴实的建筑并且带有娱乐休闲的房间、酒窖、阁楼和庄园中农业用途的附属建筑物。这样的建筑物例如帕拉迪奥设计的巴多尔别墅（*d*），将商业与娱乐的实用功能和富丽堂皇的风格结合在一起。他们的外观在耕作的土地上造成的视觉效果为无数乡村建筑物树立了新的标准。

RENAISSANCE AND BAROQUE PALACES AND VILLAS

The nobility of medieval Europe generally lived in fortified buildings for security. In Italy, a long tradition of urban life led the merchant nobility to construct urban fortresses or palaces rather than isolated castles. Facing inwards to a courtyard, which acted as the principal means of internal circulation, these large houses had a tradition of elegance and spaciousness long before the Renaissance. The new concern for Classical antiquity and proportion in the fifteenth century did not remove the need for protection. Renaissance architects, such as Benedetto da Maiano (1432-90), designer of the Palazzo Strozzi in Florence (*a*) in 1489, concentrated on creating precise Classical details and proportions on the forbidding façades and hidden courts, and organised the overall plan into a rational and symmetrical composition.

During the sixteenth century a more settled political life often allowed the nobility to move out of their castles to more comfortable accommodation. In northern Europe the aristocracy maintained their major residences at the centre of their rural territories. In France the castle, or château, was still surrounded by a moat and fronted by its customary wide court, but the principal range of buildings no longer had to serve a defensive function. Architects could now concentrate on the grandeur of the approach and interiors, and the outlook on to carefully designed views across the immediate landscape. The unrestricted space, and the contemporary Baroque preoccupation with the creation of impressive sculptural effects, allowed architects, such as Louis Le Vau (1612-70) in the principal buildings of the Château de Vaux-Le-Vicomte (*c*) of 1657, to produce dynamic compositions and interior volumes that would have a profound influence on all subsequent architectural planning.

In Italy the urban tradition continued but, inspired by accounts of the villas of the ancients, the nobility built pleasure pavilions for themselves just outside the town. One of the most influential of these was the Villa Giulia (*b*), built outside Rome by Giacomo Barozzi da Vignola (1507-73), Bartolomeo Ammanati (1511-92) and others for Pope Julius Ⅲ (1487-1555) in 1550. The outside world has no part in this design as the building looks inwards to carefully controlled views of its own internal landscape. An unadorned garden court is surrounded by an open curved façade decorated with painted ceilings and delicately sculpted arcades containing ancient sculptures of gods. This composition ends with a small arch leading to a sunken garden with fountains, known as a nymphaeum, and a walled garden beyond.

At about the same time, the aristocracy in the Venetian republic were building villas for their country estates. These often quite modest houses combined rooms for pleasure and entertainment with cellars, lofts, and outbuildings for the agricultural functions of the estate. Buildings such as the Villa Badoer (*d*) by Palladio combined business with pleasure and grandeur with practicality. Their outward perspective on the working landscape set a new standard for countless rural buildings.

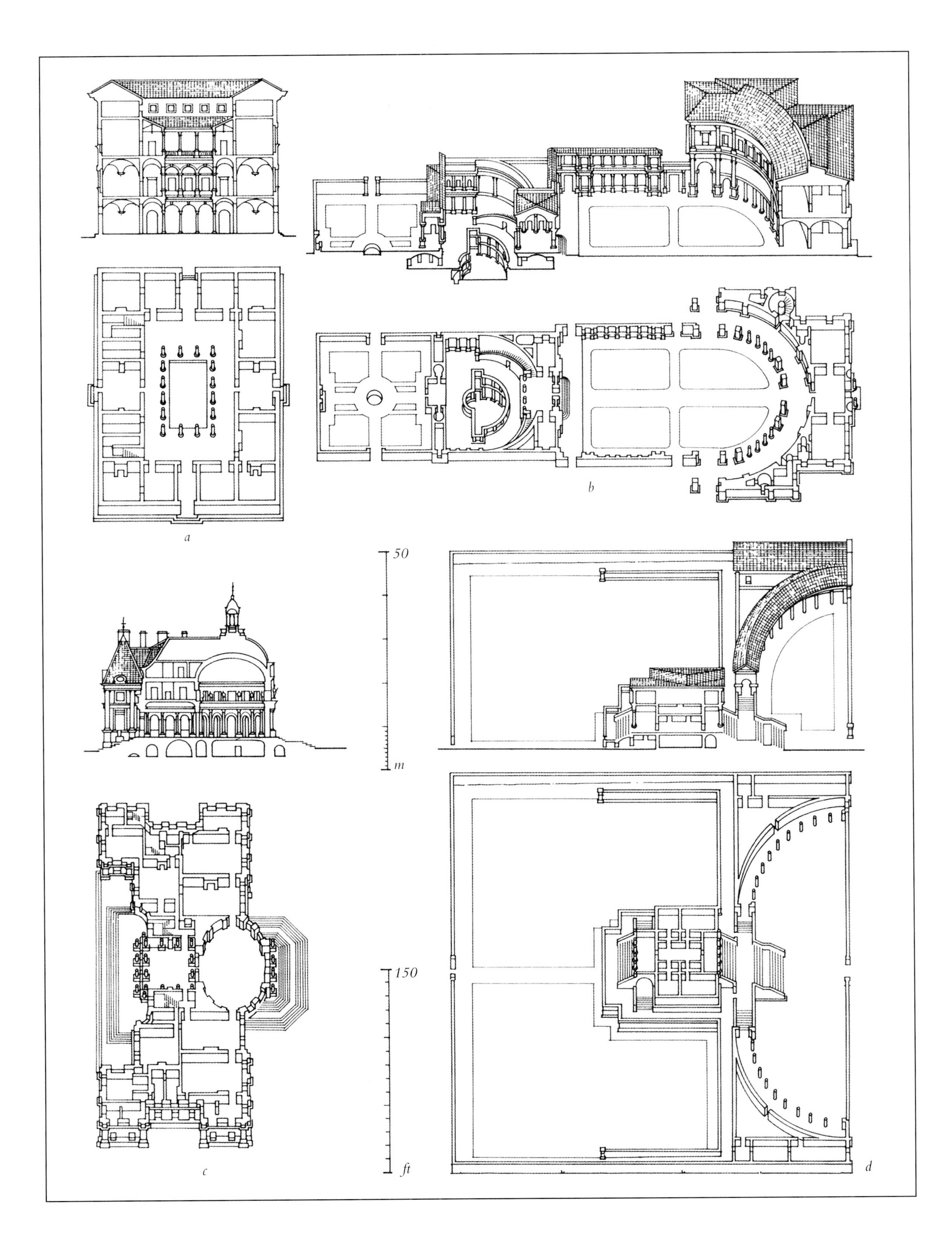
a
b
c
d
50
m
150
ft

小型城镇住宅

在城镇中，朝向街道的地段价格高昂，建筑宽度因而受到限制。这一限制导致在房屋设计中，自然光通常都是从后面和前面照射进来，楼梯设计得比较蹩脚，如果正面的房间与房子中其他房间分开，主入口及其通道就必须设置在一边。

图中（*a*）所示的是一座修建于1500年的罗马住宅。作为意大利城镇的传统，即使是深宅大院也会在一楼建有商铺。这使得建筑师有机会创造一种三条纵向分割，即三个开间的平衡建筑设计，因此需区分出偏离中心的通往上层住宅的入口。

下一个例子（*b*），取材于16世纪的匈牙利城镇比斯特里察，包括同一家庭拥有的商业经营场所和住宅。一扇巨大的门占据了一楼的正立面，通过这扇门可将大件货物送入店铺。采用这种设计，使得上层成排的对称窗户与下层开放空间互不相关，却十分合理。

一栋法国建筑（*c*），由皮埃尔·勒·米埃（1591—1669年）于1647年修建，简单的平面包括在一个内部庭院之后的马厩。这一设计成为更大型的城镇住宅的普遍特色，这些住户养得起马匹，这在后来伦敦修建的住宅中也很常见。屋顶与窗户的设计使得房屋的正立面显得集中凸出，这也弱化了大门建于不对称位置而造成的不平衡感。

低地国家有着这样的悠久传统，将狭长的房屋集中修建在一起，而获取邻街或靠近运河的宝贵商业地段。如图（*e*），取材于17世纪的阿姆斯特丹，展现出独特的巴洛克式山墙，是建筑师对古典规则与传统哥特式山墙进行调和的产物。其余的北部国家放弃了这样的山墙设计，而选择向另一方向倾斜的屋顶。

英国，比其他国家更多地发展了几乎相同的房屋设计。以社会地位区分，得到了社会各个阶级的普遍认同。图例（*d*）和（*f*）的房屋，建于1823年，分别展示了最高和最低标准。大型住宅是一位贵族的府邸，但平面与一位手工艺人的小房子几乎相同。房屋的区别在于规模以及内外部装饰的质量。

20世纪小城镇的设计问题，由于大部分的居民拥有了机动车而从根本上得以改善。只有一面邻街，有时车辆要存放在房子内。需要大面积开放空间，如图（*g*）所示，商户（*a*）和手工作坊（*b*）也面临着同样的问题。在例子中，车库大门和入口连成一条水平线，长方形的车库入口缩小了垂直比例后被复制。

THE SMALL TOWN HOUSE

In towns, the width of buildings is often limited by the high value of the land that faces on to the streets. This restricts the design of houses as daylight can often only be obtained from the back and the front, the stairs can be awkward and, if the front room is to be separated from access to the rest of the house, the main entrance and its passageway must be located to one side.

Illustrated (*a*) is a house in Rome built in 1500. It is a tradition in Italian towns that even imposing dwellings often have shops on their ground floor. This has allowed the architect to produce a balanced design by creating three vertical divisions, or bays, thereby disguising the off-centre entrance for the accommodation above.

The next example (*b*), from the sixteenth century in the Hungarian town of Bistrita, includes the common combination of business premises with substantial residential accommodation for the same family. The façade is dominated by the large door for heavy goods deliveries to the workshop and shop on the ground floor. The design is rationalised by a row of symmetrical windows on the upper floor unrelated to the openings below.

A French design (*c*), by Pierre Le Muet (1591-1669) in 1647, is a simple plan including stables beyond an internal courtyard. This became a common feature of larger town houses, where the occupants could afford horses, and was extensively used in later houses in London. The façade is given a strong central emphasis by the roof and windows. This lessens the imbalance created by the asymmetrical position of the door.

The Low Countries had a long tradition of narrow houses jostling for valuable trading frontage on their streets and canals. Example (*e*), from seventeenth-century Amsterdam, shows the distinctive Baroque gables that were created as architects tried to reconcile Classical discipline with the traditional Gothic gables. Other northern countries abandoned these gables in favour of roofs pitched in the opposite direction.

In England, more than any other country, the development of rows of virtually identical houses, grouped according to status, became widespread for all social classes. Illustrations (*d*) and (*f*) from 1823 show the highest and lowest standards respectively. The larger house could have an aristocratic occupant, but has a virtually identical plan to the small house intended for an artisan. The houses are distinguished by their size and by the quantity of internal and external decoration.

The problems of designing small town houses in the twentieth century have been changed principally by widespread ownership of motor cars. Where only one street front exists, sometimes the cars have to be accommodated within the house. The larger opening required is seen in illustration (*g*) and poses similar problems to those of the shop or workshop in (*a*) and (*b*). In this example the garage door and entrance are united by a horizontal band and the square of the garage opening is repeated in diminishing proportions vertically.

a
b
c
d
e
f
g
m
11
35
ft

小型乡村住宅

小型乡村住宅这种建筑类型可建于城镇、郊区或者乡村。这里展示的所有设计都是前门在正中并且大厅的两边各有一个房间，尽管看起来十分相似，但还是有明显的差别。

16世纪罗马郊区小型住宅，如图所示（*a*），平面简单，带有开放式门廊，或者称为凉廊，其后部可通往一楼房间和楼梯。同样简单的立面，使得上层与下层窗户之间的关系比起一层房间窗户的平衡显得更为重要。

位于伦敦郊区里士满的威克宅邸（*b*），由罗伯特·米尔恩（1733—1811年）于1775年设计。平面十分复杂，并且在有限的空间内创造出了最佳效果。楼梯移至另一侧，走廊从入口处直接通往椭圆形沙龙。正面两侧房间的窗户都位于内部中央，正立面设计得简单而且协调。工作间并不显眼地设于沙龙的各边，扇形楼梯简洁地设计于椭圆形沙龙后面。

18世纪的瑞典住宅如图所示（*c*），有着对称的正立面，隐藏不规则的房间。尽管单个房间的设计尝试将房门和窗户随意的位置变得合理，但整体布局还是缺乏主题。入口大厅中，大楼梯位于前门的一侧，在这一点上造成的不对称性由假墙弥补。

19世纪中期法国的农舍（*d*）平面奉行简朴与实用原则。狭长的大厅有足够的空间用来安放楼梯，大厅的一侧是一个大厨房，另一侧则是起居室，后面建有盥洗室。只有较少的外部设计表现出法国特征——在平面上，这一细微的改动可见于欧洲的各个国家。

德国亚琛的一座别墅（*e*），建于本世纪早些时候，在房间的分割上不那么刻板。大厅、后面的房间与前面的一间房间没有严格的分界。将入口设在楼梯平台下的大厅中，使得楼梯位于相对于前门与大厅的对称位置。

19世纪晚期的加利福尼亚住宅（*f*）平面可见于任何欧洲国家。独具特色的美式外观设计创造出穿过主要正立面的特色门廊，屋檐下的深飞檐和低矮屋顶，大部分虽为平顶，却试图创造出法式的双面坡，也称复折屋顶的效果。

THE SMALL COUNTRY HOUSE

The small country house can as a type be in a town or suburb as well as in the countryside. All of the designs illustrated here have front doors in the centre and one room on each side of a hall but, despite their apparent similarities, there are significant differences between them.

The small sixteenth-century house on the outskirts of Rome, illustrated (*a*), has a simple plan with an open porch, or loggia, at the back which gives access to the ground-floor rooms and the stairs. The equally simple elevation gives more importance to the relationship between the upper and lower windows than to the balance of windows inside the ground-floor rooms.

Wick House in Richmond, on the outskirts of London (*b*), was designed by Robert Mylne (1733-1811) in 1775. The plan is sophisticated and creates the maximum effect within the limited space available. The stair has been moved to one side and the passage from the entrance leads straight to an oval salon. Both of the front rooms have windows placed centrally on the interior and the façade is simple and balanced. Service rooms are inconspicuously fitted into each side of the salon and a quadrant stair fits neatly behind the oval.

The eighteenth-century Swedish house in illustration (*c*) has a symmetrical façade that conceals an irregular collection of rooms. Although the design of individual rooms attempts to rationalise the arbitrary positions of the doors and windows, there is no underlying theme in the layout. There is a large entrance hall containing a grand stair to one side of the front door and the consequent loss of symmetry has been disguised with a false wall.

The mid-nineteenth-century French farmhouse (*d*) has an austere and utilitarian plan. A narrow hall is just large enough for the stair and leads to a large kitchen on one side and a living-room on the other with a washroom behind. Only the sparse exterior features identify the design as French—minor variations of this plan could be found in almost any European country.

A villa in Aachen in Germany (*e*), built early this century, shows a less rigid separation of rooms. The hall, the back rooms, and one of the front rooms have no rigid divisions between them. The stairs are symmetrically positioned in relation to the front door and the hall by bringing the entrance through a small lobby under a half-landing. The plan is not contained in the simple rectangle suggested by the façade but is modified by ground-floor extensions.

The late-nineteenth-century Californian house (*f*) has a plan that might be found in any European country. The distinctive American appearance is created by the characteristic porch across the main façade, the deep cornice below the eaves and the low roof, which is largely flat intended to give the impression of a French double-pitched, or mansard roof.

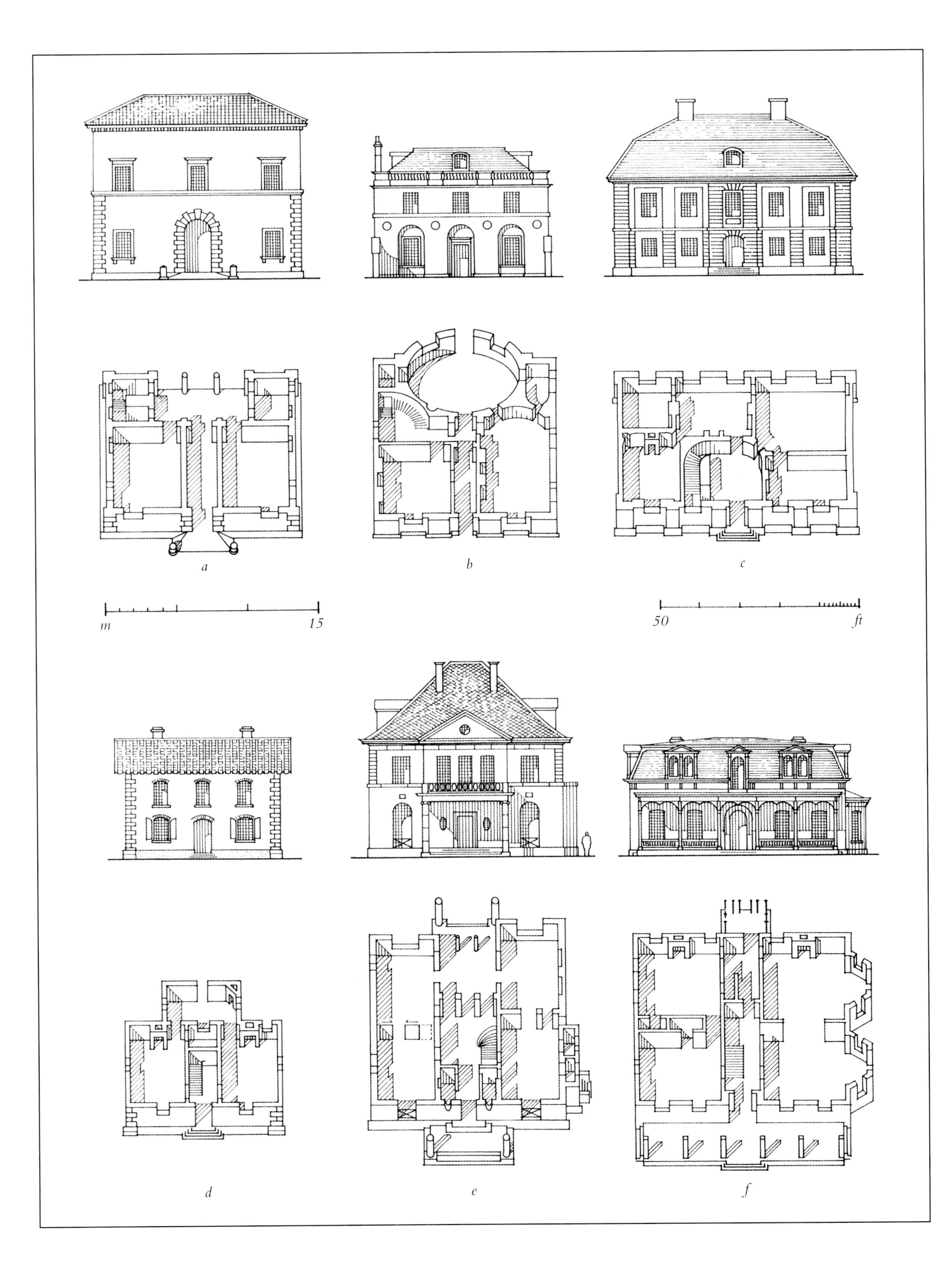
a
b
c
m
15
50
ft
d
e
f

新用途，新解读

时代变迁，建筑也随之变化。近些年来，如同以往，古典传统经采纳与发展，为人们提供了与从前大不相同的实用住宅。有时，某种古代建筑形式可以直接成为一种新的启示，新的建筑服务于不同的目的，然而形式上仍然与最初的样子极其相似。

在其他时候，对新用途的需求十分独特且空前绝后，因此创造出新的古典设计原则，如果成功，这一原则就会成为古典传统的一部分。

纽约的宾夕法尼亚火车站（*a*）由麦克金、米德&怀特设计公司于1904年设计，目的是为了让电力火车通过繁忙的曼哈顿商业区中心。这座宏伟的建筑，将人、汽车以及火车的复杂活动交会于有序的古典设计中，其间融合的各种主要元素均受罗马浴室建筑的启发。不再受火车蒸汽的熏烤，这栋建筑划分几层来区隔不同的交通方式。火车轨道与站台都设在最下面一层，通过位于大街上的主入口进入人行道，穿过商铺连廊来到候车大厅，售票处排列于高大且引人注目的中央大厅四周，这一中央大厅与罗马卡拉卡拉浴场的中央大厅极为相似。在这座大厅之上，主厅覆盖以巨型钢筋玻璃天棚，可经过楼梯通往下面的站台。两侧桥面上铺有斜坡供机动车通行，行人也可从此处通过，加上后面的办公区，创造出罗马浴场般的完整环绕效果。

英格兰利兹市政厅（*b*）由卡斯伯特·布罗德里克（1821—1905年）于1853年设计，为满足新兴的富足工业城市中公众和行政需求而修建，是公民自豪的象征。这座自由设计的古典建筑坐落在一个升高的台基之上，其中央包含能够容纳8000人的公共集会大厅和音乐厅——是当时最大的大厅。其四周环绕着法庭、会议室和办公室，以满足工业化社会不断增长的复杂需求。一座宏伟的、装饰完备的传统钟楼，由民众集资修建，耸立在整个建筑之上。

1839年，牛津的大学城，查尔斯·罗伯特·科克雷尔（1788—1863年）将博物馆与语言学院融合在同一建筑中——阿什莫尔博物馆与泰勒瑞安学院（*c*）。这是当时修建的众多特殊设计的博物馆之一，关注大众对艺术的新需求并且扩大公共教育。在坚固的雕刻正立面的后面，采用希腊复兴和文艺复兴时期的细部设计，独具匠心地将美术馆、报告厅和图书馆的不同需求融合在一起，成为一座独树一帜的精美建筑。

NEW USES, NEW INTERPRETATIONS

As times have changed, buildings have had to change. In recent times, as in the past, the Classical tradition has been adapted and developed to provide accommodation for uses quite unlike anything that has come before. Sometimes an ancient building type provides such a direct source of inspiration that the new building can serve its different purpose well, yet still closely resemble the original.

At other times the needs of the new use are so particular or unprecedented that architects have taken the principles of Classical design to create a new building type which, if it is successful, becomes in its turn a part of the Classical tradition.

Pennsylvania Railway Station in New York (*a*) was designed by McKim, Mead & White in 1904 for electric trains crossing the busy business centre of Manhattan Island. This vast building managed the complex meeting of people, motor cars, and trains within an ordered Classical design incorporating major elements inspired by Roman bath buildings. Free from the smoke of steam trains, changing levels were used to separate the different forms of transport. The railway lines and platforms were all on the lowest level. Pedestrians entered principally through a grand entrance at street level and walked along a shopping arcade to waiting-rooms and ticket-offices grouped around a high and impressive central hall resembling the central hall of the Baths of Caracalla in Rome. Beyond this hall the main concourse was covered with a large iron and glass canopy and led, by way of staircases, to the platforms below. Pedestrians could also enter from the sides on bridges over covered ramps for motor vehicles which, together with ranges of offices to the rear, gave the station an enclosure like those of Roman baths.

The Town Hall in Leeds, England (*b*) was designed by Cuthbert Broderick (1821-1905) in 1853 to satisfy both the public and administrative needs of the newly rich industrial town, and as an expression of civic pride. The freely designed Classical building was raised on a high platform and contains at its centre a public assembly room and concert hall to seat 8000 — the largest in its time. Around this are ranged the law courts, committee rooms, and offices required for the increasingly complex administration of an industrial community. A huge, wholly decorative and unorthodox tower, paid for by public subscription, rises over the building.

In 1839, in the university city of Oxford, Charles Robert Cockerell (1788-1863) combined a purpose-built museum with a language institute in a single building, the Ashmolean Museum and Taylorian Institute (*c*). It was one of a number of specially designed museums of the period, which saw a new concern for public access to the arts and an expansion in public education. Behind a strongly sculptured façade, the varying needs of galleries, lecture rooms, and libraries are brought together with an inventive combination of Greek Revival and Renaissance details to create a highly original and sophisticated building.

a
b
c
100
ft
m
30

新技术

有史以来，新的建造技法与新的材料不断被开发出来。古典建筑师们将这些创新不断发展演变成为我们熟知的古典建筑传统，并为新的技术提供发展空间。古希腊人将木质建筑改为石质；罗马人将拱券与烧制的砖块变为古典建筑语汇。

19世纪与20世纪的进步尤为迅速，工业发明与越来越复杂且人口不断增长的社会并肩发展。新的建筑材料成为可能，使得设计者能够大量建造具有特色的建筑，这样的建筑在早些年，即使单独建造，也要耗费大量的人力物力。大量使用烧制陶土作为建筑装饰的传统可追溯至古希腊时代，但到了18世纪、19世纪，更加宏伟坚固的建筑物可通过使用工业铸造的铁等材料来实现。这样就可以在工厂内完成整个正立面的建造，再将部件运到建筑工地进行组装。铸铁建筑，如纽约的豪沃特大楼（*d*），由约翰·P. 盖诺（约1826—1889年）于1856年设计，结合了精致的古典装饰与全新制造技艺所提供的先进技术条件。

工程师采用新的钢铁制造技术与硅酸盐水泥设计地基和框架，使得建筑物能够达到前所未有的高度。安全电梯技术的发展也使得建筑物能够达到这样史无前例的高度。在纽约人口密集的曼哈顿岛，新的高层建筑在20世纪早期得以发展。这些高塔的规模对于建筑师来说是新的挑战，他们一直在这一不确定风尚的时代中为创造合适的建筑风格而努力奋斗着。古典风格从不断增长的建筑高度中寻求机遇并在这一过程中有了迅速的发展。卡斯·吉尔伯特·Jr.（1894—1975年）设计的位于纽约的美国法院（*a*）于1936年落成，是这类建筑的先驱。

新的机器及其使用者也需要新的建筑形式来安置。铁路、机动车辆和飞机都需要自己的建筑。尽管机器是新的，使用者或者机器潜在的需求并不陌生，因为19世纪和20世纪古典建筑设计已经开始为其服务了。牛津著名的莫里斯车行（*b*），以名爵跑车而闻名，由哈维·帕特里奇·史密斯（1889—1964年）于1932年设计，世界第一座机场建于伦敦南部的克里登（*c*），由政府的设计师于1928年建造。

古典建筑通常是社会需求变化的反映。技术革新与建造技艺所提供的机遇通常是为了满足居住传统发生变化后所产生的新需求。

NEW TECHNOLOGIES

Throughout history new methods of building and new materials have been invented. Classical architects have taken these innovations and made them familiar by developing the Classical tradition to make room for new technologies.The ancient Greeks changed their timber buildings to stone: the Romans made the arch and the baked-brick part of the vocabulary of Classical architecture.

In the nineteenth and twentieth centuries the pace of change has been particularly fast and industrial invention has gone hand-in-hand with the growth of a more complex and crowded society. New building materials have become available, enabling designers to mass produce features that in an earlier age would have been made individually at great expense. The mass production of architectural decoration in fired clay goes back to ancient Greece, but in the eighteenth and nineteenth centuries much larger and stronger building components could be made by using new industrial processes in the casting of iron. This made it possible to produce whole façades and even buildings in a factory, and then transport the component to the site for assembly. Cast-iron buildings such as the Haughwout Building in New York City (1856) (*d*), by John Plant Gaynor (*c.*1826-89), combined an intricacy of Classical decoration with the latest opportunities offered by new manufacturing techniques.

New methods of steel production and the invention of Portland cement enabled engineers to design foundations and frameworks to carry buildings of unprecedented height. The development of the safety lift allowed easy access to these heights. In the crowded island of Manhattan, in New York, a new architecture of tall buildings evolved in the early twentieth century. The scale of these towers was so new that architects struggled in an age of stylistic uncertainty to create an appropriate architecture. Great advances were made in the development of the Classical tradition to take advantage of the opportunities of increased height. The US Courthouse in New York (*a*) by Cass Gilbert Jr. (1894-1975) completed in 1936, is one of the last of these pioneering buildings.

New building types were also required to house new machinery and accommodate its users. Railways, motor cars, and aeroplanes in their turn have required their own buildings. Although the machinery is new, the underlying needs of the users or the machines are often familiar and, as with many new uses of the nineteenth and twentieth centuries, Classical buildings have been designed to service them. The famous Morris Garage in Oxford (*b*), which gave its name to the MG sports car, was designed by Harvey Partridge Smith (1889-1964) in 1932, and the world's first purpose-built airport at Croydon, south of London (*c*), was designed by government architects in 1928.

Classical architecture has always responded to the changing needs of society. The opportunities offered by technological innovation and construction techniques are just some of the new demands that bring about changes in a living tradition.

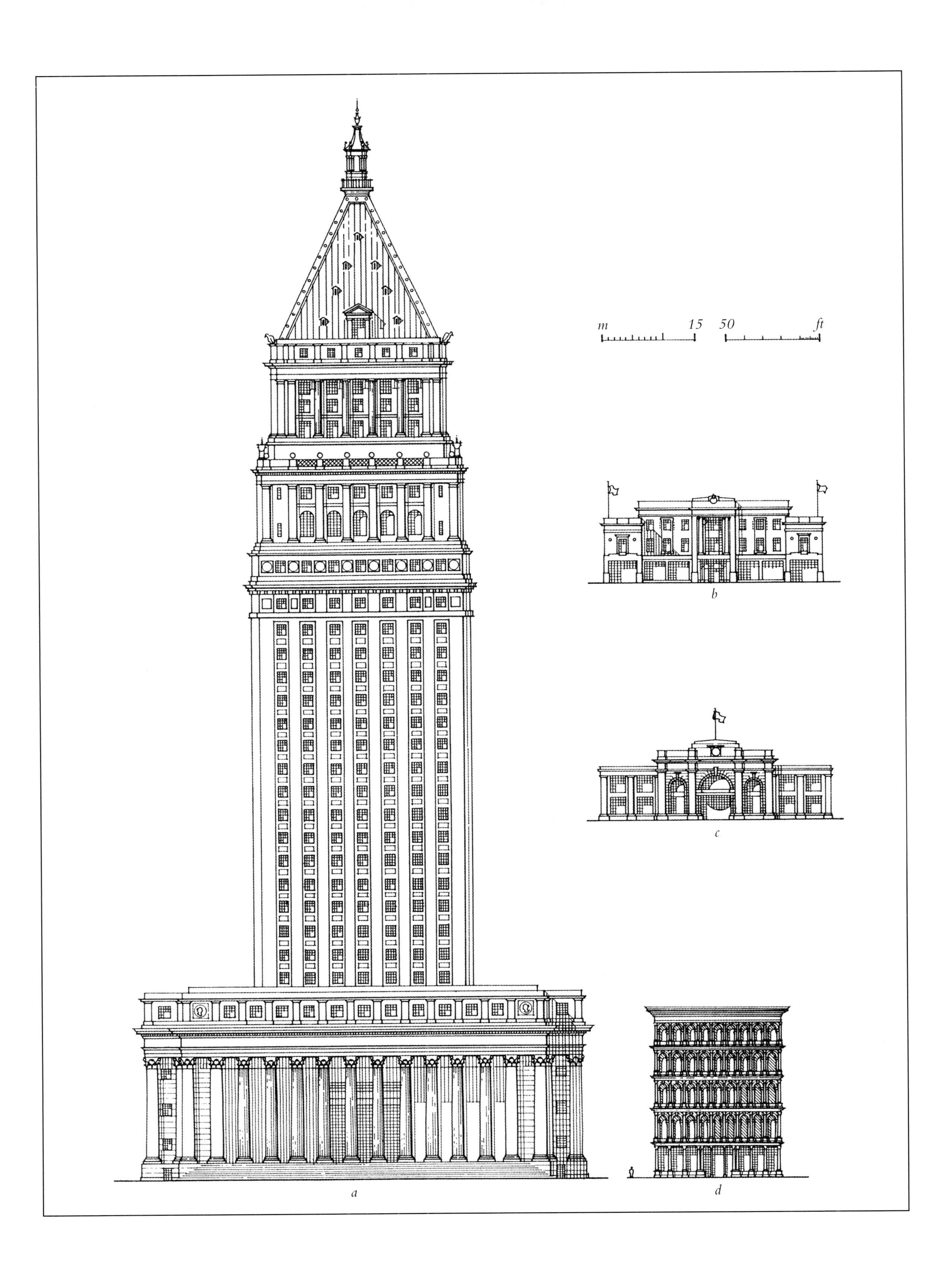
m
15
50
ft
a
b
c
d

奇异特征

古典建筑不一定非要严谨、保守或死板。过去进行的尝试与奢侈的风格都成为这一传统永远的附属物。

在古代，人们竖起独立的柱子，用于纪念或者宗教目的，上面支撑着献祭给上帝的礼物、奖杯或者雕像。作为独立式的纪念碑，单独的柱子通常比建筑物中的柱子更有创意。最著名的高柱（*d*）就是图拉真皇帝于113年在罗马城中央竖起的，作为其论坛一部分的柱子。连续的带状雕刻呈螺旋状缠绕在柱子上，好似一个巨大的卷轴，讲述着图拉真皇帝征服达契亚（罗马尼亚） 的故事。在其顶端，距离地面41米的高度，矗立着皇帝的雕像。图拉真广场其余的部分已经毁坏，可柱子依然保存完好，直至今日更是成为许多设计的模板，从烛台到同样大小的复制品。

罗马人在设计他们的坟墓上十分富有想象力。墓地以及他们的大型纪念建筑都需要遵循严格的设计法则，置于城墙之外。这样的独立设计使其得已保存。盖尤斯·克斯提乌斯的坟墓（*e*）建于公元前12年，是一座高达29米的水泥金字塔，正立面为白色石材，修建于罗马城外，后期与扩建的城墙相连。它必然地反映了一些个人兴趣并且与其征服埃及的事件有关，因此演变出了早期的增加了埃及风格的罗马古典传统。尤利乌斯的坟墓（*c*）位于法国阿尔勒附近，建于公元前40年，是一座风格独特的塔式纪念碑。它由雕刻的底座、开放式方形二层结构和带有逝者雕像的圆形塔顶构成。这是墓碑众多形式中的一种，其对建筑的影响从中世纪一直延续到今天。

大型住宅建在入口车道处的大门，自17世纪以来，通常包括雇工们居住的小房子即门房。通往主要建筑物的道路两旁，为衬托主人的地位，设计中注入了大量的精力与原创。在英格兰的汉普顿，斯特拉顿公园（*a*）的一座大门由小乔治·当斯（1741—1825年）设计。两个小型希腊复兴式门房经弧形墙与一座奇特的大门相连，乍看上去是中世纪风格，但其装饰却带有古希腊特征。

这些奇异特征的变化是无穷的，甚至还包括人们建造的用来提升18世纪英式花园浪漫气氛的人造古典废墟（*b*）。尽管将这些奇异的特征与所有的古典建筑形式进行分类是可能的，但采用有创意的古典设计来满足不同需求的机会却十分有限。

CURIOSITIES

Classical architecture does not have to be serious, constrained, or inflexible. Experiments and extravagances of the past have become permanent additions to the tradition.

Single columns were often erected in antiquity for commemorative or ritual purposes and supported gifts to the gods, trophies, and statues. As free-standing monuments, single columns were often more inventive than those used for buildings. One of the most famous was the tall column (*d*) that the Emperor Trajan put up in AD 113 as a part of his Forum in the centre of Rome. A continuous strip of sculpture is wrapped around the column in a spiral, like a huge scroll, telling the story of Trajan's conquest of Dacia (Romania). On the top, 41 metres above the ground, stood a statue of the emperor. While the rest of Trajan's Forum was destroyed, the column has been preserved to this day and in the succeeding centuries has become the model for many designs, from candlesticks to full-size copies.

The Romans were very imaginative in the design of their tombs. Strict laws required that graves, and their often large commemorative structures, be located on sites outside the city walls. This isolation often ensured their preservation. The Tomb of Gaius Cestius (*e*), of 12 BC, is a 29-metre-high concrete pyramid faced in white stone, built just outside the city of Rome and later incorporated into the enlarged city walls. It must have reflected some personal interest or involvement in the recent conquest of Egypt and is an early example of the Egyptian taste that the Romans added to the Classical tradition. The Tomb of the Julii near Arles in France (*c*) was built in 40 BC and is an example of a particular type of tower-shaped monument. It has a sculptured base with an open, square, second level, and a circular top containing statues of the deceased. This is just one of many styles of tomb whose influence can be seen in buildings from the Middle Ages to this century.

The gates on the entrance drives of large houses from the seventeenth century onwards often included small houses, or lodges, to be occupied by employees. As the approach to an important building and the roadside expression of the status of the occupier, a great deal of attention and originality could be devoted to their design. In Hampshire in England one of the gates to Stratton Park (*a*) was designed by George Dance the Younger (1741-1825). Two small Greek Revival lodges are joined by curved walls to a curious gate, almost medieval at first sight, but decorated with ancient Greek features.

The variety of these curiosities is endless and even includes artificial Classical ruins (*b*) to enhance the romantic atmosphere of an eighteenth-century English garden. Although it is possible to divide these curiosities and all Classical building types into categories, there is no limit to the opportunities to serve different functions with inventive Classical designs.

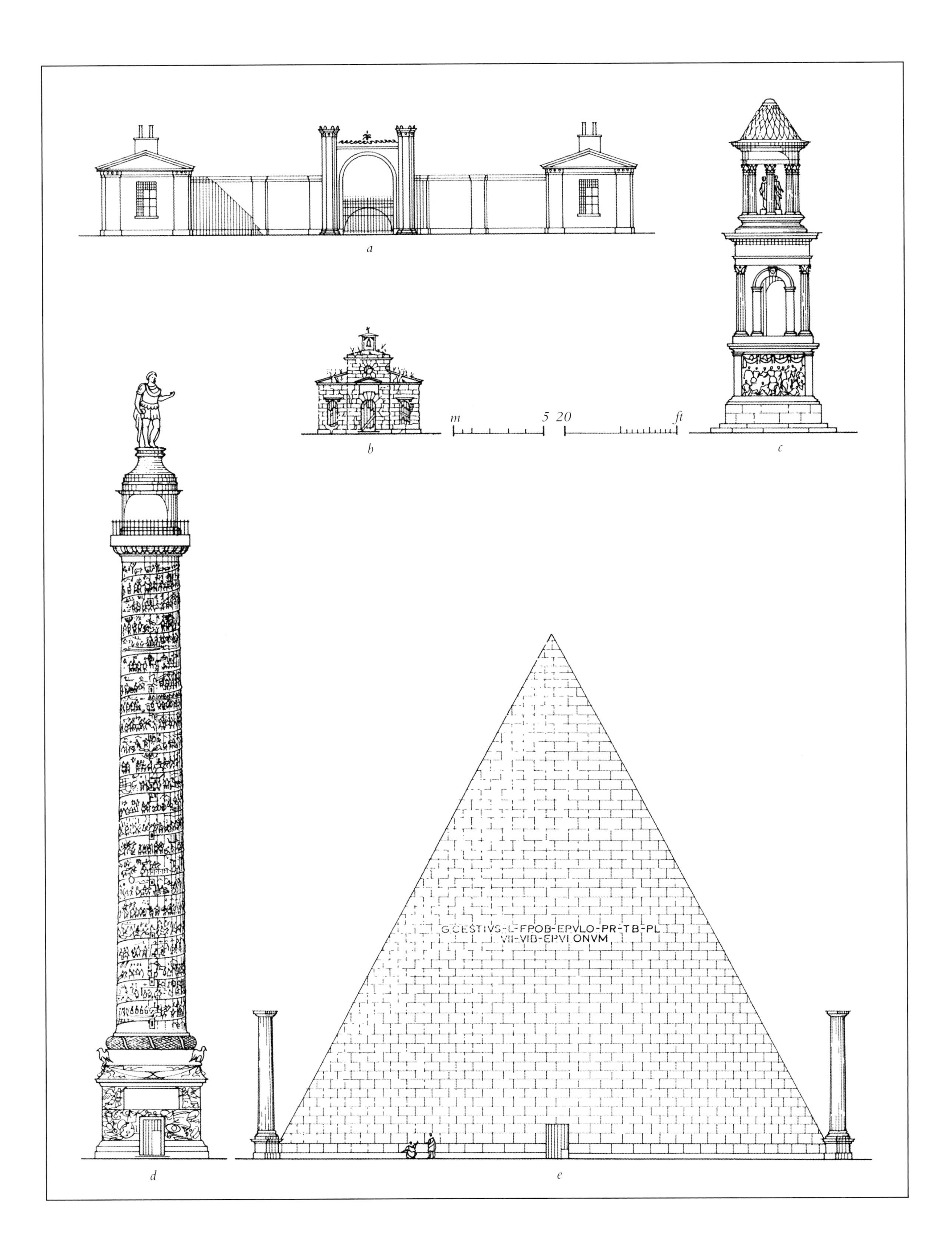

a
b
m
5
20
ft
c
d
G.CESTIVS-L-F.POB-EPVLO-PR-TR-PL
VII-VIR-EPVLONVM
e

3. 柱式

柱式

古典建筑的核心就是柱式。五种区别明显且已形成程式化体系的柱子与其上的水平支撑，称为柱式。从古至今，柱式一直引领古典建筑师。不了解柱式就不能深刻理解古典设计。

下面阐释三种主要柱式，它们是多立克柱式、爱奥尼亚柱式和科林斯柱式。另外两种，托斯卡纳柱式与混合柱式将在本章节的后面加以描述，这两种柱式变体分别是从多立克柱式和科林斯柱式中演变出来的。每一种柱式都可以很容易地从柱子顶端——柱头的装饰辨别出来。多立克柱式有一个简单的垫状细部，称为钟形圆饰，其上是一个方形板，称为冠板。爱奥尼亚柱式有两个涡旋状卷曲，或称螺旋饰，置于方形但却是模制的冠板之下。科林斯柱式有三层柱头，叶形的装饰取材于地中海植物莨苕，包裹着花瓶状核心，置于模制冠板之下，这一冠板四面均向内呈弧线。

然而，柱头通常仅仅是区分各个柱式的最基本方式。所有的柱式都分成几部分，每一部分都有自己的名字。这些在下一页中有图解。同样的名称与划分可用于不同柱式，每部分在不同柱式中都各不相同，并且可通过它们中的任何一个来区分一种柱式。

柱式基本上是成比例的体系，而每个柱式各部分的比例都不相同。为方便起见，两千年来人们一直沿袭着柱式的比例应当以逐渐变窄的柱身最底部直径的倍数或分割来衡量的传统。这种度量称为模数。

几个世纪以来，作家们发布每种柱式的理想比例与细部设计。三种柱式在帕拉迪奥于1570年著成的《建筑四书》中都有图解。这些出版物试图掩饰由于建筑物规模和视觉内涵所造成的比例上常见的变化，以及历史上发现的细部设计与比例上的变体。像帕拉迪奥这样的作家为古典设计提供了实用且权威的基础，缺乏了解很难成功设计出变体，却可能误导人们，错误地认为古典建筑仅仅是古代法则的应用而且缺乏创造潜力。

3. THE ORDERS

THE ORDERS

At the heart of all Classical architecture lie the Orders. It is impossible to understand Classical design in any depth without a knowledge of the five distinct and formalised systems of columns and horizontal supports, called the Orders, that have guided Classical architects from antiquity to the present.

The three principal Orders are illustrated here; these are the Doric, Ionic, and Corinthian. There are two others which are discussed later in this chapter, Tuscan, and Composite, but these are derivatives of the Doric and Corinthian Orders respectively. Each Order is most readily identified by the decoration at the top of the column-the capital. Doric has a plain cushion-shaped detail, or echinus, below a square plate, or abacus. Ionic has two spiral curls, or volutes, below a square but modelled abacus. Corinthian has a three-tiered capital of stylised leaves from a Mediterranean plant, the acanthus, gathered around a vase-shaped core below a modelled abacus that curves inwards on all four sides.

The capital is, however, only the most elementary way of identifying each Order. All the Orders are divided into parts, each of which has its own name. These are noted on the illustration opposite. The same names and divisions apply to each Order. Each part differs in each Order and it is often possible to identify an Order from any one of them.

The Orders are essentially a proportioning system and the proportion of each part is different for each Order. For convenience it has been accepted for two thousand years that the proportions of the Orders should be measured in multiples or divisions of the diameter of the lowest part of the relevant tapering column shaft. This is called the module.

For centuries authors have published ideal proportions and details for each Order. The three illustrated are from Palladio's *Four Books of Architecture* of 1570. These publications have tended to disguise both the common variation of proportions, according to the size of building and visual context, and the variety of details and proportions to be found historically. Authors such as Palladio have provided a useful and authoritative basis for Classical design where a lack of familiarity makes successful variation unlikely, but they can give the misleading impression that Classical architecture is merely the application of old rules and devoid of creative potential.

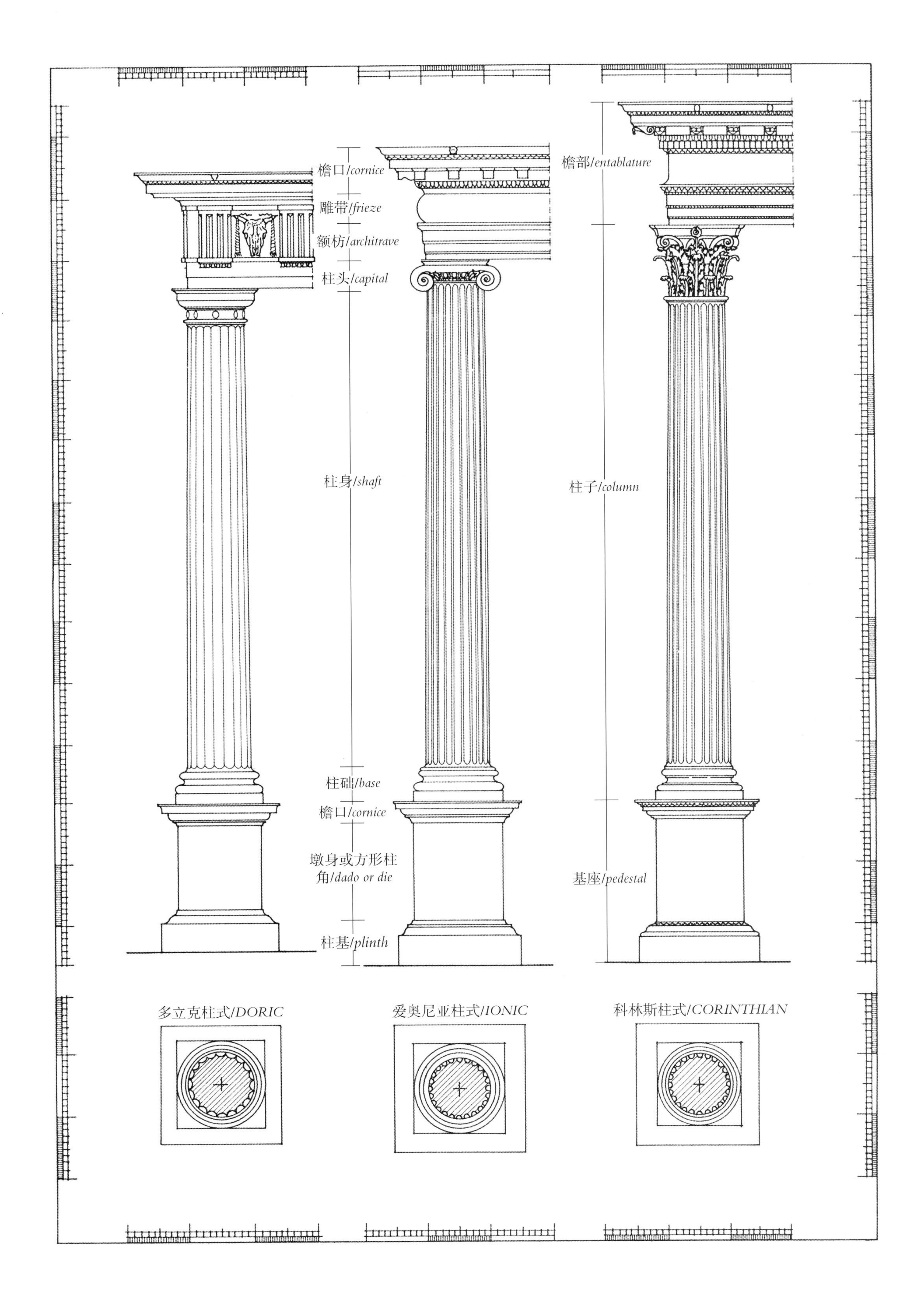
檐口/cornice
雕带/frieze
额枋/architrave
柱头/capital
柱身/shaft
柱础/base
檐口/cornice
墩身或方形柱
角/dado or die
柱基/plinth
檐部/entablature
柱子/column
基座/pedestal
多立克柱式/DORIC
爱奥尼亚柱式/IONIC
科林斯柱式/CORINTHIAN

多立克柱式：起源

多立克柱式是所有柱式的基础。它的起源难以追溯，但是还能够通过合理的判断分辨出其早期历史的某些方面。

多立克柱式大部分的石质装饰特征与木制结构有着明显的差别。这一差别可见于古代某时期一些保存下来的木制寺庙。图例（*b*）为重新修复的公元前6世纪弗朗索瓦陶瓶，展现了早期泥制屋顶的庙宇。可以确定，这座庙宇有泥砖墙和两根木柱支撑着木头的横梁。这座建筑物体现了晚期多立克柱式大部分细部设计，但是很明显一些细部设计与合理的木制结构要求并不相符。公元前7世纪木制多立克柱式起源于希腊，这一看似简单的源头逐渐变得复杂起来，可见于图解（*a*），展现出的是埃及哈塞布苏女王（在位期间为公元前1473—1458年）陵庙柱廊，建于公元前1480年。这一实例与希腊多立克柱式有着惊人的相似之处，这绝非偶然。这些实例证明，试图追溯多立克柱式精美的细部设计源于木造结构是无效的。即使是木制结构也可能是受埃及石造结构的影响，而埃及石造结构是由埃及木造结构演变而来的。

认为人们以现代的眼光将结构与其装饰区分对待，这种观点是危险的。他们更可能用美学的随性来装饰木制建筑。因此，不能从表面上来解读图例（*c*）。这是众多尝试中的一例，将木制多立克建筑合理重建，包含了一些猜想，并没有直接的证据。这或多或少地解释了为什么不仅是多立克柱式，还有其他柱式都是彼此相关联的。

在木制柱子上放置木板或石板，称为冠板。将其延伸成为横梁，用短板即方嵌条相连，方嵌条上方的小型凸出为束带饰，作为延长平板的收边。这些横梁与其连接的嵌板就构成了额枋。屋顶的横梁置于额枋之上，下方用销钉又称圆锥饰固定。这些横梁裸露的末端为三陇板，其间有缺口，称为柱间壁，这就构成了带状雕刻。在带状雕刻之上，木板上支撑着椽即飞檐托块，平铺并且牢牢地用更多的圆锥饰固定在木制屋檐即挑檐滴水板之上。挑檐滴水板必须固定在大型黏土屋瓦末端，防止其余的屋瓦从浅屋顶上滑落。最后一块屋瓦呈沟槽状称为波纹线脚或者波状花边。檐口、椽与沟槽共同构成飞檐。

THE DORIC ORDER: ORIGINS

The Doric Order is the foundation of all the Orders. Its origins are obscure, but there are some aspects of its early history that we can identify with reasonable certainty.

Much of the characteristic stone decoration of Doric is clearly derived from timber construction. This was recognised in antiquity at a time when some examples of timber temples survived. Illustration (*b*) is reconstructed from a painting on the François Vase from the sixth century BC and shows an early temple with a mud roof. The temple almost certainly had mud-brick walls and two timber columns supporting timber beams. This building features most of the details of the later Doric Order, but significantly some details do not correspond to the requirements of a rational timber structure. The apparently simple picture of the timber origin of Doric in the seventh century BC in Greece is further complicated, see illustration (*a*), which shows part of a colonnade from the funerary temple of Queen Hatshepsut (reigned *c.*1473-58 BC) in Egypt from 1480 BC. This has such a remarkable resemblance to Greek Doric that it cannot be accidental. These examples demonstrate that attempts to explain the fine details of the Doric Order strictly according to a timber origin are futile. Even in the original timber structures there may have been the influence of a much older Egyptian translation of timber to stone.

It is dangerous to assume that these people saw structure and its decoration in a modern sense as separate; they are more likely to have decorated their timber buildings with aesthetic caprice. Illustration (*c*) must not, therefore, be taken too literally. It is one of many attempts to reconstruct a timber Doric building rationally and includes a certain amount of guesswork for which there is no direct evidence. It does, none the less, help to explain not just the Doric Order, but the way the parts of all the Orders relate to one another.

The wooden columns are capped by a board or stone slab, the abacus. Spanning them are beams which are bound together by a short board, or regula, and over the regula is a small projection, the tenia, which seems to be the edge of a continuous flat board. These beams and their connecting boards make up the architrave. The roof beams rest on the architrave and are fixed to it from below by pegs, called guttae. The exposed ends of these beams are triglyphs and together with the gaps between them, the metopes, make up the frieze. Over the frieze a timber plate supports the rafters, or mutules, which are laid flat and are firmly secured to overhanging wooden eaves, the corona, by more pegs. This corona must be fixed to the last of the large clay tiles in order to prevent the rest of them slipping off the shallow roof. This final tile sometimes takes the form of a gutter called the cyma, or wave, after its shape. The plate, rafters and gutter are together called the cornice.

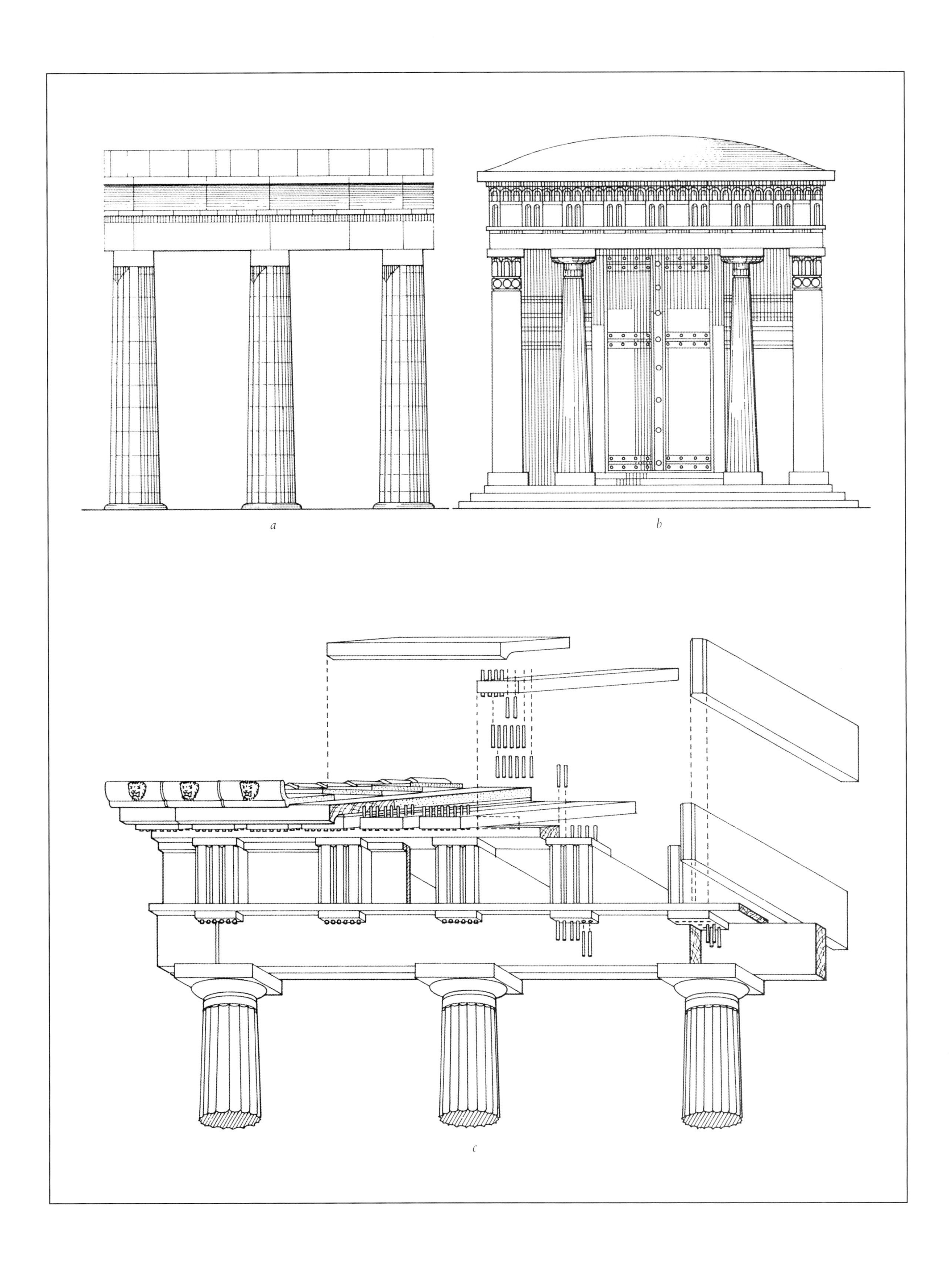
a
b
c

希腊多立克柱式

早期的希腊建筑师修建他们最早的石造建筑时谨慎地模仿木造建筑，改良比例来适应石造建筑更加沉重和欠佳的结构特征。同样的谨慎手法也明显见于多立克柱式的视觉效果发展方面。在不同的细部设计和排列上有过不少尝试，尤其是在早期的建筑中，但所有这些都建立在一个早已清晰确立的、对传统元素进行组合的基础上。细微调整的成功实例被采纳，差强人意的想法被抛弃，但几个世纪以来总体趋势是使柱体越来越纤细，这大概是对石造建筑的支撑力有信心促成的。

位于那不勒斯南部的意大利殖民地帕埃斯图姆的波塞冬神庙（*b*），可追溯至公元前460年，展现出早期石造建筑坚实的比例。柱子粗壮，直径为4.25倍柱底径, 即模数，檐部为1.75倍柱底径高。这些与公元前95年提洛岛的建筑（*c*）形成对比，这些建筑的柱子高度为7.25倍柱底径，檐部为1.85倍柱底径。

雅典的帕提农神庙（*a*和*d*）在古代世界享有很高声誉，18世纪中叶以来，通常被认定为即使不是所有希腊建筑，也是希腊多立克柱式中的最佳典范。它建于雅典极度繁荣与富于创新的时代，成为关于比例的深入研究范例。这一研究成为可能，部分原因在于人们对柱式设计传统持续的关注。柱子高度为5.5倍柱底径，檐部为两倍柱底径高。额枋与带状雕刻大概等高，飞檐稍超过0.5倍柱底径高。

希腊多立克柱式一直忠于其木造起源，可能是因为当时同时期存在着大量木造建筑的缘故吧。屋檐檐口下的飞檐托块通常呈斜坡状与屋顶的沟槽相匹配，同时在横梁末端使用三陇板是一种普遍现象。这种柱式的特征是柱身上有浅浅的沟槽，可能是起源于木材纹理或者工具修整木材时留下的痕迹。通常根据木造柱式细部设计，位于角落的最后一块三陇板通常被移置飞檐托块的一角上，而不是放在其通常所在的柱子中间位置上，因此带状雕刻上最后一块柱间壁要比其他的长。

在希腊世界，木造建筑与石造建筑的相似性由于它们都有色彩明艳的装饰而达到了一定的高度。浅浮雕中的雕刻装饰仅限于应用在单个凸出的装饰线脚，以及像柱间壁这样非木造基本结构上。对这些古典细部设计历史的肯定体现在始终不渝的熟练应用中。

THE GREEK DORIC ORDER

Early Greek architects built their first stone buildings cautiously in imitation of their timber buildings, modifying the proportions to allow for the heavier weight and poorer structural qualities of stone. This same caution is evident in the visual development of the Doric Order. There were a number of experiments with different details and arrangements, particularly in the earlier buildings, but all such variations took place within a very clearly established traditional composition of elements. Minor successful modifications were adopted and unsatisfactory ideas rejected, but the general tendency over the centuries was to make the Order more slender, perhaps encouraged by a greater confidence in the strength of stone.

The Temple of Poseidon in the Italian colony of Paestum, south of Naples (*b*), dates from about 460 BC and shows the stout proportions of the early stone temples. The columns are broad and 4.25 diameters, or modules, high, and the entablature 1.75 diameters high. This can be contrasted with a building from the island of Delos (*c*) of about 95 BC where the column is 7.25 diameters and the entablature 1.85 diameters.

The Parthenon in Athens, (*a*) and (*d*), was highly respected in antiquity and since the middle of the eighteenth century has been generally regarded as the finest of the Greek Doric, if not of all Greek, buildings. It was designed in a particularly prosperous and creative period of Athenian history and is often considered to be the model example of a refined study of proportions made possible in part by the continued adherence to the traditional arrangement of the Order. The column is 5.5 diameters high and the entablature 2 diameters high. The architrave and frieze are of approximately equal height and the cornice a little over 0.5 diameters high.

The Greek Doric Order is consistently faithful to its timber origins, probably because of the parallel existence of significant numbers of surviving wooden buildings. The mutules under the eaves in the cornice always slope to match the pitch of the roof and the use of triglyph beam-ends is universal. Columns almost always have the shallow fluting characteristic of the Order which seems to derive either from the timber grain or more likely from the cuts of the metal tool used to trim the log. The last triglyph on a corner is always moved to the corner of the entablature away from its usual position on the centre of the column, perhaps following a timber detail, thereby making the last metope rather longer than those on the rest of the frieze.

In the Greek world, the similarity between timber and stone buildings was heightened by the bright painted decorations they shared. Sculptural decoration in relief was limited to individual projecting mouldings and places such as the metopes where there would have been no essential timber structure. Recognition of the history of such Classical details as these has always given their knowledgeable use a satisfying consistency.

a
b
c
d

罗马和文艺复兴多立克柱式

希腊对于柱式的应用与罗马人对其采纳从未间断。罗马人的伊特鲁里亚祖先早在罗马人征服希腊在意大利南部殖民地之前就采用本土版本的希腊柱式，同时逐渐侵蚀了罗马本民族的建筑传统。但是长达4个世纪以来，罗马人为其使用的柱式带来一些变化，其中最为重要的就是多立克柱式的改变，这是希腊柱式中最古老也是最保守的一种柱式。

罗马人延续着希腊后期的发展趋势，让柱式变得纤细，罗马的马塞勒斯剧院（*d*）建于公元前13年，柱子高7.65倍柱底径。人们认为希腊式边角三陇板所处的特殊位置是不合理的，应当重新放回柱子的中央，与带状雕刻的其他部分相匹配。受当地传统的影响，罗马人引入了其他的改变。偶尔给柱子增加装饰线脚柱础或者装饰与柱头之下更常见、更复杂的希腊圆环，现在已经简化成增加一个小圆环串珠装饰线脚，为可有可无的柱槽收边。檐部与传统的希腊木造建筑上的结构往往大不相同，这种木造建筑十分少见，并已成为历史。马塞勒斯剧院的檐部长度大约为两倍柱底径，并带有简化的额枋。小型齿状凸起，即齿状装饰，在檐口或飞檐托块上，藏于挑檐滴水板之下。在后期的实例中，修建于305年的戴克里先浴场（*b*）檐部上的飞檐托块（没有图示）完全被省略，丰富的雕刻应用于所有细部设计，整个柱式通过这一手法展现出罗马的华美盛况。

文艺复兴再现了罗马风格，但对希腊风格知之甚少。15世纪、16世纪的意大利建筑师研究各种各样的罗马废墟，但是总体上他们还是更喜欢将古典建筑传统融入新的设计理念中去。印刷的发明将这些理念带到更多观众面前。

许多文艺复兴时期的作家发表了自己设计的柱式。结果是普遍增加了罗马柱式的柱础（*a*）和爱奥尼亚柱式的柱础（*c*），并且创造出一种新的柱基使得多立克柱式与其他柱式相匹配，例如文森佐·斯卡莫齐（1548—1616年）多立克柱式（*c*）。几乎所有的出版物中的柱子都是8倍柱底径高，但是柱础、柱头和檐部等细部设计因作者不同而风格各异。一些书中展示了在檐口处应用齿状装饰或爱奥尼亚风格柱础的设计。飞檐托块带有销钉，又称圆锥饰，或者成为凸出的挑檐滴水板下面的浅层装饰，或者像1562年维尼奥拉多立克（*a*）柱式中，不再是椽而变成了水平装饰托架。

1750年重现希腊建筑之前，这些合理的罗马多立克版本是普遍存在的。它们在文艺复兴之后得以广泛应用，既是确立的古典传统的变体，也是文艺复兴时期出版物持续影响下的产物。

ROMAN AND THE RENAISSANCE DORIC

There was no break between the Greek use of the Orders and their adoption by the Romans. Rome's Etruscan had used provincial versions of the Greek Orders long before the Roman conquest of the Greek colonies of southern Italy eclipsed Rome's native architectural traditions. However, more than four centuries of Roman use brought about some changes and none was so significant as the modifications to the Doric, the oldest and most conservative of the Greek Orders.

The Romans continued with the later Greek tendency to make the column more slender and on the Theatre of Marcellus in Rome (*d*) of 13 BC the column is 7.65 diameters high. The eccentric position of the Greek corner triglyph was thought to be irrational and it was brought back to the centre of the column to match the others on the frieze. Influenced by local traditions, the Romans introduced other modifications. Occasionally a moulded base was added to the column and more commonly the complex Greek system of rings below the capital was simplified by the addition of a small, round, astragal moulding which marked the end of the now optional fluting. The entablature was often quite different from that of traditional Greek timber buildings, which were by now rare and historic. The Theatre of Marcellus has an entablature of about 2 diameters with a reduced architrave. There are small tooth-like projections, or dentils, in the cornice and the mutules are concealed under the corona. On later examples, such as the Baths of Diocletian (*b*) of AD 305, the mutules on the entablature (not illustrated) are omitted altogether and the whole Order takes on a typically Roman opulence with rich carving applied to all the details.

The Renaissance rediscovered Rome but knew little of Greece. Fifteenth-and sixteenth-century Italian architects studied the diverse ruins of Rome but, above all, they were interested in fitting ancient architecture into new theories of design. The invention of printing brought these theories to a wide audience.

Many Renaissance authors published their own versions of the Orders. This resulted in the universal addition of a Roman base (*a*) or an Ionic base (*c*) and the creation of a plinth in order to match Doric with the other Orders as in the Doric Order of Vincenzo Scamozzi (1548-1616) (*c*). Almost all publications made the column 8 diameters high, but details of the base, capital, and entablature varied according to the author. Some books offered dentils in the cornice or an Ionic style of base. Where mutules were included the pegs, or guttae, either became a shallow decoration on the underside of the projecting corona or, as in Vignola's Doric (*a*) of 1562, they cease to be rafters and became decorative horizontal brackets.

Until the rediscovery of Greek architecture in about 1750 these rationalised versions of Roman Doric were almost universal. Since the Renaissance they have been used regularly, both as an established variation in the Classical tradition and as a consequence of the continued influence of the Renaissance publications.

a
b
c
d

多立克柱式的应用

多立克柱式之所以得名是因为这一建筑风格源于希腊多立安人占领希腊大陆和殖民地时期。早期的希腊共和国使用这一柱式标志这一建筑与多立安人有关。我们也从罗马建筑师兼作家马库斯·维特鲁威·波利奥（死于15年）那里了解了柱子的比例，即后来的柱式是从男人的身体比例中衍生而来的。这种柱子粗壮的比例与其他柱子相比显得更加粗犷是由于多立安人比亚细亚的希腊人身材更加魁梧的缘故，但维特鲁威提出神庙柱式的选择要与所供之神性别相同，这一建议似乎并未被采纳。

这一柱式的男性象征意义得以保存，公元前6世纪西西里的塞利纳斯的柱子与同一时期的一座运动员雕像都对这一意义进行了很好的诠释。

在随后的几个世纪中，这一柱式有了更多的内涵。作为柱式中装饰最少的一种，多立克柱式已经在古代后期与近现代和造价不高且实用的建筑紧密结合在一起了。人们认为这一柱式所带有的男性内涵使其特别适用于军用建筑、监狱和其他这样刻意创造出来的坚固且粗犷的建筑物。

作为升序排列中最下面的一种柱式（详见P128），这些特征可同时表达下面的含义，例如防卫特征以及文艺复兴时期宫殿低层的服务型特征。在这种情况下，为强化其坚固的特征经常使用粗琢工艺。

然而多立克柱式也能够表达和实用与雄壮的隐喻无关的优雅与繁复。1761年，修建于爱尔兰马里诺的威廉·钱伯（1723—1796年）住宅（*a*），采用多立克柱式效仿古罗马风格，创造了一座极其高雅繁复的小型建筑。

希腊文艺复兴为这一柱式带来了新的解读方式。钱伯住宅与同时期克劳德·勒杜在巴黎设计的一座大门（*b*）形成对比。在后者中，多立克柱式的使用与大胆的设计，象征着刻意回归原始的纯净。新的关于希腊纪念碑的研究激发了人们更加热切的崇敬之心，古希腊人对比例精熟的运用，使得希腊多立克柱式超越了其最基本的起源，在诸如申克尔早期于19世纪教堂（*c*）这样的实例中，精心设计出了优雅的氛围。在同一世纪的早些时候，当人们聚焦于装饰最少的柱式比例问题上时，曾尝试展现其脱离传统装饰的潜在经典比例。20世纪多立克建筑，例如1924年建于芬兰于伐斯居拉的雨果·阿尔瓦·亨里克·阿尔托（1898—1976年）工人俱乐部（*d*），展现出新的抽象纯粹之美。

THE USE OF THE DORIC ORDER

The Doric Order is so called because it was the architectural style of the Dorian Greeks who occupied the Greek mainland and its colonies. In the early Greek republics the use of this Order signified a building associated with the Dorians. We also know from the Roman architectural author Marcus Vitruvius Pollio (*d.*15 AD) that the proportions of the column, and hence the Order, were traditionally thought to be derived from the proportions of a man. The broader proportions of the column in comparison with the other Orders and the more rugged reputation of the Dorians in relation to the Asian Greeks lend weight to this tradition, but Vitruvius' suggestion that a temple Order was chosen to match the sex of the god does not seem to have been followed.

The male association of the Order has remained and it is here illustrated by a sixth-century-BC column from Selinus in Sicily and a statue of an athlete from about the same period.

In succeeding centuries the Order has come to take on other connotations. As the least decorative of the Orders it has been associated both in later antiquity and in more recent times with inexpensive and utilitarian buildings. When this is combined with the masculinity of the Order it can be seen as particularly suitable for military architecture, prisons or other buildings that are deliberately robust or aggressive.

As the lowest of an ascending tier of Order (see page 128) these characteristics can simultaneously express, for example, the defensive character and the service functions of the lower floor of a Renaissance palace. In this context, and as a reinforcement to its robust character, the Order is often accompanied by rustication.

Doric can, however, also express elegance and sophistication unconnected with any allusion to practicality or sturdiness. William Chambers' (1723-1796) Casino, at Marino in Ireland (*a*), built in 1761, employs a Roman Doric Order in emulation of ancient Rome to create a small building of surprising grace and complexity.

The Greek Revival of the eighteenth century introduces a new range of interpretations of the Order. Chambers' Casino can be contrasted with the design by Ledoux for a Paris gateway (*b*) of about the same date, where the use of a Greek Doric Order and the bold design represent a deliberate return to a primitive purity. New studies of Greek monuments engendered a profound respect for the proportional mastery of the ancient Greeks and the use of the Greek Doric Order went beyond an association with fundamental origins to allow designs such as Schinkel's early-nineteenth-century church design (*c*) to express an air of calculated elegance. Concentration on the proportions of the least decorated Order led, in the first part of this century, to an attempt to show underlying Classical proportions stripped of much of their historic decoration. Twentieth-century Doric buildings such as Hugo Avar Henrik Aalto's (1898-1976) Worker's Club in Jyvaskyla in Finland (*d*) of 1924 suggest a new abstract purity.

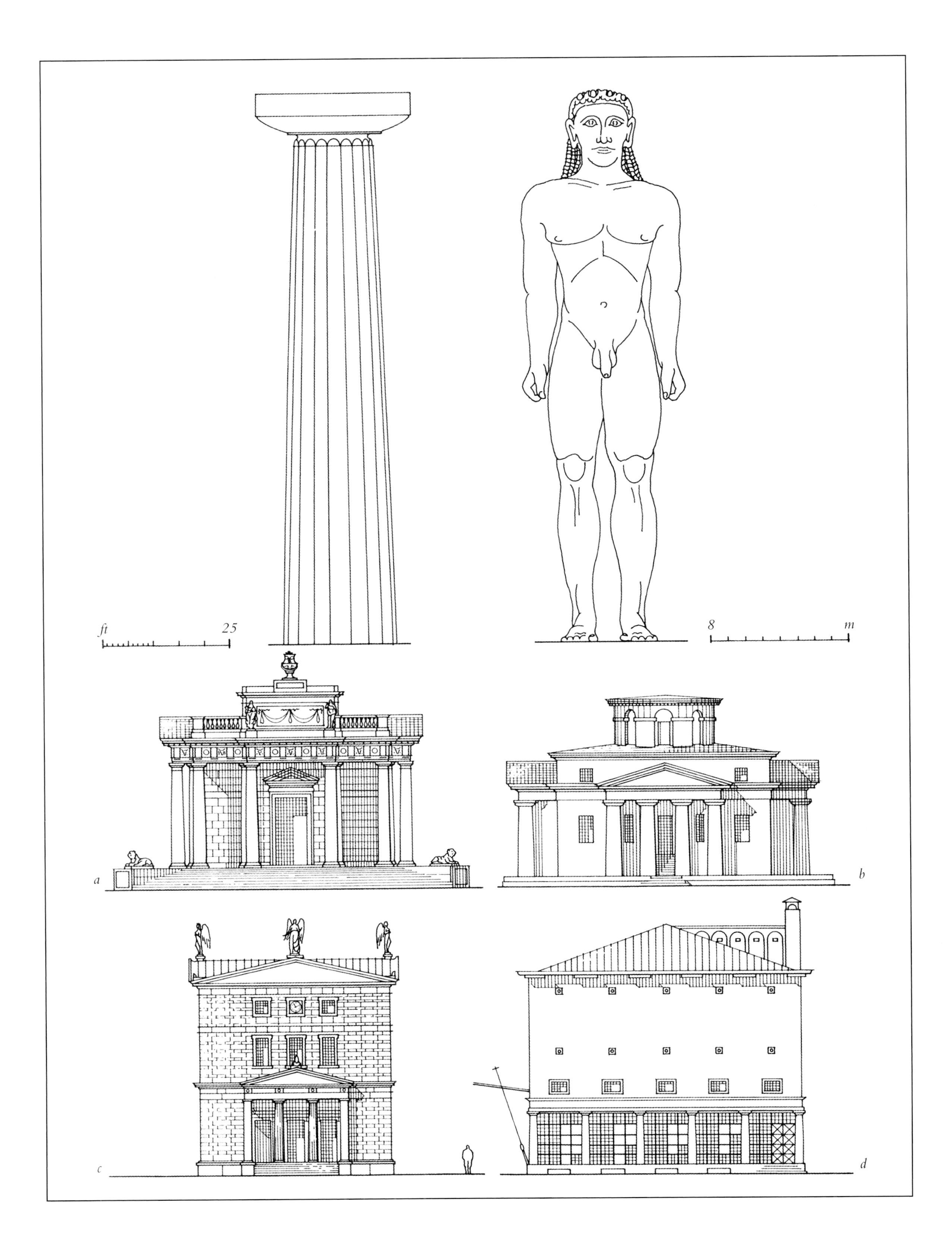
ft
25
8
m
a
b
c
d

托斯卡纳柱式

按照通常的柱式排列顺序，将托斯卡纳柱式放在前面是因为这是一种最宽大的柱式。有一种错误的印象认为托斯卡纳柱式是多立克柱式的一个分支。它应用于罗马共和国的早期，发生在希腊殖民向南部发展并将罗马建筑紧密结合于古典建筑主流之前。考古学上几乎没有托斯卡纳和伊特鲁斯坎神庙的柱身以及其他元素如何组装的实例。意大利本国的多立克柱式或者托斯卡纳柱式的遗迹可见于这些建筑中，但是多立克柱式可追溯至相似元素出现于意大利中部之前的一个世纪。

公元前1世纪，维特鲁威在描述多立克、爱奥尼亚和科林斯柱式的附录中首次提到了托斯卡纳柱式，或多或少地从希腊作家的作品中抄录了一些。毫无疑问的是，这种描述带有一种爱国情结，赋予建筑物以本国古代崇高的神庙同等的智慧与神秘的崇敬之心，正如希腊柱式所承载的是同样的内涵。我们可以复原一座早期的罗马或伊特鲁斯坎神庙（*b*），这些神庙始建于公元前150年，从考古发现可推断。这些建筑物置于石造墩座之上，墙体却是木造或泥制的，屋顶为带有陶土装饰的木造结构。有很深的凸出屋檐，可能是为了保护泥墙，图例中所示的还带有黏土屋瓦。茅草屋顶神庙，必须建有陡屋顶，然而这样的建筑只发展到了1世纪。

罗马皇宫建筑没有采用任何托斯卡纳柱式的设计。这种本土化的柱式的贡献主要体现在多立克柱式在罗马的发展上，与伊特鲁里亚柱式同样的特征融合，例如柱础、光滑的柱身和更加复杂的柱头。当本国的神庙年久失修，人们便采用当时新颖的多立克柱式进行重建，虽然保留了一些传统的平面来维持罗马神庙的独特原貌。能将古老的建筑轻而易举地重新设计，原因不仅在于对希腊柱式传统的崇敬，还因为原有风格的具有派生性并缺乏成形的体系。

我们所熟知的托斯卡纳柱式是16世纪理论家们从维特鲁威的描述中复原的产物，也是简化版的多立克柱式，使得人们对托斯卡纳柱式的描绘符合逻辑的需要。这种柱式在古典建筑中保持了近五百年的稳固地位。

托斯卡纳柱式（*a*和*c*），取材于帕拉迪奥设计，与其他版本相似，符合维特鲁威柱高为柱底直径7倍的描述。柱础也是同样的模式，与简单的罗马多立克风格相似，尽管根据维特鲁威的叙述与考古证明应当是完全呈圆形的。但为与其他柱式相匹配而增加了基座。

THE TUSCAN ORDER

In the usual sequence of Orders the Tuscan is placed first because it has the widest column. This can give a misleading impression as the Tuscan Order is, in fact, a provincial form of the Doric. It was used in the early years of the Roman republic before closer contact with Greek colonies to the south brought Roman architecture into the mainstream of the Classical tradition. Archaeologically there is little evidence of the way the columns and other elements of Tuscan or Etruscan temples were assembled. Remains of native Italian Doric or Tuscan columns have been found from these buildings but the Doric Order can be traced to at least a century before anything similar appears in central Italy.

The Tuscan Order is first mentioned by Vitruvius in the first century BC as an addendum to a description of the Doric, Ionic, and Corinthian Orders, more or less copied from Greek authors. There can be little doubt that this description was a patriotic attempt to give the architecture of ancient and revered native temples the same intellectual and mythological respectability as was attached to the Greek Orders. We can reconstruct one of these early Roman or Etruscan temples (*b*), from about 150 BC, from archaeological evidence. These buildings were raised on a stone podium but the walls were probably made of timber and mud, and the roof was timber with terracotta decoration. They had very deep projecting eaves, probably to protect the mud walls, and the example illustrated has a clay tile roof. Thatched temples, which must have had steeper roofs, did, however, survive until the first century AD.

No Tuscan Order was formalised in Roman imperial architecture. The principal legacy of this indigenous style has been the Roman development of the Doric Order, which incorporated some features of Etruscan columns such as the base, smooth shaft, and more complicated capital. When native temples fell into disrepair they were rebuilt in the up-to-date Doric Order although the traditional plans were retained to create a distinct Roman temple type. The case with which the old buildings were so redesigned is witness not just to respect for Orders of Greek ancestry, but to the derivative nature of the original style and its lack of a systematic tradition.

The Tuscan Order as we now know it is the result both of sixteenth-century theorists drawing their evidence from Vitruvius' description and of the need to reduce the Doric Order to make their description of the Tuscan Order fit in with a logical sequence. In this form it has become established for nearly five hundred years as a part of the Classical tradition.

The Tuscan Order shown, (*a*) and (*c*), is from Palladio and, in common with all other version, follows Vitruvius in giving a column height of 7 diameters. The base is also of a consistent pattern, resembling the simple Roman Doric type, although to accord both with Vitruvius and archaeological evidence it should be totally circular. A pedestal has been added to match the other Orders.

b
ft 20
6 m
a
c

托斯卡纳柱式的应用

托斯卡纳柱式因维特鲁威1世纪的著作得以延续，并于文艺复兴时期有了确定的形式。维特鲁威努力使得托斯卡纳成为一种正规的柱式，可以轻而易举地在与同时期的意大利中部的拉丁和托斯卡纳民族聚居地发现这样的本土风格。16世纪的理论家确立了托斯卡纳柱式的形式，可能更多的是将这一古典设计看作多立克柱式的一种变体，而不是一种独立的柱式。直至今天，托斯卡纳与多立克柱式相似的比例与细部设计，以及在多立克檐部上自由地运用托斯卡纳的设计通常让人们很难将两者区分开来。在古代与文艺复兴之后，托斯卡纳风格即柱式都或多或少地有了独立的特征。

托斯卡纳柱式没有与特定的人类外形相关的古典传统，但在文艺复兴之后的一些出版物中，作家赋予这种柱式以人的特征，与三种主要柱式并称。因此，由于托斯卡纳柱式的宽度和它更加本土以及不那么复杂的起源，这一柱式表现的是强壮魁梧的男性。为保持一贯性并表达对以往作家们对柱式特征描述的认可，托斯卡纳柱式就由托斯卡纳运动员的形象来代表。他的胡须赋予其粗犷的性格，与不蓄须的多立克男子形成对比；更进一步的是这一雕塑与多立克人有关，象征着托斯卡纳柱式与多立克柱式的延续关系，由于托斯卡纳与多立克人都是运动员，这就使人联想到了两种柱式之间的关联。

托斯卡纳柱式有许多多立克柱式的特征。由于其极简的细部设计，使用起来比多立克柱式更加经济。伊尼戈·琼斯（1573—1652年）于1631年修建的伦敦考文特花园圣保罗教堂（*d*），特别设计成十分经济的建筑，同时也是罕见的真正意义上采用维特鲁威描述的托斯卡纳柱式的实例。

这一柱式的比例与其原始的象征意义使其特别适合用于表达粗犷的风格。伦敦的约克府水门（*a*）修建于1626年，设计者可能是巴尔萨泽·热尔比耶（1592—1663年）；牛津大学通往皇后学院的霍克斯莫尔之门（*c*）建于1709年，都装饰有厚重粗犷的托斯卡纳柱式。皇后学院开放式穹顶展现出托斯卡纳建筑常见的优雅，并暗示着这种柱式的比例具有开发潜力，应当与其他柱式同样地采用精心设计。

总而言之，托斯卡纳柱式表达的是简洁的概念。这种简洁是建筑物功能的产物，或者正如威廉·托马斯（逝于1800年）于1781年设计的墓室（*b*）一样，可以创造出庄严肃穆的气氛和稳固的几何形制。

THE USE OF THE TUSCAN ORDER

The Tuscan Order owes its continued use and the form it took after the Renaissance to the first-century writings of Vitruvius. The Etruscan architecture that Vitruvius tried to make into a formal Order would have been identified immediately by his contemporaries as the native style of the Latin and Etruscan communities of central Italy. The Tuscan Order formulated by sixteenth-century theorists would probably have been regarded in antiquity as a variant of Doric rather than as a separate Order, and even today the similarity of proportions and details on Tuscan and Doric columns and the freedom of design often used on Doric entablatures can make identification difficult. Both in antiquity and after the Renaissance there has, none the less, been a style, or Order, which has a separate identity.

There is no ancient tradition associating the Tuscan Order with a specific human form, but in publications after the Renaissance some authors chose to give the Order a human character in line with the range of characters given to the three principal Orders. Therefore, due to the Tuscan column's breadth and its more native and apparently less sophisticated origin, the Order was sometimes given the character of a thickset rustic man. In the interests of consistency and in recognition of the tradition of the attribution of a character by past authors, the Tuscan Order is here represented by the figure of an Etruscan athlete. His beard gives him something of a rustic character in contrast to the clean-shaven Doric figure; the more advanced representation of the sculpture in relation to the Doric man indicates that the origin of the Tuscan Order post-dates the Doric and, as both the Doric man and the Tuscan man are athletes, there is a reminder of the association of the two Orders.

The Tuscan Order shares many of the attributes of the Doric. It is even more economical than Doric to use, due to its very simple detailing. St Paul's Church in Covent Garden in London (*d*), by Inigo Jones (1573-1652) in 1631, was specifically designed to be economic to construct and is also a rare attempt to represent literally Vitruvius' description of the Order.

The proportions of the Order and its primitive associations make for a particularly suitable relationship with rustication. Both York Water Gate in London (*a*) of 1626, probably by Balthazar Gerbier (1592-1663), and Hawksmoor's gate to Queen's College, Oxford (*c*) of 1709 have heavily rusticated Orders. The open dome on Queen's College shows an elegance unusual on Tuscan buildings and indicates that the Order can be used to exploit its proportions but otherwise be treated with the same finesse as the other Orders.

Above all, the Tuscan Order expresses simplicity. The simplicity can be a consequence of the function of the building or, as with the mausoleum design (*b*) by William Thomas (*d*.1800), of 1781, can intentionally create an atmosphere of sombre severity or geometric solidity.

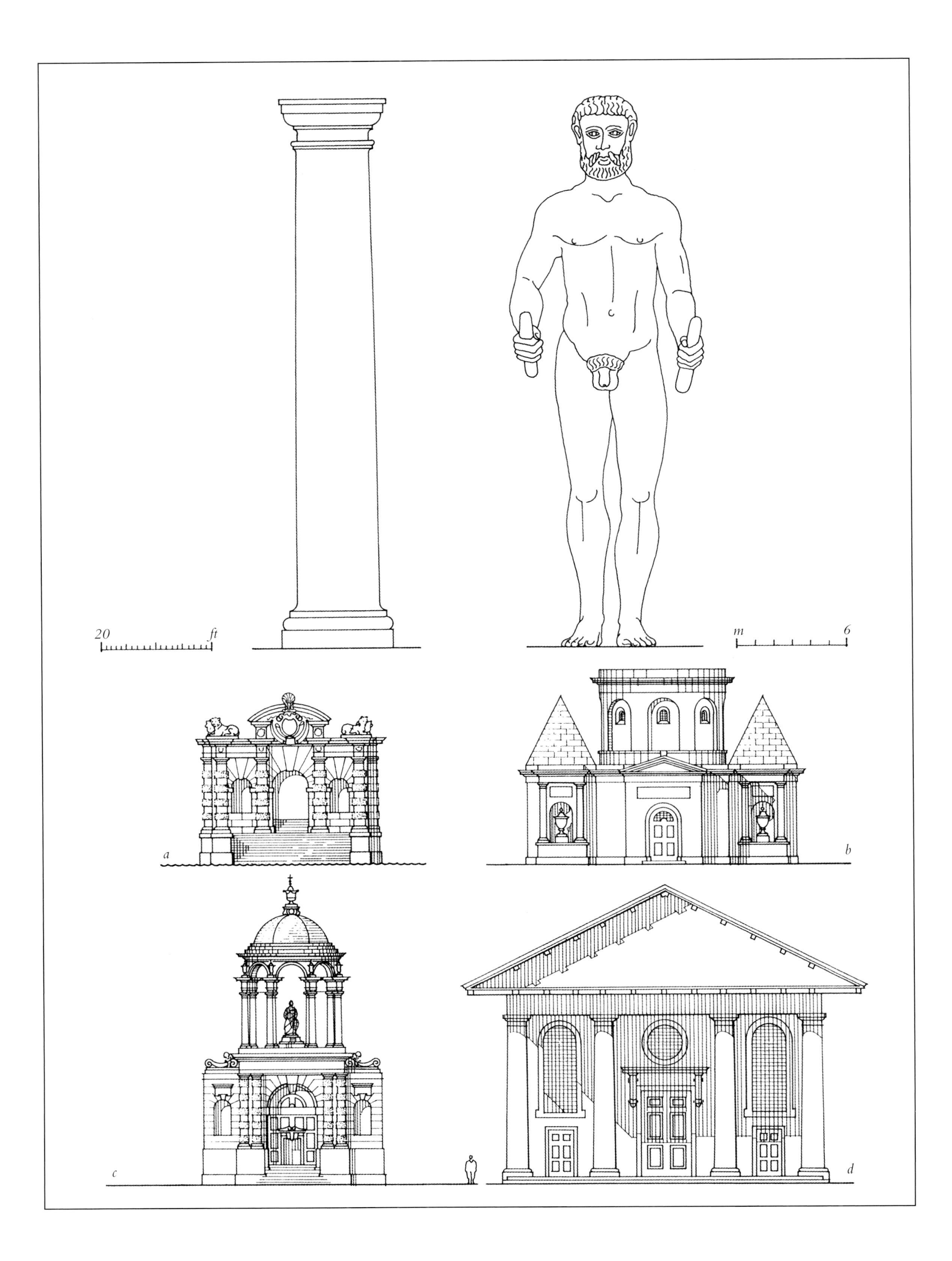
20
ft
m
6
a
b
c
d

爱奥尼亚柱式：起源

爱奥尼亚柱式起源于希腊爱奥尼亚地区，也就是今天地理上土耳其的西海岸邻近希腊群岛的地方。由于这一独特的建筑形式首次出现的证据在公元前6世纪中叶，因此，几乎可以肯定地说，这一柱式是多立克柱式的发展。

这一柱式的细部设计与多立克柱式有着明显的差别，最突出的就在于纤细的柱体与柱头上独具特色的螺旋纹案、卷涡。根据古代作家维特鲁威描述，这些卷涡与女性的发卷有着传奇般的关联，并且柱子本身就是从女性身体的比例中演化而来的。柱子与人物之间的关联在古代与埃及十分常见，赋予神庙柱式以所供之神的个性，在哈索尔女神神庙的柱式中可见一斑，如图（*a*）所示。早期的爱奥尼亚建筑同样真正采用人像柱子，或称为女像柱，建于圣地特尔斐的，象征希腊民族的6世纪的宝库（*c*）中。不难理解，最早出现的爱奥尼亚柱式，在萨摩斯岛奉献给女神赫拉的神庙和以弗所城中的阿尔忒弥斯神庙（*b*），可见代表女神本尊的女像柱。

柱头的起源令人费解。在公元前10世纪，近东地区的某些建筑物中就有柱子的柱头上带有螺旋柱式装饰，并且同样的设计也应用于腓尼基和叙利亚的象牙、金属制品上。公元前7世纪，爱奥尼亚建筑使用一种独特的柱头（*d*和*e*），这可能是受到了更古老建筑的影响，并且可能是早期爱奥尼亚柱头（*g*）的起源。这一进化理论与卷曲的传说，与其所供奉神庙而自然产生的联想意义并不矛盾。

这一柱式在其他方面也与多立克柱式不同。柱础一词是首先通过这一柱式进入建筑语汇的，可能纯粹是为了装饰效果。额枋由叠加的部分构成，可能是模仿层叠的板子。在早期的例子中，屋檐下建有一系列大型长方形的凸出，称为齿状装饰，由于其类似牙齿的形态，可能象征着紧密排列的过梁末端。齿状装饰的起源可更早地追溯至早期缺乏带状雕刻而造成檐部（*f*）较浅这一因素。

爱奥尼亚柱式在5世纪中叶进入希腊本土，多立克建筑师们使其与多立克柱式有了更加直接的理性关联。雅典伊利索斯的神庙（*h*）建有多立克比例、深带状雕刻和简单柱础的柱子。后期的建筑师重新启用了齿状装饰和阶梯状额枋的设计。爱奥尼亚柱式的比例与细部设计已经确立，保留着独特、优雅的风格和早期柱式的装饰特征。

THE IONIC ORDER: ORIGINS

The Ionic Order originated in the Greek communities of Ionia, which is today the west coast of Turkey and the adjacent Greek islands. The Order is almost certainly a later development than Doric as the first evidence of architecture of this distinctive type dates from about the middle of the sixth century BC.

The details of the Order differ noticeably from the Doric, the most apparent differences being the slender column and the distinctive curls, the volutes, on the capital. According to the ancient author Vitruvius these volutes had a legendary association with curls of female hair and the column itself was similarly thought to have derived from the proportions of a female figure. Such association of columns and figures was common in antiquity and the Egyptians particularly gave temple columns some of the personality of the relevant deity, as can be seen in the column from a temple to the goddess Hathor, illustrated (*a*). Early Ionian buildings also had such literal column figures, or caryatids, on sixth-century treasuries (*c*) built to represent individual Greek communities at the sacred site of Delphi. It is not hard to understand that the first Ionic columns, which were built to the goddess Hera, on the island of Samos, and Artemis, in the mainland city of Ephesus (*b*), could have been seen to represent the goddesses themselves.

The origins of the capital itself are more confusing. There were column capitals with spiral decoration on near-eastern buildings in the tenth century BC and similar designs were used on Phoenician and Syrian ivory and metalwork. Buildings of the seventh century BC in Ionia used a distinctive capital, (*d*) and (*e*), that was probably influenced by these older examples and is likely to have been the source of the first Ionic capitals (*g*). This theory of evolution does not necessarily contradict the legend of the curls as such an association could have been suggested simultaneously by the dedication of the temple.

The Order differs in other respects from Doric. The column base was first introduced to the Classical vocabulary through this Order, probably for purely decorative effect. The architrave is made up of overlapping sections, perhaps imitating over lapping boards. In early examples a series of large rectangular projections below the cornice, called dentils due to their resemblance to teeth, are likely to represent the exposed ends of closely spaced beams. This origin for the dentils is further supported by the early absence of a frieze and the consequently shallower entablature (*f*).

The Ionic Order was rationalised to relate more directly to the Doric Order when it was introduced to the Doric builders on the Greek mainland in the middle of the fifth century. The Temple on the Ilissus in Athens (*h*) includes a deep frieze of Doric proportions and a simplified base. Later builders reintroduced the dentils and stepped architrave. The proportions and details of the Ionic Order now became established, retaining the distinctive, elegant, and decorative character of the early Order.

a
b
c
d
e
f
g
h

爱奥尼亚柱式

爱奥尼亚柱式的部分与比例的设计，于公元前5世纪相对稳定下来。这种柱式，少有例外，大部分是柱底直径的8～9.5倍高，从希腊共和国时代直至今天都是如此。尽管罗马建筑师有时将其削减至7倍柱底直径，自从文艺复兴时期开始，高度为柱底直径的8～9倍已经成为规则。

早期柱子的实例中，柱身上纵向的沟槽似乎是有意为之，即使不是经常采用。萨摩斯岛和以弗所早期柱子的实例中柱身上有40条沟槽，有着和多立克柱身上一样浅浅的轮廓。这些沟槽后来演变成24条更深的半圆形刻沟，沟槽间有窄窄的纵向修边，或称平缘。一些罗马与文艺复兴时期的柱式没有沟槽，有些罗马式建筑的柱子上只有20条，但广为接受的爱奥尼亚柱式带有24条凹槽。

柱础有着复杂的发展历史。首先它是一个凸出的圆盘，与柱身末端以凹面装饰线脚相连。这些有时置于方形平板上，通常装饰复杂的水平刻槽，或者将向外凸出部分的轮廓设计成好似弧线形，或者将其划分为一系列较深的凹槽与凸面曲线的组合。这类柱础，可见于普里埃内的雅典娜波丽亚丝神殿（*d*）建于公元前334年，称为以弗所式。更多的例子表现出大量的变化。当爱奥尼亚建筑在希腊本土修建时，这种设计得以简化，两个或者更多的凸出柱脚圆盘线脚与凹弧装饰线脚清晰地区分开来。柱脚圆盘线脚可能采用传统水平沟槽装饰，或者其他方法装饰，称为雅典式柱础。可见于公元前5世纪末修建的雅典厄里希翁神殿（*c*）。这一柱础在罗马成为标准模板，后来在文艺复兴时期，也可以选择增加有特色的基座（*a*）。

檐部一直保持一贯的风格。早期并没有带状雕刻这一设计，也未见于普里埃内建筑（*d*），后来也没能延续到罗马时期。带状雕刻作为附属品，没有任何结构基础，作为一个整体装饰物以浮雕的形式出现。额枋通常有三道，其中两个可能划分为逐渐向外的形态，呈阶梯状，有时每一层都有装饰。飞檐檐口上通常有凸出的齿状装饰，尽管在厄里希翁神殿中被省略，但在罗马和文艺复兴的建筑中能够见到。16世纪帕拉迪奥和斯卡莫齐（*b*）提出用成排的小托架代替齿状装饰。

柱头比其他元素更能代表柱式的特征。尽管卷涡应用普遍，在其设计中有很多变体，卷涡复杂的几何特征以及其各种变化需要单独进行描述。

THE IONIC ORDER

The arrangement of the parts and proportions of the Ionic Order became relatively consistent in the fifth century BC. The column, with few exceptions, has remained between 8 and 9.5 diameters high from Greek republican examples to the present day. Although Roman architects sometimes reduced this to 7, since the Renaissance a height of between 8 and 9 diameters has been the general rule.

Vertical fluting on the column shaft seems always to have been intended, if not always executed, on early examples of the Order. The earliest examples of columns from Samos and Ephesus have forty flutes, with the same shallow profile as Doric columns. These flutes soon reduced in number to twenty-four and became deeper, semicircular grooves with narrow vertical bands, or fillets, between them. Some Roman and Renaissance columns have no fluting and the occasional Roman building has only twenty flutes, but it is generally accepted that the Ionic column is fluted twenty-four times.

The column base has a complicated history. At first it was a swollen disc joined to the lower end of the shaft by a small concave moulding. This sometimes sat on a square slab and was usually decorated with a complex series of horizontal grooves which either left the outline of the outward swelling visible as a simple curve or divided it up into a series of deep concave grooves and convex curves. Bases of this type, such as on the Temple of Athena Polias at Priene (*d*) of 334 BC are called Ephesian. Extant examples show considerable variety. When Ionic buildings came to be constructed on the Greek mainland this arrangement was simplified, two or more swollen torus discs being clearly separated by concave scotia mouldings. The torus discs could be decorated with the traditional horizontal grooves or in other ways. This is known as the Attic base. It is found on the Erechtheion in Athens (*c*) of the late fifth century BC. This base became the standard type in Rome and then in the Renaissance when, characteristically, an optional pedestal (*a*) was added.

The entablature has remained quite constant. The early omission of a frieze, which may have been absent at Priene (*d*), did not survive into the Roman period. As an addition without any structural origin, the frieze was often decorated in its entirety with relief sculpture. The architrave generally has three, but occasionally two, divisions stepping progressively outwards and sometimes decorated at each step. The cornice usually has a row of projecting dentils which, although they were omitted on the Erechtheion, are generally found on both Roman and Renaissance buildings. In the sixteenth century both Palladio and Scamozzi (*b*) suggested a variation which puts a row of small brackets in place of the dentils.

The capital is, more than any other element, the defining feature of the Order and, in spite of the universal use of the volute, there is great variety in its design. The complex geometry of the volute and these variations require their own description.

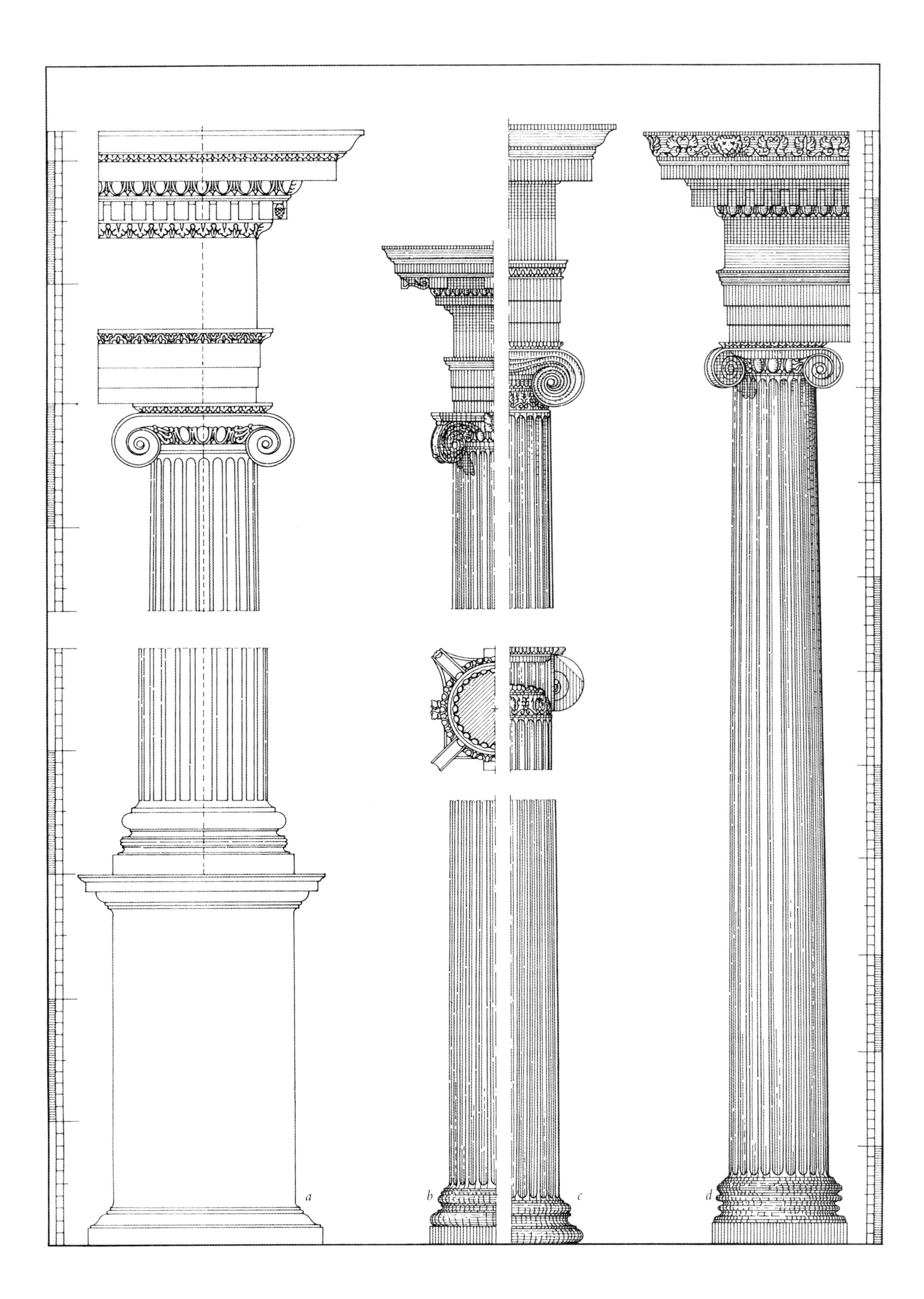

a
b
c
d

爱奥尼亚柱头

卷涡装饰，作为爱奥尼亚柱头的突出特征，经常以各种形式出现于经典装饰中。他们可见于许多自然形态中，但是当设计者们再现这一形态时，需要精确的圆形及方形的几何构图依据。

尽管我们了解卷涡最初是如何计算的，文艺复兴时期之后演变出几个简单的方法。它们都具备以下这一基本特征：卷涡的画法是画一系列的1/4圆，在中间的小正方形上逐渐移动这些1/4圆的圆心，将圆弧逐步缩小。尼古拉斯·高曼（1611—1665年）的卷涡画法如右面图中所示。

卷涡装饰的旋涡眼在距离柱子中心一个半径的距离处，并从卷涡顶端向下延伸1/4直径的距离。旋涡眼本身为柱底直径的0.045倍。作为涡旋绘制基准的正方形的边长为旋涡眼直径的一半。线1～4构成的正方形，其中一边置于旋涡眼的垂直直径的正中间。将位于涡旋中心的正方形一边切分成六个部分，形成等分点1、5、9、*X*、12、8和4，在这些点上画水平延长线，连接2和3与旋涡眼中心*X*，形成2*X*和3*X*线，上述等分点水平延长线与2*X*和3*X*相交（形成其余点）。将这一系列的1/4圆周的中心，沿着正方形上数字的顺序移动，正方形逐渐缩小，即可绘制成卷涡。第一个1/4圆周的中心始于冠板下*A*点的垂直线上，结束于点1、2水平延长线上的*B*点；然后将圆心移置点2，半径缩短交于*B*点，下一个1/4圆周移至*C*点，以此类推，直至圆心移至点12并最终汇入旋涡眼。当卷涡为双边时，应重复这一过程，并且每一步都稍向内移动。

爱奥尼亚柱头的设计，都遵循这种几何绘制方式，却可以创造出许多变化。实例（*a*）取材于公元前5世纪雅典的厄里希翁神殿。实例（*b*）、（*c*）和（*d*）全部来自罗马，展现出罗马人的不同理解。简化版的实例（*d*）来自罗马圆形大剧场，而（*c*）尤为重要，位于转角的柱子上面的卷涡经旋转后，朝向两个方向，这一事例展现了罗马人是如何理性地解决角柱柱头这一问题的。罗马人发明了这种柱头，四边相同并且每个柱子上卷涡的方向都旋转了。最后的实例（*e*），由米开朗琪罗设计，修建于1568年的罗马保守宫，展现出的对这一柱式独创的细节修饰，后来盛行于巴洛克建筑中。

THE IONIC CAPITAL

Volutes, which are the distinguishing feature of the Ionic capital, appear frequently in various forms on Classical decoration. They are found in a number of natural forms, but when they are reproduced they require the application of a precise geometry based on circles and squares.

Although we have evidence of how volutes were originally calculated, since the Renaissance several simple methods have been developed. They all share one essential characteristic: the spiral is created by drawing a series of quarter-circles and gradually diminishing their size by moving the centres of the circles progressively around a small central square. Nikolaus Goldman's (1611-1665) method for drawing a volute is illustrated opposite.

The centre of the eye of the volute is half a diameter out from the centre of the column and a quarter of a diameter down from the top of the volute. The eye itself is 0.045 diameters. The square for setting up the spiral has sides half the length of the diameter of the eye. One side of the square, the line 1 to 4, is centrally placed on the vertical diameter of the eye. Two smaller internal squares are created by joining the two outer corners 2 and 3 to the centre of the eye, dividing the side of the square on the centre of the volute into six equal parts to give points 1, 5, 9, *X*, 12, 8, and 4 and then extending these points horizontally to meet the lines 2*X* and 3*X*. The spiral can now be drawn by moving the centre point of a series of quadrants progressively round the diminishing squares in the numbered order shown. The circumference of the first quadrant starts vertically above its centre at point *A* below the abacus and stops at point *B* on the horizontal extension of line 1, 2. The centre is moved to point 2 and the radius reduced to meet point *B* and the next quadrant drawn to point *C*, and so on until the quadrant centred on point 12 finally merges into the eye. When the volute line has two edges these should gradually converge by repeating the process but moving each point slightly inwards.

The design of Ionic capitals, while sharing this geometry, can vary considerably. Example (*a*) is from the fifth-century-BC Erechtheion in Athens. Examples (*b*), (*c*) and (*d*) are all from Rome and show the variety of Roman interpretations. The simplified version (*d*) is from the Colosseum and (*c*) is of particular importance as an example of how the Romans rationalised the capital to overcome the problem that occurred on a corner column where one volute of the capital had to be twisted round to face both ways. The Romans invented a capital where all four sides were identical and all volutes twisted round on every column. Finally, example (*e*) by Michelangelo from the Palazzo dei Conservatori in Rome, built in 1568, shows the inventive embellishment of the Order that became popular in Baroque architecture.

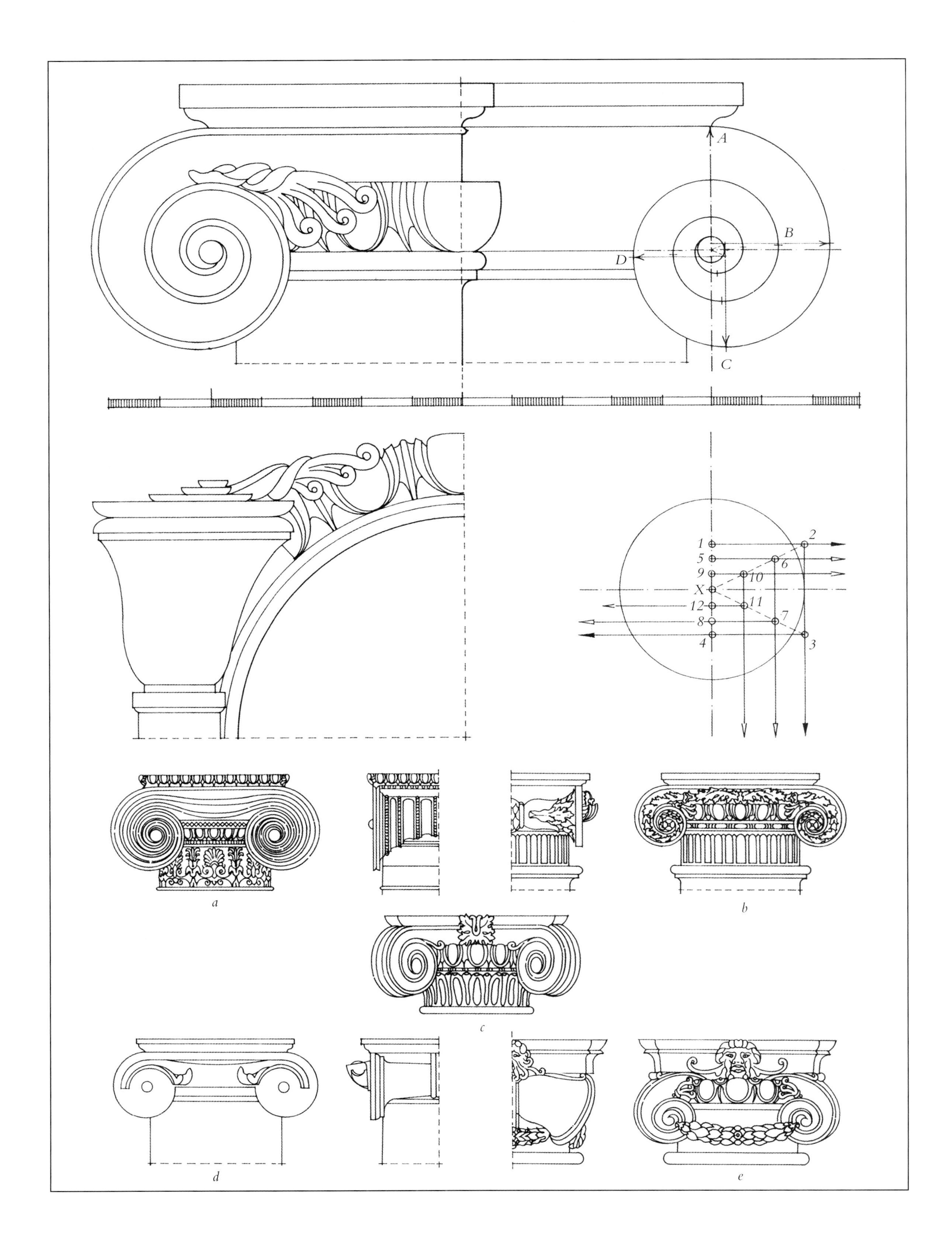
A
B
D
C
1
2
5
6
9
10
X
12
11
8
7
4
3
a
b
c
d
e

爱奥尼亚柱式的应用

早期希腊城邦认为爱奥尼亚柱式与今天的土耳其西部及其殖民地有关。这一柱式及其装饰据称源于成熟女性的身体比例与特征，与代表男性的多立克柱式形成对比。古代爱奥尼亚人，与更加富有活力的多立安人比起来，事实上的确生活得更加奢侈、悠闲，因此而显得女性化。这一柱式的发源地之一——萨摩斯岛上公元前6世纪的雕塑，更进一步证明了这一传说。柱式上的卷涡好似妇女头上的发卷，竖直的沟槽好似妇女衣裙上的褶皱，正如上文中提到用于膜拜的雕塑的图例展示出来的一样。

爱奥尼亚柱式既不像多立克柱式一般简单也不似科林斯柱式般变化万千。因而其使用好似局限于古代和文艺复兴之后的时期。尽管罗马人应用它，但爱奥尼亚柱式却不像科林斯柱式或者罗马混合柱式那样受欢迎。文艺复兴早期，爱奥尼亚柱头应用于建筑物的部分结构中，表达了当时人们对柱式流行趋势的各种不同理解，但少有整个建筑采用爱奥尼亚设计的实例。

维特鲁威关于这一柱式女性起源的叙述，带来了深远的影响，因而爱奥尼亚式一直与女性相关联。人们认为它与静坐活动以及不活跃的人个性十分吻合。16世纪确立了五个主要柱式递升的排列顺序，这一传统赋予爱奥尼亚柱式新的声望，并且使其成为多立克柱式与科林斯柱式之间的过渡。基于对这一折中比例的兴趣，1559年帕拉迪奥设计了佛斯卡里别墅（*c*），这是一栋爱奥尼亚式建筑。

巴洛克建筑师似乎对爱奥尼亚柱式颇感兴趣，在17世纪以及随后的世纪交替时期广泛应用。进入18世纪纯粹古典时代后，巴洛克的变体赋予这一柱式新的重要地位。约翰·索恩（1753—1837年）历史性地对希腊爱奥尼亚柱式的精准运用，于1800年设计伦敦郊外的皮特香格庄园（*a*）唤起了古典简洁风尚。在美国，1847年俄亥俄州代顿蒙哥马利郡法院（*b*）的设计，严格再现希腊爱奥尼亚神庙建筑，使用新的考古知识表现对英雄民主主义下的公平性的信仰。卡雷拉与黑斯廷斯选择精巧的巴洛克爱奥尼亚式，于1894年设计了纽约的斯隆府邸（*d*）。这种爱奥尼亚形式比考古学演变而来的柱式，甚至是已经具有丰富装饰的科林斯柱式和混合柱式，都更具备广阔的创造空间。

THE USE OF THE IONIC ORDER

The early Greek city-states identified the Ionic Order with the peoples of what is today western Turkey and with their colonies. The column and its decoration are said to be derived from the proportions and features of a mature woman, in contrast to the masculine Order of the Dorians. In antiquity the Ionians were, indeed, contrasted with the more vigorous Dorians by a luxurious and leisurely lifestyle that was then associated with femininity. Statues from the island of Samos in the sixth century BC, one of the places of origin of the Ionic Order, further support the myth. The column was said to have curls like those of a woman's hair and the vertical fluting was said to represent the folds of a woman's dress, like the dress of the statue of the worshipper illustrated.

The Ionic Order has neither the simplicity of the Doric nor the versatility of the Corinthian. Its use seems consequently to have been limited both in antiquity and since the Renaissance. Although used by the Romans, it was less favoured than the Corinthian Order or their own Composite Order. In the early Renaissance, Ionic capitals were used on parts of buildings to express the various interpretations of the Order current at the time, but there were few examples of wholly Ionic architecture.

The influence of Vitruvius' story of the female origin of the Order seems to have persisted and Ionic continues to be associated with the female sex. It is thought suitable for sedentary activities or individuals of an inactive nature. The establishment in the sixteenth century of a tradition of placing the five Orders in an ascending sequence gave Ionic a new respectability and an intermediate position between Doric and Corinthian. It may be nothing more than an interest in an intermediate set of proportions that led Palladio to make the Villa Foscari near Venice (*c*) of 1559 an Ionic building.

Baroque architects seemed to find the Ionic Order particularly interesting and in the seventeenth century, and subsequently at the turn of this century, it was widely used. Baroque distortion gave the Order a new significance in the return to Classical purity in the eighteenth century. John Soane's (1753-1837) use of an historically accurate Greek Ionic Order at Pitzhanger Manor outside London (*a*) in 1800 could be seen to evoke and archaic simplicity. In the USA the rigid reproduction of a Greek Ionic temple for the Montgomery Country Courthouse, Dayton, Ohio (*b*) of 1847 uses this new archaeological knowledge to represent a belief in the quality of justice in an heroic democracy. Carrére and Hastings chose an elaborate Baroque Ionic for the Sloane House in New York (*d*) of 1894. This form of Ionic offered greater opportunities for creative embellishment than archaeologically derived examples, or even the already rich Corinthian and Composite forms.

20
ft
m
6
a
b
c
d

科林斯柱式：起源

与多立克和爱奥尼亚柱式相比，科林斯柱式发展得相对较晚。它以科林斯城的名字命名，但其起源与重要性湮灭于神话传说中。第一个科林斯柱式的实例可追溯至大约公元前400年。

罗马作家维特鲁威，可能是重新讲述一个早期希腊神话，提到了这一柱式的起源。著名的雅典雕刻家卡利马科斯（约公元前210/205—240年），曾经过一座墓碑，属于一个未到婚嫁年龄少女，上面放着一篮子物品。篮子上盖有屋瓦，置于莨苕植物的根茎上，因此莨苕叶子向上包裹着篮子生长。卡利马科斯受到启发，将其作为一种新柱头的设计基础。对这一传说的解读见图例（*e*）。

这个传说解释了这一柱式纤细的比例源自年轻女性的身体比例，但不止于此。卡利马科斯以其青铜艺术而闻名，科林斯城也以青铜制品著称，这些被称为“科林斯艺术”。青铜的起源可以解释为什么叶子装饰的命名与排列都好似中心带有内核的铸件。死亡与死后拥有财产的关联也许是使用莨苕装饰的原因。莨苕叶子与古代希腊葬礼遗迹、墓碑又称石碑相关联，为了与柱子比较，图（*c*）为放大了的石碑，通常使用莨苕叶饰或叶丛状装饰，与早期的科林斯柱式相似。

柱子上使用植物装饰并不陌生，柱子（*a*）为方便比较而简化了，是来自埃及的众多实例之一。最初了解的来自希腊的实例（*b*和*d*），取材于巴塞建于公元前400年的阿波罗神庙，独立安放在爱奥尼亚柱阵的内部。这个位置一般用来放置神像，因此柱子本身可能就是膜拜的对象，大概象征着阿波罗的众多神力之一——猝死之神。

科林斯柱式位于爱奥尼柱子中央，这种布局令人大开眼界。科林斯柱式的檐部，并不似多立克与爱奥尼亚柱式的檐部那样，随着柱式的发展而演变，而是采用爱奥尼亚柱式的檐部，仅在罗马时期发展出了显著的特征。这一柱式在早期的发展过程中，也仅用于建筑内部，通常外部为多立克式，例如建于公元前375年特尔斐的圆形建筑物（*f*）。直到公元前174年在雅典的奥林匹亚宙斯神殿，科林斯柱式才应用于主建筑的外部。

THE CORINTHIAN ORDER: ORIGINS

By comparison with the Doric and Ionic Orders, the Corinthian Order is a relative newcomer. It is named after the city of Corinth, but its origin and significance are shrouded in mystery and myth. The first example dates from about 400 BC.

The Roman author Vitruvius, probably repeating an earlier Greek legend, tells a story of the origin of the Order. The famous Athenian sculptor Callimachus (*c*.210/205-240 BC) is said to have passed by a basket of possessions placed on the grave of a Corinthian girl who had died on reaching marriageable age. This basket had a roof-tile over it and had been placed over the root of an acanthus plant which had then grown up around the basket. Callimachus; attracted by this, used it as the basis for a design for a new type of column capital. An interpretation of this legend is shown in illustration (*e*).

The myth explains the slender proportions of the column, which are said to derive from those of a young woman, but it may tell us more. Callimachus was famous for his bronzework and the city of Corinth was so well known for its bronzes that these could be referred to merely as 'Corinthian work'. A bronze original could provide an explanation both for the name and way the leaf decoration seems to be applied like repeated castings to a central core. The connection with death and possessions for the afterlife may be the reason for the acanthus decoration. The acanthus was associated with funeral celebration in Greek antiquity and gravestones, or stele, such as the one illustrated (*c*), enlarged for comparison with the column, often had acanthus or anthemion decoration similar to the first known Corinthian capitals.

The use of plant decoration for columns was not new and column (*a*), reduced for comparison, is one of many from Egypt. The first known Greek example, however, from the Temple of Apollo at Bassai, (*b*) and (*d*), of 400 BC was centrally placed in significant isolation in an internal Ionic colonnade. This would normally have been the position of a cult statue, so the column may itself have been the object of veneration, perhaps a reference to one of Apollo's many attributes as the god of sudden death.

The internal position of this Corinthian column among Ionic columns is also informative. The Corinthian entablature did not develop with the column, as with the Doric and Ionic. Rather, the column at first tended to be added to an Ionic entablature, only developing distinct features in Rome. In its early development it was also used exclusively internally, and usually on buildings with Doric exteriors such as the Tholos at Delphi (*f*) of about 375 BC. It did not make an appearance on the exterior of a major building until 174 BC in the Temple of Zeus Olympius in Athens.

a
b
c

d

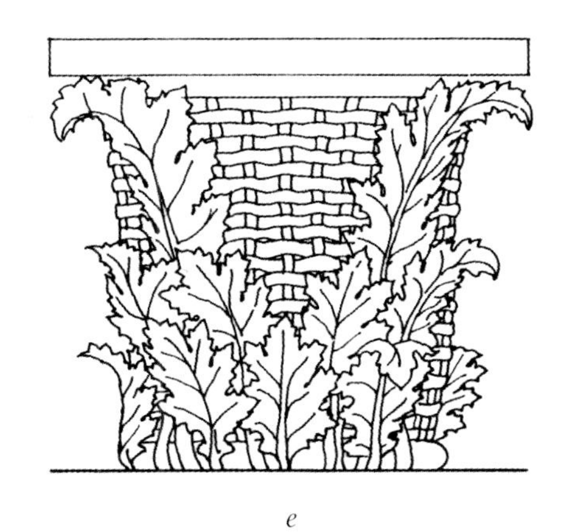
e

f

科林斯柱式

科林斯柱式在2世纪之后应用于主建筑物中，才被认定为是一种发展完善的柱式。因此，在古代世界，这一柱式的主要应用受罗马的影响，或者由罗马的建筑师使用。

科林斯柱式源于爱奥尼亚柱式的一种变体，大部分科林斯柱式的细部设计都来自于爱奥尼亚先祖。科林斯柱式基本上采用的都是爱奥尼亚柱身和柱础，向上延伸，将低矮的爱奥尼亚柱头替换成高耸的科林斯式柱头。这就使得科林斯柱式的高度增加到柱底直径的9.25～10.5倍。很大一部分的古代科林斯柱式刚好是柱底直径的10倍并且当作家们于文艺复兴后期发布理想的比例时，他们倾向于选择简单的数字如9、9.5与更加精准常见的比例——10。1562年维尼奥拉设计的科林斯柱式（*a*），这种版本的大量实例和建于215年罗马塞拉皮斯神庙（*b*）的实例证明，这一柱式的高度恰好为柱底径的10倍，然而建于131年的雅典哈德良凯旋门（*c*）的柱子高度为柱底径的9.5倍。

柱头之下科林斯与爱奥尼亚柱式完全相同。科林斯式柱础采用爱奥尼亚雅典式柱础，带有一条或两条凹面或凹弧边饰，上下饰有凸出的柱脚圆盘线脚。所有早期这一柱式的实例，柱身上都刻有24条半圆形垂直沟槽，爱奥尼亚柱式后期的版本也是如此。在一些罗马的实例中，柱身没有切割，或许是为了经济实用，或者是由于特殊颜色柱身的材料为有花纹的大理石、花岗岩、斑岩或者其他特殊石材的应用。文艺复兴之后，沟槽的使用就成为可选项目了。

科林斯柱式与众不同之处，除了在P94将要描述的独特柱头之外，就是罗马时代演变出的檐部设计了。高度为柱底直径的2.33倍，大体上相似，但经常比爱奥尼亚的檐部略高。额枋分为2～3个部分，与爱奥尼亚的细部设计相似，但更高一些，大概高出柱底直径的0.66倍。科林斯柱式带状雕刻与爱奥尼亚同样，或者朴素，或者由一整块雕刻装饰嵌板构成。然而，飞檐的细部设计与爱奥尼亚柱式有很大差异。一些飞檐，如塞拉皮斯神庙（*b*），保留了具有明显爱奥尼亚柱式特征的齿状装饰，但不久就开始采用一种延展且间隔较大的齿状装饰，类似托架，正如在哈德良凯旋门（*c*）所示。这一托架后期，被修饰成雕刻小卷轴的形态，或称为飞檐托饰，支撑着上面简化的凸出装饰线脚。然而齿状装饰并未被完全摒弃，或装饰于飞檐之上，或者成为类似120年建于罗马的万神殿（*d*）内部柱式上朴实的齿状装饰。这一设计即使不是普遍的，也是十分常见的罗马装饰技法，自从文艺复兴以来得到广泛的应用。

THE CORINTHIAN ORDER

Corinthian cannot really be regarded as a developed Order prior to its use on major buildings in the late second century BC. The principal use of the Order in the ancient world was, consequently, under Roman influence, or executed by Roman architects.

The Corinthian Order started as a variation of the Ionic; most Corinthian details owe their origin to their Ionic ancestry. The Corinthian column is essentially an Ionic shaft and base lengthened by substituting the low Ionic capital with a tall Corinthian capital. This gives an increased height to Corinthian columns to somewhere between 9.25 and 10.5 diameters. A significant number of ancient columns were exactly 10 diameters and when authors published ideal proportions after the Renaissance they tended to select simple figures such as 9, 9.5 or, more frequently and more accurately, 10. Both the large-scale example (*a*), a version of Vignola's Corinthian Order of 1562, and the column from the Temple of Serapis in Rome (*b*) of AD 215 are 10 diameters high, while the columns of the Arch of Hadrian in Athens (*c*) of AD 131 are 9.5 diameters high.

Below the capital the similarities between Corinthian and Ionic are exact. The Corinthian base is the Ionic Attic base with a projecting torus moulding above and below one or two concave scotia mouldings. All early examples of Order have the same twenty-four half-round vertical flutes cut into the shaft, as do later versions of Ionic columns. In several Roman examples the column shaft is left uncut, either for economy or where special coloured shafts of figured marble, granite, porphyry, or other special stones are used. After the Renaissance, the use of fluting became simply optional.

The distinctive aspect of the Corinthian Order, other than the unique capital discussed on page 94, is the entablature it developed during the Roman period. At about 2.33 diameters it is broadly the same as, but often a little higher than, an Ionic entablature. The architrave, with two or three divisions, is identical in detail to the Ionic, but rather taller, at about 0.66 of a diameter. The Corinthian frieze is also either plain or an uninterrupted panel of sculpted decoration. The details of the cornice, however, differ significantly from the Ionic. Some cornices, such as on the Temple of Serapis (*b*), retained the distinctly Ionic decoration of dentils, but soon an extended and widely spaced dentil, like a bracket, was introduced, as on the Arch of Hadrian (*c*). This bracket was later embellished by being carved into the shape of a small scroll, or modillion, supporting reduced projecting mouldings above. The dentils were not, however, abandoned. Either the cornice was modified to include dentils or, as in the interior Order of the Pantheon in Rome (*d*) of AD 120, a plain detail was put in their place. This arrangement became normal, if not universal, Roman practice and in this form has been widely adopted since the Renaissance.

a
b
c
d

科林斯式柱头

科林斯式柱头十分正式，同时以相对真实的手法表现了一种茛苕植物沿着结实内核生长的形态。这种地中海地区常见的野生植物的叶子与卷须，包裹着朴素的支柱状内核，由小的圆形串珠装饰线脚将其从柱身上分离出来。这一支柱状内核通常好似一个倒挂的钟形，上置方形冠板、四边内卷，形成4个凸出的对角喇叭。钟形内核水平分为3个部分，可能等高也可能不等高，这些分割标志着叶片的边缘。下面两层为两排简单的叶片，每排都相同，在顶部翻卷并交错排列装饰在钟形内核的周围。最顶层的每一面上，两条叶茎上萌发出的叶片与卷须，最终形成16条螺旋装饰。这些纹案在顶端会合，形成冠板对角喇叭下较大的卷涡和其间的小卷涡。小卷涡之间有花朵纹饰绘制于冠板卷曲弧线的中间。

科林斯柱头对于茛苕叶子表现的真实度各有不同，而且茛苕植物本身不同种类与叶片形态也大相径庭。修建于94年的罗马维斯帕先与提图斯神庙，使用的典型柱头如放大图例所示，但也有不少对传统设计不同的解读。

图例（*a*）是一个最早在建筑外部使用的科林斯柱式的实例，取材于建于公元前334年雅典李西克拉特合唱团纪念碑。底层的叶子十分朴素，中层的叶子精致并饰有花朵，上面的卷须似乎取材于其他植物，顶部覆有小棕榈叶。图例（*b*）取材于修建于公元前80年罗马附近蒂沃利的希贝尔神庙，中间的叶片减少，带有巨大的卷涡与花朵纹案。柱头取材自建于公元前7年的罗马卡斯托和普鲁克斯神庙（*c*），是一个更加标准的范例，有着独特、鲜活的叶子雕刻，中间缠绕着小的卷涡。建于80年的罗马圆形大剧场中的柱头（*d*）是一种普通简化版的实例，未经切割的叶子可能是出于经济适用的角度考虑，也可能是可以与精致的柱头形成对比。实例（*e*）是18世纪早期巴洛克式建筑，由丹尼斯·狄德罗（1713—1784年）与让·勒朗·达朗贝尔（1717—1783年）设计，独树一帜地在解读传统设计的基础上创造了新的风格。卷须被卷曲的叶片取代，从顶部一直垂置于中层朴素的叶饰纹案处。最后一个实例（*f*）比其他正常的柱子要小，取材于罗马的斯克罗法宫。这一文艺复兴时期的柱头展现的是一种简化装饰元素的例子，通过缩减细部设计的比例来适应较小的柱头。

THE CORINTHIAN CAPITAL

The Corinthian capital is a formal but relatively realistic representation of an acanthus plant growing around a solid core. Leaves and tendrils of the common wild Mediterranean plant are gathered around a plain cylindrical centre separated from the column shaft by a small, round, astragal moulding. This round core is usually shaped like an inverted bell and capped by a square abacus with four inward-curving sides to create four projecting diagonal horns. The bell is divided horizontally into three parts, which can be equal or vary in height, and these divisions mark the lines of the leaves. The lower two divisions have simple rows of leaves, identical in each row and turned over at the top, which alternate with one another around the bell. In the uppermost division leaves and tendrils sprout out of two stalks on each face and divide up to terminate in a total sixteen spirals. These join at their tips to create large volutes under each horn of the abacus with smaller volutes between them. Between the smaller volutes a flower rises up to sit in the centre of each inward curve of the abacus.

The accuracy of the representation of the acanthus leaves Corinthian capital can vary considerably and the acanthus plant itself has different species with different leaf forms. The large illustration shows a typical capital from the Temple of Vespasian and Titus in Rome of AD 94, but there are many different interpretations of the traditional design.

Illustration (*a*) is an example from the first known external use of Corinthian Order, the Choragic Monument of Lysicrates in Athens of 334 BC. The lower leaves are plain and the middle range delicate with flowers between them while the tendrils above seem to belong to a different plant capped by a small anthemion. Example (*b*) is from the Temple of the Sybil at Tivoli near Rome of 80 BC and has a very much reduced range of middle leaves, heavy volutes, and a very large flower. The capital from the Temple of Castor and Pollux in Rome (*c*) of 7 BC is of a more standard type, with distinctive, lively leaf-carving and intertwining smaller volutes. The capital from the Colosseum Rome (*d*) of AD 80 is an example of a common simplified form with uncut leaves either for economy, deliberate effect, or to contrast with more elaborate capitals. Example (*e*) is an early-eighteenth-century Baroque design by Diderot and design by Denis Diderot (1713-1784) and Jean-Baptiste le Rond d'Alembert (1717-1783) and is an inventive reinterpretation of the traditional form to create a new style. The tendrils are replaced by curled leaves which descend to the top of the middle row of plain leaves. The last example (*f*) would be of the smaller size than the others and is from the Palazzo della Scrofa in Rome. This Renaissance capital demonstrates a type of reduction of elements that is often used to reduce the scale of detail to a level appropriate to a smaller size of capital.

a
b
c
d
e
f

科林斯柱式的应用

科林斯柱式并不像多立克与爱奥尼亚柱式一样与种族有关，因而在这些特性消失后得以广泛应用。在古代，只有柱子才具有象征意义；独特的檐部在后期才得以发展，此外柱子演变成多立克柱式和爱奥尼亚柱式。从P90讨论的柱头产生的传说可见，这一柱式可能和女性与葬礼有关。由于这一柱式首先使用于建筑物内部，一定是人们认为更具装饰特色的柱子适合富丽堂皇的内部设计的缘故。柱子体现出女性的重要性，因为在古代，女性的居家与处于被动的社会地位与男性涉外以及活跃的社会职责形成了鲜明对比。人们一直坚持认为这一柱式与年轻女性有关联，在维特鲁威留存至今的著作中有记述，可见于公元前174年雅典奥林匹亚宙斯神殿中的一个版本，安置在希腊风格维纳斯雕像的旁边。

科林斯柱式成为罗马人极其喜爱的一种设计，他们曾一度将早期的神庙与早期的柱式改造成这一风格。极强的装饰效果与雕刻特征，使得这一柱式变得更加迷人，用于装饰的巨大花费使得神明与供奉者都倍感荣耀。

罗马人广泛使用科林斯柱式，这确保它留存于中世纪、文艺复兴早期并且事实上以不曾间断的形态进入到20世纪。这一柱式比例优雅，柱头雕刻繁复使其得以长盛不衰。

为了理解文艺复兴以来建筑师们创造出的一些特殊联想意义，有必要深入研究其装饰风格。布鲁内列斯基设计的巴齐礼拜堂（*c*）于1429年建于佛罗伦萨，采用几何形制科林斯设计，带有罗马帝国时期柱头；是文艺复兴早期建筑，极力唤醒了古典特征。1774年修建于爱丁堡的罗伯特·亚当注册办公室（*b*）同样也反映了古罗马的建筑风格，但是也叠加了文艺复兴时期确立的一些特征。托马斯·汉密尔顿皇家物理学院（*d*）建于1845年，也在爱丁堡，使用雅典风塔样式的科林斯柱式，寻求回归罗马和文艺复兴之前的古典纯粹风格。亚瑟·海盖特·马克穆多（1851—1942年）在伦敦北部设计的一座宅邸（*a*），建于1883年，有简单的柱头和低矮敦实的柱身，其灵感来自工艺美术运动和16世纪的矫饰主义。

THE USE OF THE CORINTHIAN ORDER

The Corinthian Order had none of the ethnic associations of the Doric and Ionic and was only widely used when such distinctions had lost their significance. Only the column had a meaning in antiquity; the development of a distinct entablature did not evolve until later and the column was often used with the Doric or Ionic Order. From the legend of the invention of the capital on page 90, it would appear to have had female and funerary associations. As it was also first used internally, the more decorative character of the column must have been thought appropriate for the greater opulence of the interior. This may also be linked to the female significance of the column as, in antiquity, the passive domestic role of the female was contrasted with the outdoor active role of the male. The association of the Order with a young women has persisted, preserved by the survival of the ancient writings of Vitruvius, and is represented here by a column from the Temple of Zeus Olympius in Athens of 174 BC placed alongside a version of a Hellenistic statue of Venus.

The Corinthian Order became a firm favourite of the Romans and earlier temples in earlier Orders were at times rebuilt in that style. The strongly decorative and sculptural nature of the Order was probably considered to be more attractive, and the evident expense of the decoration gave greater glory to the deity and donor alike.

The widespread use of Corinthian by the Romans ensured its survival throughout the Medieval period, early Renaissance and beyond in a virtually uninterrupted sequence until the twentieth century. The elegance of its proportions, the sculptural opportunities of the capital and prestige attached to the opulence of the carving have maintained its popularity.

In order to understand some of the more particular associations intended by architects since the Renaissance it is often necessary to examine the style of the decoration. The Pazzi chapel (*c*) by Brunelleschi was built in 1429 in Florence in a geometrically ordered Corinthian design with capitals of a Roman Imperial type; it is one of the first Renaissance buildings strongly to evoke the character of antiquity. The central section of Robert Adam's Register Office in Edinburgh (*b*) of 1774 is also intended to reflect the architecture of ancient Rome but is overlaid with features that had become established in the Renaissance. Thomas Hamilton's Royal College of Physicians (*d*) of 1845 in the same city on the other hand, by the use of a Corinthian Order from the Tower of the Winds in Athens, seeks to return to an historic purity that pre-dates both Rome and the Renaissance. A house designed by Arthur Heygate Mackmurdo (1851-1942) in north London (*a*) in 1883 has simplified capitals and squat columns in a bold and original design which draws its inspiration from the Arts and Crafts Movement and sixteenth-century Mannerism.

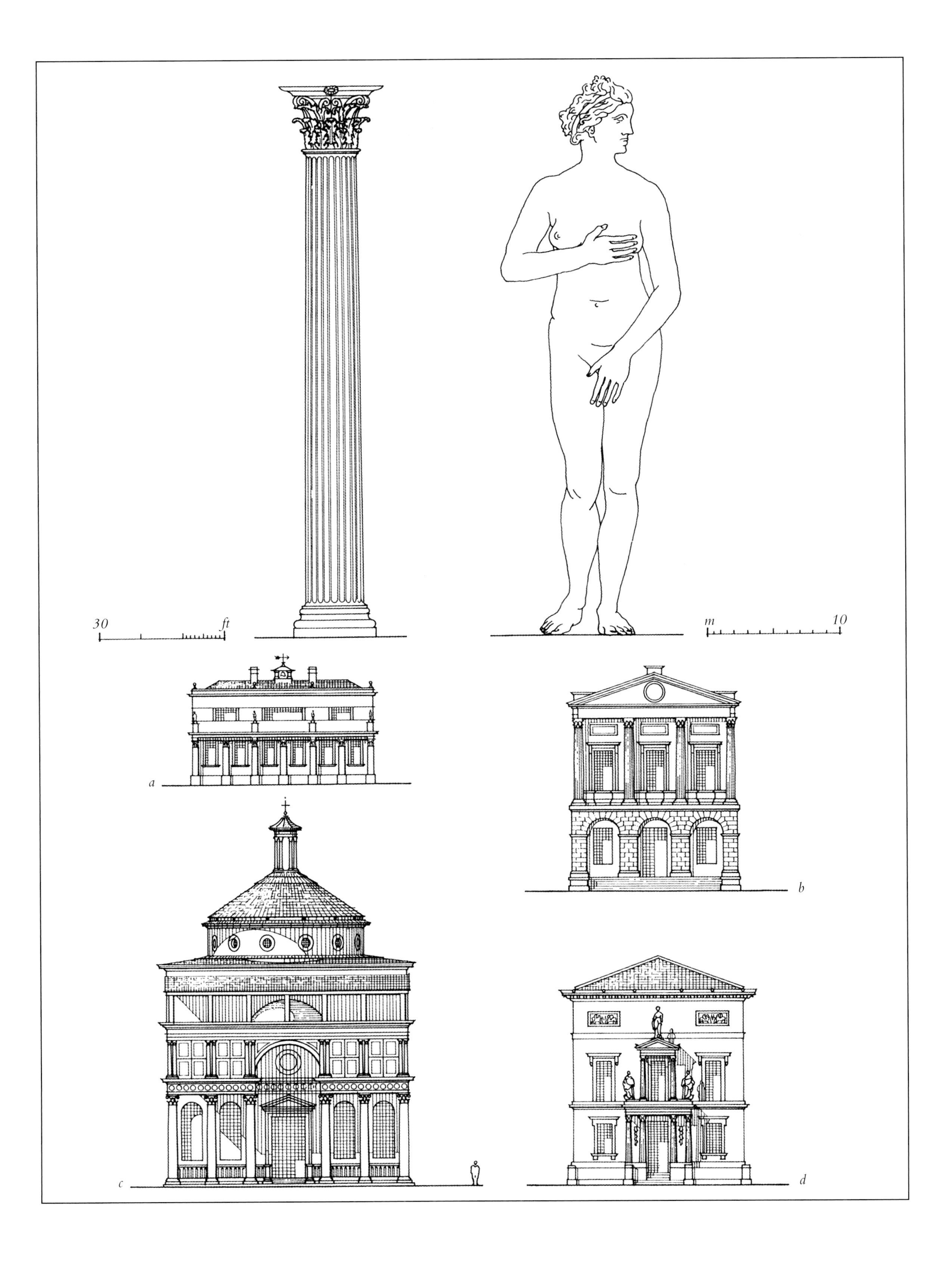
30
ft
m
10
a
b
c
d

混合柱式

混合柱式是仅有的一种纯罗马式柱式。首先见于82年古罗马广场提图斯凯旋门之上，尽管很可能是在此之前的一个世纪创造出来的。这一柱式后来又应用于大约204年的塞普蒂默斯·塞维鲁（145—211年）凯旋门（*c*），以及一系列其他凯旋门、纪念碑和浴场的设计中。这一柱式似乎是从科林斯柱式演变而来的，是当时罗马人的最爱，同时也广泛应用于凯旋门的建造。在很多方面，它与科林斯柱式相似，柱头有所不同，在文艺复兴时期被认为是爱奥尼亚与科林斯柱头的组合，因此得名混合式。

其柱头的下面两层设计与科林斯柱头相似，饰有莨菪叶片，4个对角则装饰罗马爱奥尼亚柱式的卷涡。因此，究其起源，这一柱式受爱奥尼亚的影响要大于科林斯柱式。早期爱奥尼亚柱头通常有较深的棕叶带状纹案，装饰于卷涡之下。罗马时期的实例则为各种不同的纹案装饰在这一位置，并且带有对角卷涡。由于科林斯式的檐部源于爱奥尼亚式，这一演变的起源对于混合式来说也是一种可能。

在1世纪，柱头的种类一经确立，其显著特征和柱身与檐部的关系就保留下来了。因此不需要古典文献就能够从文艺复兴时期的实例中认定这是一种确立的柱式。

柱头高度通常和柱底直径相等，在冠板下分成3个区块，通常是等高的。下面两层为两排莨菪叶饰，这与科林斯柱头是一样的。顶层的每一面，叶片纹案之上都有两朵小花向上攀升，并穿过冠板的底部。爱奥尼亚柱式的典型柱头纹案——卵箭装饰线脚位于两层科林斯式冠板之下，同时柱头的卷涡穿过冠板的底层。花朵纹案从卵箭饰中间升起，覆盖了两层冠板的中间部分。

整个柱高为柱底直径的10倍，但塞普蒂默斯·塞维鲁凯旋门（*c*）为9.66 倍，而建于305年的戴克里先浴场（*b*）则为10.5倍。文艺复兴时期的作家，将混合柱式看作是柱式发展链条上的最后一环，认为其高度与科林斯柱式有关。帕拉迪奥（*a*）设计混合柱式比例为10，使其成为排序中的最后一个，与科林斯的9.5倍形成对比。另一方面，斯卡莫齐（*d*）设计的混合柱式为9.75倍柱底直径高，按照发展演变关系，比他设计的科林斯柱式的10倍要矮一些。

混合柱式的所有其他细部设计都参照科林斯柱式，包括齿状装饰的组合、飞檐托饰、其他装饰线脚和檐部组合元素的变化等方面。

THE COMPOSITE ORDER

The Composite Order is the only wholly Roman Order. It is first known in the Forum in Rome on the Arch of Titus of AD 82 although it may have been invented about a century earlier. The Order was subsequently used on the Arch of Septimus Severus (AD 145- AD 211) (*c*) in about AD 204, and on a number of other triumphal arches, monuments, and baths. It seems to be a derivative of the Corinthian Order, which was a favorite of the Romans and also used extensively on triumphal arches. In many respects it is identical to the Corinthian Order, but it is distinguished by its column capital which was considered in the Renaissance to be a combination of Ionic and Corinthian and hence called Composite.

The capital shares the two lower rows of acanthus leaves of the Corinthian Order with the four diagonally placed, decorated volutes of the Roman Ionic Order. It may therefore owe its origin more to a development of the Ionic Order than the Corinthian. Early Ionic capitals often have a deep decorative anthemion band below the volutes and Roman examples exist with various kinds of decoration in this position combined with diagonal volutes. As the Corinthian entablature is also derived from the Ionic, evolution from this source must remain a possibility.

Once the capital type was established in the first century AD both its distinguishing characteristics and its relationship with the column and entablature were retained. It was therefore sufficiently established as an Order to be recognised as such in the Renaissance without any ancient literary references.

The capital is generally 1 diameter high and below abacus is divided into three parts of approximately equal height. The lower two parts consist of two rows of acanthus leaves, as on Corinthian capitals. On each face two small flowers rise up out of the top row of leaves to penetrate the upper part of the capital. The egg and dart moulding of a typical Ionic capital sits immediately below a two-part Corinthian abacus and the volutes of the capital always penetrate the lower part of this abacus. A flower rises directly out of the egg and dart and covers the centre of both parts of the abacus.

The total column height is generally about 10 diameters, but on the Arch of Septimus Severus (*c*) the columns are 9.66, while in the Baths of Diocletian (*b*) of AD 305 they are 10.5. Renaissance authors, when placing the Composite as the last of a sequence of all the Orders, related the height to the Corinthian. Palladio (*a*) made the Composite 10 as the last of an ascending sequence, compared to Corinthian at 9.5. Scamozzi (*d*), on the other hand, made the Composite Order 9.75 diameters high - lower than, and a subservient derivative of, his Corinthian Order of 10.

All the other details of the Order follow the Corinthian example, both in the variety of their combination of dentils, modillion brackets, and other mouldings and in the varied proportion of the elements in the entablature.

a
b
c
d

混合柱式的应用

古代没有关于混合柱式的文献记载，这一名称是文艺复兴时期创造出来的。罗马作家维特鲁威提到了公元前1世纪科林斯柱头的各种创新设计，但这样的创造隶属于科林斯标准柱头的范畴，我们不能确定这些是记录混合式柱头的文献。一个世纪之后，小普林尼（61—约113年）关于柱式的记录也未提及任何新的样式。然而毫无疑问的是，从建筑所提供的证据以及3个多世纪的设计保持一贯性的角度考虑，混合柱式除了作为科林斯柱式的变体，在使用上一定具备一些其他重要特性。

这一柱式与凯旋门有关联，例如203年的塞普蒂默斯·塞维鲁凯旋门（*c*）。一些皇室家族的庙宇以及象征皇权的纪念碑，都显示出混合柱式与罗马以征服为荣耀的传统有着特殊的关系。庆祝胜利与羞辱被征服者在罗马人的生活中是极其严肃的事情，使用一种并非直接从被征服国家复制而来的建筑设计来纪念他们的胜利，对于纪念碑的创造者来说很有吸引力。

正如这里提到的所有其他的柱式都采用雕像的形式，文艺复兴时期的作家也采用具有象征意义的雕像来设计混合柱式，提图斯凯旋门的一个柱子上刻有罗马胜利女神雕像，呈现出为皇帝戴上桂冠的姿态。将象征女性的爱奥尼亚柱式与科林斯柱式结合构成混合柱式，采用传统上的女性形象来塑造有翼胜利之神雕像，来代表这一柱式显得十分恰当。

胜利的象征手法源于基督教会的传统，代表中世纪基督的胜利。然而在文艺复兴之后，混合柱式很少应用于整栋建筑中。缺乏文献资料的证明，科林斯柱式本身具有的魅力，加之混合柱式作为科林斯柱式分支的观点，都限制了这种柱式的影响力。

1565年帕拉迪奥在威尼斯附近维琴察，使用混合柱式修建他未完成的卡皮塔妮凉廊（*a*）。这一华丽的柱式及其胜利的寓意，成为威尼斯的象征。霍克斯莫尔在1695年之前设计的伊斯顿·内斯顿府邸（*b*）是一座位于英格兰北部的住宅，选择这一柱式可能是比较随意的。罗马的圣马塞洛教堂（*d*）由卡洛·方塔纳于1682年设计，坐落在离塞普蒂默斯·塞维鲁凯旋门不远的地方。使用混合柱式可能是象征中世纪基督的胜利回归。巴洛克建筑经常使用这种柱式，但它们同样装饰爱奥尼亚柱式，与混合柱式极其相似，因此强化了爱奥尼亚、科林斯与混合柱式之间本来就已经存在且模糊不清的关系。

THE USE OF THE COMPOSITE ORDER

There are no certain literary references to the Composite Order from antiquity and the name was not created until the Renaissance. The Roman author Vitruvius mentions the existence of different inventions in the design of Corinthian column capitals in the first century BC but, as such inventions existed within the range of standard Corinthian capitals, we cannot tell if this is a reference to the Corinthian capital. A century later a reference by Pliny the Younger (AD61 – *c.*113) to the Orders makes no mention of anything new of this type. There is, however, little doubt from the evidence of buildings and the consistency of design over some three centuries that the use of the Order must have had some significance other than as a variant of Corinthian.

The use of the Order in association with triumphal arches, such as that of Septimus Severus (*c*) of AD 203, with some temples to the cult of the imperial family and with other monuments to imperial power, tends to suggest a specific connection with the conquering glory of the Roman state. The celebration of victory and the humiliation of the vanquished were highly formalised events in Roman life and the use of an architectural Order not directly copied from a conquered nation may have had some attraction for the creators of the monuments that recorded these victories.

As all the other Orders have been represented here by figures and as Renaissance authors created symbolic figures for the Composite Order, a column from the Arch of Titus is here placed alongside a Roman figure of Victory poised to crown an emperor with a victor's laurel wreath. The combination of the female Ionic and Corinthian in the Composite Order makes the choice of a traditional female figure of winged Victory to represent the Order seem particularly appropriate.

The symbolism of victory was inherited by the Christian Church and used to signify the victory of Christ in the Middle Ages. However, Composite was little used for complete buildings after the Renaissance. The lack of literary reference, the attraction of the Corinthian, and the view of the Order as a derivative of Corinthian seem to have limited its appeal.

Palladio used the Composite Order for his incomplete Loggia dei Capitani in Vicenza near Venice (*a*) (1565). Here the opulence of the Order and its association with victory may be intended as a reference to the Venetian state. The choice of the Order for Easton Neston, a house in northern England (*b*) by Hawksmoor before 1695, is more arbitrary. The church of S. Marcello in Rome (*d*), by Carlo Fontana (1682) is located a short distance from the Arch of Septimus Severus where the use of the Composite Order may be a return to medieval symbolism of the victory of Christ. Baroque architects made most frequent use of the Order, but they also embellished the Ionic Order in a way that closely resembled the Composite thereby reinforcing the ambiguous relationship that already existed between the Ionic, Corinthian, and Composite Orders.

30 ft
m 10
a
b
c
d

变体与创新：柱子

柱子的功能不局限于用来支撑建筑物。垂直并逐渐变窄的形态不仅代表人类的不同身形，也可以追溯回树干与植物的形态，这正是柱子的外形与装饰的起源。在过去的几个世纪里，柱子不时被装饰或扭曲，或者象征着其自然的鼻祖或者给柱式增加装饰元素，代表某些建筑、顾主或者建筑师的一些特殊想法。

希腊特尔斐的神殿内独立式柱子（*f*）大约建于公元前3世纪，模仿一种莨菪植物茎的形态。在罗马柱式（*g*）上能够看到装饰叶片在古代的回归。这一主题被16世纪法国建筑师菲利贝·德·洛梅（1510—1570年）重新采纳（*i*），并有19世纪和20世纪的法国文艺复兴追随者们，如纽约的麦克金、米德&怀特设计公司（*j*）等继承。柱子显然是纪念原始神庙中以树干为柱的传统。古代与文艺复兴时期对这一起源有文献记载。布拉曼特1477年为米兰圣安布罗吉奥教堂设计的回廊（*h*）是15世纪有力的证明。

当装饰延伸至整个柱身（*k*），就成为单纯的丰富建筑艺术的表现手法。装饰可仅限于柱子上部（*l*）和下部（*m*），分区通常用于柱身下1/3的顶部。

在罗马时代经常使用扭曲的柱子，这一设计在文艺复兴时期得以复活。狭窄的螺旋纹案（*e*）十分常见，创造出只有沟槽是扭曲的印象。大部分后期的柱式都带有宽的或者花环螺旋纹案，起源于罗马皇室，并在罗马圣彼得大教堂（*a*）中重新使用，一度被认为是来自所罗门神庙。17世纪巴洛克建筑师们充满激情地复制这些形态，其中最著名的就是贝尔尼尼在圣彼得神坛（*b*）上为支撑天棚（或称华盖）而设计的柱子。

螺旋纹案的设定方法是将柱身水平分为48个部分并在中间画小型支柱（*d*）。柱子圆周上取8个点（*c*），按顺序连接这些点形成48段，在柱体上投射出螺旋环绕的支柱形态。柱子的直径按照螺旋上每一点构成的柱子表面支柱来测量。柱子的顶部与底部逐渐并入柱础和柱头，通过4个相等阶段*A*、*B*、*C*、*D*（*c*）回归真正圆心。

1524年佛罗伦萨的劳伦森图书馆（*n*）设计中，米开朗琪罗仿造罗马家具倒转了逐渐变窄的柱身。这种新形态与赫耳墨斯柱、半人半柱的古代雕刻有关，后来成为广为接受的形制（*o*），经常应用于较小的元素中，如窗户、壁炉和家具上。

VARIATIONS AND INVENTIONS: COLUMNS

The column is much more than a means of supporting a building. The vertical tapering shape not only represents different human figures, but refers back to the tree trunks and plants that were the origin of the form and its decoration. Over the centuries the column itself has from time to time been embellished and distorted, either to suggest this natural ancestry or to add more decoration to an Order or to refer to some other idea specific to the building, patron, or architect.

A free-standing column from the Greek sanctuary at Delphi (*f*), probably from the third century BC, takes the form of an acanthus stalk. Decorations of leaves recurred in antiquity and can be seen on Roman columns (*g*). This theme was taken up by the sixteenth-century French architect Philibert de l'Orme (1510-70) (*i*) and passed on to nineteenth- and twentieth-century followers of the French Renaissance such as McKim, Mead & White in New York (*j*). The column is most obviously reminiscent of the tree-trunk columns of primitive temples. Literal references to this origin are found in antiquity and the Renaissance. Bramante's cloister for the church of S. Ambrogio, Milan (*h*) of 1477 is a powerful fifteenth-century example.

When decoration extends over the whole shaft of the column (*k*) this is only an expression of architectural exuberance. The decoration can be limited to the upper (*l*) or the lower (*m*) part of the column, the division normally coming at the top of the lower third of the shaft.

Twisted shafts were often used by the Romans and were revived in the Renaissance. A narrow spiral (*e*) is quite common and creates the impression that only the fluting has been twisted. Most later columns with a broader, or wreathed spiral owe their origin to imperial Roman examples re-used in St Peter's in Rome (*a*), once thought to have come from the Temple of Solomon. These complex shapes were enthusiastically copied by Baroque architects in the seventeenth century, most notably by Bernini to support the canopy, or baldacchino, over the altar in the rebuilt St Peter's (*b*).

The spiral is set out by dividing the shaft horizontally into forty-eight parts and drawing a small cylinder in the centre (*d*). Eight points on the circumference of the cylinder (*c*) are joined progressively up the column in forty-eight stages to create a spiral on the surface of the cylinder which is projected on to the surface of the column. The diameter of the column is measured from the surface of the cylinder at each point on the spiral. At the top and bottom the spiral is eased into the base and capital by returning to the true centre in four equal stages, *A*, *B*, *C*, and *D* at (*c*).

In 1524 in the Laurentian Library in Florence (*n*) Michelangelo reversed the taper of some columns in a manner reminiscent of Roman furniture. This new shape associated the column with herms, the half-human, half-pillar sculptures of antiquity, and has since become an accepted variant form (*o*) most frequently used on smaller elements such as windows, fireplaces, and furniture.

1
2
3
4
5
6
7
8
A
B
C
D
48
0
a
b
c
d
e
f
g
h
i
j
k
l
m
n
o

变体与发明：柱头与柱础

设计者与工匠们对雕刻的创新常常集中于柱头。即使是最传统的柱头也能够反映出建筑师的意图。而传统之外有很多的变化。

多立克柱头简单的形态通常被赋予额外的装饰，例如在建于1486年的罗马坎榭列利亚宫（*a*）以及罗伯特・亚当于1762年在英格兰修建的西翁府邸的拱门（*b*）。另一方面，卷涡作为一种极端的爱奥尼亚柱头装饰变体，十分特别，相对少见。罗马城市庞贝中一座住宅（*c*）上装饰的不同寻常的卷涡纹案，展现出古代本土建筑师通常能够设计出柱式的原创版本。

科林斯与混合柱式复杂的雕刻引领了大量的创新与实验。早期科林斯式变体，例如建于公元前40年雅典风塔上（*d*）的柱式经常被复制。建于公元前170年土耳其海岸米利都议会大厦中的科林斯柱头（*e*），展现出如何减少成排的叶饰来适合小规模建筑比例。古罗马广场上的混合柱式（*f*）与爱奥尼亚柱头有着相似的装饰。后期的混合式柱头发明了稀有的花饰柱头（*g*），为威廉・钱伯与18世纪后期设计，独特的玉米柱头象征着新世界，由本杰明・拉特罗布（1764—1820年）于1815年为美国国会大厦（*h*）设计。

柱头上的人像与动物通常代表着建筑物的属权与用途。罗马实例的特色是胜利的象征，如图（*i* 和 *l*）所示，然而建于10年的协和神殿（*j*）的柱式则表现的是献祭的公羊。这些动物形象赋予哥特式柱头设计以灵感，早期的文艺复兴时期实例，如佛罗伦萨的巴齐礼拜堂（*k*），结合了中世纪理念与古典灵感。

扁平的柱子，又称壁柱，迫使科林斯柱式或者混合柱式发生改变，以适应形态的变化。罗马时期的实例（*m*），科林斯式卷涡被有翼马取代，实例（*p*）上反转的卷涡取材于坎榭列利亚宫，是典型的早期文艺复兴式设计。冠板下的卵箭饰来自博洛尼亚（*n*）与布雷西亚（*o*）表明这些柱头都是混合式的变体。

柱础同样可以有变化。一个早期的实例，源于土耳其迪迪玛的阿波罗希腊神庙（*q*），装饰的柱础成为入口处柱子的标志，它们可以追溯至公元前1世纪。繁复的装饰如实例（*r*）经常应用于罗马柱础，然而附加叶子装饰的柱础如实例（*s*），取材于罗马的康斯坦丁教堂，十分常见。有时也有例如图例（*t*）这样的发明，由16世纪法国建筑师让・古戎（1510—1572年）设计，取材于巴黎卢浮宫，文艺复兴之后也有沿用。

VARIATIONS AND INVENTIONS: THE CAPITAL AND BASE

The sculptural inventiveness of the designer and craftsman are often concentrated on the column capital. The most conventional column capital can reveal the intentions of the architect. There is more variety than the conventions suggest.

The simple form of the Doric capital has often been given additional decoration, such as on the Cancelleria in Rome (*a*) of 1486 and on the gate arch at Syon House in England (*b*), by Robert Adam in 1762. The volute, on the other hand, is so specific that extreme variations to the Ionic capital are relatively rare. An unusual arrangement of the volute from a house in the Roman city of Pompeii (*c*) shows how domestic architecture in antiquity often featured quite original versions of the Orders.

The complex sculpture of the Corinthian and Composite Orders has led to a large number of inventions and experiments. Early Corinthian varieties such as the Order on the Tower of the Winds in Athens (*d*) of about 40 BC have been copied many times. The Corinthian capitals from the Council House at Miletus on the Turkish coast (*e*) of 170 BC, show how a reduction in the rows of leaves can accommodate the design to suit smaller-scale buildings. A Composite capital from the Forum in Rome (*f*) has a marked similarity with decorated Roman Ionic capitals. Later Composite inventions include a fanciful floral capital (*g*), by Chambers in the late eighteenth century, and a special maize capital symbolizing the New World for the United States Capitol (*h*) by Benjamin Latrobe (1764-1820) in 1815.

Figures and animals in capitals often refer to the use or ownership of the building. Allusions to victory characterise such Roman examples as (*i*) and (*l*), while a capital from the Temple (*j*) of Concord of AD 10 depicts sacrificial rams. Such figured designs inspired Gothic capitals, and early Renaissance examples such as on the Palazzo Pazzi in Florence (*k*) unite medieval concepts with Classical inspiration.

The flat column, or pilaster, necessitated a modification to the Corinthian or Composite capitals to accommodate the change of shape. In a Roman example (*m*) the Corinthian volutes are replaced by winged horses, while the reversed volutes on example (*p*) from the Cancelleria are typical of the early Renaissance. The egg and dart below the abacus on examples (*n*) from Bologna and (*o*) from Brescia suggest that these capitals were intended as variants of the Composite Order.

Column bases can also be varied. An early example is from the Greek Temple of Apollo at Didyma in Turkey (*q*) where the decorated bases distinguish the entrance columns; they probably date from the first century BC. Heavy decoration such as that on example (*r*) was often applied to Roman bases while the addition of leaves on (*s*), from the Baptistry of Constantine in Rome, is unusual. Occasional inventions such as the base (*t*), by the sixteenth-century French architect Jean Goujon (1510-1572), from the Louvre in Paris, have continued since the Renaissance.

a
b
c
d
e
f
g
h
i
j
k
l
m
n
o
p
q
r
s
t

柱子：细化

从早期的古典建筑开始，柱子就设计成从柱底直径部分向上至柱头逐渐变窄的形式，整根柱子在不同高度呈不均匀的锥形分布。这种锥形或者膨胀，称为支柱收分曲线，可能反映了早期用树干作为柱子的传统，柱身直径的自然收缩。

早期希腊柱式在柱础部分比膨胀的最宽处要窄。古代设计支柱收分曲线时采用不同的体系，文艺复兴时期演变出更多的方法，总体保持柱子总体宽度处于柱底直径宽度之内。

文艺复兴时期的计算方法始于计算柱础之上，柱底直径与柱头下面柱子顶端直径之差，尽管有所差别，但大部分柱子柱顶直径都是柱底直径的0.85倍。

实例（*a*）为从柱础向上呈锥形。连接柱顶直径上*A*点与*B*点，其延长线与柱底直径水平延长线交于*X*点。从柱子顶端用圆规确定*B*点，*B*点为*A*点至柱子中心线柱底半径长度。将柱身等分，从*X*点画延长线，与柱中心线交于等分点*C*。从每个*C*点向上量柱底半径长度，画出一系列*D*点，就形成了支柱收分曲线的弧度。

实例（*b*）收缩了柱身上面的2/3部分。柱顶直径等宽向下延伸至距离柱子根部1/3处，确定中心线上的*A*点。*A*点到*B*点的圆形周长为柱子周长，将这一圆周等分，数量与柱顶部分的等分相同。这一圆周上的等分点，垂直向上延伸，与柱顶等分线交于一系列*C*点，连接这些*C*点就形成了逐渐向上收缩的弧度。

对柱子添加沟槽的不同方法如图所示。多立克柱式有20条浅槽，交会处为尖角，其他柱式通常有24条深槽，可以是复杂的弧线，或者如图所示的简单半圆形。尖锐的边缘容易损坏，柱子底部通常都是平滑的，如图例（*c*），取材于庞贝古城，建于1世纪，或者采用特殊的弧形细部设计填充，称为卷绳状雕饰，如图例（*d*）同样取材于庞贝古城。卷绳状雕饰本身可装饰为绳子状或者丝带状，精美的例子如图例（*e*），由法国建筑师菲利伯特·德·洛梅于16世纪设计。

THE COLUMN: REFINEMENTS

From the earliest Classical buildings onwards, columns have been reduced in diameter towards the capital with a taper unevenly distributed over the height of the column. This tapering, or bulging, effect is known as entasis: it probably echoes the natural reduction in the diameter of the tree trunks first used as columns.

Early Greek columns could also be narrower at the base than at the widest point of the swelling. Various systems were used in antiquity for setting out entasis, and further methods have been developed since the Renaissance, which generally keep the width of the column within the width of the diameter at its lowest point.

The starting point for Renaissance calculations is the difference between the lower diameter of the column measured just above the base and the upper diameter measured just below the capital. Although there is some variety in this difference the upper diameter is usually about 0.85 of the lower diameter in all the Orders.

Example (*a*) tapers from the base. *A* setting-out point *X* is found by extending a line from the outside of the upper diameter, at point *A*, to the level of the lower diameter, through point *B*. Point *B* is found by extending half the width of the lower diameter down from the top of the shaft with compasses to meet the centre line of the column. The column shaft is then divided into equal parts and, from point *X*, lines are extended to cross the points *C* where these divisions meet the centre line. From point *C* each line has half the diameter of the column added to it to produce a series of points *D* which give the curve of the entasis.

Example (*b*) tapers in the upper two thirds of the shaft. The upper diameter is extended downwards to meet a circle of the same diameter as the lower third of the column at point *A*. The circumference of the circle from point *A* to point *B*, outside the column, is divided into the same number of equal parts as the upper portion of the column. The points on the circumference of the circle are extended vertically to meet the horizontal divisions of the upper part of the column at points *C* which are connected to give a gradual curve.

The different methods of fluting columns are also illustrated. Doric columns have twenty shallow flutes, meeting at sharp edges, and the other Orders usually have twenty-four deeper channels which can have complex curves or the simpler semicircles shown. Where the sharp edges are vulnerable to damage, the lower part of the column can be left unfluted, as in example (*c*) from Pompeii, first century AD, or can be partially filled with curved detail known as cabling, as in example (*d*), also from Pompeii. Cabling can itself be decorated like a rope or a ribbon and very elaborate examples such as (*e*) were devised by the French architect Philibert de l'Orme in the sixteenth century.

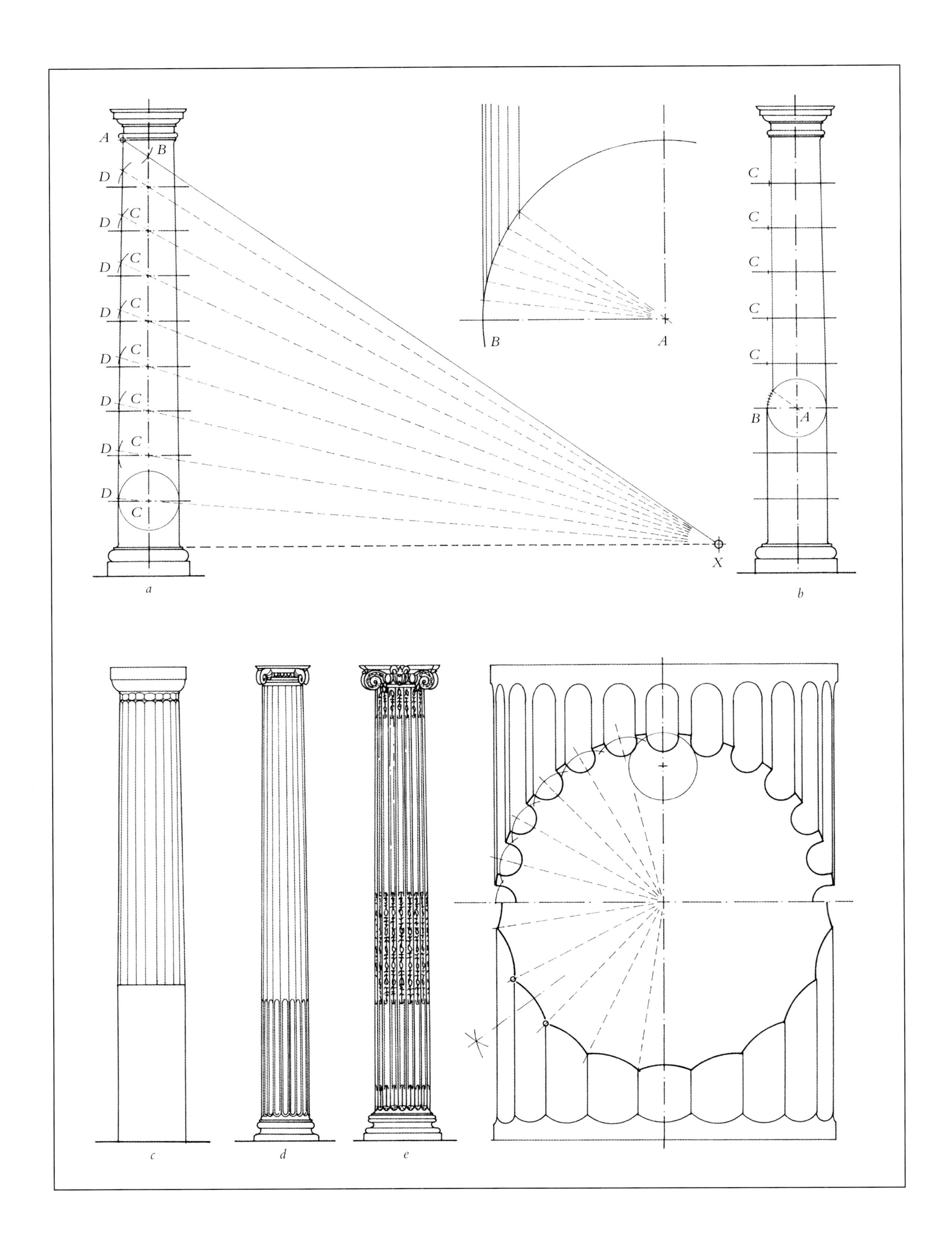
A
B
D
C
X
a
b
c
d
e

壁柱

建筑物中独立式的柱子既不实用也不能达到令人满意的效果。尽管柱式起源于独立式柱子，但很快人们开始模仿这些柱廊，修建带有同样装饰系统的墙面上的凸出物。这些较浅的凸出物称为壁柱，但凸出超过半个柱身时就称为附墙柱。

早期引入壁柱设计，装饰朴素的墙面，使其与建筑物其他部分的独立柱式相呼应，并且使得柱式能够应用于任何一种建筑。

壁柱的表面与后面的墙面之间的距离各不相同。图例（*b*）为附墙柱，与墙面连接的地方隐藏在柱身内部弧线的下面，因此给人一种独立式柱子的感觉。如果改成方形截面如图（*c*），这种效果就会消失，长方形的界面使得附墙柱看上去更粗壮。如果方形附墙柱进深与下部的柱子宽度相等，就增加了柱身的宽度，由于在与后面的墙体连接时支柱收分曲线消失了。图例（*d*）是一个半支柱进深的壁柱，这是圆形柱身最小的深度了。壁柱的弧度削减了凸出的视觉效果。方形形制的使用如图（*e*）所示，可弥补这种缺憾。由于壁柱的深度较浅（*f*），因而使用长方形，如果有支柱收分曲线，也仅限于收回到墙体的两个面。

当独立式柱子邻近墙面时，通常墙上会有相应的壁柱，如图（*a*）所示。罗马人进一步在朴素墙体前演变出了具有装饰效果的柱廊，通过将独立式柱子列于墙体前，将檐部作为独立部分延伸出来，称为每一个柱顶上的墙面凸出部分。

壁柱通常是长方形的，并且这一特色使得壁柱有了一系列独特的细部设计。实例（*g*和*k*）没有任何支柱收分曲线。实例（*g*）取材于公元前4世纪迪迪玛的阿波罗神庙，比同一时期的任何独立式柱子都纤细，并且柱头的样式也仅使用于壁柱之上。另外两个壁柱实例（*h*和*j*）展现出了外面镶有嵌板的两个壁柱版本，其上可镶嵌大理石或其他表面装饰材料。这种流行的变体和嵌板，如实例（*k*）所示，可装饰塑像或其他自然或几何浮雕纹案。同样有可能将沟槽与柱头设计应用于长方形的形制中。305年修建的戴克里先浴场（*i*）中的壁柱上，沟槽的数量减少了。实例（*j*和*k*）展现出文艺复兴时期的柱头，是一种简化的科林斯柱式，改良后适用于方形截面。

THE PILASTER

It is not always practical or desirable to use free-standing columns in a building. Although the Orders originated with independent columns, it was not long before imitations of these colonnades were built by applying the same decorative system to projections in walls. These projections are called pilasters when they are shallow, and engaged columns when they project more than half a column width.

The early introduction of the pilaster allowed plain walls to be decorated in a way that related to free-standing columns on other parts of the building and gave the opportunity to apply the Orders to any kind of structure.

The face of a pilaster can be set at different distances from the wall behind. Example (*b*) is an engaged column. The connection with the wall is concealed by the inward curve of the shaft, thereby giving the impression of a free-standing column. If this is changed to a square profile as in (*c*) the effect is lost and the rectangular section of the engaged column gives the impression of greater bulk. If a square engaged column is the full depth of the lower column width the apparent bulk is further increased, since the entasis of the column is lost as it connects to the wall behind. Example (*d*) is a pilaster half a column deep and is the minimum depth possible for a circular shaft. The curve of the pilaster diminishes the visual effect of the projection. The use of a rectangular form as in (*e*) can compensate for this. As the depth of the pilaster becomes shallower (*f*) a rectangular form has to be used. If there is to be any entasis this can be restricted to the two faces that return into the wall.

When a free-standing column sits adjacent to a wall, there is often a corresponding pilaster on the wall as in example (*a*). The Romans further developed the decorative effect of colonnades against plain walls by setting free-standing columns against the wall and bringing the entablature out in individual sections, called ressauts, over each column.

Pilasters are more often than not rectangular and this has resulted in a series of details specific to pilasters. None of the examples (*g*) to (*k*) has any entasis. Example (*g*) from the fourth-century BC Temple of Apollo at Didyma is very much more slender than any contemporary free-standing columns and has a capital of a type that is only found on pilasters. The two pilasters (*h*) and (*j*) illustrate two versions of a panelled outer column face which can have inset marble or other decorative finishes. This is a popular variation and the panels can, as in example (*k*), be decorated with figures or other naturalistic or geometric decoration in relief. It is also possible to adapt fluting and capitals to the rectangular form. In the Baths of Diocletian (*i*) of AD 305 the flutes are reduced in number. Examples (*j*) and (*k*) show Renaissance capitals of a simplified Corinthian type modified to accommodate the square profile.

a
b
c
d
e
f
g
h
i
j
k

柱子间距

柱子之间的距离，或称柱间距，在柱式的应用过程中，是需要考虑的重要因素。不仅要通过变换柱子之间的距离来达到不同的效果，同时还引入了交错与柱群间距，尤其是在文艺复兴之后，为建筑师们带来了几乎是无限的选择与组合方式。每一种间距本身，正如选择细部设计一样，表达了设计者个人的风格喜好。

罗马作家维特鲁威为我们留下了现存最早的基于神庙正面的柱子间距分类标准。这是依照两柱之间的距离为柱径的倍数来测量。他将其分为五类：密柱式（*a*）柱间距为柱径的1.5倍；双径柱距式（*c*）柱间距为柱径的2倍；宽柱式（*d*）为3倍；离柱式（*e*）为4倍；伏排式（*b*）——他的最爱——柱间距为柱径的2.25倍，除了入口对面的中央间距，此处为3倍。维特鲁威更进一步评论4倍柱径柱距，即离柱式，看起来太宽阔，只适合带有木制额枋的托斯卡纳神庙。当然如果细部设计从石造建筑物中继承而来，那么独立式柱子的柱间距比这个大，看上去就太过宽阔了。

维特鲁威主要对描述这一体系的原则十分感兴趣。我们知道古代柱间距的变化远比他让我们了解到的要多得多。在同一建筑物上也会有变体，像缩短角柱间距这样的调整屡见不鲜。

中央入口处宽阔的间距十分普遍，并且容易分辨。这一布局在实例（*f*）中所示，中间间距为3倍柱径的间距，两边为1.5倍。削减柱间距来强化通道两侧，使得中间4倍柱径的间距看起来更能够让人接受，如图（*g*和*h*）所示。两侧密集间距（*h*）为0.5倍柱径，因此有必要修建2倍宽的人行道，这可以在实例（*f*和*g*）中以令人满意的方式实现。

在柱廊的末端设计双柱（*i*），呈现出明显的视觉效果，并且可以与不规则的柱间距相结合，如图（*j*）所示，柱间距为4倍和2倍柱径交错排列。当需要在末端或边角处呈现更强的视觉效果和结构支撑时，柱子可联合在一起。实例（*k*）展现出两个连接在一起的柱子，形成了一个心形柱，尽管这种不太常见，但在古代和文艺复兴时期内角与外角设计中都能找到实例。更常见的方法是使用一个更明确的终止，将一个柱子连接到厚重的长方形设计中，如（*l*）所示，为一组相互连接的方柱或圆柱。所有这些变体通常结合于巴洛克建筑内，形成复杂与躁动的印象，如模拟实例（*m*）所示，掩饰了其作为结构支撑的复杂柱群。

COLUMN SPACING

The spacing of columns, or intercolumniation, is an important consideration in the application of the Orders. Not only are there different effects to be gained by varying the spaces between columns, but the introduction of alternating and clustered spacing, particularly after the Renaissance, has given architects an almost unlimited range of options and combinations. Each type of spacing will itself, as with any choice of detail, express the stylistic preferences of the designer.

The Roman author Vitruvius has left us with the earliest categorisation of column spacing, based on the fronts of temples. This is measured by the number of column diameters that can be fitted between two columns. He gives five classes: pycnostyle (*a*) with 1.5 diameters spacing; systyle (*c*) with 2 diameters; diastyle (*d*) with 3; araeostyle (*e*) with 4; and eustyle (*b*)—his favourite—with 2.25 diameters between the columns, except for the central space opposite the entrance, which has 3. Vitruvius further comments that a spacing of 4 diameters, the araeostyle, looks too wide and is only suitable for native Tuscan temples with their timber architraves. Certainly free-standing columns with a spacing any wider than this can look too wide if accompanied by details directly derived from stone buildings.

Vitruvius was primarily interested in describing a system of rules. We know that the variety of column spacing in antiquity was much more diverse than he would have us believe. Variations occurred even in the same building—refinements such as closer spacing at corners are not unusual.

Wider spacing for a central entrance is common and can be quite pronounced. The arrangement in (*f*) shows a centre spacing of 3 column diameters with 1.5 at the sides. This strengthening of the sides of openings with closer spacing can make a central spacing of 4 look quite acceptable as in (*g*) and (*h*). The close spacing of 0.5 diameters on the sides of (*h*) makes a double pedestal necessary although this could also be satisfactorily used on (*f*) or (*g*).

The visual termination of a colonnade with coupled columns (*i*) can be effective and it can be combined with irregular column spacing between, as in (*j*) where widths of 4 and 2 diameters between columns alternate. Where greater visual or structural strength is required on ends or corners, columns can be joined together. Example (*k*) shows two columns linked to form a single heart-shaped column and, although this is quite unusual, it is found both in antiquity and the Renaissance at external and internal corners. It is more usual to form a more positive stop by attaching a column to a heavier rectangular block which can, as in (*l*), be expressed as a group of linked square and circular columns. All of these variations and more were often combined in Baroque buildings to give the sort of complex and restless impression seen in the hypothetical example (*m*) which also disguises in its complicated clusters of columns quite large structural supports.

a
b
c
d
e
f
g
h
i
j
k
l
m

4. 比例的理论

比例体系

通过确立在二维或更多维度空间中的数学关系，相同的形状与形制能够以不同大小重复呈现。这种方式使得一间1m × 2m的房间，在平面上以同样的比例变为2m × 4m，或者30m × 60m。这种情况下的数学关系为1：2的比率，房间可根据这一比率分配。

过去，缺乏尺寸标准和精确工具使得人们在测量时采用比例关系，例如使用3：4：5的三角形来画直角，依靠比率而不是尺寸来转换建筑信息。这一习惯可见于塞巴斯蒂安・塞里奥（1475—1554年）1575年写成的《建筑五书》中的大门图解（*a*）。尺寸最初源自建筑师的人体结构的比例，并且直到度量标准出现之前，一直是人类身体的理想比例。这种信仰认为人的身体是按照上帝的形象创造出来的，与古代信仰结合时，数学证据就有了神圣庄严的规定，赋予比例以神秘或宗教的重要性。

古代人认为美不是任意的，而是神的想象力，近来人们相信可以用科学分析的方法，形成了关于比例的理论和理想体系，例如弗朗西斯科・迪・乔吉奥・马蒂尼（1439—1501年）在15世纪末设计的教堂平面（*b*）。

古典柱式在过去的几个世纪中演变，构成了复杂的比例体系。经过几代人的实践，建筑物可以通过比例来进行设计，没有模仿之嫌。最小的部分可能与最大的部分以柱宽的模数相关联，菲利贝・德・洛梅（*c*）在其1567年写成的《建筑》一书中提到。这一传统可能得以改进，而不失其完整性，并且这一体系也不拒绝进一步的发展。比例家族、柱式家族，每一种都有其特有的身份、变体和传说。对柱式和比例的理解是每一个人解读古代密码的入口，却从未有人真正彻底了解过。

柱式只是众多建筑方法中的一种，为了在复杂的世界当中创造出和谐。附加的比率体系能够应用于柱式当中。这些体系可以给设计以额外的统一性，在不考虑而存在于其中的幻想哲学理论的情况下，使它们变得更加庄重。

4. THEORIES OF PROPORTION

PROPORTIONAL SYSTEMS

By establishing the mathematical relationship between two or more dimensions, similar shapes and forms can be repeated in different sizes. By this means a room of 1 metre by 2 metres, for example, will be the same proportion in plan as a room of 2 metres by 4 metres or 30 metres by 60 metres. The mathematical relationship in this case is expressed as the ratio 1:2 and all these rooms will be proportioned according to this ratio.

In the past the absence of standard dimensions and the lack of accurate instruments led to the use of proportional methods of surveying, such as the use of 3:4:5 triangles for creating right-angles, and the reliance on proportion rather than dimension for the transfer of information for construction. This can be seen in Sebastian Serlio's (1475-1554) diagram for a door (*a*) in his *Five Books of Architecture* (1575). Dimensions were at first derived from the physique of the architect and then, until the advent of the metric system, from idealised proportions of the human body. The belief that the human body was created by and even in the image of a god, when combined with the ancient conviction that mathematical proof was divinely ordained, gave proportions a mystical or religious significance.

The ancient idea that beauty is not arbitrary but a vision of the divine, and the more recent conviction that it is subject to scientific analysis, has led to the formation of theories about proportions and ideal systems such as Francesco di Giorgio Martini's (*bap.* 1439–1501) church plan (*b*) from the late fifteenth century.

The Classical Orders have evolved over the centuries to constitute a sophisticated proportional system. Buildings can be designed according to proportions established by generations of usage without resorting to imitation. The smallest part can be related to the largest through the module of the column width, shown by Philibert de l'Orme (*c*) in his book *Architecture* of 1567. The conventions can be modified without losing their integrity and the system is not closed to further development. There are families of proportions, or Orders, each with their own identity, variations, and mythology. An understanding of the Orders and their proportions is an entry into an ancestral code appreciated by everyone but never fully understood by anyone.

The Orders are only one architectural method of creating harmony in a confusing world. Additional systems of proportions can be applied to the Orders. These systems can give an added unity to a design almost regardless of the sometimes fanciful philosophies that lie behind them and seem to give them an enhanced gravity.

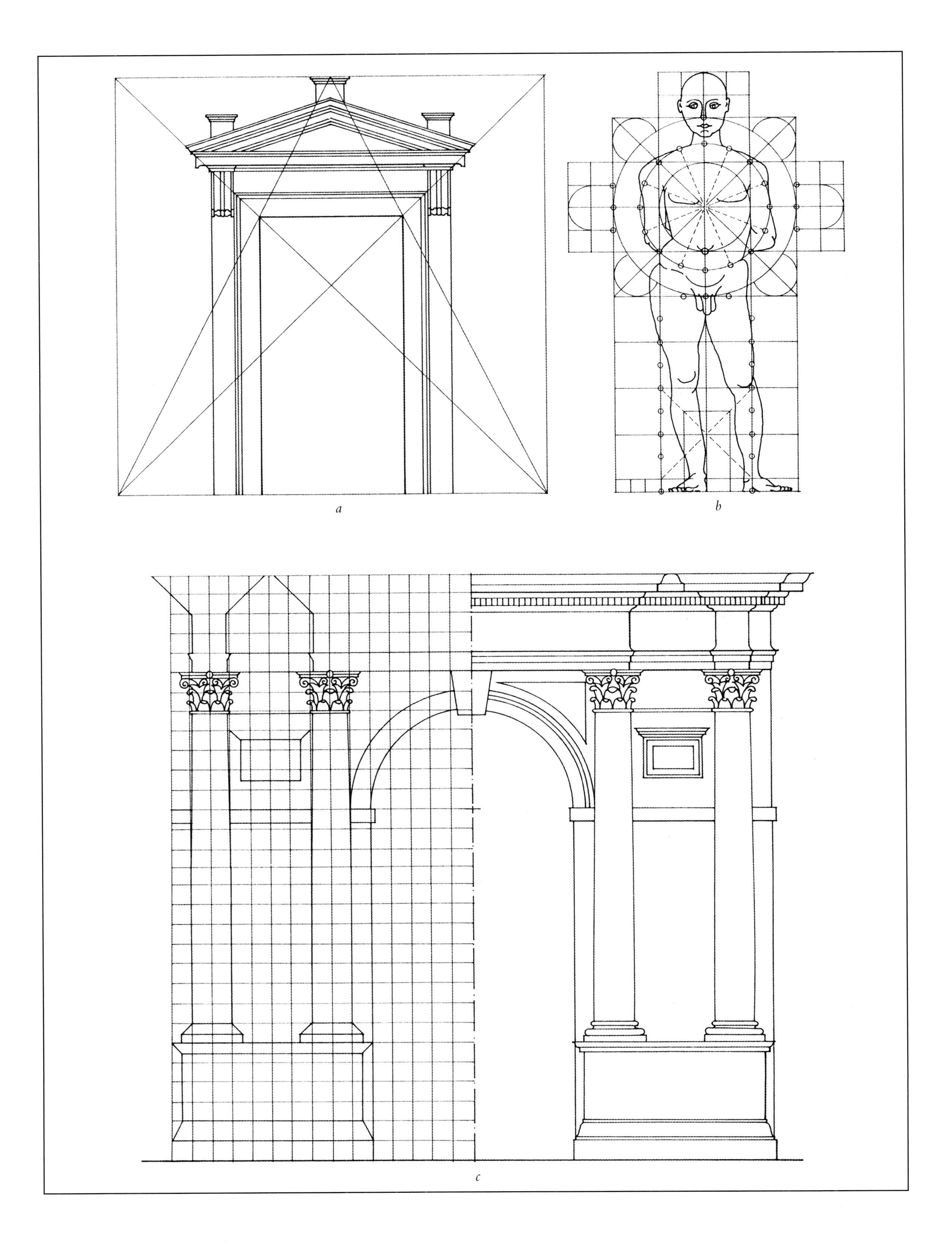
a
b
c

简单比例

圆形与球形（c）以及正方形与立方体（a），通常代表理想形态。等边三角形与金字塔形（d）和其他形态有时也包括在内。这些形状很容易使用直角尺或者圆规来构建，并且被赋予神秘的特征。正方形与圆形彼此关联，在罗马作家维特鲁威笔下，都有人体比例有关系。这种简单几何图形与自然的联合在文艺复兴时期影响极其巨大。许多像里奥纳多·达·芬奇（1452—1519年）这样的作家与艺术家纷纷展示这样的设计（b）。

人们相信圆形与球形是完美的，这一信仰建立在圆周距离圆心绝对等距的基础之上。这种完美被与永生以及神圣关联起来。圆形与球形的形制可见于所有建筑时期，但是其使用仅限于独立的几何学。经常根据圆形创造出变体的几何形态，例如八边形、六边形以及更少见的五边形（e）。

正方形和立方体不仅各部分相等，而且比圆形更容易掌握和分割。古代与文艺复兴时期作家描述房间平面经常与正方形进行参照。因此，屋子的边比例为4：5是$1^1/_4$个正方形，如果是3：5则是$1^2/_3$个正方形（f）。平面、正立面和体量可以划分为一系列环环相扣的形状，可还原成一系列小正方形或整数（h）。

在公元前6世纪毕达哥拉斯发现了和弦上最初的4个音，与弹拨时紧绷的弦上精确的长度划分有关，这些划分由一系列简单的比率所控制，即1：2：3：4。柏拉图将这些数字发展成为两个序列，一个是依次乘以2，变成了1-2-4-8，另一个是依次乘以3，1-3-9-27，通常集合在一起用希腊字母Λ（g）来表示。直到18世纪，人们一直认为这些数字代表着宇宙的和谐，一种统一宗教、数学和音乐的哲学。人们断定和谐音乐的数学原理能够创造出视觉和谐，文艺复兴时期的建筑师们使用这些规则发展出了相关的比例体系，创造出了中级的第三种符号，因此在维度上超越了二维空间。有三种方法能够找到中间或均分符号：算数、几何和调和。算数平均（A）通过以下公式计算，$M-A=A-N$，或者$A=1/2(M+N)$。因此5和9的算数平均数为7，因为5-7=7-9。几何平均（G）通过以下公式计算，$\frac{M}{G}=\frac{G}{N}$，或者$G=\sqrt{(M\times N)}$。因此4与9的几何平均数为6，因为$6=\sqrt{(4\times 9)}$。调和平均方法（H）为以下公式$\frac{1}{M}-\frac{1}{H}=\frac{1}{H}-\frac{1}{N}$，简化成$H=\frac{2MN}{M+N}$。因此12和6的调和平均数为8，因为$8=\frac{2\times 12\times 6}{12+6}$。

SIMPLE PROPORTIONS

The circle and sphere (*c*) and the square and cube (*a*) are often represented as ideal forms. The equilateral triangle and the pyramid (*d*) and other figures are sometimes included. These shapes lend themselves to easy construction with the square and compasses and have been given mystical properties. The square and the circle can be related to one another and were both associated with the proportions of the human form by the Roman author Vitruvius. This unification of simple geometry and nature was particularly influential in the Renaissance. It has been illustrated by many authors and artists including Leonardo da Vinci (1452-1519) (*b*).

Belief in the perfection of the circle and the sphere is based on the absolute regularity of the distance of the perimeter from the centre. This perfection has been associated with eternity and divinity. Circular and spherical forms are found in all architectural periods, but their use is limited by their self-contained geometry. Variations are often created with geometric forms related to the circle such as the octagon, hexagon, and, more rarely, the pentagon (*e*).

The square and cube not only have regularity in the equality of their parts but are also easier to manipulate and subdivide than circular figures. Ancient and Renaissance authors described room plans by their relationship to a square. So, a ratio of the sides of a room of 4:5 was a square and quarter and 3 : 5 a square and two thirds and so on (*f*). Plans, façades, and volumes could be divided into interlocking series of shapes reducible to a series of small squares or whole numbers (*h*).

In the sixth century BC Pythagoras discovered that the first four notes of the harmonic series could be related to precise divisions of the length of a taut string when plucked and that these divisions were governed by a series of simple ratios 1 : 2 : 3 : 4. Plato developed these numbers into two series, one which multiplies by two, 1-2-4-8, and the other by three, 1-3-9-27, usually shown together in a figure based on the Greek letter Λ (*g*). These figure were thought to represent the harmony of the universe, a philosophy that unified religion, mathematics and music until the eighteenth century. It was assumed that the mathematics of musical harmony would create visual harmony and Renaissance architects used these principles to develop a coherent proportional system which could generate an intermediate third figure, and therefore dimension, out of two. There were three methods of finding this intermediate or mean figure: arithmetic, geometric, and harmonic. The arithmetic mean (A) is found by the equation $M - A = A - N$, or $A = 1/2(M+N)$. So the arithmetic mean of 5 and 9 is 7 since $5 - 7 = 7 - 9$. The geometric mean (G) is found by the equation $\frac{M}{G} = \frac{G}{N}$, or $G = \sqrt{(M \times N)}$. So the geometric mean of 4 and 9 is 6 since $6 = \sqrt{(4 \times 9)}$. The harmonic mean (H) is found by the equation $\frac{1}{M} - \frac{1}{H} = \frac{1}{H} - \frac{1}{N}$ which simplifies to $H = \frac{2MN}{M+N}$. So the harmonic mean of 12 and 6 is 8 since $8 = \frac{2 \times 12 \times 6}{12+6}$.

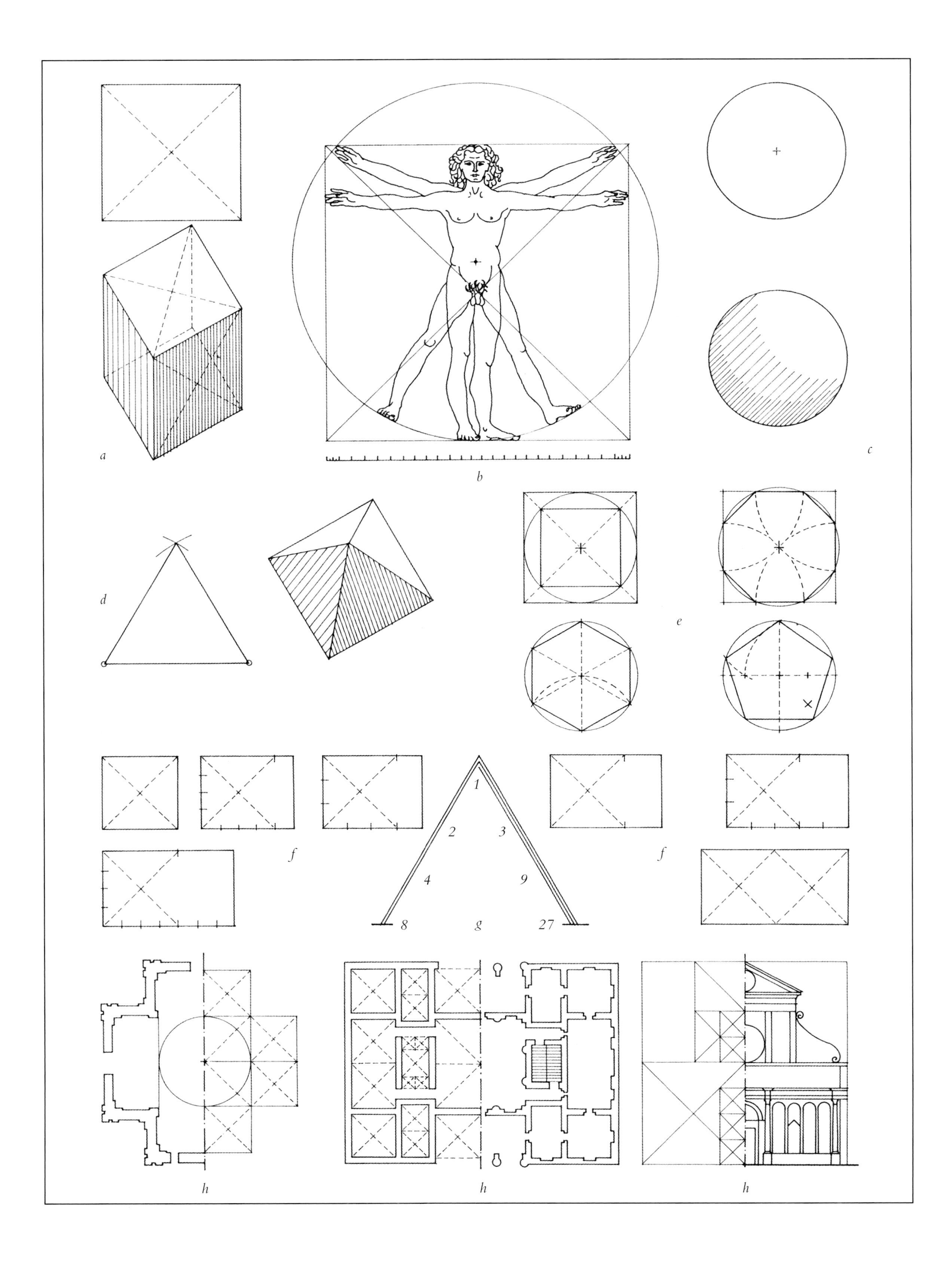

a
b
c
d
e
f
1
2
3
4
9
8
g
27
f
h
h
h

复杂比例

古代与文艺复兴时期的比例体系通常是以比率和数字为基础的。当建筑需要达到高度精准的水平时，这种理论不仅是实用的，而且复杂的理论需要用相对简单的数学、几何以及音乐理论来表达。

文艺复兴时期只倡导一种基于复杂比率的长方形构图，源自罗马作家维特鲁威。$\sqrt{2}$长方形（*a*）结合了正方形和圆形的几何形制。尽管长方形两边的比例为复杂的$1:\sqrt{2}$，约等于1：1.414，很容易使用正方形的对角线和圆规画图，并且有一些实用功能。

正方形可以划分成具有相同比率的对称部分，$\sqrt{2}$长方形可以用同样方式来划分，分割后包含一系列不同的$\sqrt{2}$长方形和正方形（*b*）。八边形作为一种几何图形，介于圆形与正方形之间，能够用来构建并扩大八边形与八角星形（*c*）。

另一种长方图形是一个带有复杂比例的简单几何图案，称为黄金矩形（*d*）。正方形划分成两部分，以其中一半正方形的对角线为半径，以正方形等分线的端点为圆心，向正方形外部画弧，与正方形一边的延长线相交，这样就构建了一个长方形，长宽比例约为1：1.618。黄金分割的比例是独特的，并具有明显特征。古代与文艺复兴时期的数学家对这些特征十分了解，但并无史料记载19世纪以前的建筑师对此感兴趣。当时人们试图通过早期建造实践去寻求数学的灵活性与美学标准以及其使用的间接证据。然而没有证据证明这个数字或其在任何时期刻意的应用，象征着永恒之美，但在最近一段历史中，发现了其特征中有趣的地方。

黄金矩形能够分割为正方形和另一个黄金矩形，一连串逐渐变小的正方形与黄金矩形或逐渐缩小的正方形与黄金矩形（*e*）。这种排列看似无穷，基于各部分相同的比率（*d*）例如1：0.618…，*BD*：*CD*，同样1.618…:1，*BD*：*AB*。这一黄金分割比率也可见于五角星和逐渐缩小的五角星（*f*）。12世纪在比萨，意大利数学家列昂纳多·斐波那契（约1170—1250年）也用一个数列记录了一些有趣的关系。在斐波那契数列中每个数字都是前两个数字之和，1-1-2-3-5-8-13等。自然界按照这一原则不断增长的现象十分常见，这一数列中相邻的两个数字的比率也逐渐接近1：1.618。

COMPLEX PROPORTIONS

Proportional system in antiquity and the Renaissance were most frequently based on ratios of whole numbers. Not only was this practical when accuracy in construction was difficult, but even sophisticated theories could be expressed in terms of relatively simple mathematical, geometric, and musical principles.

Only one rectangular figure with proportions based on a complex ratio was advocated in the Renaissance, taken from the Roman author Vitruvius. This figure was the root-two rectangle (*a*) which combined the geometry of the square with the circle. Although the sides of the rectangle have a complex ratio $1:\sqrt{2}$, or approximately 1 : 1.414, it is very simple to set up geometrically with the diagonal of a square and compasses and has some useful characteristics.

The root-two rectangle can be divided up in the same way as a square to give symmetrical parts of equal proportion and the subdivisions can be varied to include a mixture of root-two rectangles and squares (*b*). It is also a part of the geometry of the octagon, a common intermediary between the circle and the square, and can be used for setting up and expanding an octagon or a star octagon (*c*).

Another rectangular figure with a simple geometry and a complex ratio is the golden rectangle (*d*). A square is divided into two and a diagonal of one of the half-squares becomes a radius with its centre on the central division of the square. This radius is swung outwards to extend to the side of the square, creating a rectangle with a ratio between the sides of approximately 1 : 1.618. The ratio of the golden section has some unique and remarkable characteristics. Many of these were known by mathematicians in antiquity and the Renaissance, but there is no documentary evidence of any architectural interest before the nineteenth century. At that time experiments sought to link its mathematical versatility with a scientific standard of beauty and circumstantial evidence of its use was claimed from earlier buildings. While no proof of the eternal beauty of this figure or its conscious use in any period but the recent past can be found, its characteristics are of some interest.

The golden rectangle can be divided up to give a square and another golden rectangle, a series of interlocking golden rectangles and squares or diminishing patterns of squares and golden rectangles (*e*). The permutations are seemingly endless and are based on the equality of the ratios of the parts (*d*) such that 1:0.618..., *BD*:*CD*. is the same as 1.618...:1, *BD*:*AB*. The ratio of the golden section is found in the star pentagon and diminishing star pentagons (*f*). It also has an interesting relationship with a series of numbers set down by the Italian mathematician Leonardo Fibonacci of Pisa (*c.*1170–*c.*1250) in the twelfth century. In the Fibonacci series each successive number is the sum of the two previous numbers, 1-1-2-3-5-8-13 and so on. Growth according to this principle is often found in nature and the ratio between any two adjacent numbers gradually approaches 1 : 1.618 as the series progresses.

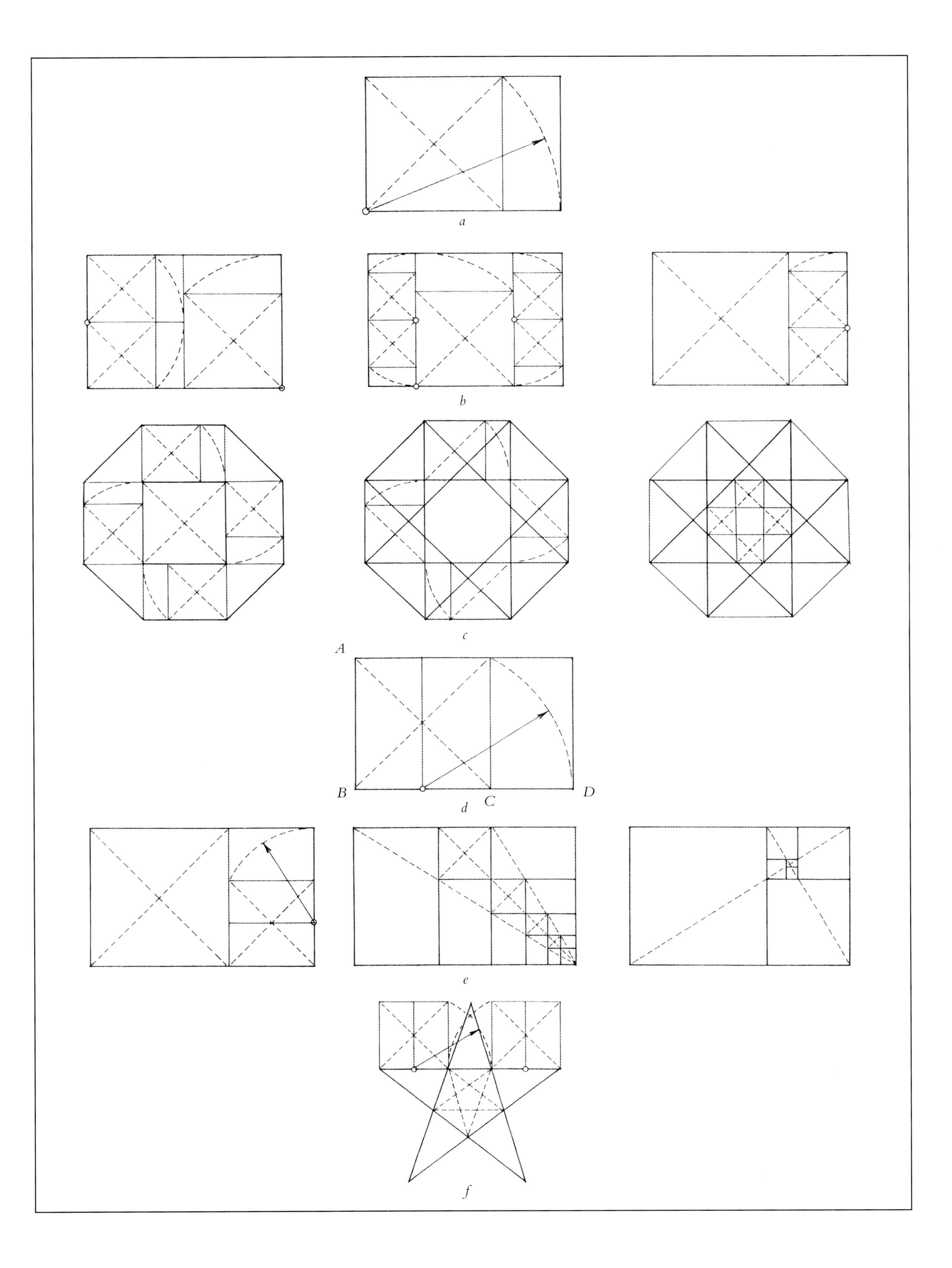
a
b
c
A
B
C
D
d
e
f

5. 装饰线脚

简单装饰线脚

柱式的细部设计包括一系列的直线形和弧形凸出与凹陷装饰。这些称为轮廓或装饰线脚。尽管其中一些成为独特古典柱式的标志，同一范畴内的装饰线脚以一种特定的方式组合在一起，构成每种柱式的细部设计。它们的形制、位置和彼此之间的关系在过去的几个世纪中逐渐演变成为规则，松散地规范着柱式的构成与装饰。对装饰线脚的了解，对于全面掌握古典建筑来说十分必要。

装饰线脚可能是简单形状，或者带有雕刻、绘画的装饰。在古典建筑物上，装饰线脚与建筑其他部分一样覆盖着鲜艳的彩绘装饰，很多早期的雕刻装饰可能也源于绘画纹案。

平缘（*i*）仅仅是小型的方形凸出物，如图所示，一面或者两面弧度向外，或者仅仅是两面水平向外凸出。它很少有装饰，但有时会有波浪卷纹案设计，如图（*ii*）和（*iii*）。平缘通常紧贴着置于扁平的垂直挑檐滴水板之上（*iv*），这在所有柱式的飞檐中都是一样的。在装饰更加复杂的罗马及其后时代的飞檐中，挑檐滴水板通常雕刻有垂直的沟槽纹案（*v*），通常可以用僵硬的叶饰（*vi*）代替。

串珠饰（*vii*）比其他装饰线脚更常用。这是一种小的半圆形凸出，可见于柱头上，在垂直立面的额枋上，或飞檐大型线脚上，在托架涡卷花饰的外侧面以及柱式的其他部分。串珠饰线脚通常带有装饰。珠缘（*viii*）通常是各种珠子以及卷状装饰线脚，（*ix*和*xiv*）原来是串珠饰特有的，但随着时间推移，逐渐增加了缆绳或绳索，如（*xv*）、（*xvi*）、（*xvii*）、（*xix*）和（*xxi*）以及其他的扭索形态，例如丝带装饰线脚，（*xviii*）和（*xx*）。许多这些装饰不断地应用于罗曼式与哥特式建筑。串珠饰设计变换的可能性极大，（*xxii*）到（*xxvii*）仅仅展示了一些过去对自然形态及其混合形态的应用。带有方向性的自然形态例如（*xxiii*）和（*xxvi*）通常在中央部分逆转方向，呈对称纹案。

5. MOULDINGS

SIMPLE MOULDINGS

The details of the Orders are made up of a series of straight and curved projections and recesses. These are called profiles, or mouldings. Although some are identified with a particular Classical Order, the same range of mouldings are put together in a specific way to make up the details of each Order. Their form, location, and relationship to one another have developed over the centuries to create principles which loosely regulate their composition and decoration. A knowledge of the mouldings is essential to a full understanding of Classical architecture.

Mouldings can be left as simple shapes or decorated with carving, painting or both. On ancient buildings, mouldings, in common with the rest of the structure, were covered with brightly painted decorations and the origin of many of the early forms of sculpted decoration probably lay in painted patterns.

The fillet (*i*) is no more than a small square projection which can, as indicated, curve outwards to its face on one or both sides or merely project horizontally on both faces. It is rarely decorated but it sometimes has a simple wave-scroll design, (*ii*) and (*iii*). The fillet often sits immediately above the flat vertical face of the corona (*iv*) which is universal to the cornice of all the Orders. In more heavily decorated Roman or later cornices, the corona is occasionally carved with a vertical fluted pattern (*v*) which can in turn be embellished with stiff leaves (*vi*).

The astragal (*vii*) is used more frequently than any other moulding. It is a small semicircular projection and is found on column capitals, between vertical faces on architraves, between larger mouldings on cornices, in the outer face of scrolled brackets and elsewhere in the Orders. Astragal mouldings are often decorated. Beading (*viii*) and various bead and reel mouldings, (*ix*) and (*xiv*), were originally specifically associated with the astragal, but in time it was also decorated with cable or rope designs, (*xv*), (*xvi*), (*xvii*), (*xix*) and (*xxi*), and other twisted forms such as ribbon moulding, (*xviii*) and (xx). Many of these decorations continued to be used in Romanesque and Gothic architecture. The variety of possible astragal designs is huge and (*xxii*) to (*xxvii*) illustrate only a few of the natural forms and mixtures of natural forms that have been used in the past. Directional natural forms such as (*xxiii*) and (*xxvi*) often reverse at the centre of their length to make the pattern symmetrical.

i
ii
iii
iv
v
vi
vii
viii
ix
x
xi
xii
xiii
xiv
xv
xvi
xvii
xviii
xix
xx
xxi
xxii
xxiii
xxiv
xxv
xxvi
xxvii

更加简化的装饰线脚

柱脚圆盘线脚有着与串珠饰同样的半圆形剖面，但在同一柱式中通常比串珠饰大一些。柱脚圆盘线脚在任何一种柱式中通常都作为柱础的一部分并且很少有装饰。然而，柱脚圆盘线脚起源于萨摩斯岛和以弗所的第一个爱奥尼亚柱式上作为柱础的扁平凸出圆盘，并带有不同宽度的凹槽，盘绕于圆周上（*a*）。这些沟槽的频度、间距和数量差别很大。柱式形制转变为科林斯式，在罗马成为托斯卡纳、多立克和混合式，但是沟槽装饰（*a*）从爱奥尼亚柱式延续下来。通常情况下，保留了柱础的装饰并且这种连续的带状装饰能够延伸到整个建筑物的基础。

使用一连串的好似一捆绳索（*b*）的凸面曲线来代替一系列凹槽，这种手法可见于早期的爱奥尼亚建筑；一种交织的扭索状装饰纹案（*c*和*d*），可见于公元前5世纪的爱奥尼亚柱础。以成捆的杆子作为装饰的形制是模仿罗马时代司法权威的象征——束棒，丝带饰则强化了这一象征（*b*）。

自罗马时期，一些自然与规范的装饰应用于柱脚圆盘线脚装饰上。其中一些实例如图（*e*和*g*）所示。另外两个实例可见于P273。柱脚圆盘线脚装饰可增加一些其他装饰，如卵箭饰和叶丛状装饰。

装饰柱脚圆盘线脚所采用的许多手法可应用于垫状的，即鼓突的带状雕刻上，见P99的混合柱式。然而，鼓突的带状雕刻少有完整的半圆形，并且其装饰范围也不包含爱奥尼亚柱础（*a*）上的水平凹槽。

齿状装饰，（*i*）~（*n*）是成排的齿状凸出装饰，起源于早期爱奥尼亚式建筑中飞檐下层部分，可能象征着横梁的末端。从爱奥尼亚柱式开始，他们逐渐演变为科林斯柱式的一部分，随后又变为混合式。在罗马，它们经常被应用于多立克柱式，文艺复兴时期演变成为一种独特的锯齿状多立克飞檐。

在其最简单形制中（*i*），齿状装饰不仅是一连串的垂直长方形凸出。在罗马爱奥尼亚式建筑中，一个独立的边角齿状装饰有时悬挂在底层凸出（*j*）之下。这种不合逻辑的安排对于横梁底部来说，一般会通过在边角处悬垂松果装饰来解决，参见实例（*k*和*l*）。在科林斯式和混合式飞檐中，底层凸出有时会采用特殊手法加以装饰（*l*）。在后期的建筑中，所有与木制结构起源有关的元素都不复存在，齿状装饰也只是一种装饰特征（*m*和*n*）。

FURTHER SIMPLE MOULDINGS

The torus moulding has the same semicircular profile as an astragal, but is consistently larger than an astragal on the same Order. Torus mouldings are most commonly found as a part of the base of a column of any Order and are often left undecorated. The origin of the torus moulding was, however, as a bulging flat disc at the base of the first Ionic columns in Samos and Ephesus and had a series of concave channels of different widths running around its circumference (*a*). The frequency, spacing, and number of these channels could vary considerably. The form was transferred to the Corinthian Order, and in Rome to the Tuscan, Doric, and Composite Orders, but the channelled decoration (*a*) was reserved for the base of the Ionic Order. Generally it has remained a base moulding and can be used as a continuous band at the base of buildings.

The substitution of a series of concave channels with a series of convex curves like a bundle of rods (*b*), but without ribbons, is found on early Ionic buildings; an interlacing guilloche pattern, (*c*) and (*d*), can be seen on Ionic bases of the fifth century BC. Decoration in the form of bundles of robs resembled the Roman symbol of judicial authority, the fasces, and ribbons were added to reinforce the association (*b*).

Since the Roman period a number of natural and formal decorations have been applied to the torus moulding. Some of these are illustrated in (*e*) to (*g*). Two further examples can be seen on page 273. The torus has received other types of decoration which are illustrated elsewhere, such as the egg and dart, and anthemion.

Many of the devices used to decorate the torus are also used for the cushion, or pulvinated, frieze seen on the Composite Order on page 99. The pulvinated frieze, however, is rarely a complete semicircle and its decorative range does not include the horizontal grooves found on the Ionic base (*a*).

Dentils, (*i*) to (*n*), are a row of tooth-like projections that originated in the lower part of the cornice of early Ionic buildings and probably represented beam-ends. From the Ionic Order they became a part of the Corinthian Order and hence the Composite Order. In Rome they were occasionally used on the Doric Order and in the Renaissance a specific 'denticular' Doric cornice was evolved.

In their most simple form (*i*), dentils are no more than a series of vertical rectangular projections. On Roman Ionic buildings an isolated corner dentil was sometimes suspended from an underlying projection (*j*). This illogical arrangement for a beam-end was often resolved by suspending a purely decorative pine-cone in the corner, (*k*) and (*l*). On Corinthian and Composite cornices the underlying projection was at times itself decorated with a curiously specific device (*l*). On later buildings any association that might have remained with a timber structural origin was lost and dentils were treated as a solely decorative feature, (*m*) and (*n*).

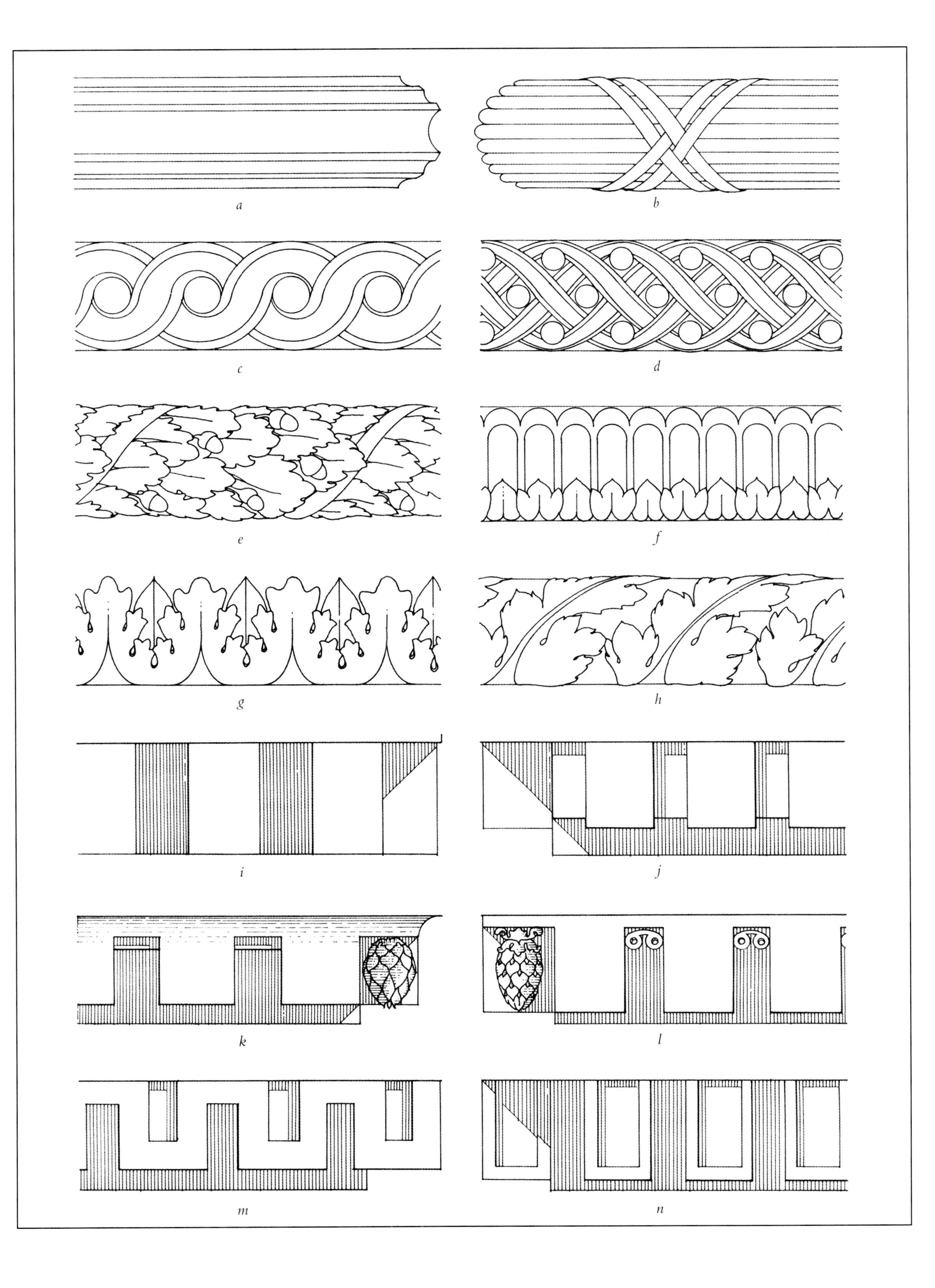

a
b
c
d
e
f
g
h
i
j
k
l
m
n

复杂装饰线脚

一系列的复杂装饰线脚作为古典装饰的特征逐渐演变出来。这些复杂的轮廓，每一种都有各自的名字和几何结构，但大部分的变化都属于这一框架之中。

正波纹线脚，或称双弯曲线，装饰线脚（*a*～*f*），是双弧线，上凹下凸。除了一些多立克柱式实例之外，正波纹线脚是其他柱式中檐口最上方的装饰线脚。它常见于壁柱的柱头、基座檐口，倒置于壁柱和柱础底座以及其他柱式各个部分之上。

正波纹线脚似乎并无特殊的装饰形式。通常为较浅的形制（*a*、*d*和*e*），为希腊人所喜爱，较深的坚固的形态为罗马人喜爱。实例（*d*～*f*）凹弧大于凸弧，半径发生变化时也符合这个关系。

反波纹线脚，或称反向弯曲，为上凸下凹的反曲线（*g*～*l*）。这一形状逐渐演变，融入了舌箭装饰纹案，应用于所有柱式的檐口上：爱奥尼亚柱式、科林斯柱式和混合柱式的额枋顶部，某些爱奥尼亚柱头冠板处以及其他柱式的某些部位。在同一柱式的挑檐滴水板上，它通常比正波纹装饰线脚要小一些。

凸圆形装饰线脚（*m*～*o*），与卵箭装饰纹案有着特殊关联，这是最具特色也是最为广泛应用的一种古典装饰细部设计。它凸出装饰于檐口、额枋、爱奥尼亚和混合柱式的柱头上，并且其简单的形制（*o*）或只有1/4圆的形态，能用于罗马多立克与托斯卡纳柱头的卵形花边上。

希腊多立克卵形花边（*p*～*r*），按照其演变顺序分别从早期寺庙（*p*），演变到古代雅典形制（*q*），再到后期希腊风格（*r*）。卵形花边一词，意为海胆，特指多立克和爱奥尼亚柱式柱头，尽管同样轮廓有时作为少量的凸圆形装饰线脚变体，出现在一些希腊檐部设计上或充当早期多立克檐口最上部的装饰线脚。

凹弧饰（*s*～*u*），是一种实用并且常见的无装饰轮廓，可独立充当主要装饰线脚或者扮演较宽或较窄装饰特征之间缓冲连接的角色。凹弧边饰装饰线脚或者向上延伸（*v*～*x*），或者单独或成对出现，但要保持其较长半径处于常规位置之上，分割某些多立克柱式，所有爱奥尼亚、科林斯和混合柱式柱础上的柱脚圆盘线脚。

所有这些实例都设定精准，并且经过筛选，展示出几何构图的改变在不同形状上造成的光影变化。拥有自信的笔画和经验，无须圆规便可以画出更加细微的改变。

COMPLEX MOULDINGS

A series of complex mouldings has evolved that is specifically identified with Classical design. Each of these complex profiles has its own name and geometric structure, but there is a great deal of variety within this framework.

The cyma recta, or ogee, moulding, (*a*) to (*f*), is a double curve, concave above and convex below. With the exception of some Doric examples, the cyma recta is the uppermost moulding on the cornices of all the Orders. It is also found on pilaster capitals, on the cornices of pedestals, and upside-down on pilaster and pedestal bases as well as in other parts of some Orders.

The cyma recta seems to be associated with no specific form of decoration. Generally the shallower forms (*a*), (*d*), and (*e*), were favoured by the Greeks and the deeper, stronger forms by the Romans. Examples (*d*) to (*f*) have larger concave than convex curves, and when the radius varies it is always in this relationship.

The cyma reversa, or reverse ogee, has the convex curve uppermost and the concave curve below, (*g*) to (*l*). The shape evolved to accept tongue and dart decoration and is found in the cornices of all the Orders: crowning the architrave of the Ionic, Corinthian, and Composite Orders, on the abacus of some Ionic capitals and in other parts of some Orders. It is usually smaller in scale than the cyma recta moulding above the corona on the same Order.

The ovolo moulding, (*m*) to (*o*), is specifically associated with egg and dart decoration and in this form is one of the most distinctive and widespread Classical ornamental details. It is prominent in cornices, architraves, Ionic, and Composite capitals and, in its simpler form (*o*) or even as a quarter-circle, can be used as the echinus of a Roman Doric or a Tuscan capital.

The Greek Doric echinus, (*p*) to (*r*), is shown in the sequence of its evolution from early temples (*p*), through the classic Athenian form (*q*), to the late Hellenistic type (*r*). The name echinus, meaning sea-urchin, refers exclusively to Doric and Ionic column capitals, although a similar profile sometimes appears as a very small scale variant of ovolo in some Greek entablatures and as the uppermost moulding on early Doric cornices.

The cavetto moulding, (*s*) to (*u*), is a useful and often undecorated profile which can be a principal moulding on its own or can perform the function of easing the junction between wider and narrower features. The scotia moulding can be used either way up, (*v*) to (*x*), but keeps the longer radius below in its most common position separating, either singly or in pairs, the torus mouldings of some Doric and all Ionic, Corinthian, and Composite bases.

All these examples have been set out with precision and selected to show how changes in geometry alter the fall of light on the different shapes. With a sure hand and experience, more subtle variations can be drawn without the aid of compasses.

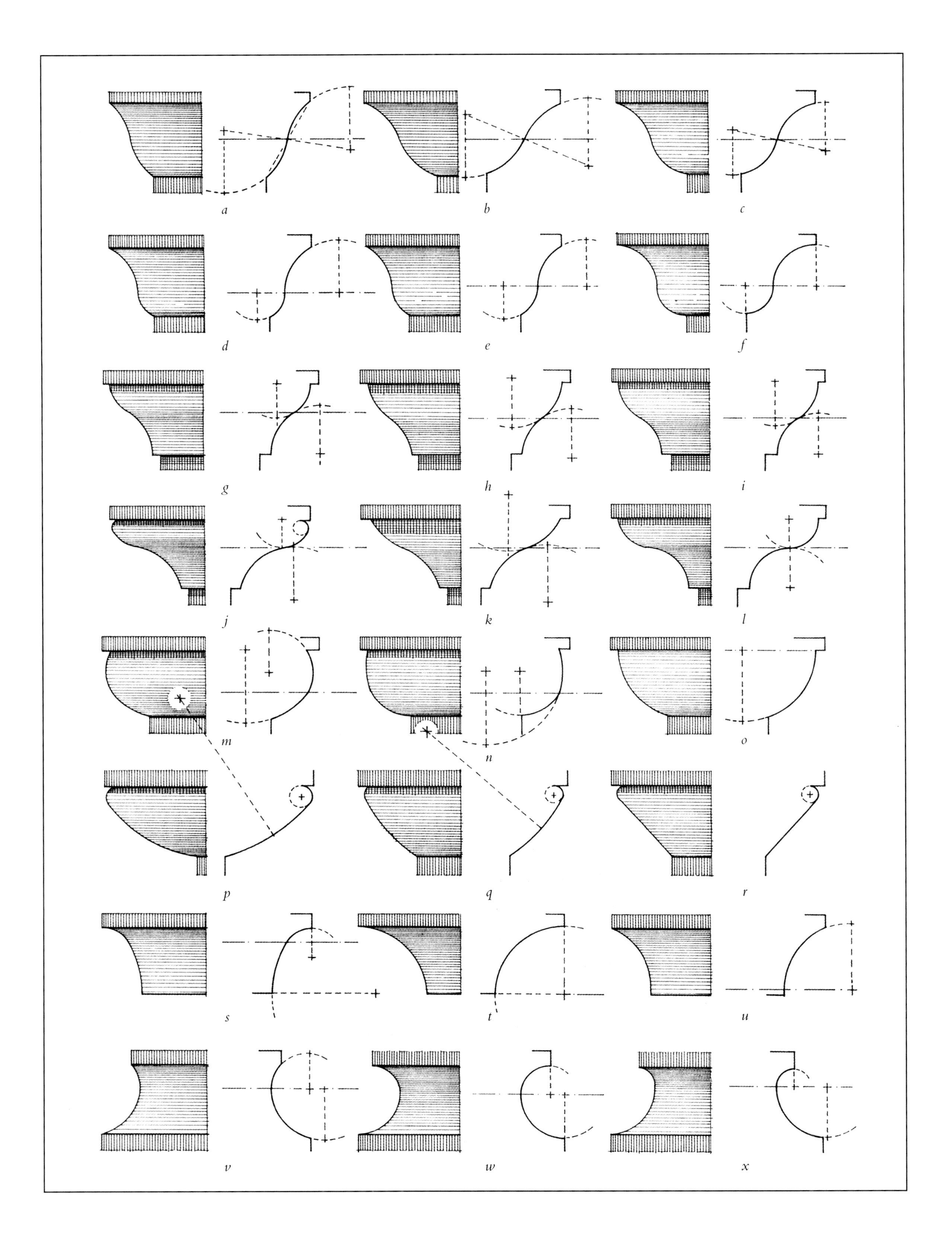
a
b
c
d
e
f
g
h
i
j
k
l
m
n
o
p
q
r
s
t
u
v
w
x

卵箭饰与舌箭饰

两种装饰线脚与两种装饰形制有着特殊的关联。凸圆形装饰线脚与卵箭饰纹案密切相关，装饰纹案本身就称为凸圆形线脚装饰。正波纹装饰线脚通常装饰有叶子，称为舌箭饰或者叶舌饰。很有可能这两种装饰线脚都是专门发展出来适用于这种类型的装饰，或者绘画或者雕刻。当然这两种装饰纹案与其各自的装饰线脚在过去几个世纪中有着贯穿始终的关联，使其具备了独立的特征并与古典设计密切相关。

卵箭饰是古典设计中最为广泛应用的一种纹案。其个性鲜明，正如其名字所表达的，似一排悬垂的卵形。早期的例子来自公元前7世纪原始爱奥尼亚柱头装饰（*a*），然而，这一装饰尚未显示出与鸡卵有明显的关联。但当其开始出现在公元前5世纪爱奥尼亚建筑（*b*）中时，便开始出现在随后几个世纪中所展现出的特征。

可以找到许多变体，但是，正如取材于意大利的希腊殖民地的实例（*c*）所示，这些形状更像卵，因此强化了其关联性。罗马人在采用这一设计的时候通常很少进行改变（*d*），但在希腊实例中更像叶子茎部的箭形，（在罗马时代）成为箭头或飞镖形，现代名称由此而来。文艺复兴时期，箭饰通常得以强化，由于卵形中间较大的间距（*e*）。同样宽敞的间距可以融合为一种简单的形制（*f*）适用于小型装饰线脚，所有的细部设计都毫无偏差地包含其中。

反波纹装饰线脚中的卵箭装饰纹案（*h*）可见于公元前5世纪的建筑，同时还有鸟喙凸雕装饰线脚（*g*），这种装饰线脚逐渐不再使用，直到18世纪后期才得以复兴。对鸟喙凸雕装饰线脚彩绘的研究发现，它被装饰成一种风格独特的下垂叶饰。卵箭饰中的尖叶饰纹案有着同样的设计基础，不过叶子纹案转向线脚底部，并且削减了凸雕。罗马时期，沿用希腊时代的形制（*h*），时常加以改良。在一些实例中，直接对原来翻转叶子的概念（*i*）加以细化，而有时，尤其在大尺度上，原来设计的形状可以仅仅是雕塑装饰的背景（*j*）。罗马实例在文艺复兴时期（*k*）得以延续，当应用于小型装饰线脚（*l*）时，单个叶片变宽，为了适应足够的细部设计使得设计具备可识别性。

卵箭饰与舌箭饰在转角处都采用特殊叶片纹案或同样形状的无装饰线脚。

EGG AND DART AND TONGUE AND DART

Two mouldings are particularly associated with two forms of decoration. The ovolo moulding is so closely associated with egg and dart pattern that the decoration itself is also called ovolo. Cyma recta moulding is usually decorated with a form of leaf design called tongue and dart or leaf and tongue. It is quite possible that both of these mouldings developed specially to accommodate these types of decoration, either in painted or carved form. Certainly the consistent association of these two forms of embellishment with their respective mouldings over the centuries has given them a linked identity and a particular association with Classical design.

Egg and dart decoration is the most widespread decorative form in Classical design. It is most clearly identified, as its name suggests, by its row of hanging egg shapes. Early examples of the decoration on primitive Ionic capitals of the seventh century BC (*a*), however, do not have such an obvious association with the form of eggs. But by the time it appears on Ionic buildings in the fifth century BC (*b*), it has developed the characteristics that it will display in the succeeding centuries.

Many variations are to be found, but, as example (*c*) from Greek colonies in Italy demonstrates, these often display more of the egg shape and so serve to reinforce the association. The Romans adopted the design with unusually little alteration (*d*), but the dart, which in Greek examples was more like the stem of a leaf, takes on the arrowhead or dart shape of its modern name. In the Renaissance the dart was often further emphasised by a wider spacing between the eggs (*e*). A similar wider spacing may be incorporated into a simplified form (*f*) which is suitable for smaller mouldings where the inclusion of all the details is impractical.

Tongue and dart decoration on cyma reversa mouldings (*h*) appears on fifth-century BC buildings together with an undercut bird's beak moulding (*g*) which fell out of use until revived in the late eighteenth century. Investigation of traces of paint on the bird's beak mouldings indicate that it was decorated as a stylised drooping leaf. The pointed-leaf decoration of tongue and dart has the same fundamental design except that the leaf has turned down to the bottom of the moulding and elaborated the undercut. The Romans, while adopting the Greek form (*h*), at times elaborated the design. In some cases the elaboration derived directly from the original concept of turned-down leaves (*i*) while, particularly at a large scale, the shape of the original design could be taken as little more than background for sculptural enrichment (*j*). The Roman forms were adopted in the Renaissance (*k*). When used for very small mouldings (*l*), the individual leaves can be made very much broader in order to accommodate sufficient detail to make the design recognisable.

Both egg and dart and tongue and dart decorations are taken around corners by the application of a special leaf or an undecorated moulding of a similar shape.

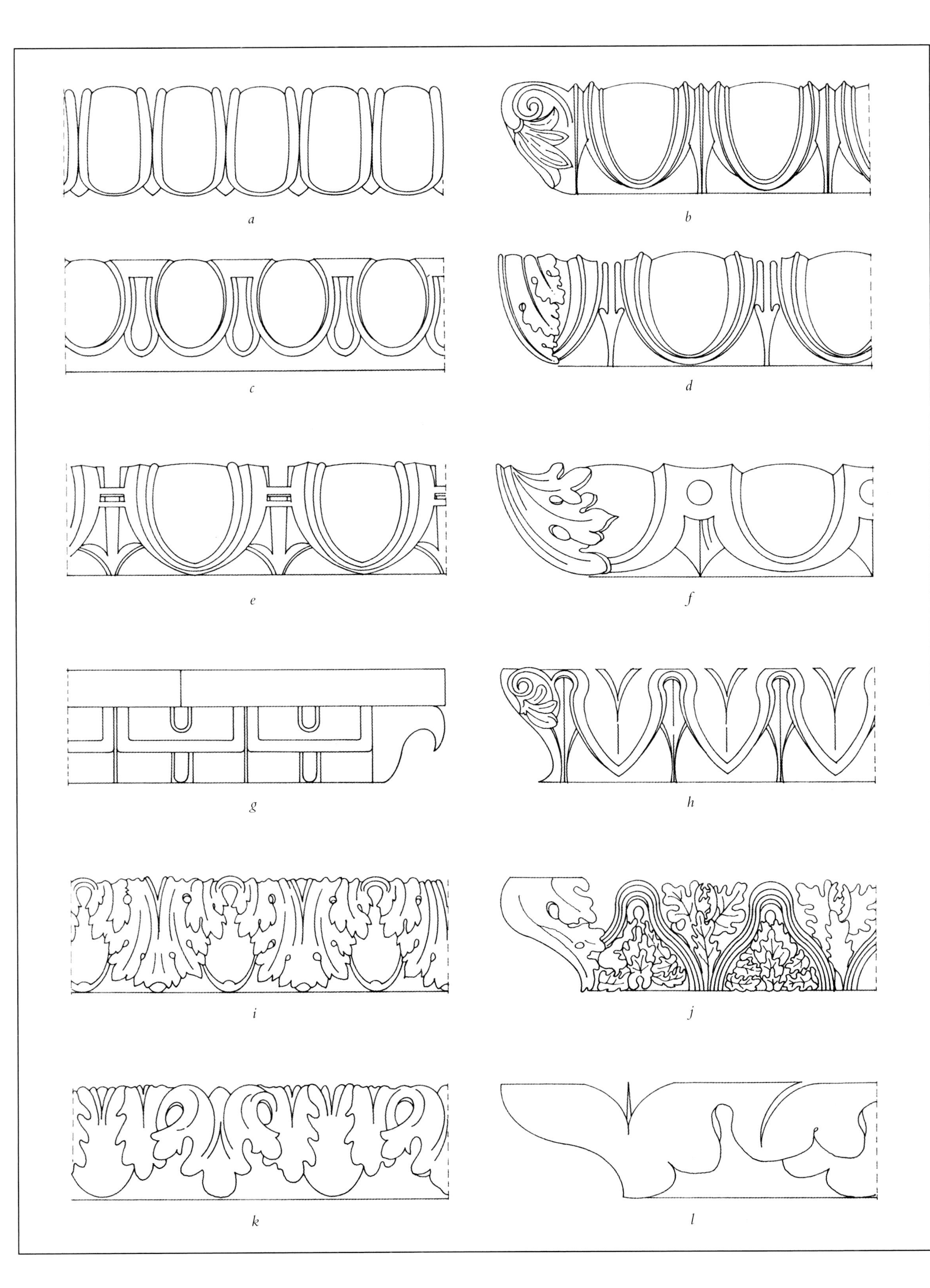

a
b
c
d
e
f
g
h
i
j
k
l

更加复杂的装饰线脚

当凸圆形装饰线脚和反波纹装饰线脚有了相对稳定的装饰语汇，其他复杂的装饰线脚也就具备了同样的特征。这些线脚上增加的装饰似乎并未遵循某种可识别的模式。通常情况下，古典建筑所有的细部设计、所有的轮廓都精细地加以丰富的雕刻、彩绘或者鎏金。只有凹弧边饰除外，这种线脚很少带有其他装饰。

从文艺复兴时期开始，关于古典建筑的研究就是针对废墟或者历经变革保留下来的建筑物。在这些建筑物上仅存的装饰就是雕刻，通常是作为彩绘装饰的基底或者强化手法。然而经历了几个世纪的变迁，原有建筑上的彩绘大多都消失了，古典建筑是无色彩的观点开始滋长。在文艺复兴早期，彩绘建筑十分常见。但19世纪末期，在对洛可可富丽堂皇的装饰进行严肃抵制的基础上，建立起了希腊复兴主义。而研究揭示出早期希腊建筑是色彩明艳的结果令人感到惊讶。罗马建筑同样依赖彩绘进行装饰，在此基础上还增加了金色的叶子和色彩斑斓的天然大理石。

当线脚的演变不再与单个装饰形态相关，雕刻中应用的装饰便与彩绘装饰有相似之处。早期希腊神庙中卵形花边上的舌箭饰（*i*）是彩绘的，但是舌形装饰、飞镖饰与卵箭饰通常是雕刻于卵形花边之上的。考虑何种形态最适合某种装饰线脚，因而限制了装饰范围。能够见到各种流行的抽象古典叶片纹案设计并不奇怪，例如叶丛状装饰绘制在希腊卵形花边上（*j*），雕刻在19世纪卵形花边上（*k*），并且雕刻于两个正波纹装饰线脚（*e*和*g*）之间。同样的，僵直的叶片可见于另一种19世纪卵形花边（*l*），以及帕拉迪奥于16世纪设计的正波纹装饰线脚（*d*）上。

其他的实例展现出装饰可能覆盖的范围。凹弧饰中的圆花饰（*a*）仅仅是针对相对简单的装饰表面，众多设计变体中的一种。在正波纹装饰线脚上，莨苕叶设计（*f*）以及重复的海豚纹案（*h*）源自罗马建筑，这两个实例只是众多罗马变体设计手法中的两种，垂直沟槽设计常见于法国18世纪纹案书籍（*c*）。最后月桂树叶凸圆形装饰线脚取材于罗马后期建筑物（*b*），展现出保留了最原始设计的卵箭饰线脚装饰，也可以采用其他变换的手法。

FURTHER COMPLEX MOULDINGS DECORATED

While ovolo and cyma reversa mouldings have a relatively consistent decorative vocabulary the same cannot be said for the other complex mouldings. The addition of decoration to these mouldings seems to have followed no discernible pattern. In common with all details of Classical buildings, however, all these profiles were often elaborately enriched with carving, painting, or gilding. The only exception is the scotia moulding which is rarely decorated.

Since the Renaissance, the study of the buildings of antiquity has been the study of ruins or much-altered survivals. The only remaining enrichments on these buildings are carvings, which were often originally intended as a foundation and enhancement for painted decoration. The passage of many centuries, however, removed much of the original paint and the idea grew up that the buildings of antiquity were designed without colour. In the early Renaissance, the painting of buildings was common, but the Greek Revival of the late nineteenth century was founded on a desire for sober refinement to counteract the decorative exuberance of the Rococo. It came as a great shock when research revealed that early Greek architecture was a riot of bright colour. Roman architecture, too, relied on painted colour and to this the Romans added gold leaf and the natural colours of bright marbles.

When the evolution of a moulding is not related to a single form of embellishment, the decoration that is applied in carving is likely to be the same as might be found in painting. The tongue and dart on the echinus of an early Greek temple (*i*) is painted, but tongue and dart and, more often, egg and dart are found carved on echinus mouldings. The range of embellishment is only limited by what is suitable for the shape of the moulding. It is, therefore, not surprising to see variants of the popular Classical abstracted leaf designs, such as the anthemion, painted on a Greek echinus (*j*), carved on a nineteenth-century echinus (*k*), and carved on two cyma recta mouldings, (*e*) and (*g*). Equally, versions of the stiff leaf can be found on another nineteenth-century echinus (*l*) and on a cyma recta moulding (*d*) by Palladio in the sixteenth century.

Other examples are shown to illustrate the possible range of embellishment. The rosette pattern on a cavetto (*a*) is only one of many alternatives for this relatively simple decorative surface. On a cyma recta, the acanthus design (*f*) and repeating dolphin design (*h*) from Roman buildings are only two examples from a wide variety of Roman treatments, while the vertical channels from an eighteenth-century French pattern-book (*c*) are relatively unusual. Finally, the bay-leaf ovolo decoration from a late Roman building (*b*) demonstrates that even the moulding normally reserved for egg and dart embellishment can receive a varied treatment.

a
b
c
d
e
f
g
h
i
j
k
l

6. 增加与削减

柱式之上的柱式

古典建筑物不仅局限于在固定形制与大小建筑物中使用柱式。从古代开始，柱式就被设计成适用于建筑物，这些建筑与作为其基础的简单寺庙并不相似。长达几个世纪，解决建造不同建筑物的设计问题而演变出一些惯例，已经成为古典传统的一部分。有大量这样的手法，一些特别应用于某一特定历史时期并且被其他时代排斥，但由于他们对传统的贡献，还是得以成功再现。这些创新解决了许多设计问题，但是永远不能达到完美的水平。

为了使高度增加到一层之上，需将柱子一层层叠起来。在早期希腊神庙内部，例如建于公元前438年的雅典帕提农神庙（*a*），一排小型多立克柱子直接叠加在下面大柱子的额枋之上。下面柱子逐渐向上变窄的趋势延续至上层柱子，因此上层柱子明显变小。

后期两层建筑，例如建于公元前150年的雅典阿塔罗斯柱廊（*b*），上层使用了更加纤细的爱奥尼亚柱式。设计原则是将比较纤细的柱子置于比较粗壮的柱子之上，自然削减重量，符合结构需要。80年修建罗马圆形大剧场时，将这一手法发挥到极致，多立克、爱奥尼亚、科林斯和混合柱式按照升序依次排列。尽管布局在古代或者文艺复兴早期并没有一定之规，但罗马圆形大剧场（*d*）的保存以及文艺复兴时期对柱式的分类，带来了这种柱式垂直顺序的普遍应用。

罗马作家维特鲁威建议，上层柱子为下层的3/4大小。他的建议似乎适用于两层的建筑中，能够达到令人满意的效果，但是如图（*c*）所示，如果继续向上延伸就变得有些极端了。通常采用文艺复兴时期作家们推荐的做法，即上层柱底直径为下层柱顶直径，如图（*e*）所示。此图也包含了所有柱式的基座，尽管也可以采用其他设计。

6. ADDITION AND SUBTRACTION

THE ORDERS IN ORDER

Classical architecture is not restricted by the use of the Orders to buildings of a limited size or form. Since antiquity the Orders have been manipulated to accommodate buildings quite unlike the simple temples that are their foundation. Over the centuries conventions have evolved for solving the design problems created by different buildings and these have become part of the Classical tradition. There are a huge number of these devices. Some are specific to certain historical periods and rejected by others, but where they have been successful they have been repeated as contributions to the development of the tradition. Many design problems have been solved by these inventions, but the list can never be complete.

In order to achieve a height greater than a single storey the Orders can be placed on top of one another. Early examples are in the interiors of Greek temples, such as the Parthenon in Athens (*a*) of 438 BC, where a smaller row of Doric columns was placed directly on the architrave of larger columns below. The taper of the lower columns is continued through to the upper columns so that the upper columns are significantly smaller.

In later two-storey buildings, such as the Stoa of Attalos in Athens (*b*) of 150 BC, the more slender Ionic Order was used for the upper floor. The principle of placing more slender Orders above stouter Orders give a natural reduction in weight which corresponded to structural requirements. It was taken to its fullest extent in the construction of the Colosseum in Rome (*d*) in AD 80, where Doric, Ionic, Corinthian, and Composite are set in an ascending sequence. Although this arrangement was not universal in antiquity or the early Renaissance, the survival of the Colosseum and the categorisation of the Orders in the Renaissance led to a general adoption of this vertical sequence of the Orders.

The Roman author Vitruvius recommends that upper columns should be three quarters of the size of those below. His recommendation seems to apply to two-storey buildings where it can be quite satisfactory but, as illustrated in (*c*), if continued upwards it can become quite extreme. It is more usual to follow the recommendation of Renaissance authors and make the upper diameter of the lower Order equal the lower diameter of the upper Order, as show in (*e*). This illustration includes pedestals for all the Orders although other arrangements can be used.

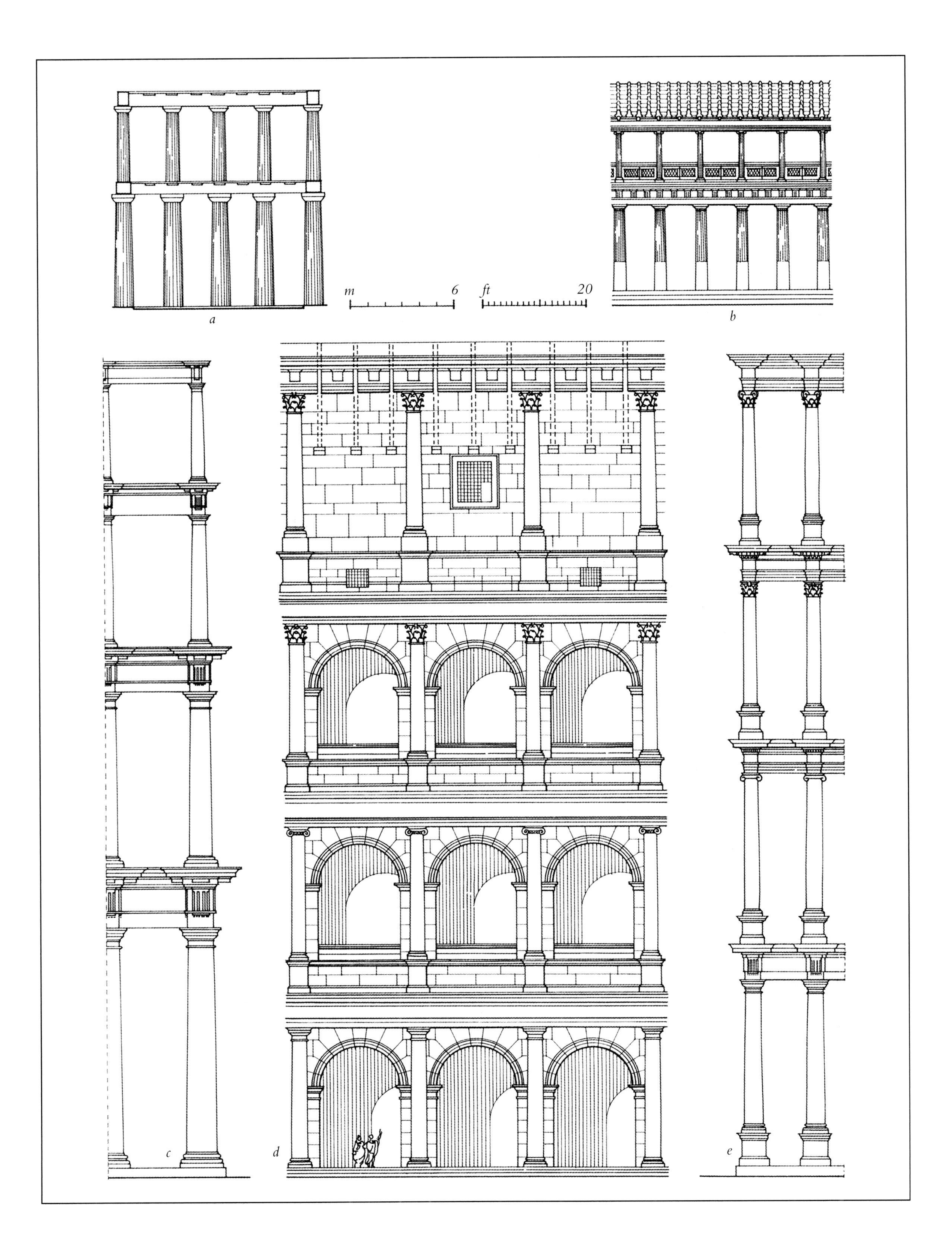

a
m 6
ft 20
b
c
d
e

柱式的巧妙应用

应用简单柱式的基本形态有时不能达到令人满意的建筑物视觉和功能性需求。

檐部的高度能够通过增加小垫块（*a*），或者通过在此垫块上增加细部设计（*b*）来进一步提升，细部设计与假想的上层柱子的柱础成比例。基座能够有更加直观的表达，如图（*c*）所示，使得上面的建筑更富有意趣。

实例（*c*）有足够的高度，能够成为独立一层，因此称为阁楼。罗伯特·亚当建于1767年的卢顿花园府邸（*d*）设计了带有窗户的阁楼。帕拉迪奥于1565年设计的瓦尔马拉那宫（*e*）也是同样的，但是阁楼层基座上的雕塑，创造出更加高大的形象。将栏杆置于阁楼上层，例如乔治·丹斯1739年在伦敦设计老翁府邸（*f*），更能够突出高度。

法国奥朗日的罗马凯旋门（*h*）的中心桥跨有一个较低的阁楼区块，为一个实用的山花充当背景。上面安放雕像基座。建于1841年，牛津的阿什莫尔博物馆（*g*），设计师为查尔斯·罗伯特·科克雷尔，中央柱子居于大型檐口之下，与整个建筑的高度成比例，阁楼位于两个檐口之间。在伦敦威斯敏斯特保险公司办公大楼（*i*）设计中，查尔斯·罗伯特·科克雷尔将阁楼层及其基座改为矮粗的上层柱式与三角墙。这些柱式有着短胖的柱身和特殊柱头。

在底层，通过基座来增加高度。帕拉迪奥的迪亚沃罗府邸（*j*）将柱子安放在一个高基座上，并在其间设计了窗户。菲利贝·德·洛梅于1547年设计的阿内府邸（*k*）中，上面两层门廊带有厚重的基座来控制向上柱式的高度。帕拉迪奥设计的波尔图费斯塔宫（*l*），展现了普通设计手法，将整个底层变成粗琢基座。

檐部的高度可能带来窗户安放位置的问题。实例（*j*）将底层窗户设计在雕带之间，但是法国森斯建于1535年的主教宫（*m*）的设计师，采用了前卫但并非独一无二的手法，增加底层雕带的高度，削减上面的檐部设计。独特的巴洛克细部设计可见于1704年英国的康德府（*n*），将柱子之上的雕带和额枋限定于垂直区块，即柱顶石中，以便额枋能够穿过窗户。像霍克斯莫尔这样的设计师，在1712年修建牛津大学拉德克里夫图书馆（*o*）时，通过简化并且修饰柱式的手法完全避免了此类问题，其中保留下来的细部设计并未显现出古典元素传统地位。

MANIPULATING THE ORDERS

The visual or functional needs of a building may not be satisfactorily answered by a simple application of the Orders in their basic form.

The height of an entablature can be raised by adding a small block (*a*), or further raised by adding details to this block (*b*) that correspond to the proportions of a column pedestal of an imaginary upper Order. The pedestals can be expressed more literally, as in (*c*), and provide more interest at the top of the building.

Example (*c*) has sufficient height to become a separate storey, known as an attic storey. Robert Adam's Luton Park House (*d*) of 1767 shows an attic story with windows. Palladio's design for Palazzo Valmarana (*e*) of 1565 is similar, but statues on the attic pedestals create an appearance of greater height. Placing a balustrade on top of the attic storey, as on George Dance the elder's Mansion House in London (*f*) of 1739, gives a still greater impression of height.

The central bay of the Roman triumphal arch (*h*) in Orange, France has a lower attic block that acts as backdrop to an applied pediment. The pedestal for the statue sits above. In the Ashmolean Museum in Oxford (*g*) in 1841, by C.R. Cockerell, the principal Order sits below a large cornice, proportioned according to the total height of the building, and the attic sits between the two cornices. On the Westminster Insurance Office in London (*i*) C.R. Cockerell turned the attic storey and its pedestals into a squat upper Order with its own pediment. These attic Orders can have their own short columns and special capitals.

At a low level, height can be gained by pedestals. On the Casa del Diavolo (*j*) Palladio raised the columns on high pedestals and fitted windows between them. The upper two storeys of the doorway of Philibert de l'Orme's Château d'Anet (*k*) of 1547 have very heavy pedestal blocks to control the heights of the ascending Orders. Palladio's Palazzo da Porto Festa (*l*) illustrates the common device of turning the whole lower storey into a rusticated pedestal.

The height of the entablature can cause problems with the location of the windows. Example (*j*) has low windows in the frieze but the designer of the Archbishop's Palace in Sens, France (*m*) of 1535 has taken the radical, but not unique, step of increasing the height of the lower frieze and reducing the height of the upper entablature. A distinctively Baroque detail at Cound Hall, England (*n*) of 1704 limits the frieze and architrave to a vertical block, called a dosseret, above the column in order to allow windows to pass by the entablature. Architects such as Hawksmoor in his unexecuted design of 1712 for the Radcliffe Camera in Oxford (*o*) avoided the problem altogether by so simplifying and modifying the Order that the remaining details gave little clue as to conventional position of the Classical elements.

a
b
c
d
e
f
g
h
i
j
k
l
m
n
o

高柱式

层与层之间的高度适当，一系列的垂直柱式中单个柱子很小，总体看来使得建筑不那么宏伟。为了使两层或更高的建筑物带有与古代神庙同样的效果，就要使用完整的古典柱式，柱子达到标准高度。这就称为高柱式，或大型柱式。

尽管高柱式从16世纪才开始使用，罗马建筑通常混合使用大型柱子和特征并不突出的小柱子。建于216年的罗马卡拉卡拉浴场（*a*），内部宽阔的水泥屋顶圆拱由高达11.5米的科林斯式花岗石柱支撑，中间带有一排小型科林斯柱式。

一个早期影响巨大的高柱式应用实例，可见于罗马议会大厦建筑群（*b*）中，由米开朗琪罗于1546年设计。图中展示了三栋建筑物之一的两个开间。正立面为统一整体，由位于基座上的巨大科林斯式壁柱连接。底层高柱式间嵌有辅助的爱奥尼亚柱式，柱高为一层层高。

罗马拉特拉诺圣乔凡尼教堂（*c*）由亚历山德罗·加利莱（1691—1737年）于1733年修建，其正立面以高达20米的混合柱式为基础。在高柱式之内有两层高科林斯柱式。底层的辅柱比上层的高，并支撑着檐部。上层的柱式支撑着圆拱，避免大小柱式檐部之间产生冲突。

帕拉迪奥1565年在威尼斯修建的圣乔吉奥·马吉奥莱教堂（*d*），实际上是两个神庙式的正立面交织在一起。巨大的混合式圆形壁柱安置在基座之上，支撑巨大的中央山花，小型科林斯式方形壁柱支撑底层山花，分割后成为边廊的底层天花板。科林斯柱式作为1/4柱再次出现，成为中央门的框架，混合柱式的基座成为科林斯柱式之间壁龛的底座。

伦敦修建于1866年的共济会大厦，其中一部分如图所示（*e*），弗雷德里克·佩皮斯·科克雷尔（1833—1878年）避免了高柱式和辅柱在混合使用时所造成的檐部潜在的冲突。高柱式在粗琢的压抑之下，带有一个粗壮的科林斯式柱头，柱头的高度与科林斯辅柱的柱头高度一样，辅柱与粗琢高底座等高。

格拉斯哥成排的住宅（*f*）建于1859年，亚历山大“希腊人”·汤姆森（1817—1875年）使用同一檐部结合了高柱式与辅柱，通过使用对于高柱式来说极为纤细的柱子，使得檐部高度对于高柱式与辅柱看起来都很恰当。他也移除了高柱式之间的小型上层柱式，因此避免了檐部的视觉冲突。

GIANT ORDERS

When floor-to-floor heights are relatively modest, a vertical series of Orders can be individually small and collectively make the whole building appear insignificant. To give the buildings of two or more storeys the same impact as the large temples of antiquity, a complete Classical Order, with columns the full height of the building, was used. This is called a colossal, or giant Order.

Although giant Orders did not come into use until the sixteenth century. Roman architecture often combined columns of great size with smaller columns for lesser features. In the interior of the Baths of Caracalla in Rome (*a*) of AD 216 the wide concrete roof-vault is carried on granite Corinthian columns 11.5 metres high with screens of a smaller Corinthian Order between them.

One of the early and influential uses of a giant Order was in the group of buildings on the Capital in Rome (*b*), designed by Michelangelo in 1546. Two bays of one of the three buildings are illustrated. The façades are unified by pilasters of a giant Corinthian Order on pedestals. Inset between the giant Orders on the ground floor is a subsidiary Ionic Order with full columns one storey high.

The façade of S. Giovanni in Laterano, Rome (*c*), designed by Alessandro Galilei (1691-1737) in 1733, is based on a giant Composite Order with columns 20 metres high. Inside the giant Order are two one-storey tiers of Corinthian columns. The ground-floor subsidiary Order is higher than that on the upper floor and carries an entablature. On the upper floor the columns carry an arch so avoiding any conflict between the entablatures of large and small Orders.

Palladio's church of S. Giorgio Maggiore in Venice (*d*) of 1565 is, in effect, two temple fronts interwoven with one another. A giant Composite Order of round pilasters on pedestals supports the larger central pediment and a smaller Corinthian Order of square pilasters supports a lower pediment, which is split to provide the lower roofs for the side-aisles. The Corinthian Order reappears as quarter-columns framing the central door and the pedestals of the Composite Order become the bases for small niches between the Corinthian columns.

With the Freemason's Hall, in London, of 1866, part of which is illustrated (*e*), Frederick Pepys Cockerell (1833-1878) avoids a potential conflict between the entablatures of the giant and subsidiary Orders by combining them. The giant Order is suppressed with rustication and given a squat Corinthian capital the same height as the capital of the Corinthian subsidiary Order, which is brought up to the same level on a high rusticated plinth.

On a row of houses in Glasgow (*f*), built in 1859, Alexander "Greek" Thomson (1817-1875) combined a giant and subsidiary Order with the same entablature but, by using exceptionally slender columns for the giant Order, made the entablature seem to be an appropriate height for both. He also removed the small upper columns from between the giant Order columns, so avoiding any visual conflict at entablature level.

a
b
c
d
e
f

比例变化

人们可以采用不同的方式来应用柱式以改变建筑的外观。为了展示这些设计手法，针对同一建筑物（*a*）采用一系列不同的柱式设计。实例有着相同的比例，但在实际应用中，有可能增加设计来创造更加优雅的效果。

由于单个柱子比较纤细，简单的应用为升序排列的多立克、爱奥尼亚和科林斯柱式，高度依次递减（*b*），创造出精美特征。檐部坚实的水平分割使得建筑物看起来比较低矮。当第一组柱式抬高到底座（*c*）上时，其上高大的窗户看起来十分凸出，整栋建筑也更加气派。将脆弱的底层建筑及装饰设计得更加粗壮，这同样也是更加实用的手法。

实例（*b*）和（*c*）是十分典型的文艺复兴早期和兴盛期的柱式应用。在两种情况中，最好使用更多的柱子来削减柱子之间的距离，并且交叉变换它们的间距。

引入两层高柱式，置于粗琢底座（*d*）之上，能够改变建筑物的特征。使其变得更加有气势，单个柱式也更加凸出。使用单一柱式赋予建筑物以该柱式的特征。将粗琢的一层换成高柱基座（*e*），使建筑物看起来更高，成为与街道等高的建筑物的重要设计元素。

实例（*d*）和（*e*）增加了檐部的高度，刚好包含在这个正立面上层的窗户空间当中。凸出的檐口和檐部的细部设计成为重要元素。这两种高柱式的设计从16世纪沿用至今，帕拉迪奥使其流行起来。在低调的建筑中，檐部的高度会带来问题，因此省去雕带或者额枋的做法并不罕见。

当高柱式延伸至三层（*f*），建筑看起来更高、更富戏剧效果。宽阔的柱身将建筑物强有力地分割，窗户的水平分割不复存在，它们变成垂直的分组。这种典型的巴洛克式高柱式的应用，使得采用完整的檐部成为可能，并且省去了雕带。

这种高柱式粗壮的檐部可结合阁楼使用，以削减建筑物的视觉高度。将檐口置于上层窗户的窗台处，下面的一层建筑朴素无装饰（*g*），就创造出了两层建筑物的视觉效果。通过阁楼更具装饰性的处理（*h*）对其加以强调，但沉重的檐部将立面分割成两部分。

CHANGING SCALE

The Orders can be applied in a number of different ways to alter the appearance of a building. To illustrate this, a series of alternative applications of the Orders to the same underlying three-storey building (*a*) are shown together. The examples all have the same proportions, but in practice it would be possible to manipulate the design to produce more refined results.

The simple application of ascending Doric, Ionic, and Corinthian Orders of diminishing height (*b*) produces a delicate character due to the small size of the individual Orders. The strong horizontal divisions of the entablatures give the building a low appearance. When the first Order is raised on a plinth (*c*) the tall windows above become more important and the whole building tends to look more imposing. This can also be a more practical arrangement as the decoration on the vulnerable ground floor is more robust.

Examples (*b*) and (*c*) are typical early and High Renaissance applications of the Orders. In both cases the distance between the columns might be better reduced by introducing more columns and spacing them at alternate widths.

The introduction of a two-storey giant Order on the rusticated plinth (*d*) transforms the character of the building. It becomes more impressive and the individual columns are given greater importance. The use of a single Order allows the whole building to assume the character of that Order. The replacement of the rusticated lower floor with tall column pedestals (*e*) gives the feeling of greater height, and the pedestals become important design elements at street level.

The increased height of the entablature in examples (*d*) and (*e*) can just be contained in the space over the top window on this façade. The projection of the cornice and the detail of the entablature now become significant. Both of these giant Order arrangements were used from the sixteenth century onwards and popularised by Palladio. In modest buildings the height of the entablature caused problems and it is not uncommon to find the frieze or the architrave omitted.

When giant Order columns extend to three storeys (*f*) the building looks both taller and more dramatic. The wide columns divide the building so forcefully that the horizontal arrangement of the windows is lost and they are seen as vertical groups. This typical Baroque use of a giant Order makes it impossible to accommodate a full entablature so the frieze has been omitted.

The strong entablature of a giant Order can be used with an attic storey to reduce the apparent height of a building. By taking the cornice to the sill of the upper windows and leaving the upper floor plain (*g*) the impression of a two-storey building is created. A more decorative treatment of the attic floor (*h*) will give it more emphasis, but the heavy entablature will divide the façade into two parts.

a
b
c
d
e
f
g
h

7. 隐藏柱式

渐进式省略：柱式

柱式的全部装饰并不总是适用。装饰的程度能在保留柱式原有的特征与比例的基础上拓展或缩小，可以通过选择使用装饰元素来描绘潜在的全部装饰。以某种方式掌控这些柱式的能力是几个世纪以来人们对古典传统设计熟练掌握的产物，是古典设计的基本组成部分。

罗马多立克柱式，如图（*a*）所示包含了传统上与此柱式相关的所有细部设计。当以较低成本应用于较小柱式上时，或者需要创造出比较大胆的效果，许多这些细部设计就不太适合。

图例（*b*）展示的是柱上无沟槽设计，大部分的三陇板也被省去。可将光滑柱身设计应用于任何一种柱式上，但多立克柱式上三陇板的渐进式省略具有其独到的特征。保留了柱子上单一的三陇板以及额枋上独特的相关细部设计，作为雕带水平配比的标准。

在下一个阶段（*c*）中设计方形柱子、无三陇板或带状雕刻上的柱间壁。柱子保留支柱收曲分线并且带有完整的檐口。檐口上精美的细部设计能够在檐口本身得以简化之前逐渐削减，正如图例（*d*）所示。这一实例中省略了柱子上的柱间壁。简化的檐口与雕带模糊了多立克与托斯卡纳柱式之间的差别。

进一步削减细部设计，如实例（*e*）和（*f*）所示，创造出的柱式非常简单地集合了各个部分，保留了相应比例，仍然是一个完整柱式。在简单的柱式（*f*）表达中，柱式的细部装饰都消失无踪，檐口成为已不复存在的细部设计的最后见证。

这里仅仅展现了一个柱式细部设计的渐进式削减。有可能完整保留一些元素而省略所有其他的装饰。在实例（*g*）中只有雕带和额枋被省略为平面。檐口和方形柱础（*h*）已经简化，但是保留了光滑的三陇板。完整的檐口（*i*）置于最朴素的柱式之上，只有少量组合设计，但仍存在更多的可能。

7. HIDDEN ORDERS

PROGRESSIVE OMISSIONS: THE ORDER

The full range of decoration associated with the Orders is not always appropriate. The extent of decoration can be increased or decreased while retaining the character and proportion of the Orders; the potential presence of a full decorative scheme can be suggested by introducing selected decorated elements. The ability to manipulate the Orders in this way is the product of centuries of familiarity with the Classical tradition and is an essential part of Classical design.

The Roman Doric Order illustrated (*a*) includes the complete range of details traditionally associated with the Order. When this is to be constructed in a small size, at a low cost or to create a bold impression, many of these details can be inappropriate.

Illustration (*b*) shows the fluting on the column removed and most of the triglyphs omitted. It is possible to use a smooth column shaft on any of the Orders, but the progressive omission of the triglyphs on the Doric Order has some unique characteristics. The retention of the single triglyphs over the columns and the rather peculiar survival of the associated details on the architrave indicate the basis for the horizontal proportioning of the frieze.

The next stage (*c*) has square columns and no triglyphs or metopes in the frieze. The column retains its entasis and there is a full cornice. The fine detailing on the cornice can also be progressively reduced before the cornice itself is simplified as in (*d*). This example also shows the omission of the entasis on the columns. The simplified cornice and frieze blur the distinction between Doric and Tuscan.

Reducing the details further, as in examples (*e*) and (*f*), produces a very simple assembly of parts while retaining the proportions and a suggestion of the full Order. In the simplest expression of the Order (*f*) there is no trace of the column details and the cornice acts as the last reminder of the details that have been lost.

This is only an illustrative progression of reducing an Order. It is equally possible to retain some elements in full detail while reducing others. In example (*g*) only the frieze and architrave have been reduced to a plain face. The cornice and the square column base in (*h*) have been simplified, but bare versions of triglyphs remain. A full cornice in (*i*) sits over the most unadorned expression of the proportions of the Order. These are just a few combinations, many more are possible.

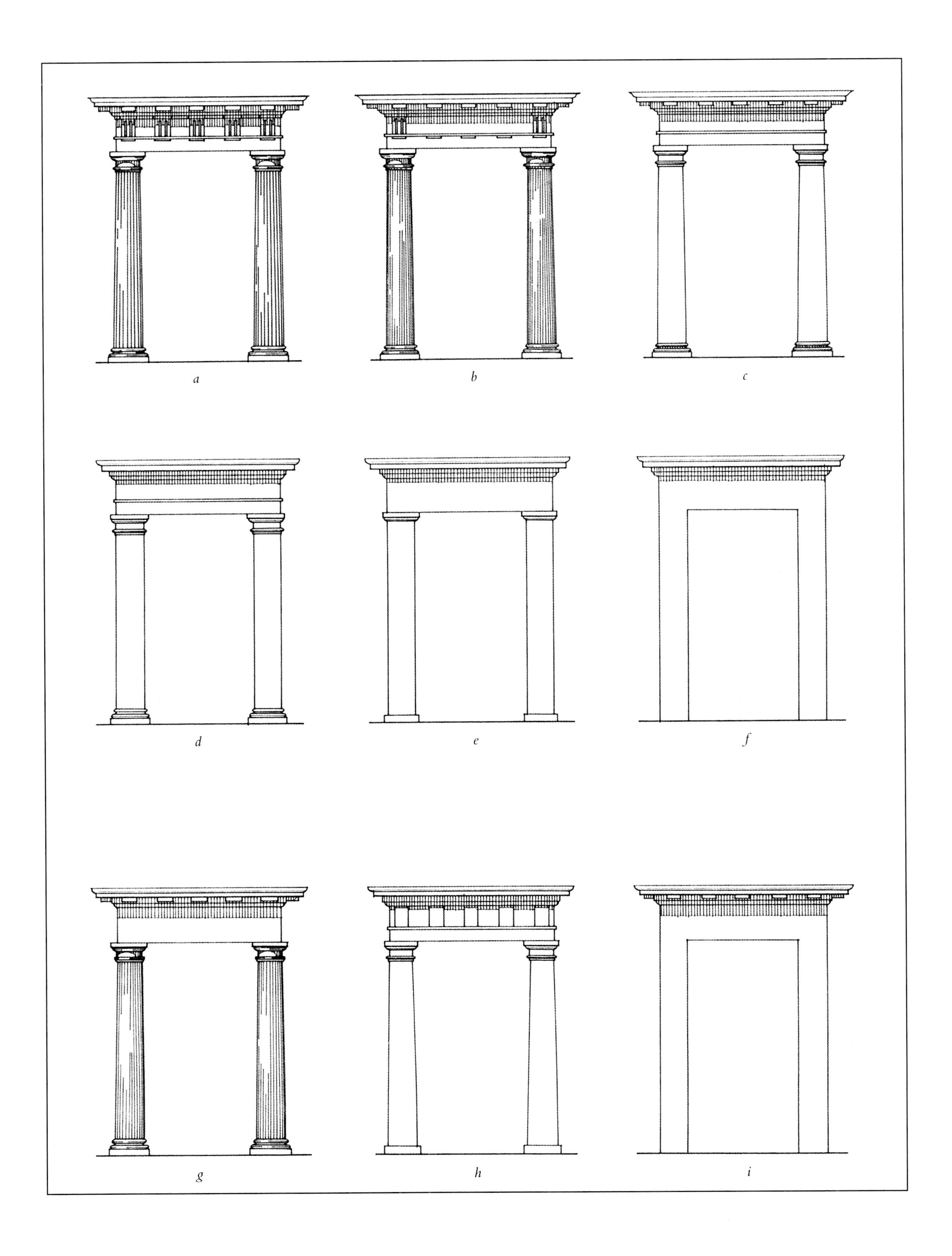
a
b
c
d
e
f
g
h
i

渐进式省略：建筑物

对古典外立面细部设计的削减，创造出一些极简建筑物，其古典起源难以追溯。许多18世纪末英国住宅的细部设计仅限于其大门，用以展示其设计中包含的古典设计原理。这些建筑物得到赞赏是由于它们依然沿用古典柱式比例作为其设计基础。古典建筑中的柱式即使在不能够直接看到的情况下也总是存在的。

这里举出了渐进式省略古典正立面的实例，是4个稍作改动的联排住宅，位于伦敦贝德福德广场，由威廉·斯科特（活跃于18世纪70年代）和罗伯特·格鲁（活跃于18世纪70年代）于1776年设计。

第一个实例（*a*）是一座完整的神庙式砖砌正立面。粗琢并涂以灰泥的底座包含了正门并且支撑着4个爱奥尼亚壁柱，上面省略了带有檐部的额枋设计。一个完整的山花掩盖了部分屋顶。在第二个实例中（*b*），柱式与山花消失不见了。设计更加简朴，保留了第一个实例中的粗琢地面，并且削减了檐部设计。栏杆取代了山花，坚实的柱基取代了省略了的柱子位置。在进一步省略的实例（*c*）中粗琢手法消失了，显露出了下面的砖石。拱门开口处仍采用粗琢工艺，但十分简朴；水平雕带标志着柱子的柱础线，确定了柱子的比例以及保留下来的檐部上方檐口的大小。最后的实例（*d*）是贝德福德广场建筑中细部设计最少的，但却代表了该时代进一步削减所采用的普遍建筑手法所达到的水平。完全采用砖砌面，两条水平雕带代表柱础和檐口，仅有两个1/4柱暗示这一设计的古典传统。

这些实例源自18世纪伦敦，但其设计原理可应用于任何时代任何地点的任何古典建筑上。这种可能性意味着完整的古典正立面可能依附于最朴实的正立面。有可能在基本的建筑物上采用与装饰精美的古典手法所特有的相同比例设计。

装饰与建设成本息息相关，现有的建筑能够通过增加其缺失的隐藏柱式而得以改进。成排的房屋，正如我们所见的贝德福德广场的建筑，可以结合起来形成统一体，集合成更大的建筑；大部分的精美版本（*a*）置于中间，稍朴素一些的（*b*）放在末端作为亭子，其间穿插着装饰更少的版本（*c*）。总之，最朴素的建筑也能够从本质上采用与尊贵建筑同样的设计。

PROGRESSIVE OMISSIONS: THE BUILDING

The reduction of detail on Classical exteriors has created some buildings of such simplicity that their Classical origin is not always obvious. Many English houses from the late eighteenth century have little more than the details of the door to reveal the Classical principles that underlie their design. The elegance which has made these buildings so admired is, none the less, derived from the Classical Orders that are the foundation of their proportions. In a Classical building the Orders are always present even if they are not immediately visible.

A progressive reduction of a Classical façade is illustrated here with four slightly modified examples taken from a row of houses in Bedford Square in London, designed by William Scott (*fl.*1770s) and Robert Grews (*fl.*1770s) in 1776.

The first example (*a*) has a complete temple front applied to a brick face. A rusticated base in painted stucco contains the front door and supports four Ionic pilasters below an entablature with the architrave omitted. A full pediment hides part of the roof. In the second example (*b*) the columns have gone and with them the pediment. The design is much more modest but retains the rusticated ground floor of the first example, and the reduced entablature. A balustrade has replaced the pediment and its solid plinths correspond to the positions of the lost columns. In the next stage (*c*) the rustication has gone, to reveal the brickwork below. The arched door-opening is still rusticated, but in a modified way; otherwise a horizontal band marks the line of the base of the columns which would set the proportions of the Order and the size of cornice on the remaining reduced entablature. The final example (*d*) has less detail than any of the buildings in Bedford Square, but represents a level of further reduction that was common at this time. It has a sheer brick face with only two horizontal bands to mark the position of the column bases and cornice. In the simple door, the same height and width as the adjacent windows, only two quarter-columns give a clue to the Classical basis of the design.

These examples are taken from eighteenth-century London, but the principle could apply to many Classical buildings from any period and in any place. The possibility of suggesting that a complete Classical front lies just beneath the barest façade and the ability to proportion the most rudimentary building in the same way as the most elaborate are uniquely fertile aspects of Classical design.

Decoration can be closely matched to construction budgets and existing buildings can be improved by adding missing parts of their hidden Orders. Rows of houses can, as in our example in Bedford Square, be combined to form unified compositions that collectively resemble larger buildings: the most elaborate version (*a*) is placed in the centre, the next level (*b*) as pavilions at the ends and multiples of less decorated examples (*c*) between. Above all, the most humble building can share the underlying design of the most exalted.

a
b
c
d

渐进式省略：内部设计

古典室内细部设计是我们日常生活的一部分。许多人的房间里带有檐口、雕带和额枋设计，却未能意识到这些建筑语汇的起源，这些元素在古典柱式中的作用，以及柱式中这些熟悉元素的成比方法。

罗伯特·亚当1773年设计的伦敦德比府邸可展现这样的内部设计，这些设计原理可用于几乎所有的古典室内装饰。

房间墙面的设定首先以完整的古典柱式（*a*）为基底，在这里使用的是科林斯式。窗台仍位于基座的顶端，从这条线开始，柱子上升至适合檐部比例的高度，檐部上面紧接着屋顶。这种柱式以房间高度为基准，确立天花板上作为檐部一部分的檐口与雕带大小。基座的顶端成为护墙板扶手又称椅子扶手；其轮廓通常是基座上檐口的扁平版本。墙裙，又称壁脚板，是柱础的柱基，其轮廓构成相同。和基座一样，其间的墙面称为护墙板。门上带有独自的檐部，在这种情况下与主檐部大小相同。檐部底层装饰线脚、额枋，环绕大门后与壁脚板相接。窗户带有同样的细部设计，但没有雕带和檐口。

这种布局可以变化。柱子通常起于地面，细部设计、檐口的比例、雕带和壁脚板也发生相应变化。护墙板不是必要的元素，但也经常包含在设计之中，当柱础起于地面时，可充当柱子间坚固的扶手。门窗通常有其独立的古典柱式，有时带有各自的柱子与山花，细部设计会相应变小。在高大的房间里，檐口位于天花板水平线的下方，两者由较大的连续向内翻卷的弧线（又称凹圆线）相连，或者是穹顶能够从一个房间横跨至另一个房间。一个高大的房屋可能有两种柱式，一种叠于另一种之上，或由一个高柱式和一个辅柱构成。

这些细部设计只有在最华丽的房间中全部展现。精简的版本也很常见。实例（*b*）保留了一些显著的特征。有装饰完整的檐口、门上的檐部以及护墙板。实例（*c*）是人们所熟知的简单房间设计，保留了天花板檐口、门窗边的额枋以及墙裙，最后一个实例虽然简单，却与第一个例子比例相同，保留了改进的潜力以及两个内部设计中反映古典原则的装饰。

PROGRESSIVE OMISSIONS: THE INTERIOR

The details of the Classical interior are part of our everyday lives. Many people live in rooms with cornices, friezes, and architraves without realising the origin of these words, the place of these elements in the Classical Orders and the way in which the Orders set the proportions of such familiar features.

The example illustrated is from the interior of Derby House in London, designed by Robert Adam in 1773, but the principles of the scheme could apply to almost any Classical interior.

The walls of the room are set out to accommodate a complete Classical Order (*a*), in this case Corinthian. The window-sills are at the level of the top of the pedestal and from this line the columns rise to fix the proportions of the entablature, which sits immediately below the ceiling. This Order, based on the height of the room, establishes the size of the cornice and frieze at ceiling level as parts of that entablature. The top of the pedestal becomes the dado rail or chair rail; its profile is generally a flat version of the cornice of a pedestal. The skirting, or baseboard, is the plinth of the pedestal and its profile is similarly composed. As with the pedestal, the wall between is called the dado. The door has its own entablature, in this case equal in size to the principal entablature. The lower moulding of the entablature, the architrave, is carried around the door to meet the baseboard. The windows have a similar detail, but with no frieze or cornice.

This arrangement can differ. The columns often start at floor level and the details and proportions of the cornice, frieze, and baseboard change accordingly. A dado is not necessary, but it can still be included when the column bases start on the floor and should be seen as a solid balustrade between columns. The doors and windows frequently have their own independent Classical Order, sometimes with their own columns and pediment, and the details are correspondingly smaller. In a tall room, the cornice may sit below the ceiling level and be joined to it by a large continuous inward curve, called coving, or a vault may span from one side of the room to another. In a very tall room it is even possible to have two Orders, one above the other, or a giant Order and subsidiary Orders.

These details only appear in full in the most splendid rooms. It is much more usual to find a reduced version. Example (*b*) retains a number of impressive features. There is a full decorated frieze, a full entablature over the door and a dado. Example (*c*) is the familiar simple room which retains only the ceiling cornice, the architraves around the door and windows and the skirting. This last example, however simple, is set out according to the same proportions as the first and retains the potential to be improved or decorated to reflect the Classical principles shared by both interiors.

a

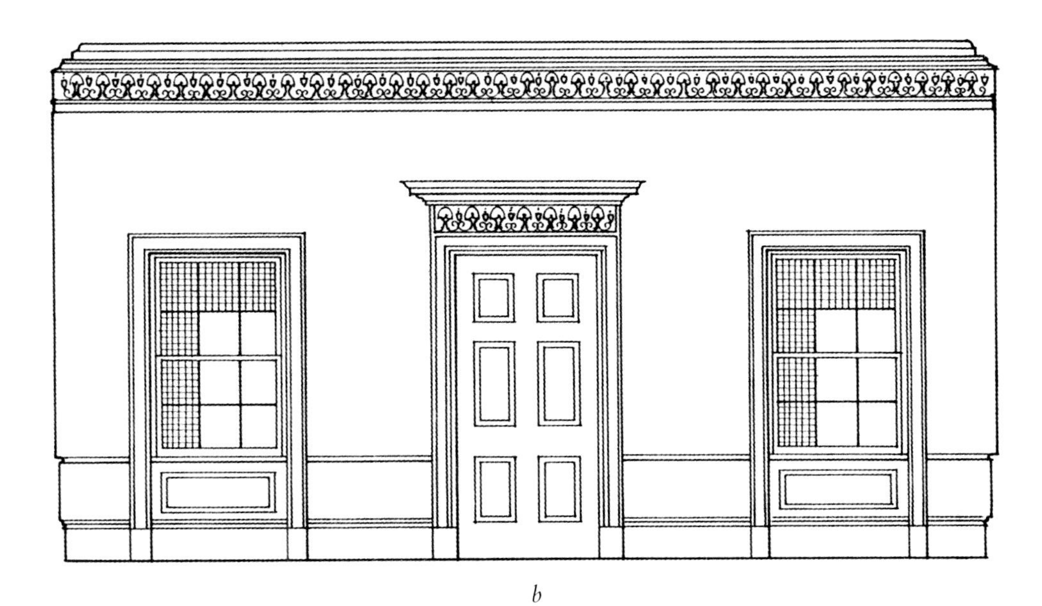

b

c

外观细部设计的选择

古典设计的一大特征就是，选择完整的古典细部设计却不带有完整的古典柱式。柱式的比例与装饰体系使其能够通过采用一些装饰元素来象征着柱式完整装饰的应用。这种设计的成功，依赖我们对常见古典建筑的熟悉，但是在不了解所选择的细部设计起源及其柱式的情况下，而对其进行重复使用，可能会造成设计失败。

这里列举了一些著名建筑上单个细部设计的实例（*a*）。在这些实例中，为使得比较效果更加直观，因此精准地复制了每栋建筑物的比例。在现实中，每个实例都可能为了适应特殊的细部设计而进行细微的调整。

在实例（*b*）中，升序排列的柱式——多立克式、爱奥尼亚式、科林斯式——保留了其整体或局部檐部设计而能够分辨清楚。下面的多立克柱式檐部完整保留下来，然而上面的爱奥尼亚柱式只保留了檐口。科林斯柱式的屋檐有完整的檐部设计。这一特殊的安排由窗户的位置决定，只看到每种柱式完整的檐部或者檐口是十分常见的。这种在装饰雕带上单独应用一系列以升序排列的各种柱式水平特征的手法，广泛应用于各时期的古典建筑中。

下面的两个例子，展现了柱式在以升序排列情况下，其他的选择性应用方法。在实例（*c*）中，底层与屋檐部分保留了柱式的装饰痕迹；底层多立克式柱式是完整的。多立克柱式与上层形成对比，赋予该层与其功能息息相关的重要性。作为对比实例（*d*），拥有完整的顶层科林斯柱式。通过使用粗琢手法，采用多立克柱式，赋予下层以厚重感，达成正立面的平衡。这种手法常见于上层需要更好的视野或者其他特殊重要意义的建筑物。

在实例（*e*）中，唯一保留下来的是科林斯柱式的檐口与雕带，其比例调整为适合整个正立面的高度。装饰丰富的檐口，不论是否带有檐部的其他元素，或者不带有柱式的其他部分，通常可见于古典建筑中，使得简单的正立面变得独具特色。可能通过改变设计的平衡来创造类似的效果。在实例（*f*）中，通过增加全高度壁柱来强调垂直效果。尽管保留了檐口与雕带，壁柱上大部分的细部设计都被省略。完整的柱头本身赋予整栋建筑以科林斯柱式的装饰特征。

SELECTIVE EXTERIOR DETAILS

The selective addition of full Classical details without the complete Classical Order from which they derive is a particular feature of Classical design. The proportional and decorative systems of the Orders make it possible to suggest the full ornament of the Order by the presence of a few decorative elements. The success of this type of design relies on our everyday familiarity with Classical buildings, but can fail when selective details are repeated without an underlying knowledge of their origin and of the Orders.

Some examples of the selective use of individual details on an identical building (*a*) are illustrated here. In these examples the proportions of each building are repeated exactly, in order to make the comparison direct. In reality, minor adjustments would be made to accommodate the particular details of each illustration.

In example (*b*) an ascending series of Orders—Doric, Ionic, Corinthian—is suggested by the retention of all or part of the entablature of each Order. The lower Doric entablature is retained in full while on the Ionic Order above, only the cornice remains. The eaves have a complete Corinthian entablature. This particular arrangement is dictated by the window positions and it is quite usual to see all the entablatures or just the cornices of each Order. This isolated application of the horizontal feature of a series of ascending Orders in bands is a very widely used design feature in Classical architecture of all periods.

The following two examples illustrate other selective applications of the Orders when they are arranged in an ascending sequence. In example (*c*) only the lower floor and the eaves retain any evidence of the decoration of the Orders; on the lower floor the Doric Order is complete. The Doric Order, by its contrast with the upper floors, gives an importance to this floor that would usually be associated with its function. As a contrast, example (*d*) has a full Corinthian Order on the top floor. The balance of the façade is maintained by giving weight to the lower floor with rustication, which in itself can suggest the Doric Order. This type of design is normally found where the upper floor commands a fine view or has some other special significance.

The only elements that remain on example (*e*) are the cornice and frieze of the Corinthian other, proportioned to the full height of the façade. A rich cornice, with or without other elements of the entablature and without any other parts of the Order, is found frequently on Classical buildings and gives distinction to an otherwise simple façade. It is possible to achieve a similar impression while changing the balance of the design. A more vertical emphasis is given to example (*f*) by the addition of full-height pilasters. Although the cornice and frieze remain, like the pilasters they are stripped of much of their detail. Full capitals alone give the building the decorative character of the Corinthian Order.

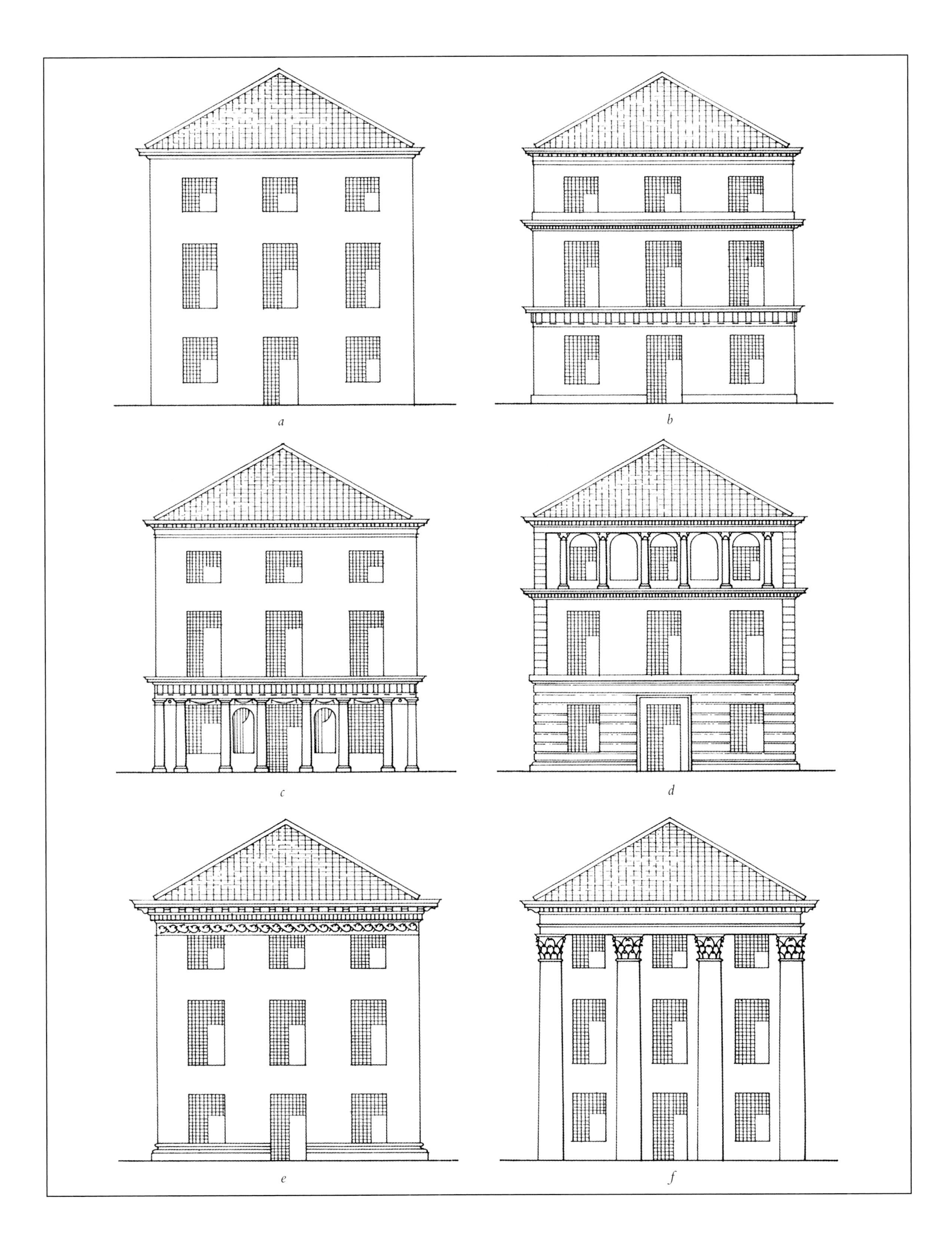
a
b
c
d
e
f

8. 山花

神庙屋顶

山花是希腊神庙（*a*）屋顶的山墙端。希腊神庙从其山墙端进入，使得山花成为重要的建筑特征。

通常古典山花屋顶的斜度较小。希腊人于公元前700年发明了一种特殊的黏土屋瓦，用于修建神庙屋顶的斜坡。这种屋瓦沿用至今，适合大面积、水平且四边翻卷的屋顶。屋瓦的连接处有较小的半圆屋瓦。新的屋顶体系十分有效，取代了早期的泥制平屋顶和陡斜的茅草屋顶。屋瓦铺设得比较松散，为了防止最外层的屋瓦滑落，设计了特殊的装饰黏土支架，又称装饰屋瓦，固定于屋顶横梁上，每行末端都以半圆屋瓦钩住。

山花通常装饰华丽。内部三角形的空间，即拱楣，有时带有雕刻群，山花上方位于末端与中央的雕刻及其置于恰当位置的底座，统称为山花雕像座。

罗马作家维特鲁威设定山花时，通过将最长水平线划分为9个部分，以便找到斜屋顶上（*c*）最低装饰线脚的高度。16世纪，塞里奥将中心线向下延长至最长水平线长度的一半，以此点为弧线的中心，从山花外端置顶点（*b*）作弧。然而，这些规则绝不是通用的。一些罗马屋顶的斜度可达30°（*g*）。北欧文艺复兴时期，一些建筑物中的屋顶斜度可达45°（*f*），以便适应当地的屋顶覆面形制。

山花的一些细部设计要遵循严格的法则。山花墙置于水平檐部的檐口之上，其顶部设计与之相同，但是檐部设计省略了顶部装饰线脚，通常是波纹线脚部分，因为这部分通常是檐沟，在这种情况下就显多余了。下面的挑檐滴水板形成了斜坡与水平部分的连接。

当檐口包含凸出部分时，例如齿状装饰（*d*）、飞檐托块（*e*）或者涡卷花饰支架，这些通常不沿着山花的斜坡排列而是发生变形，使得各边垂直，其上部和底部与山花形成一定的角度。这些特征间隔分布使其与下方相同的装饰细部设计垂直保持在同一延长线上。

8. THE PEDIMENT

THE TEMPLE ROOF

A Pediment is the gable-end of the roof of a Greek temple (*a*). Greek temples were entered by their gable-end, making the pediment an important architectural feature.

Classical pediments usually have a shallow pitch. The Greek invention of a special type of clay roof-tile in about 700 BC established the pitch of temple roofs. This roof-tile, still in use today, was large, flat and turned up at the sides. The joints between the tiles were covered by smaller half-round tiles. The new roofing system was very efficient and replaced earlier flat mud roofs and steep thatched roofs. The tiles were laid loose and, to stop the outermost tiles falling off, special decorated clay brackets, or antefixa, were secured to the roof beams and hooked up over the end of each row of half-round tiles.

Pediments were often lavishly decorated. The triangular space inside, the tympanum, at times contained sculptural groups and the top of the pediment could have sculpture at the ends and centre called, with their supporting bases where appropriate, acroteria.

The Roman author Vitruvius set out pediments by dividing the maximum horizontal dimension into nine to find the height to the lowest moulding at the top of the pitch (*c*). In the sixteenth century, Serlio took a distance down from the centre equal to half the maximum horizontal dimension and made this the centre of an arc from the outside of the pediment to its uppermost point (*b*). These rules are, however, far from universal. Some Roman roofs had a thirty-degree pitch (*g*) and in buildings of the north European Renaissance, pitches of up to forty-five degrees (*f*) accommodated local forms of roof covering.

Some details of pediments obey strict rules. The top of the pediment is the same as the cornice of the horizontal entablature on which it rests, but on the entablature the cornice loses its top moulding, usually a cyma, as this was originally the gutter and would be redundant in this position. The corona, below, forms the junction between the pitch and the horizontal.

When the cornice contains projecting features, such as dentils (*d*), mutules (*e*) or scrolled brackets, these usually do not tilt over to follow the slope of the pediment but are distorted, so that the sides are vertical while the top and bottom of the features stay at the angle of the pediment. These features are spaced so that they are vertically in line with the equivalent detail below.

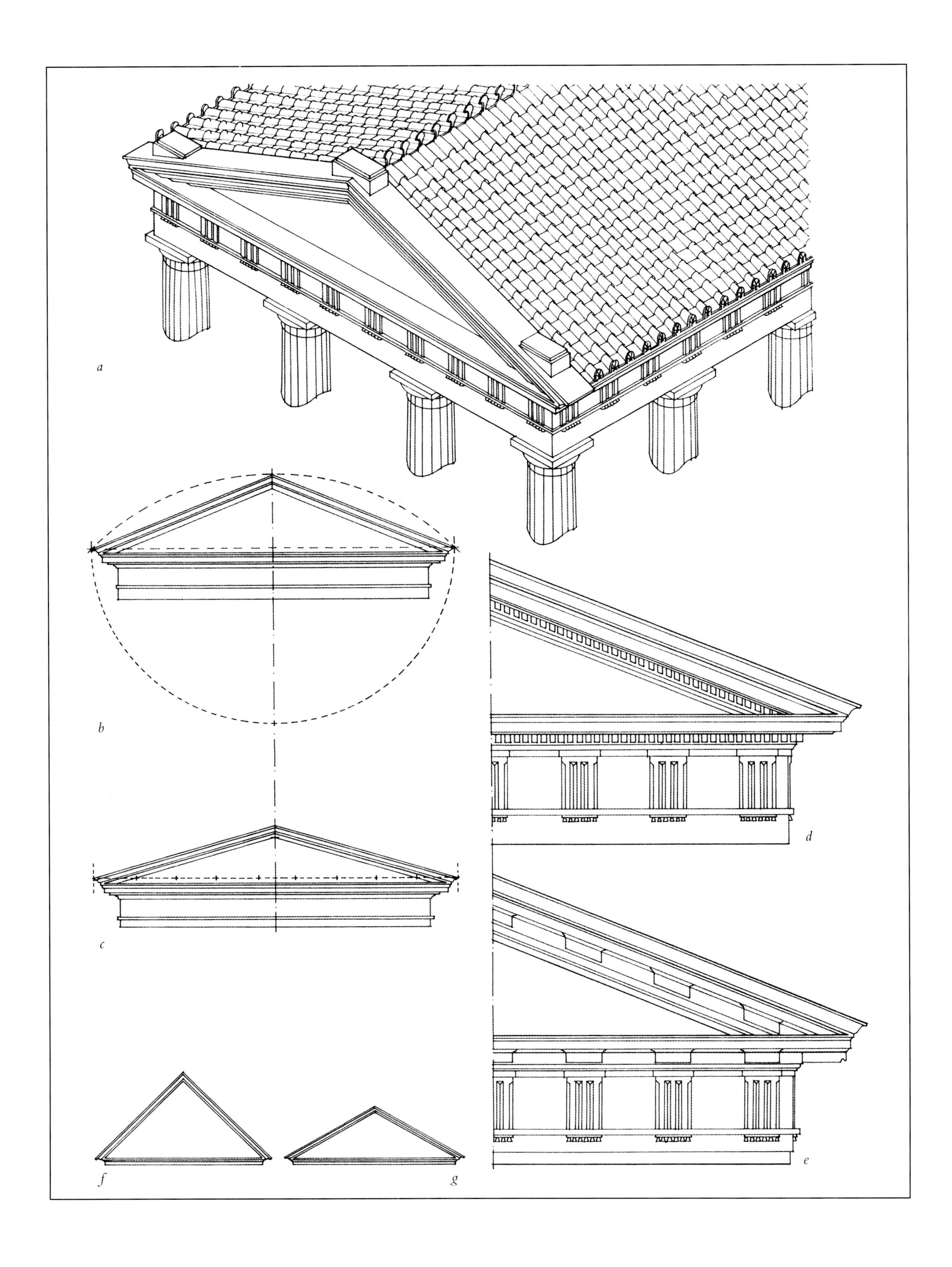

a
b
c
d
e
f
g

山花的种类

从罗马时代起，当山花所应用的建筑部分不具备神庙形态的山墙时，其用法发生了一系列的改变。山花的重要性被如此牢固地确定，以至于即便简化为变形的残余，其身份仍得以保留。

大部分山花的变体涉及拆除不同的部分。这些改变所产生的术语使人困惑。山花的顶部省略有时称为“开放式”，有时则称为“残缺式”；山花底部省略设计，产生的术语也同样让人困惑，并无普遍接受的术语体系。

山花的一部分能够凸出于柱子或支架之上，使得水平檐部的中心位于其后（*a*）。檐部的中心位置能够完全省略，创造出山花顶点（*b*）下方的开放空间。这一布局使得拱券或其他垂直构造能够向上不受阻碍地穿过柱头。

山花的斜坡可能中断并向前延伸。残破的檐部和斜坡能够同时向前（*d*）或者斜坡本身可能中断（*c*）。另外，斜坡中段部分可以省略，使得斜坡屋顶上部山花完全开放（*e*）。产生的间隙（*e*）取决于柱上檐部垂直部分，又称柱顶石的宽度，但也可削减。也可省略嵌入式檐部与嵌入式斜屋顶，只保留斜屋顶的两个垂直残部（*f*）。这一细部设计与矫饰主义、巴洛克式以及洛可可式建筑风格有着特殊关联。

山花不一定是三角形的，其顶部可能是弧形的。这些通常为低矮节段性曲线（*g*），高度与三角形墙相同。山花的弧形顶端（*g*）从最外部水平点到对应的顶点由一条分割线分成两部分，这一中心点在直角处与中心线相交。它们等高使得弓形和三角形的山墙能够混合形成交替模板，或者通过改变其类型来强调其中一个或者多个特征。弧形山花可与三角形山花进行同样方式的改进（*h*）。

山花也可以是个完整的半圆形，但最上层的波纹线脚终结时会切面朝下。有时为避免这种情况，将波纹线脚底部与柱子或托架中心对齐，然后向外折叠，终结时呈水平状（*i*）。

有许多弧形山花有更加高级的变体。斜坡屋顶能够变成两个涡卷花饰，交会于中央或者保持分开状态（*j*）。斜坡屋顶可有双弧线（*k*），或分割为一系列的曲线，或者曲线与直线的斜坡屋顶（*l*）。这些变体通常与巴洛克或洛可可建筑风格相关联。

TYPES OF PEDIMENT

The application of the pediment to parts of buildings which do not resemble the gable-end of temple has, from the Roman period onwards, led to a number of modifications. The importance of the pediment is so well established that its identity survives even when reduced to little more than distorted remnants.

Many of the variations of the pediment involve the removal of different parts. The terminology for these modifications has become confusing. Pediments with part of the top omitted are sometimes called 'open' and sometimes 'broken'; for pediments with part of the bottom omitted, the terminology is similarly confused. There is no universally accepted system of terms.

Part of a pediment can be brought forward on columns or brackets, leaving the central part of the horizontal entablature behind (*a*). The central portion of the entablature can be omitted altogether, creating an open space beneath the apex of the pediment (*b*). This arrangement allows an arch or some other vertical feature to rise unhindered past the column capitals.

The pitch of the pediment can be broken and brought forward. The broken entablature and pitch can be brought forward together (*d*) or the pitch alone can be broken (*c*). Again, the central portion of the pitch can be omitted, leaving the top of the pediment completely open (*e*). The gap created in (*e*) is set by the width of the vertical section of entablature above the column, or dosseret, but it can be reduced. It is also possible to omit both the recessed entablature and the recessed pitch leaving only two vertical remnants of pediment (*f*). This detail tends to be specifically associated with Mannerist, Baroque, and Rococo architecture.

Pediments do not have to be triangular, they can have a curved top. These are generally low segmental curves (*g*) and of the same height as a triangular pediment. The centre of the arc of the pediment (*g*) is found by dividing a line from the outermost horizontal point to the equivalent point at the apex into two parts and projecting this centre point at right-angles to meet the centre line. Their equal height allows segmental and triangular pediments to be mixed in an alternating pattern or to give emphasis to one or more features by changing the type. Curved pediments can be varied in the same way as triangular pediments (*h*).

The pediment can also be a complete semicircle although the uppermost cyma moulding terminates facing downwards. This was sometimes avoided by bringing the bottom of the cyma moulding down in line with the centre of the column or bracket and turning it outward to terminate horizontally (*i*).

There are many further variations of curved pediments. The pitch can be turned into two scrolls which can meet at the centre or remain apart (*j*). The pitch can have a double curve (*k*) or be split into a series of curves, or curves and straight pitches (*l*). These variations are generally associated with Baroque and Rococo architecture.

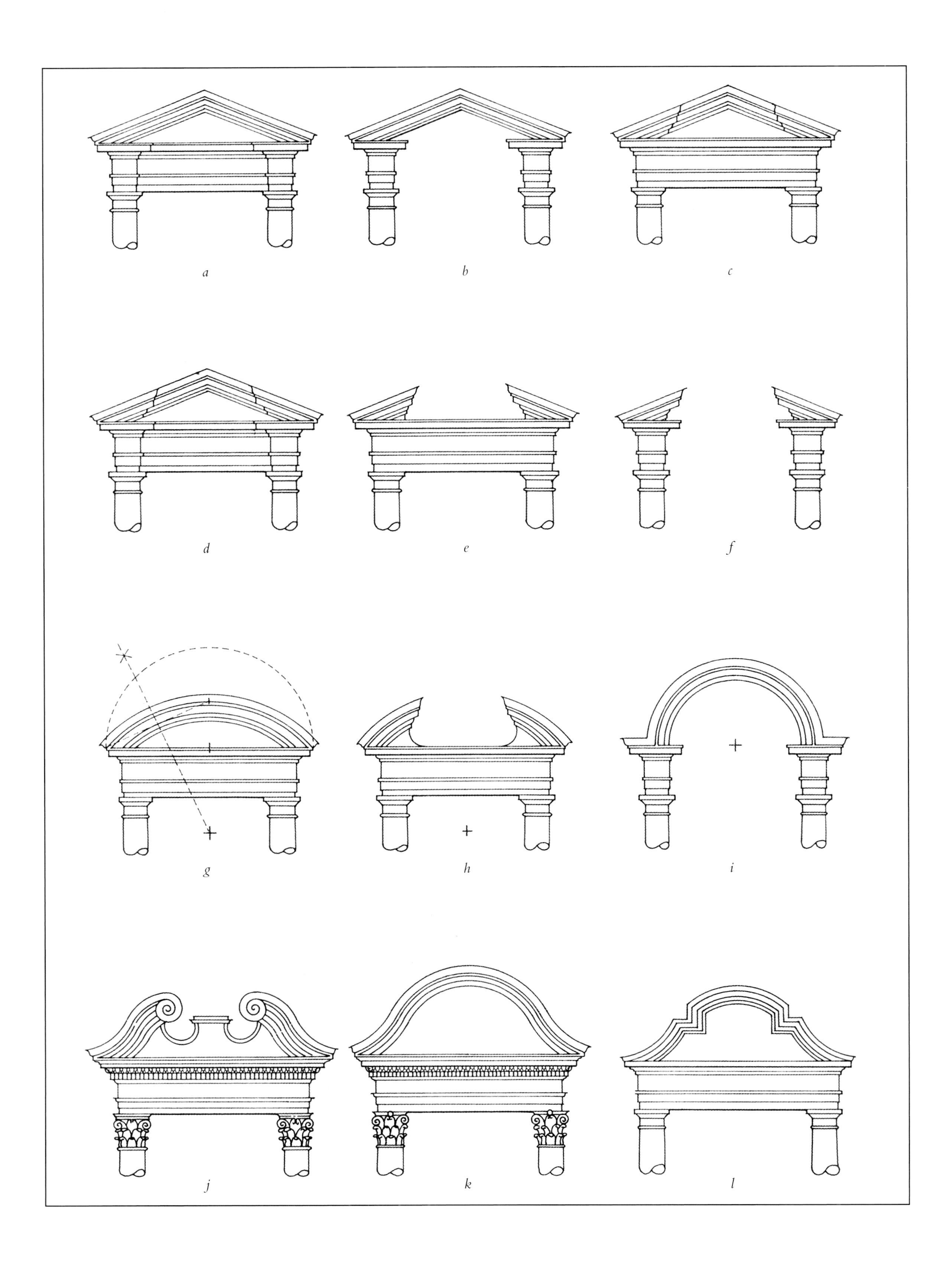
a
b
c
d
e
f
g
h
i
j
k
l

山花的各种应用

山花作为古典建筑设计元素的重要性，激励着各个时期的建筑师们发明新的组合与变体。这一进程在巴洛克与洛可可设计中达到顶峰。在新古典主义节制之风的影响下，回归简单的形制。

山花通常带有开放式顶部设计，目的是为了容纳特殊的物体。这是半身像（*a*）、大型雕像、古瓮或其他装饰类或具有象征意义雕塑的理想放置处。当山花位于门上或者两层之间时，可安装窗户（*b*）。窗户本身成为所在山花（*c*）的重要元素，并且能在铭文、壁龛或雕塑上使用相似的特征。

一系列的罗马建筑拥有一分为二的山花，每个带有单独并相对的斜坡分别置于中央装饰特征两侧，这一中央装饰特征可能是弧形山花。这一设计可见于100年图拉真市场（*d*）的外部正立面。其他罗马建筑山花中包含一个不同类型的山花，例如建于1世纪后期，维罗纳的波萨瑞之门（*e*）。

巴洛克建筑师发明了更多的变体。彼得罗·达·科尔托纳（1596—1669年），于1660年在罗马圣母玛利亚大教堂（*f*）的设计中，通过将内外山花置于同一水平线上的手法，使得一个山花叠加于另一个山花之上。1693年，波特菲斯在罗马修建的圣安多尼堂（*g*），马丁诺·隆吉（1534—1591年）设计内部山花穿过中间断开的外部山花，在同样位于罗马的建于1664年的圣文森佐教堂中，他设计了三层重叠的山花。

在随后的一个世纪中，这些变体进一步得以发展，1733年建于德国慕尼黑的圣约翰教堂（*i*），埃吉德·奎林·阿萨姆（1692—1750年）扭曲了一系列的重叠山花，形成众多曲线，随着正立面形成波浪起伏状。后期的巴洛克设计例如圣路易吉·贡扎加（*j*），由安德烈·波佐（1642—1709年）于1700年设计，最终达到了如此复杂精妙的程度，以至于其山花在模糊组合构成中失去了其重要性。

建筑物的山墙端自然是安置其山花的所在。一个完整的山花位于斜坡屋顶的末端（*k*），可以省略其不同的部分，直至最终只剩下檐口装饰线脚和较短的檐部檐口转角线脚（*l*），以此象征着其细部设计的古典起源。北部欧洲国家，本土的小型平顶黏土屋瓦，多层层叠，使得陡屋顶成为迫切的需求。17世纪，建筑师将这些建筑中传统的凸出山墙正立面改良为以南部欧洲浅屋顶为基础的古典山花，因而产生了大量的原创设计。这些复杂的巴洛克设计不同寻常的变化，（*m*和*n*），在北部国家如荷兰，拥挤的商业街正面竞相吸引人们的目光，也因此而得名——荷兰山墙。

DIFFERENT USES OF THE PEDIMENT

The significance of the pediment as an element of Classical design has encouraged architects of many periods to invent new combinations and variations. This process reached its peak in Baroque and Rococo design, before an atmosphere of Neo-Classical sobriety encouraged a return to simple forms.

An open top was often created on a pediment in order to contain a specific object. It was an ideal location for a bust (*a*), a larger statue, an urn, or other decorative or symbolic sculpture. When a pediment was placed over a door or filled a space between floors, a window could be fitted into it (*b*). The window could itself become an important element with its own pediment (*c*) and a similar feature could be used for an inscription, niche or statue.

A number of Roman buildings have a pediment split into two, each with a single and opposite pitch on either side of a central feature such as a curved pediment. This can be seen on the outer façade of Trajan's Markets (*d*) of AD 100. Other Roman buildings had pediments inside pediments of a different type, as on the Porta Dei Borsari in Verona (*e*) of the late first century AD.

Baroque architects invented further variations. Pietro da Cortona, (1596-1669) in S. Maria della Pace in Rome (*f*) of 1660, superimposed one pediment on another by bringing the inner and outer pediments up to the same level. In S. Antonio dei Portoghesi in Rome (*g*) of 1693, Martino Lunghi (1534-91) allowed the inner pediment to penetrate the open top of the outer pediment, while in SS.Vincenzo ed Anastasio also in Rome (h) of 1664, the same architect superimposed three layers of pediments.

In the following century these distortions developed still further and in the Johanneskirche, Munich, Germany (*i*) (1733), Egid Quirin Asam (1692-1750) twisted a series of overlapping pediments into a multitude of curves which in turn undulated with the sweep of the façade. Late Baroque designs such as the altar to S. Luigi Gonzaga (*j*), by Andrea Pozzo (1642-1709) in 1700, finally achieve such a complexity that the pediment loses much of its significance in the ambiguity of the composition.

The gable-ends of buildings are natural places for pediments. A full pediment located at the end of a pitched roof (*k*) can have various parts omitted until little more than some cornice mouldings and a short return of the entablature cornice (*l*) suggest the Classical origin of the details. In north European countries the indigenous small flat clay roof-tiles, overlapping in several layers, necessitated steep roofs. In the seventeenth century a great deal of ingenuity was devoted to adapting the traditionally prominent gable façades of these buildings to Classical pediments based on the shallow pitch of south European roofs. There is a remarkable variety of these complicated Baroque designs, (*m*) and (*n*), which jostle for attention in the crowded commercial street fronts of northern states like the Netherlands, from where they received their English name—Dutch gables.

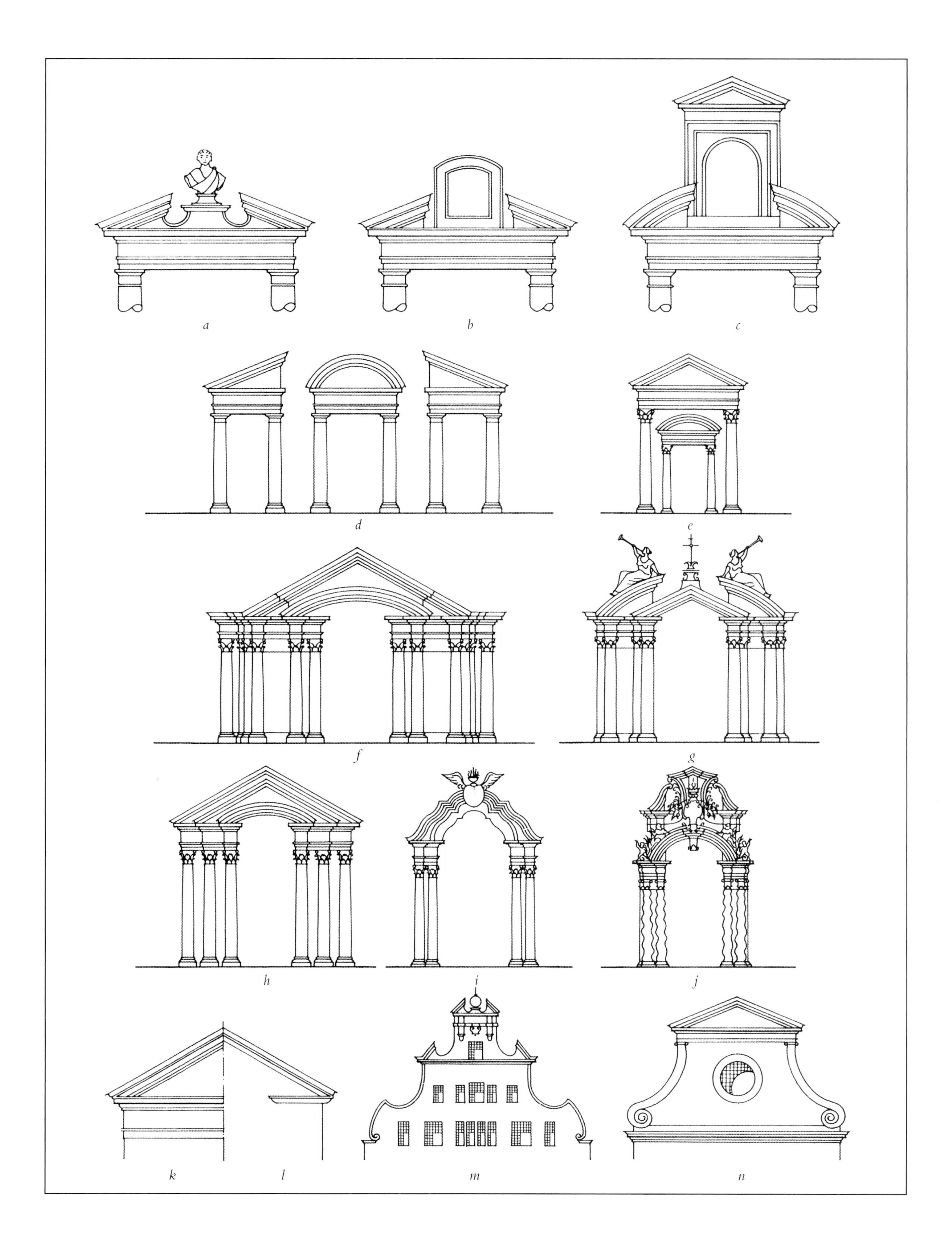

a
b
c
d
e
f
g
h
i
j
k
l
m
n

山花及其设计

山花的起源赋予其永远不可剥夺的重要地位。从早期的神庙时代直至今日，山花在古典设计中的应用既强化整个建筑，又强化建筑局部。

古希腊神庙中的山花占据支配地位，作为屋顶山墙端，占据了整个末端或者正立面的入口，其配对物位于后方。公元前430年建于雅典的帕提农神庙（*a*）装饰华丽，带有精美的雕刻，位于末端的鼓室内安放着精雕细琢的、讲述着神话故事的雕像，与女神雅典娜有关。

柱子与山花的组合与崇拜有关，当削减至两个较小的柱子用以支撑一个小型山花时，这一关联意义仍得以保留。这种简单的安排——神龛，可见于雅典哈德良凯旋门（*g*）雕像设计四周，象征着帝王的宗教信仰。小型的山花神龛，构成了雕像的壁龛或框架，或者是中空的，也同样是古典设计中不断重复的主题。

在罗马时代及其后期、文艺复兴时期，山花应用于缺乏延续的屋顶或者成排柱子的建筑中，赋予其最原始的形态。意大利北部曼图亚的圣·塞巴斯蒂亚诺教堂（*b*），建于1460年，据推测重建于此地。阿尔伯蒂首先将神庙正面设计应用于教堂，将山花的正立面从排他的异教徒设计中解放出来。在罗马圣·安德烈教堂（*e*），由维尼奥拉于1550年设计。维尼奥拉遵循2世纪万神殿的设计（P13），将柱子与山花置于穹顶结构前面。他的小教堂设计，呈椭圆形，柱子简化为扁平壁柱。传统的教堂正立面山花入口一直沿用下来。在佛罗伦萨的圣·翡冷翠大教堂（*f*），由格拉多·西尔瓦尼（1579—1675年）和费迪南多·鲁杰里（1691—1741年）设计，于1790年完工，有三个山花：第一个为截断山花；第二个为逆转山花；第三个为弧形山花，悬挂并简化为尖顶饰。

16世纪中期，威尼斯建筑师帕拉迪奥将山花应用于一栋住宅的设计之中，他相信罗马人赋予住宅以神庙般的尊贵。他的乡村住宅设计，也就是别墅，通常包括大型实用门廊，即柱廊，带有山花。但是在威尼斯附近昆托·维琴蒂诺建于1545年的蒂内别墅中（*d*），山花占据主导地位，成为整个房子的设计。

山花，应用于山墙或者假山墙之上，是小型开间以及神龛的标志，也是古典设计中最常见的元素之一。它既不需要柱子支撑也不需要依附于整个正立面。伦敦附近的一座邱宫（*c*），由塞缪尔·福特雷（死于1647年或更早）于1631年设计，山花设计应用于几个荷兰式山墙之上，强调了单个或成对的窗户。

PEDIMENTS AND DESIGN

The origin of the pediment gave it an importance that it has never lost. From the temples of earliest times to the present day, pediments have been used in Classical design to give emphasis both to whole buildings and to parts of buildings.

The Greek temple was dominated by its pediment which, as the gable-end of the roof, occupied the whole end or entrance façade, and its counterpart at the rear. The Parthenon in Athens (*a*) of 430 BC was heavily decorated with rich sculpture and the tympana at each end were filled with elaborate groups of figures telling mythological stories associated with the life of the goodness Athena.

The combination of columns and pediment was associated with reverence and could retain this association when reduced to two small columns supporting a small pediment. This simple arrangement, the aedicule, can be seen around the statue on Hadrian's Arch in Athens (*g*) of AD 131, suggesting the religious devotion due to the Emperor. The small pedimented aedicule, forming a niche or frame for a statue, or even empty, has become a recurring theme in Classical design.

In Rome and later, in the Renaissance, the pediment was applied to buildings that lacked either the continuous roof or the row of columns that had given it its original form. With his church of S. Sebastiano in Mantua, northern Italy (*b*), built in 1460 and conjecturally reconstructed here, Alberti was the first to apply the temple front to a church, freeing the pedimented façade from its previous exclusively pagan associations. S. Andrea in Rome (*e*) was designed by Vignola in about 1550. Vignola follows the example of the second-century Pantheon (page 13) by placing columns and a pediment in front of a domed structure. His little church, however, is oval and the columns are reduced to flat pilasters. The tradition of pedimented entrance façades for churches has continued. In S. Firenze in Florence (*f*) by Gherardo Silvani (1579-1675) and Ferdinando Ruggieri (1691-1741), completed in 1790, there are three pediments: one segmental, one reversed, and one curved, suspended, and reduced to a finial.

In the middle of the sixteenth century the Venetian architect Palladio applied the pediment to houses, believing that the Romans would have given an equal dignity to houses as to temples. His designs for country houses, or villas, usually included a large applied porch, or portico, with a pediment, but on the Villa Thiene in Quinto Vincentino near Venice (*d*) of about 1545, the pediment took over and became the design of the whole house.

The pediment, applied to a gable or false gable and marking small openings and aedicules, is one of the most familiar elements in Classical design. It needs neither columns nor whole façades. On Kew Palace near London (*c*), built by Samuel Fortrey (*d.*in or before 1647) in 1631, pediments are used on several gables in a Dutch fashion and to emphasise windows both individually and in pairs.

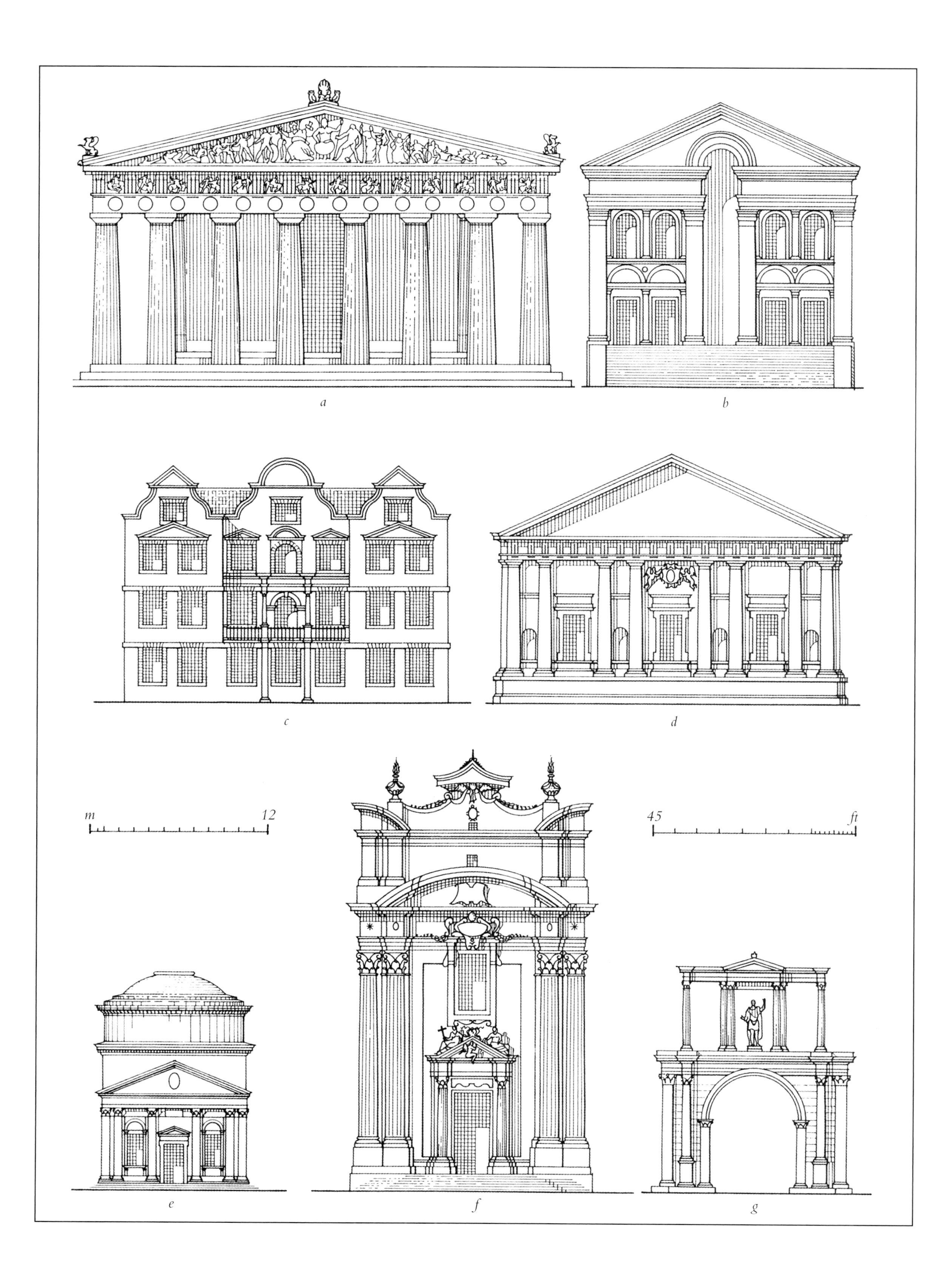

a
b
c
d
m
12
45
ft
e
f
g

9. 拱券

古代拱券

拱券的发明是重大技术革新。将楔入的楔形石环绕成半圆形，使用石头或其他高密度建筑材料，来获得比采用同样材料设计扁平的横梁更大跨度。埃及人已经掌握这一技术，早在公元前1400年就使用完整的拱券结构。希腊人也使用拱券，但主要应用于实用结构中。早期用于装饰的实例为公元前156年，在现代土耳其所在位置的普南城市场大门（*a*）。埃及人与希腊人都不使用拱券作为尊贵建筑的一部分，明显是因其风格而产生厌恶之感造成的。

伊特鲁里亚人首先在欧洲的建筑中发展了拱券结构。正如后来的罗马人一样，伊特鲁里亚人修建了城市之间永久的道路，并将水源从遥远的地方引入。拱形桥梁结构通常用于土木工程计划，同时拱券也用于具有观赏特征的重要建筑中，如城市大门。位于意大利中部的罗马殖民地法莱里（*b*），建于公元前3世纪，带有所有拱券结构设计的基本元素，成为后来4个世纪中古典建筑语汇不可分割的一部分。

罗马继续发展拱券的应用，在不断扩张的帝国范围内竖起高架桥和高架渠，以满足人们迁移、货物运输、水源输送的需求，并且将陡峭的山峰夷平或造为梯田。最令人震撼的现存的业绩就是建于公元前1世纪后期法国南部的嘉德水道桥（*c*），展现出前无古人的建筑工程技艺与勇气，不仅在嘉德河上支撑起高49米的高架水渠，通往尼姆城，同时其下层也是公路桥。

罗马人将拱券发展为建筑元素，具有重要意义。其中一个主要演变就是罗马凯旋门。这些大型的纪念结构，其建造目的是记录胜利，人们为凯旋的将军举办梦寐以求的公开胜利庆典。其形制源于为行军临时搭建的木制大门，这样的结构，如提图斯凯旋门（*d*），建于大约82年，成为罗马帝国荣耀的重要象征。

9. THE ARCH

THE ARCH IN ANTIQUITY

The invention of the arch was an important technical innovation. By wedging a series of tapering stones together around a semicircle it is possible to span much greater distances using stone or other dense building materials than is possible with flat beams in the same materials. This was known by the Egyptians, who were using full structural arches as early as 1400 BC. The Greeks also used arches, but principally for utilitarian structures. The earliest known decorative example is a market-place gate in the city of Priene in modern Turkey (*a*) from 156 BC. Neither the Egyptians nor the Greeks used arches for parts of buildings that had any prestige, evidently being averse to the form for stylistic reasons.

It was the Etruscans who first began to develop the arch in European architecture. Like the Romans after them, the Etruscans built permanent roads between cities and brought their water supplies from remote sources. Arched bridging structures were often used for these civil engineering schemes and arches were also used for important visual features such as city gates. A gate in the Roman colony of Falleri in central Italy (*b*) from the third century BC has all the essential elements of arch design that became an integral part of the Classical vocabulary some four centuries later.

The Romans continued to develop the use of the arch, erecting viaducts and aqueducts throughout their growing empire for the movement of people and goods, the supply of water and the leveling and terracing of steep hills. The most impressive survival of these unprecedented feats of engineering skill and daring is the Pont du Gard in southern France (*c*) from the late first century BC, which not only carried water in an aqueduct 49 metres above the River Gardon to the city of Nimes, but also a road bridge at a lower level.

The Roman development of the arch as an architectural element is of major significance. One of the principal factors in this evolution was the Roman triumphal arch. These large commemorative structures were built to record a triumph, the coveted public celebration of a victory awarded to a returning general. The form may have originated in temporary wooden gates erected specially for procession, and structures such as the Arch of Titus (*d*) of about AD 82 become important symbols of the glory of the Roman state.

a
b
c
SENATVS
POPVLVS QVE ROMANVS
DIVO TITO DIVI VESPIANI F
VESPASIANO AVGVSTO
m
6
ft
20
d

拱的构建

希腊人使用扁平的石质横梁，作为建筑开间的跨度，这是对早期木造结构的模仿。然而，石头不适合充当横梁的材料，由于这一建材在没有支撑的情况下承受向下压力时，易碎易折断。而另一方面，坚硬的石头在来自反方向支撑的情况下却对压迫以及折断有着相当大的抵御能力。为了构建拱形结构，楔形石块或砖块呈放射状，拼凑于开间上方，因此每个石块上方的重量所产生的压力使其能够彼此紧密结合。石块之间挤压所产生的抗力使得整个拱结构不会瓦解。同样的原理也可应用于扁平或者较浅的拱结构。圆拱是常见的古典设计类型，尖拱，即哥特式拱却是最有效的。

圆拱有一个圆心（*a*）位于圆拱开始即起拱的直线（*b*）上。从这一中心开始，在其起拱的直线之上，以固定的半径（*c*）画弧线。拱券起拱于两个石块，又称拱墩（*d*）之上，通常为凸出设计并雕刻有装饰线脚。拱的直径通常为两柱或支墩之间的距离（*e*），即跨度（*f*）。拱由一系列楔形砖块，或称拱砌砖（*g*）构建而成，这些石块可多可少，依据石块大小及拱的跨度而定。中央的拱砌砖叫作拱顶石（*h*），通常通过凸出及/或装饰的手法得以强化。拱的底部为拱腋（*i*），其上为冠部（*j*）。整个拱的外表面或者拱的边缘称为拱背（*k*），位于拱下的内表面称为拱腹或者内弧面（*l*）。拱一侧的石雕工艺为扶壁（*m*），两个相邻拱之间的区域称为拱肩（*n*）。拱内的空间称为山墙饰内三角面（*o*）。

拱在不完整的情况下无法支撑自己。它们必须在临时支撑结构即拱架（*p*）上进行构建。这一结构通常由下面带支撑的木材搭建。拱墩可能源于拱架支撑结构。拱的形状在拱架中成形，将一系列小型木造压条（*q*）固定起来，数量足够支撑每块拱砌砖。拱砌砖从两侧拱墩开始，逐渐叠加，直至交会，加入拱顶石完成整个拱结构。因此，拱顶石将整个结构固定。大型的早期罗马石拱有时由两个穿堂各据一边构成，这赋予拱顶石极其特殊的重要意义。

任何一种横跨开间结构的建造目的都是将上部墙壁重量转移至两侧，最终垂直落于地面。圆拱对于这种目的来说不算完美，重量倾向于将圆拱向外推。由于这一原因，拱或者成排的拱两边通常需要实体墙壁部分。

THE CONSTRUCTION OF ARCHES

The Greeks used flat stone beams to span openings in buildings, in imitation of earlier wooden structures. Stone is, however, an unsuitable material for beams as it is brittle and breaks under the stress of downward pressure when unsupported. Hard stone will, on the other hand, offer considerable resistance to being crushed and broken while supported on the opposite side. To make an arch, wedge-shaped pieces of stone or brick are set together over an opening in a radiating pattern so that the weight above each of the pieces pushes them against one another. The resistance of the stone to being crushed prevents the collapse of the arch. The same principle applies to the construction of flat and shallow arches. Round arches are the common Classical type, while pointed, Gothic, arches, are the most efficient.

A round arch has a single centre (*a*) which sits on a line (*b*) from which the arch rises, or springs. From the centre, a constant radius (*c*) describes the line of the arch from the line of springing upwards. The arch spring off two blocks, or imposts (*d*), which often project and can be carved with mouldings. The diameter of the arch, the distance between the columns, or piers (*e*), is the span (*f*). The arch is made up of a series of tapering blocks, or voussoirs (*g*), which can be many or few according to the size of stone or span of the arch. The central voussoir, which binds the whole arch together, is called the keystone (*h*); it is often emphasised through projection and/or decoration. The lower part of the arch is the haunch (*i*) while the upper part is the crown (*j*). The entire outer surface or edge of the arch is known as the extrados (*k*) and the inner surface below the arch as the soffit or intrados (*l*). The stonework to one side of an arch is the abutment (*m*) and the area between two adjacent arches is the spandrel (*n*). The space inside the arch is the tympanum (*o*).

Arches cannot support themselves until they are complete. They must be constructed on a temporary support known as centering (*p*). This is usually made of timber and is supported from below. It seems likely that imposts originated as supports for centering. The shape of the arch is formed in the centering and a series of small timber battens (*q*) is fixed in place in sufficient number to support each voussoir. The voussoirs are laid progressively up from the imposts on each side until they meet and the arch is completed by the keystone. The keystone, therefore, 'locks' the design into place. Large early Roman stone arches were sometimes laid by two gangs, one on each side, and this must have given the keystone particular significance.

The purpose of any structure that spans an opening is to transfer the weight above to the walls on either side and thence down vertically to the ground. The round arch is not the perfect shape for this, the weight tending to push the side of the arch outwards. For this reason arches or rows of arches usually have substantial section of wall at each end.

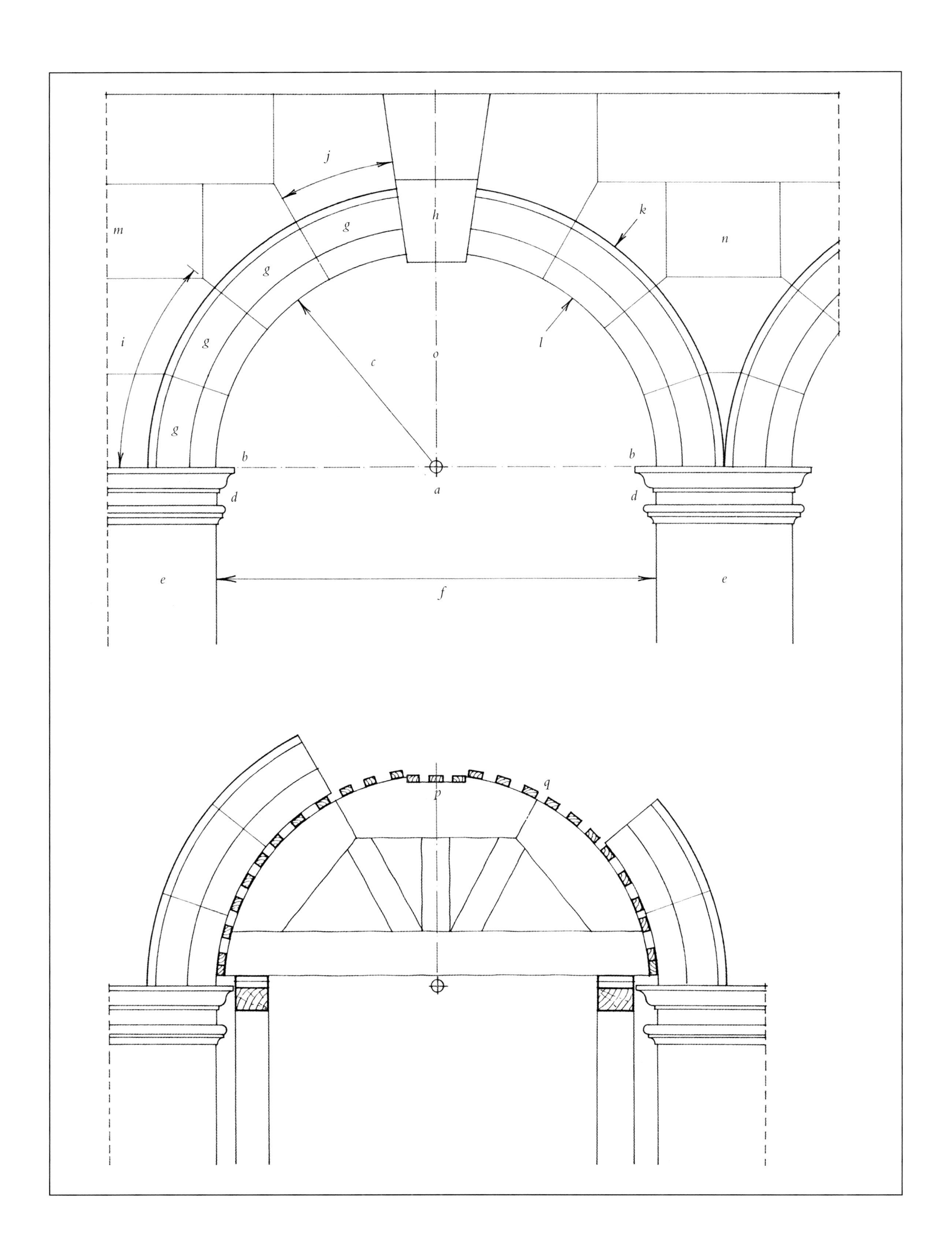

j
h
m
k
g
n
g
i
g
o
l
c
g
b
b
a
d
d
e
e
f
p
q

拱券与柱式

将拱券与古典柱式组合，以引入与单个柱式相关联的新建筑特征。这些就是拱墩与拱门饰。拱墩是恰好位于拱券起拱处下方的细部设计。它与柱头或檐口相似。拱门饰为沿拱券轮廓，呈弧形的额枋部分。

拱券与柱式相结合位于两个柱子之间，拱券的比例决定柱子之间的距离。最常见的拱券的比例为高是宽度的两倍，（*a*和*b*）。如果柱子与拱券的位置尽量接近并且不带基座，柱础与拱门饰就能确定拱券的宽度为5倍柱底径左右（*a*）。如果有基座，就必须增加支墩宽度，使得柱础的檐口能够凸出来，但是由于柱身会变小，空间就增加至7倍柱底径（*b*）。如果从拱门饰顶端至额枋底部有0.5倍柱底径的富余空间，即可固定拱券与柱式组合的比例。

罗马人在相对较晚时期引入了拱券与柱式的组合，并没有给已确立地位的传统系列形制即拱墩与拱门饰，带来新的发展。为了保证柱式的发展进程，文艺复兴时期的作家保持托斯卡纳拱券细部设计简洁。由维尼奥拉（*c*）和斯卡莫齐（*d*）设计的实例都来自16世纪。多立克式也同样简洁，维尼奥拉的设计（*e*）和帕拉迪奥的设计（*f*）均展现了多立克式柱头版本的拱墩。

当柱式的柱头变得复杂时，就不便将其改造为拱墩了。维尼奥拉的作品（*g*）和帕拉迪奥的作品（*h*）在爱奥尼亚式拱券中继续沿用多立克式柱头版本的设计，但帕拉迪奥增加了装饰来反映这一柱式极富装饰效果的特征。

科林斯柱式在古代罗马最为流行，因此在拱券细部设计中最为多变。维尼奥拉的作品（*i*）通过使用多立克式柱头与拱墩创造出了拱券的对比效果，而帕拉迪奥的设计（*j*）则在科林斯式额枋之上增加了卵箭装饰线脚的拱门饰，并且采用在多立克式柱头上增加装饰的手法设计了拱墩。斯卡莫齐的设计（*k*）更为传统，使用罗马后期柱头与2/3未经改动的科林斯额枋。康斯坦丁凯旋门（*l*）提供了产生完全反差效果的实例。在这一凯旋门中，拱墩是位于朴素串珠饰之上完整的檐口部分。同样的对比也可见于混合柱式中。维尼奥拉的设计（*m*）与其科林斯式拱券设计仅有细微差别，但是在建于203年的塞普蒂默斯·塞维鲁凯旋门（*n*）上，拱墩与变化的檐口部分相似，用于装饰华丽的柱头之上，拱门饰为混合式额枋增加了卵箭装饰线脚。

ARCHES AND THE ORDERS

The combination of the arch and the Classical Orders led to the introductions of new features that could be associated individually with each Order. These are the impost and the archivolt. The impost is the detail just below the point where the arch springs. It resembles either a column capital or a piece of cornice. The archivolt is the profile that follows the line of the arch and is a curved architrave.

As the arch is combined with the Orders by sitting between two columns, the proportions of the arch will determine the spacing of the columns. One of the most commonly used proportions has an arch with a height twice its width, (*a*) and (*b*). If the columns are to be placed as close to the arch as possible and there is no pedestal, then the column base and the archivolt will establish the width of the arch and this will be about 5 diameters (*a*). If there is a pedestal, the piers must increase in width to allow for the projection of the cornice of the pedestal, but as the columns will become smaller, the spacing will increase to about 7 diameters (*b*). If a space of 0.5 diameters is left between the top of the archivolt and the bottom of the architrave, the proportions of the combined arch and Order will be fixed.

The relatively late introduction of combined arches and Orders by the Romans did not give rise to a clearly established series of conventional forms for imposts and archivolts. In order to maintain the progression of the Orders, Renaissance authors keep their Tuscan arch details simple. Examples (*c*) by Vignola and (*d*) by Scamozzi are both from sixteenth century. The Doric is also simple and both Vignola (*e*) and Palladio (*f*) show imposts that are versions of Doric capitals.

When the capital for the Order becomes more complex it cannot be so readily adapted as an impost. Vignola (*g*) and Palladio (*h*) when illustrating Ionic arches continue to use modified Doric capitals, but Palladio adds further decoration to reflect the more decorative character of the Order.

The Corinthian Order was the most popular in ancient Rome and, perhaps as a consequence, has the most varied arch details. Vignola (*i*) contrasts the arch with the Order by using a Doric capital and archivolt, while Palladio (*j*) makes the archivolt by adding an egg and dart moulding to a Corinthian architrave and the impost by adding further decoration to a Doric capital. Scamozzi (*k*) is more conventional, using a late Roman capital and two-part Corinthian architrave with no modifications. The Arch of Constantine (*l*) offers a complete contrast. Here, the impost is a whole section of cornice above a plain astragal moulding. Similar contrasts are to be found in Composite arches. Vignola (*m*) only varies his details slightly from his Corinthian arches, while on the Arch of Septimus Severus (*n*) of AD 203 the impost resembles a section of cornice modified to make an ornate capital and the archivolt is a Composite architrave enriched with an egg and dart moulding.

a
b
c
d
e
f
g
h
i
j
k
l
m
n

拱券与柱式：变体

尽管内拱的宽度为内部高度的一半（比例为1：2）是一个常见的经验法则，但使用这一精确比例有时不太可能或并不恰当。支墩与柱子关系，拱券上的冠饰以及额枋、拱顶石均可改变。

托斯卡纳柱式比例规则由于缺乏古代的证据，因而仅仅建于理论基础之上。维尼奥拉在1562年展示了1：2 的比例，在柱子不带基座的情况下，不设计拱顶石也没有拱门饰（*a*）。另一方面，帕拉迪奥于1570年记录了1：1.65的比例，带有拱门饰和拱顶石（*b*）。

维尼奥拉的多立克式拱券（*c*）同样为1：2的比例。其设计带有拱门饰，但是再次省略了拱顶石。只有在没有基座的情况下，支墩才有其独立的底座，他引入了一种装饰线脚，置于柱头底部，用以延长柱子。这与古代流传下来的最早的实例形成对比，这一实例取材于公元前13年建的罗马马塞勒斯剧院（*d*），其比例为1：2.4，没有额枋或者凸出的拱顶石，拱券恰好置于柱式的框架之中。两个文艺复兴多立克拱券（*e*和*f*），比例接近1：2，但其他的细部设计有所不同。在实例（*e*）中，这种柱式被大大简化，拱门饰与额枋相接，而实例（*f*）却带有拱顶石和狭窄的支墩。

维尼奥拉的爱奥尼亚拱券（*g*）再次采用了1：2的比率，尽管他的拱券有带基座的柱式，并且也有拱顶石，但没有一个拱顶石是带有简单底座的。马塞勒斯剧院的上层（*h*）有着非常相似的拱券比例，但是缺乏拱内饰以及较宽的支墩，创造出了极其不同印象。取材于罗马建于1550年的朱利亚别墅的一个带有基座的爱奥尼亚式拱券（*i*）中，拥有更加常见的文艺复兴构造，但是拱券更低矮，比例为1：1.7。

一个不寻常的文艺复兴时期拱券（*j*）置于结合了托斯卡纳和科林斯细部设计的柱式上，值得注意的是，其拱内饰被拱墩隔断。更加传统的科林斯式拱券源于维尼奥拉设计（*k*），分配的比例为1：2，而他所设计的其他柱式现在增加了拱顶石。这一比例在帕拉迪奥的科林斯式带有基座的拱券（*m*）设计中重复使用。建于公元前9世纪，位于意大利北部苏萨的奥古斯特凯旋门（*l*），与古代很多实例相似，十分独特。这一比例为1：1.5的拱券，置于主要柱式范围内，带有小壁柱支撑拱门饰。

维尼奥拉的混合式拱券（*n*）与其科林斯式拱券相似并且再次使用1：2的比例。在罗马建于203年的塞普蒂默斯·塞维鲁凯旋门（*o*）上，几乎相同的比例，为1：1.8，但柱子位于很高的基座之上，改变了柱式与拱券的关系。

ARCHES AND THE ORDERS: VARIATIONS

Although an internal arch width of half the internal height (a ratio of 1:2) is a useful rule of thumb, it is not always possible or desirable to use this precise proportion. The relationship between the piers and the columns, the crown of the arch and the architrave, and keystone can also vary.

Guides for the proportions of the Tuscan Order are based solely on theory, due to the lack of evidence from antiquity. Vignola in 1562 shows a ratio of 1:2, with no keystone and no form of archivolt when the columns have no pedestal (*a*). Palladio, on the other hand, writing in 1570, has a ratio of 1:1.65, an archivolt and a keystone (*b*).

Vignola's Doric arch (*c*) is also 1:2. It has an archivolt, but again lacks a keystone. The pier has its own base and, only when there is no pedestal, he introduces a moulding on the line of the bottom of the capital to lengthen the columns. This is to be contrasted with one of the earliest examples from antiquity, from the Theatre of Marcellus in Rome (*d*) of 13 BC, where the ratio is 1:2.4, there is no architrave or projecting keystone and the arch sits well within the frame of the Order. Two further Renaissance Doric arches, (*e*) and (*f*), have ratios very close to 1:2, but other details vary. The Order is much simplified in (*e*) with the archivolt touching the architrave, while (*f*) has a keystone and narrow piers.

Vignola's Ionic arch (*g*) again has a ratio of 1:2 and, although his arch with an Order on a pedestal does have a keystone, no keystone is shown for a simple base. The upper storey of the Theatre of Marcellus (*h*) has a very similar arch proportion, but the lack of an archivolt and the width of the piers create a very different impression. An Ionic arch with a pedestal from the Villa Giulia in Rome (*i*) of 1550 has the more usual Renaissance configuration, but the arch is lower with a ratio of 1:1.7.

An unusual Renaissance arch (*j*) sits within an Order that combines Tuscan and Corinthian details; the arch is notable for the archivolt unbroken by an impost. The more conventional Corinthian arch from Vignola (*k*) shares the ratio of 1:2 with his other Orders and now has a keystone. This ratio is repeated in Palladio's Corinthian arch with a pedestal (*m*). The arch of Augustus in the north Italian town of Susa (*l*) of 9 BC, in common with many examples from antiquity, is much more individual. The ratio is 1:1.5 and the arch sits so far within the principal Order that it has its own small pilasters supporting the archivolt.

Vignola's Composite arch (*n*) is similar to his Corinthian arch and again has a ratio of 1:2. The Composite Arch of Septimus Severus in Rome (*o*) of AD 203 is of approximately the same ratio at 1:1.8, but the columns of the Order sit on very high pedestals, changing the relationship between the Order and the arch.

a
b
c
d
e
f
g
h
i
j
k
l
m
n
o

拱券与柱式：更多变体

古典拱券并非总是位于柱子和柱式额枋构成的框架之中。尽管这是拱券最初融入建筑语汇的方式，但是在罗马建筑中对其广泛的应用带来了柱式与拱券组合方式的多样化。

1世纪时期，拱券置于额枋之上，额枋中间断开，形成一个较高的开口。在随后的年代中，同样的原理应用于成排的拱券，即拱形游廊中。拱券置于独立垂直的额枋之上，称为柱顶石，使得柱式看起来更高（*a*）。在罗马帝国东部省份，演变出另一种变体形制（*b*），形成成排的柱子支撑拱券。在这一设计中，拱券及其拱门饰直接从柱头向上起拱，额枋被完全省略。这种拱廊设计在罗马帝国后期传播到各处，并且影响了罗马式建筑的发展。这些形式的建筑在文艺复兴早期也十分盛行。

在文艺复兴后期，拱券直接从柱子上起拱，并成功地与山花组合，山花使得额枋中间部分敞开（*c*），以便柱头或拱券的拱墩与柱式的柱头处于同一水平线上。这与罗马后期另一种细部设计（*d*）有同样的效果，整个额枋呈弧形置于柱子之上。这一设计被称为拱式过梁，该发明同样从东部省份传至罗马帝国后期。它可以被看作是山花的核心特征，或者拱廊（*e*）内额枋水平部分的变体。一个不太重要且短期应用的设计手法（*f*），从檐口中部成拱，并使其与柱头保持在同一水平线上。

从文艺复兴时期开始，演变出一种特殊的拱形开口（*h*），与水平额枋和拱券组合。这是布拉曼特的发明，但是塞里奥和帕拉迪奥经常使用这种设计，并创造出极佳的效果，被命名为威尼斯式、塞里奥或帕拉迪奥式窗户。在18世纪帕拉迪奥式复兴时期十分流行。

柱式可以包含在拱券中。柱式能够独立设计于拱券内（*i*）或者从拱的内侧面延续下来，可带或不带水平额枋，为边缘开口增加了错综复杂的装饰效果（*g*）。在更大的体量上，不同大小的拱券内安置柱式的方法有很多。罗伯特·亚当使用简单成排的柱子支撑一个花瓶，在建于1792年苏格兰巴思盖特的巴尔巴迪府邸（*k*）中以此充当入口。巨大的拱券内有帕拉迪奥式窗户（*j*），窗户缩小，新月形朴素的墙面可带有装饰图案。

ARCHES AND THE ORDERS: FURTHER VARIATIONS

Classical arches do not always sit within a frame created by the columns and entablature of an Order. Although this is how they first became assimilated into the Classical vocabulary, the widespread use of the arch in Roman architecture led to an increase in the number of ways of combing the Orders with the arch.

Arches were placed above entablatures in the first century AD and the entablature was interrupted to form a high opening. In later years the same principle was applied to continuous rows of arches, or arcades. The arches then sat on isolated vertical pieces of entablature called dosserets, which gave the Order the appearance of greater height (*a*). In the eastern provinces of the Roman Empire another variant form (*b*) evolved for continuous rows of columns supporting arches. Here, the arch and its archivolt sprang directly from the column capital and the entablature was omitted altogether. This type of arcade spread to all parts of the later Roman Empire and was to influence the development of Romanesque architecture. Both these forms of arch were popular in the early Renaissance.

In the later Renaissance, arches springing directly from columns were combined very successfully with pediments that had the central part of the entablature left open (*c*) making the column capital or impost of the arch level with the column capital of the Order. This achieved a similar effect to another late Roman detail (*d*) in which the whole entablature arched over the column. Known as an arcuated lintel, this innovation also reached the late Roman Empire from its eastern provinces. It can be seen either as a central feature in a pediment or alternating with horizontal sections of entablature in an arcade (*e*). A less pronounced but contemporary treatment (*f*) involved arching the cornice only from a centre point level with the column capitals.

In the Renaissance a particular kind of arched opening (*h*) was developed which combined the horizontal entablature and arch. It was invented by Bramante, but both Serlio and Palladio used it frequently to great effect and it has become known as a Venetian, Serlian, or Palladio window. It became very popular in the eighteenth-century Palladian revival.

The Orders can also be contained within arches. The Order can sit independently inside the arch (*i*) or can line the inner face of the arch, with or without a horizontal entablature, giving added intricacy to the edge of the opening (*g*). On a larger scale there are many ways of placing the Orders inside arches of various sizes. It is, for example, possible to use the Orders as a screen across an arched opening. Robert Adam used a simple row of columns supporting a vase to act as the entrance to Balbardie House in Bathgate, Scotland (*k*) in 1792 and a Palladian window inside a larger arch (*j*) gives a reduced window size and a crescent of plain wall which can receive a decorative pattern.

a
b
c
d
e
f
g
h
i
j
k

拱楣和拱肩

半圆形的拱券与长方形的柱式组合创造出两个中间界面，并演变出其特有的装饰语汇。拱券内部从开始起拱的底边至拱顶石区域，称为门楣中心。在长方形内部开口处设计拱券，开口并未起拱至拱券，这样可以用开口部分纹案加以装饰或填充，创造出门楣中心。位于拱券两端的类似三角形状的区块，处于长方形开口空间之内，称为拱肩，这些部分经常带有装饰。

罗马凯旋门装饰华丽，有雕像和铭文以及塑像，布满了拱肩嵌板。建于203年的塞普蒂默斯·塞维鲁凯旋门（*c*），其中央开口上，与其他凯旋门相似，有两个带翼的胜利之神雕像，带着俘获的武器盔甲作为得胜的纪念，飞向中央的拱顶石。在同一座凯旋门上，有带拱肩的侧拱（*a*），拱肩上雕刻着斜倚的河神。这种拱肩装饰由于对这些古代纪念碑具有重要意义而几经复制，它们几乎完整地保留至今。英格兰，利物浦的圣乔治礼拜堂（*b*），由哈维·朗斯代尔·埃尔姆斯（1814—1847年）和查尔斯·罗伯特·科克雷尔于1839年设计，这一雕塑主题经改良以适应柱式上方升起的拱。寓言故事中的人物填充T形拱肩，其两翼向拱顶石处延伸。

简单的拱肩装饰（*d*）带有圆形的细部设计，环绕在最宽处四周。这些可以添加装饰或者较小的雕像，也能够置于拱廊中较大拱肩（*e*）之上，这些拱廊围绕着下面的柱子。更多变幻莫测的雕塑主题，如莨菪叶子或其他自然纹案设计（*f*），经常应用于拱肩之上，且通常围绕着一个居于最宽部位的主要几何纹案。

有了更加宽阔且更少限制的装饰嵌板，山墙饰内三角面的装饰变得更加丰富多样。半圆形通常建议采用放射状设计，最常见的主题为放射状沟槽图案，这是一种抽象的纹案（*g*），可能取材于贝壳。还可增加其他几何或者自然纹案（*i*），这些只是受到传统习俗与设计者的想象力限制。雕刻嵌板通常置于入口处的拱券之上或其他重要位置。山墙饰内三角面雕刻群发展为其他意识形态主题（*h*），在罗马式建筑中为一大特色，并且在文艺复兴时期的教堂中仍然流行。

简单的环形装饰细部设计对于山墙饰内三角面与拱肩同样适用。圆形嵌板位于拱券中心形成装饰框架，以便改变装饰的颜色、材质或者材料，或者能够为雕刻群（*j*）做更小的分区。圆形本身通常可依据其与半圆形拱券的关系，分割成简单几何图案的窗户（*k*和*l*）。

TYMPANUM AND SPANDREL

The combination of the semicircular shape of the arch and the rectangular shape of the Orders created two intervening surfaces that have developed their own decorative vocabulary. The area inside the arch from the line of its springing to its keystone is called the tympanum. When an arch is created over a rectangular opening and the opening does not rise into the arch, the tympanum thereby created can be decorated or filled with a decorative pattern of openings. The two virtually triangular areas created on either side of an arch inside a rectangular opening are called spandrels, and these are also frequently decorated.

Roman triumphal arches were heavily decorated with statues and inscriptions and sculpture filled the spandrel panels. The central opening of the Arch of Septimus Severus (*c*) of AD 203, in common with other triumphal arches, had two figures of winged Victory flying with trophies of captured arms and armour towards the keystone. On the same arch there are also side-arches (*a*) with spandrels containing reclining river gods. This type of spandrel decoration has been repeated many times due to the importance of these monuments in antiquity and their survival virtually intact to the present day. In St George's Hall in Liverpool, England (*b*), by Harvey Lonsdale Elmes (1814-47) and C.R. Cockerell in 1839, this sculptural theme was modified to suit arches springing above the Orders. Allegorical figures fill the T-shape spandrels, their wings spreading towards the keystone.

A simple spandrel decoration (*d*) has circular details centred on the widest point. These can be decorated or filled with smaller sculptures and can also be placed in the larger spandrels of arcades (*e*) centred on the column below. More capricious sculptural themes, such as acanthus leaf or other naturalistic designs (*f*), are frequently used for spandrels, often centred on a dominant geometric form at the widest point.

The decoration of the tympanum can be more varied, due to the broader and less restricted shape of the panel. The semicircular form immediately suggests radiating designs and one of the most common themes is a pattern of radiant fluting which can be abstract (*g*) or take the form of a shell. Other geometric or natural patterns (*i*) are often added and are only restricted by convention or the imagination of the designer. Panels of sculpture are frequently placed in arches over entrances or other important positions. The development of tympanum sculptural groups for theological subjects (*h*) became a particular feature of Romanesque architecture and remained popular for churches in the Renaissance.

Simple circular details are as appropriate to the tympanum as to the spandrel. Circular panels centred on the arch can form a framework for decorative changes of colour, texture, or material, or can divide the space for smaller sculptural groups (*j*). The circles are themselves often windows grouped into simple geometric patterns, (*k*) and (*l*), based on their relationship with the semicircle of the arch.

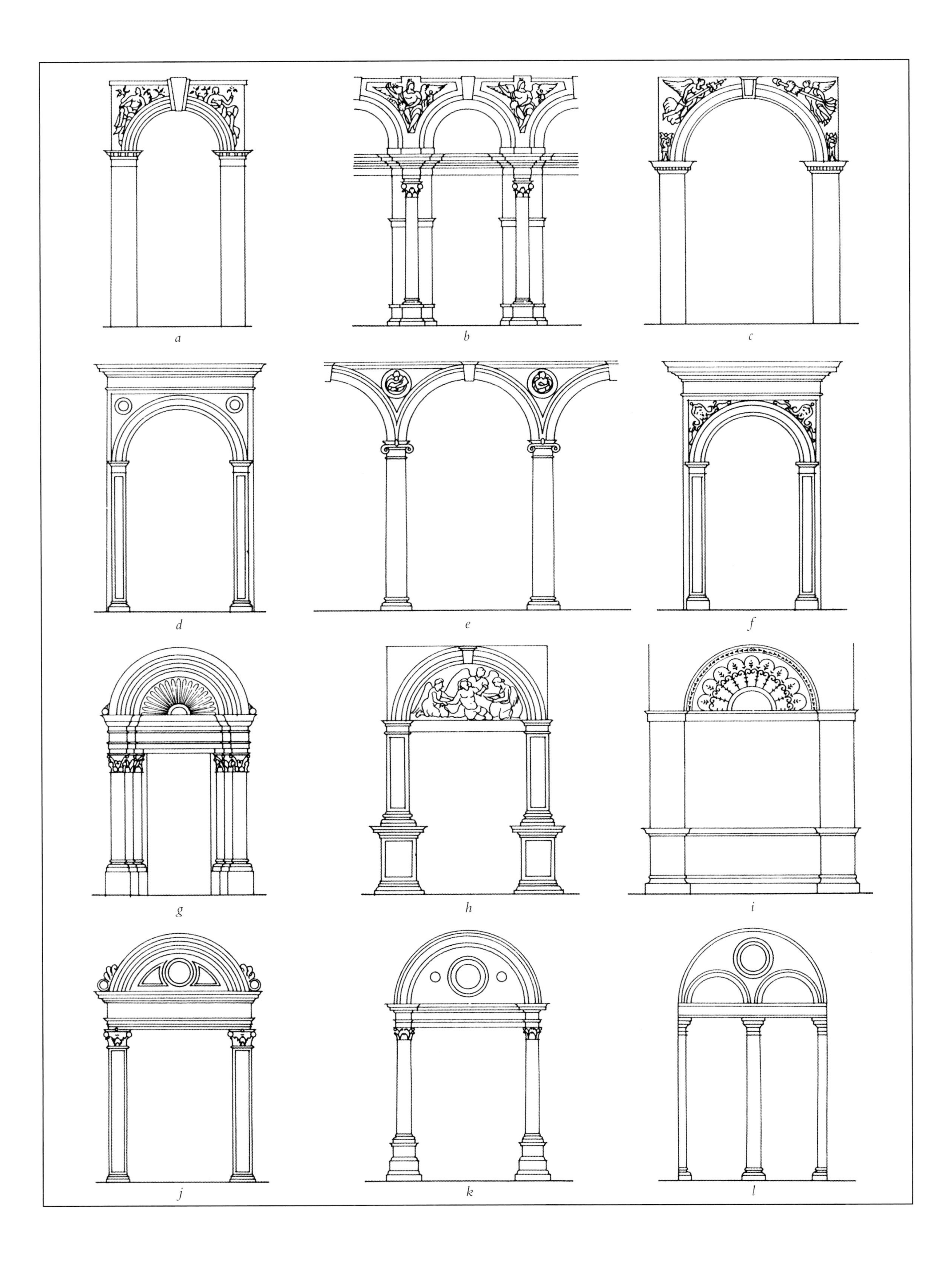
a
b
c
d
e
f
g
h
i
j
k
l

拱券与拱顶

拱券的底部，即拱腹部分加以拉伸，拱券就覆盖了一个空间而不是一个开口。拉伸的拱券变为拱顶，并且如果带有完整的半圆形拱腹，就成为圆筒形穹隆（*a*）。圆筒形穹隆，即连拱，与拱券同样古老，而且可能没有独立地位。罗马人发明了水泥，然而其结果是古老的石拱演变为三维形态，并创造出完整的结构，形成一个坚固的石质体量。这使得罗马建筑师能够相对轻松地设计出宏伟的拱顶建筑，并且促进拱顶设计的发展。水泥的流动特征使自由几何设计在这样的建筑结构中得以发挥，并且应用耐久的薄砖结构来包裹这种潮湿的材料，直至凝固，对这种自由手法几乎没有限制。在后期加入石材作为较薄的装饰覆面，加之砖石覆面的临时结构需求，因而维持了独立拱形结构外观。

圆筒形穹隆在其延长方向上可开小窗。如果这些都是拱形的（*b*），就在拱顶内部创造出一个复杂的弧形。如果两个相同的圆筒形穹隆相交（*c*），四条弧线相交点，沿对角线穿过正方形空间。这些线条称为穹棱，这种拱顶叫作穹棱拱顶。在见于罗马浴场和巴西利卡中，水泥穹棱跨度可达25米。

出于建造、强度或仅仅为了美学表现，拱顶中位于支撑点之上的拱形部分有时可通过降低拱腹的方式来表达。这就打破了圆筒形穹隆的线条（*d*）或者描绘出了穹棱拱顶（*f*）的平面。可进一步将拱顶与实际上明显用于支撑的拱形结构区分开来，柱群或混合柱式以及支墩（*e*）为每个元素提供视觉上的独立支撑。这种罗马后期的设计在中世纪得以广泛采用。

拱顶中拱腹的装饰也可区分结构中的拱形元素。位于拱顶中的简单正方形壁龛，即花格镶板（*g*），在拱券中可变为长方形。一种不同的设计，例如扭索状装饰（*i*）将增加拱顶的对比效果，正如这里的钻石形花格镶板所示。在连续的拱顶表面，设计的省略能够凸显拱顶中的拱券（*h*），这里展示的是不足半圆形的纹案。

罗马水泥结构随着西部帝国兴起而终结。拜占庭时期的建筑师使用非凡的技艺将水泥拱顶与穹隆结构变为砖石构造。在西方，罗马式和当时的哥特式建筑师，借助时尚的尖拱设计，将石造穹棱拱顶简化为斜拱即拱肋，创造出前所未有的纤细结构。圆筒形穹隆及其变体在文艺复兴时期得以复兴。

ARCHES AND VAULTS

When the underside, or soffit, of an arch is lengthened it will cover a space rather than form an opening. This lengthened arch becomes a vault and if it has an unbroken semicircular soffit it will be a barrel vault (*a*). The barrel vault, or continuous arch, is as old as the arch itself and probably had no separate identity. The Roman invention of concrete, however, resulted in the extension of the old stone arch in three dimensions to create a complete structure which set hard to form one solid mass. This enabled Roman architects to create vast, vaulted structures with relative ease and encouraged the development of the design of vaults. The fluid nature of the wet cement permitted great freedom in the geometry of these structures and the use of permanent thin brick structures to contain the wet material while it hardened did little to limit this freedom. Stone was often added later as a thin decorative surface and, with the temporary structural requirements of the brick skin, an appearance of independent arched structures was maintained.

The barrel vault can have small windows added in its length. If these are also arched (*b*), a complex, curved shape is created in the vault. If two equal barrel vaults intersect (*c*), four curved lines will mark their point of intersection, crossing the square space diagonally. These lines are groins, and this type of vault is a groin vault. Concrete groin vaults spanning up to 25 metres were constructed for Roman baths and basilicas.

In the interests of construction, strength or solely for aesthetic purposes, the arches over the point of support in vaults are sometimes expressed by dropping the soffit. This can break up the line of a barrel vault (*d*) or define the plan of a groin vault (*f*). The effect of separating the vault structure from real or apparent supporting arches can be taken further and clusters of columns or a mixture of columns and piers (*e*) provide visually independent supports for each element. This late Roman design was widely adopted in the Middle Ages.

The decoration of the soffit of vaults can also differentiate arched elements in the structure. Simple square recesses, or coffers, in the vault (*g*) can be continued as rectangles on the arch. A different design, such as guilloche (*i*), on the arch will give added contrast to the vault, shown here with diamond coffering. On an uninterrupted vault surface, just the omission of part of the design can suggest an arch in the vault (*h*) which is shown here as less than a full semicircle.

Roman concrete construction ended with the Western Empire. Byzantine architects used great skill in translating the concrete vault and dome structures into brick and stone. In the west, Romanesque and then Gothic architects, aided by the fashion for pointed arches, simplified the stone construction of groin vaults with diagonal arches, or ribs, to create structures of an unprecedented slenderness. The barrel vault and its derivatives were revived in the Renaissance.

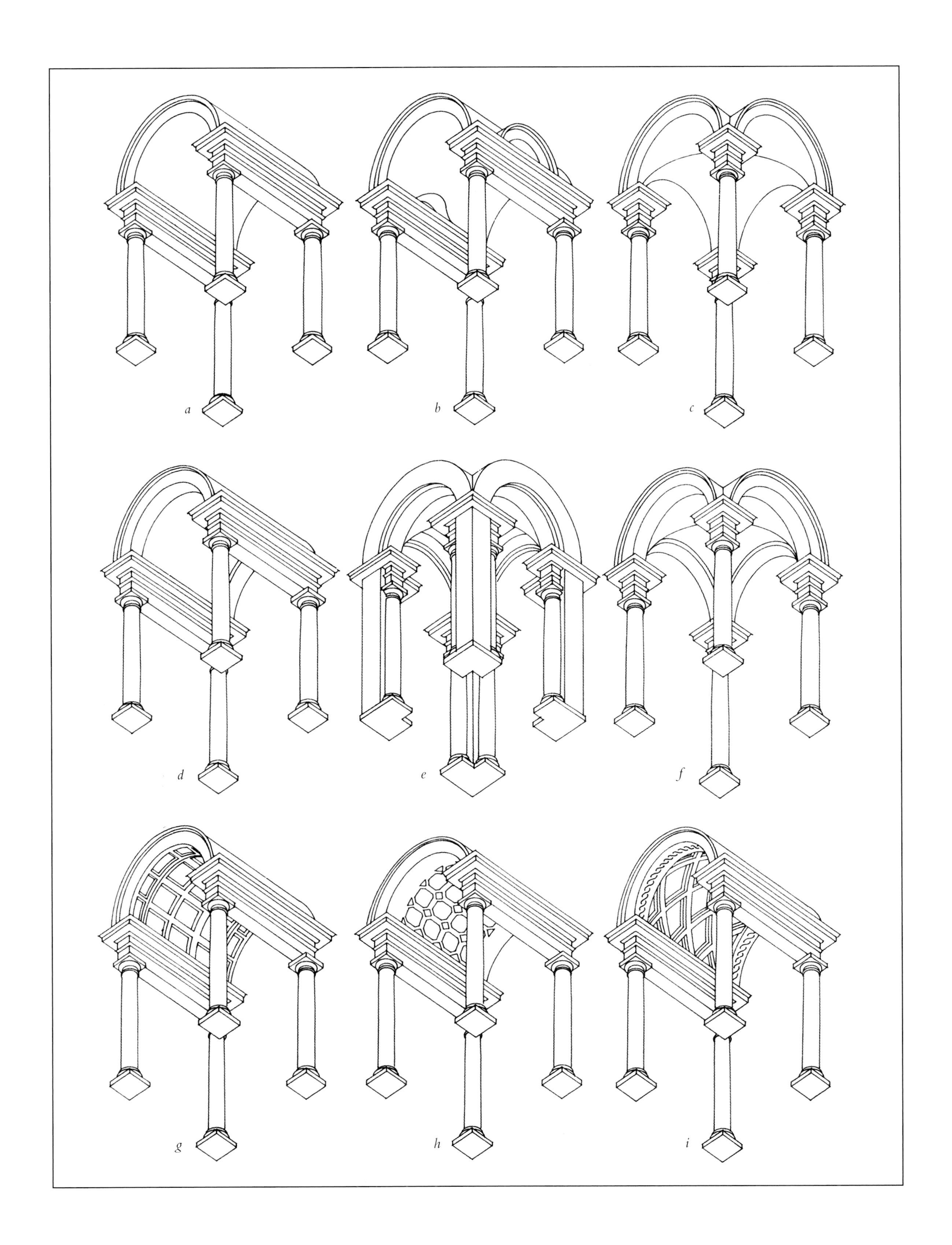
a
b
c
d
e
f
g
h
i

拱券及其设计之一

伊特鲁里亚和早期罗马的拱券是简单的砖石结构，有时带有与拱券构造相关的装饰特征，例如拱顶石与拱墩。这些拱券构成的长拱廊建于圆形剧场中并且支撑着阶梯。随着罗马向南部扩张，帝国的财富增长，与意大利城市之间的联系也增加了，罗马文明进入了古典文化的主流。正是此时，罗马为古典传统的发展做出了巨大的贡献，并且将拱券与柱式组合起来。

有一个重要的巧合，保留下来最早的实例为一座剧院——一个源于希腊的建筑模式——公元前13年的罗马马塞勒斯剧院（*a*）。尽管柱式与拱券的组合应用始于一个世纪之前，这座建筑展示了简单的延续连拱饰，这是早期实例中的特征。半圆形正立面，有着高高的上层结构，依赖重复同样的基本形制并且建有弧形墙面的手法创造出建筑效果。在柱式变换的细部设计中增加了产生对比效果的复杂装饰以及一些不同楼层间的变化。

与马塞勒斯剧院几乎是同时代的，法国南部城市欧坦的圣安德烈宫（*b*），采用相同的手法设计通道处画廊上的拱券和柱式。罗马城门的传统特征就是画廊形制的成排窗户。但是，正门并未采用柱式，而是沿用了早期大门的简单形制，但完整的檐部设计暗示了某一种柱式的存在，作为美术馆的基础并位于大门拱砌砖之上。在大部分的罗马建筑中，拱券与柱式的应用比许多后来的理论家们所论证的更加自由。

后罗马帝国时期，人们对于拱券与柱式的熟悉使得古典拱券的功能超越了一种体系对另一种体系的装饰作用，而整合成为一种建筑法则。罗马时代后期教堂内部设计，例如350年修建的城外圣埃格尼斯教堂（*c*），其内部设计有简单的拱廊柱，直接从柱头上向上起拱并且其垂直体量逐渐削减。

圣安德烈宫与圣埃格尼斯教堂这样的建筑对中世纪建筑师产生很大影响。城市大门重要的象征意义及其实用价值得以保留，在后期罗马教堂中享有无法匹敌的影响力。然而，当布斯格多在11世纪计划设计像比萨大教堂（*d*）这样的古典建筑时，古代世界的延续性已经消失，他们需要依赖来自罗马精挑细选保存下来的设计。总而言之，罗马帝国后期对拱券的应用赋予中世纪早期建筑以独特之处，即便当他们对古典传统加以错误却充满活力的应用时，也是如此。

ARCHES AND DESIGN 1

Etruscan and early Roman arches were simple masonry structures which occasionally included decorative features related to the construction of the arch, such as keystones or imposts. Long arcades of these arches were constructed on amphitheatres and to support terracing. As Rome expanded its control to the south, and both its wealth and contact with Greek cities in Italy increased, Roman civilisation entered the mainstream of Classical culture. It was at this time that the Romans made one of their major contributions to the development of the Classical tradition and united the arch with the Orders.

It is significant coincidence that one of the best preserved early examples of this combination is on a theatre—a Greek import—the Theatre of Marcellus in Rome (*a*) of 13 BC. Although the Orders and arches had been used together for a century before, this building displays the simple continuous arcading that characterised the earlier examples. The semicircular façade, which had a high upper storey, relies for its architectural effect on the repetition of the same basic form and the curve of the wall. The varied details of the Orders add a contrasting intricacy and some differentiation between the floors.

Almost contemporary with the Theatre of Marcellus, the Porte S. André in the city of Autun in southern France (*b*) has a similar treatment for the arches and Orders on the gallery over the gateway. This row of windows in form of a gallery was a traditional feature on Roman city gates. The main gates, however, have no applied columns and take the simple form of earlier gates, but the existence of an Order is implied by the full entablature that acts as base for the gallery and sits immediately above the voussoirs of the large gates. There is more freedom in the use of arches and the Orders on much Roman architecture than many later theorists have argued.

By the late Roman Empire, familiarity with the combination of the arch and the Orders had taken the Classical arch beyond the application of one system to decorate another to an integrated architectural discipline. The interiors of late Roman churches, such as S. Agnese fuori le Mura (*c*) of about 350 AD, have simple arcades of columns in an interior springing directly off column capitals and diminishing in scale vertically.

Buildings such as Porte S. André and S. Agnese had a strong influence on medieval architects. The symbolic significance of city gates together with their practical use survived, and late Roman churches enjoyed unrivalled influence. Continuity with the ancient world had, however, been lost, and, when architects like Buscheto wished to design Classical buildings such as Pisa Cathedral (*d*) in the mid-eleventh century, they relied on selective survivals from Rome. It is, above all, the late Roman use of the arch that gives these early medieval buildings their distinction, even when executed with a vigorous misunderstanding of the Classical tradition.

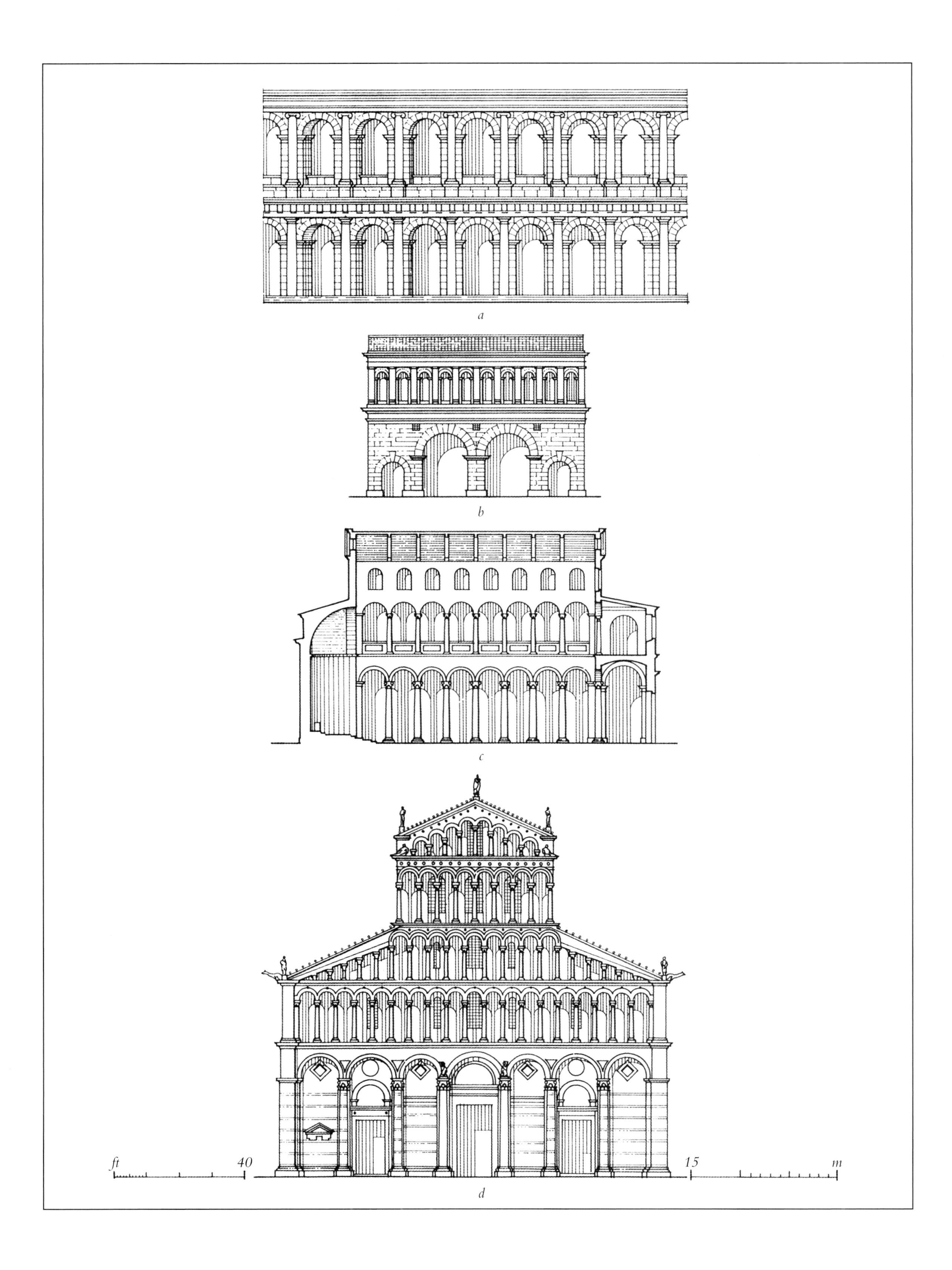
a
b
c
ft
40
15
m
d

拱券及其设计之二

文艺复兴时期建筑师，由于错误地将其与毁灭罗马帝国的哥特族人联系起来，而毅然决然地拒绝哥特式尖拱。文艺复兴早期，大部分当时的建筑都是哥特式，而圆拱是古典世界的象征。类似佛罗伦萨贡迪府邸（*b*）这样的建筑，由朱利亚诺·达·桑迦洛（1445—1516年）于1489年设计，需要用这样的观点来看待。设计依托在渐变的粗砌墙面上有序排列简单拱券的手法。顶部的檐口仅仅作为柱式的参考。

1545年，帕拉迪奥在竞赛中获胜，赢得了为意大利北部城市维琴察一座中世纪巴西利卡（*a*）修建新的正立面的机会，这时拱券成为普遍熟知的建筑语汇中的一个重要元素。帕拉迪奥的设计由两个带有威尼斯窗的拱廊构成。尽管帕拉迪奥并未发明这一细部设计，但在这里却加以最精确地运用。拱券是复杂结构的主要特征，这一结构包括多立克柱式及其上部的爱奥尼亚柱式。每一个开间处都有同样柱式的辅柱支撑拱券。

巴洛克建筑师为获得大胆的视觉效果而再次探究拱券。英国建筑师霍克斯莫尔以其惊人的组合设计而著称，使用固体墙面与开口的对比而不是精妙的细部设计。其位于伦敦斯皮塔佛德，建于1723年的塔楼以及基督教堂的门廊的混合体（*d*），这一设计几乎完全由不同大小的拱券构成，与巨大的托斯卡纳柱式形成对比。

19世纪几种历史遗迹同时复兴，赋予古典拱券以新的重要意义。哥特式建筑的复兴在将人们的注意力吸引到拱券的形状上来。对文艺复兴早期，不断增长的兴趣，加之拜占庭与罗马式建筑的复兴，在德国叫作圆拱式。圆拱包含几种受欢迎的复兴建筑风格的优势，19世纪早期的建筑师，例如申克尔拥有不止一个这样风格的设计。他在德国汉堡设计的剧院（*c*），使用大型镶嵌玻璃拱券构成的拱廊，达成了模棱两可的设计风格，成为19世纪的特征。

这一世纪的两次世界大战期间，古典建筑刻意剥除装饰，而创造出精美且简单的设计。建筑师们熟悉古典建筑语汇，以最简洁的元素阐释古典形制。古典建筑历史对于法西斯政府在罗马城外开始修建的新兴城市EUR尤为重要。不可避免且经过深思熟虑地采用与古代世界相媲美的设计。新城的中心为乔瓦尼·圭里尼（1887—1972年）、欧内斯托·拉帕杜拉（1902—1968年）和马里奥·罗曼诺（活跃于20世纪30—40年代）设计的意大利市民广场（*e*），戏称为方形角斗场，这座建筑只使用了六层拱券，简化为只保留其形状，使人们能够感受到附近首府纪念碑。

ARCHES AND DESIGN 2

Renaissance architects decisively rejected the pointed Gothic arch, incorrectly associating it with the Gothic tribes that destroyed Rome. In the early Renaissance, when the majority of existing buildings were Gothic, the round arch had a unique symbolic association with the Classical world. Building such as the Palazzo Gondi in Florence (*b*), designed by Giuliano da Sangallo (1445-1516) in 1489, must be seen in this light. The design relies on ordered rows of simple arches set in a graduated rusticated wall. The crowning cornice is the only reference to the Orders.

When Palladio won the competition to give a new façade to the medieval basilica in the city of Vicenza in northern Italy (*a*) in 1545, the arch had become just one of the elements in a universally familiar Classical vocabulary. Palladio's design is composed with two arcades of open Venetian windows. Although Palladio had not invented this detail, this is its most explicit use. The arches are a dominant feature in a complex composition which has a Doric Order below an Ionic Order. Each bay has a subsidiary column of the same Order supporting the arch.

Baroque architects again sought out the arch for bold visual effects. The English architect Hawksmoor is notable for striking compositions which use the contrast between solid wall and openings more than fine detail. In his combined tower and portico of Christ Church, Spitalfields in London (*d*) of 1723, the design is almost entirely made up of arches of different sizes contrasted with a huge Tuscan Order.

In the nineteenth century the simultaneous revival of several historic styles gave the Classical arch new significance. The revival of Gothic architecture once again drew attention to the shape of arches. An increased interest in the early Renaissance was accompanied by revivals of Byzantine and Romanesque architecture which were called in Germany the round-arch style. Round arches had the advantage of encompassing several favoured revivals and early nineteenth-century architects like Schinkel designed in more than one of these styles. His theatre design in Hamburg, Germany (*c*), by the use of arcades of large glazed arches, achieves an ambiguity of style that is characteristic of the nineteenth century.

Classical architecture between the wars in this century was deliberately stripped of much of its decoration to create refined and simplified designs. Familiarity with the Classical vocabulary allowed architects to suggest Classical form with the barest elements. The Classical past was particularly important in the new city of EUR that the Fascist government started to build outside Rome. The parallel with the ancient past was both unavoidable and deliberate and at the centre of the new city was the Palazzo della Civiltà Italica (*e*) by Giovanni Guerrini (1887-1972), Ernesto Lapadula (1902-68) and Mario Romano (*fl.*1930s-40s). Nicknamed the square Colosseum, this building uses nothing but six tiers of arches reduced to no more than their shape to evoke the monuments of the nearby capital.

m
15
40
ft
a
b
c
d
e

混合拱券及其测定

拱券并不全都是半圆形。它们可能拥有不同的比例使其能够适用于高度并非为宽度的一半的开间。

当开间的宽度不同，并且全半圆形拱券横跨其上时（*a*），拱券的高度一定是不同的。如果拱券需要是等高的设计，就有可能对中央或边拱进行调整。较窄的拱券能够升高（*b*）使其圆心高于支墩。高度就由拱券的垂直延伸部分构成。这些是上心拱也可能是简单半圆形，如左侧图例所示，也可能是右侧图例所示的弓形。大型拱券的高度可以通过两种方式来削减：三心拱（*c*）或者弓形拱（*d*）。

三心拱能够通过几种方式来绘制。当开口的宽度和拱券所需宽度达到临界尺寸（*e*），有必要了解如何设定拱券。

首先，画出底部宽度，即起拱处宽度。图示中*AB*，以及顶点的高度在中心位置*C*点，从*C*点向下画垂直线。从*C*点开始画另一条线与*AB*平行并与*B*点向上的垂直线相交于*D*点，构成长方形*CDBE*。将垂直线*CE*向下延伸，并与之等长到达*F*点，因此*CE*与*EF*等长。将*EB*划分为两部分，*EG*与*GB*等长。将*DB*划分为两部分，*DH*与*HB*等长。从*F*点开始画线穿过*G*点到达*D*点，从*H*点到*C*点画另一条线。这些线将交于*J*点，在*K*点分割*CJ*线，使得*CK*等于*KJ*，画一条穿过*K*点与*CJ*垂直的线，与*CEF*延长线交于*L*点，*L*点就是三心拱的第一个圆心，*LC*为三心拱中最大弧形的半径。从*L*点画延长线*LM*与*AB*线平行，连接*M*、*A*两点并延长至最大弧上的*N*点，连接*L*、*N*两点，并与*AB*交于*P*点，这就是三心拱的第二个圆心，第三个圆心Q点在*E*点另一侧等长处。

设定弓形弧的高度和宽度（*f*），在开口宽度处画*AB*线，*C*点为等分点，所以*AC*等于*CB*，从*C*点处画拱高交于*D*点，*E*点处划分*AD*线，*AE*等于*ED*，穿过*E*点画*AD*的垂直线，与*DC*延长线交于*F*点，*F*点就是弓形弧*ADB*的圆心。

图例（*g*～*j*）展示的是不同高度的三心拱的外观，而图例（*n*～*m*）则展示了不同的弓形弧。

MIXING ARCHES AND SETTING OUT

Arches are not always semicircular. They can have different proportions which permit them to be used in situations where a height other than half the width of the opening is appropriate.

When openings are different widths and are spanned by full semicircular arches (*a*) the height of the arches must necessarily be different. If the arches need to be the same height, then it is possible to make modifications to the central or side-arches. The narrower arches can be raised (*b*) so that their centres are above the imposts. The height is then made up by vertical extensions to the arch. These are stilted arches and can be simple semicircles, as on the left, or segmental, as on the right. The height of the large arch can be reduced in two ways: with a three-centred arch (*c*) or with a segmental arch (*d*).

The three-centred arch can be drawn in several different ways. It is most useful to know how to set out the arch when the width of the opening and the desired height of the arch are the critical dimensions (*e*).

First, draw the width at the bottom, or springing, of the arch, *AB* on the drawing, and the height at the top and centre, point *C*, and draw a vertical line from *C* downwards. Draw another line from point *C* parallel to *AB* which will meet a vertical line from *B* at point *D*, forming a rectangle *CDBE*. Extend the vertical centre line *CE* by the same distance again to point *F* so that *CE* is equal to *EF*. Divide *EB* into two so that *EG* is equal to *GB*. Divide *DB* into two so that *DH* is equal to *HB*. Draw a line from point *F* through point *G* to point *D* and draw another line from point *H* to point *C*. These lines will meet at point *J*. Divide the line *CJ* at point *K* so that *CK* is equal to *KJ* and draw another line at right-angles to *CJ* through point *K* which will cross the centre line *CEF* at point *L*. Point *L* is the first centre point of the three-centred arch. *LC* then becomes the radius of the arc of the larger of the three circles making up the arch. Extend this radius down to a line *LM* which is parallel to *AB* and draw a line from point *M* through point *A* to meet the arc of the large circle at point *N*. Connect point *N* to point *L* and where this line crosses the line *AB* and draw a line from point *M* through point *A* to meet the arc of the line *AB*, point *P*, is the second centre point of the three-centred arch. The third centre at *Q* is an equal distance, but on the other side of point *E*.

To set out a segmental arch from its height and width, (*f*), draw a line *AB* across the width of the opening and divide it at *C* so that *AC* is equal to *CB*. Draw a centre line from *C* to the height of the arch *D*. Divide *AD* at *E* so that *AE* is equal to *ED* and draw another line at right-angles to *AD* through point *E* to meet *DC* extended to point *F*. Point *F* is the centre of the segmental arch *ADB*.

The appearance of different heights of three-centred arches is shown in (*g*) to (*j*) and similar differences are shown in (*n*) to (*m*) for segmental arches.

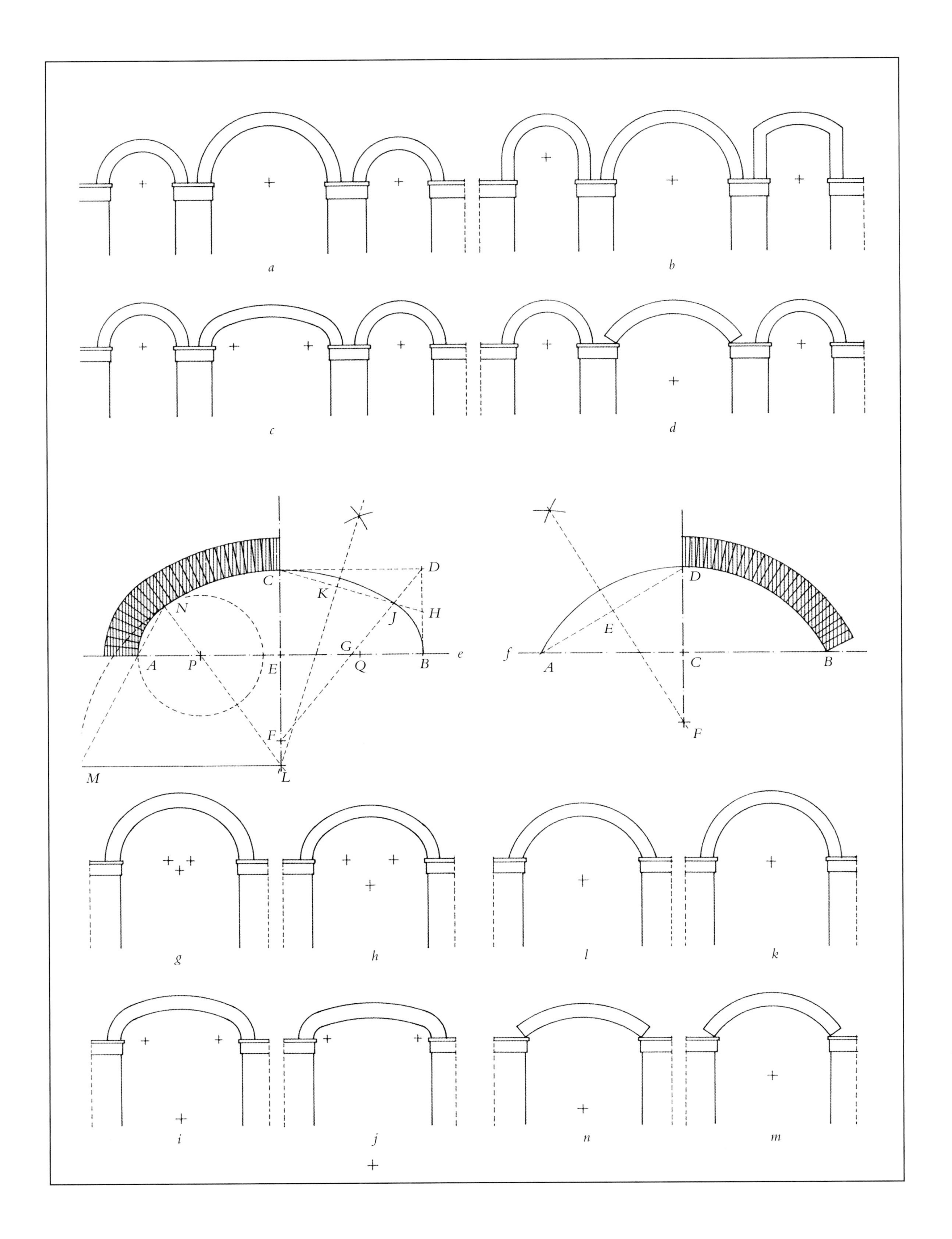
a
b
c
d
C
D
K
N
H
J
G
e
A
P
E
Q
B
F
M
L
D
E
f
A
C
B
F
g
h
l
k
i
j
n
m

10. 穹顶

古代穹顶

从史前时代到今天，人们一直都修建带有圆锥形或者圆形屋顶并且由泥或者树枝等材质构成圆形建筑。大约公元前1600年，迈锡尼人开始在希腊大陆上修建地下穹顶圆形建筑物，可能是早期传统或者宗教仪式使用的圆形建筑物所遗留下来的产物。其中最大的阿特柔斯宝库（*a*），可追溯至公元前1300年。通过将石头平放，逐渐向内叠起最后交会于中央的方式创造出石造的蜂窝形状。这种穹顶构造的原始形制随着迈锡尼文明的衰退而终结。

公元前2世纪，罗马人发现了用以制造水泥的火山砂、火山灰的独特优势，改革了建筑构造，使得建筑师放弃了古老的柱子与横梁等建筑构造，更加喜爱应用弧形屋顶，即拱顶和穹顶。穹顶是罗马人对古典建筑最伟大的贡献之一。

罗马建筑师，随着发现新材料的结构潜力，同样探索了穹顶的美学潜力。结构工程的伟大功绩与古典细部设计相结合创造出了令人敬仰的拱顶或穹顶内部设计。这些设计中最伟大的当属罗马万神殿（*c*），建于120年，哈德良皇帝统治时期。穹顶内部为半球形，置于环形等高的鼓座之上，因此43.2米的直径也是内部高度的尺寸。光线从唯一的环形开口处进入，或者称为眼洞窗，位于穹顶顶部。穹顶由坚固的罗马水泥与较轻的火山石构成；没有金属加固材料或者环形锁链来防止穹顶向外坍塌。由于这一原因罗马穹顶为碟形外观，创造出逐渐增厚的建筑结构，因此将建筑材料的重量向下转移至墙体。

穹顶结构一直沿用至罗马帝国结束。一个大型花园凉亭，人们称其为智慧之神密涅瓦·梅狄卡神庙（*b*），是从4世纪罗马保存下来的。鼓形座10个面，上层有窗户汇入穹顶。每一边由拱券穿过，除了入口，开口处进入半圆形后殿上面附有半穹顶。穹顶中有砖石拱肋，可能是修建过程中遗留下来的部分。

10. THE DOME

DOMES OF ANTIQUITY

Round buildings with conical or domed roofs and made of mud or branches have been constructed from prehistoric times to the present day. In about 1600 BC the Mycenaeans started to build domed subterranean tholos tombs on the Greek mainland, probably as a traditional or ritual survival of earlier circular buildings. One of the largest of these, the Treasury of Atreus (*a*), dates from about 1300 BC. The stone beehive shape is created by laying stones flat and stepping them gradually inwards until they meet at the top. This primitive form of dome construction ended with the decline of the Mycenaean civilisation.

The Roman discovery in the second century BC of the unique strength of a volcanic sand, pozzolana, for making concrete, revolutionised building construction, allowing architects to abandon the old post and beam form of construction in favour of curved roofs, or vaults, and domes. The dome is one of Rome's great contributions to the development of Classical architecture.

Roman architects, as they discovered the structural potential of the new material, also exploited the aesthetic potential of the dome. Great feats of structural engineering were united with traditional Classical details to create awe-inspiring vaulted and domed interiors. The greatest of these is the Pantheon in Rome (*c*), built in the reign of the Emperor Hadrian in AD 120. The inside of the dome is a hemisphere which sits on a circular drum of an equal height so that the diameter of 43.2 metres is the same dimension as the total interior height. Light is admitted only by a single circular opening, or oculus, at the top. The dome is composed of solid Roman cement mixed with light volcanic stone; no metal reinforcement or encircling chain prevents the dome from bursting outwards. For this reason all Roman domes have a saucer-shaped exterior which creates a progressive thickening of the structure, thereby transferring the great weight of the material down to the walls.

Domed structures remained in use until the end of the Roman Empire. A large garden pavilion, known as the Temple of Minerva Medica (*b*), has survived in Rome from the fourth century. The drum has ten sides, with windows at a high level which merge into the dome. Each side is pierced by an arch which, on all sides except the entrance, opens into a semicircular apse covered by a half-dome. There are brick ribs built into the dome, probably as a part of the construction process.

a
b
c
m
10
ft
30

文艺复兴后期穹顶

拜占庭时期建筑师继续发展罗马穹顶的传统，但将其从水泥结构变为砖石构造。当15世纪早期，文艺复兴对古典的激情征服了意大利城市时，穹顶的建造再次繁荣起来，意大利建筑师求助于拜占庭建筑经验而不是遗失的水泥建造工艺。

佛罗伦萨大教堂（*b*）的穹顶是站在中世纪与文艺复兴十字路口的建筑。1420年文艺复兴的艺术先驱和建筑师菲利波·布鲁内列斯基解决了看似不可能完成的任务，在建于1357年的八角形教堂神坛上修建了穹顶。已有墙体的高度和超过42米的宽度意味着应用木材作为临时支撑，即拱鹰架的传统做法是不可能的。布鲁内列斯基独具匠心地在聚拢的拱券，即拱肋之间设计了犬牙交错的薄砖层，在建造过程中将其充分交织在一起，避免了拱鹰架的使用。到1367年，哥特式的拱顶比例已经成为定式。但是布鲁内列斯基赋予这一形制以简单的线条，通过首次使用木制或金属的拱环或链条来防止穹顶向外崩塌，因此避免了石质支撑外壳和哥特式建筑师们使用的平衡块。穹顶的大小以及独创性十分闻名，其结构链条、古典细节设计以及结构分割和防风雨层被广泛复制。

1506年，开始了罗马最神圣殿堂的重建工程——圣彼得墓上的康斯坦丁巴西利卡。建造延续了一个多世纪，许多重要的意大利建筑师都参与其中。布拉曼特早期的绘图展现了大量简单的内部半球形，罗马式倒置碟形外部设计位于环形柱阵之上。然而，穹顶（*c*）与米开朗琪罗1550年设计好的平面大不相同。米开朗琪罗的穹顶为42米宽，到穹隆顶塔底部为29米高，其底座从地面算起为76米高，两层砖结构由10条铁链固定。复杂且引人注目的设计，及其凸出的轮廓和圆锥形拱肋是古典穹顶建筑的发展标尺，对无数后来的设计产生了深远影响。

圣彼得大教堂的巴洛克式辉煌，远离18世纪新古典建筑师的古典理念。深受考古学家与现存的罗马建筑影响，建筑师修建带有倒置碟形外观和半球形内部设计的简单穹顶。与古代水泥建筑不同，新古典建筑穹顶，例如建于1786年的都柏林福尔宫（*a*），由詹姆斯·冈东（1743—1823年）设计，使用木材与钢结构修建，创造出古代世界的灰泥与木材效果。

POST-RENAISSANCE DOMES

Byzantine architects continued to develop the tradition of the Roman dome, but changed the construction from cement to brick or stone. When the Renaissance passion for antiquity gripped the cities of Italy in the early fifteenth century, and the construction of domes flourished again, Italian architects turned to the experience of Byzantium rather than to the lost art of cement construction.

Standing at the crossroads of the medieval world and the Renaissance is the dome of Florence Cathedral (*b*). In 1420 the pioneering Renaissance artist and architect Brunelleschi solved the seemingly impossible task of erecting a dome over the altar of the cathedral on an existing octagonal structure designed in 1357. The height of the existing walls and a width of more than 42 metres meant that the traditional use of temporary timber supports, or centering, was impractical. Brunelleschi devised an ingenious method of building thin layers of bricks in an interlocking pattern between converging arches, or ribs, which knitted together sufficiently during construction to avoid the use of centering. The Gothic proportions of the dome had been fixed by decree in 1367, but Brunelleschi gave the form a new simplicity of line by using, for the first time, a ring, or chain, of timber and iron to prevent the dome bursting outwards, thereby avoiding the encrustation of stone supports and counterweights used by Gothic architects. The size and ingenuity of the dome became famous and its structural chain, Classical details, and separation of structural and weatherproofing layers were widely copied.

In 1506 work began on the rebuilding of the holiest shrine in Rome, Constantine's basilica over the tomb of St Peter. Building continued for more than a century and involved many leading Italian architects. Early drawings by Bramante show a huge but simple internal hemisphere with a Roman, inverted-saucer-shaped exterior above a circle of columns. The dome (*c*) was, however, built to quite different plans prepared by Michelangelo in about 1550. Michelangelo's dome is 42 metres wide, 29 metres high to the underside of the crowning lantern, and its base is 76 metres from the floor. It is constructed of two layers of brick and is bound together with ten iron chains. The complex and dramatic design, with its pointed profile and converging ribs, is a landmark in the development of the Classical dome and had a profound influence on the design of innumerable subsequent examples.

The Baroque splendour of St Peter's was too far removed from the concept of antiquity held by Neo-Classical architects at the end of the eighteenth century. Influenced by archaeology and surviving Roman buildings, architects constructed simple domes with inverted-saucer-shaped exteriors and hemispherical interiors. Unlike the concrete structures of antiquity, Neo-Classical domes, such as the double domes of the Four Courts of Dublin (*a*) of 1786, by James Gandon (1743-1823), were built with timber and iron frames to recreate the forms of the ancient world in plaster and wood.

m
25
ft
70
a
b
c

穹顶的种类

在古典建筑史上，穹顶建筑有许多不同形制，受支撑结构形状、采光方式、建造手法以及对视觉效果的追求等因素影响。

穹顶的圆形平面恰好置于环状支撑结构之上，这一结构称为鼓形座（*a*）。从古代开始，这一手法就比较常用。然而，当基础墙为方形，方形与圆形的关系就成为平面的重要特征。

穹顶可置于方形结构之上，几何差异表现得十分明显（*b*），但是这一设计并未经常使用，由于穹顶仅由墙体的四个角来支撑。为解决这一问题，穹顶能够逐渐融合变为方形（*c*）。每个边角上三角形的空间，位于拱券弧形与穹顶边缘之间，能够弯曲为拱券的起拱部分，称为穹隅。拜占庭时期建筑师十分喜爱穹隅设计。穹隅结构无须直接融入穹顶圆周中，却能够支撑中间拱形座（*d*）。如果穹顶不足完整的半球形，最终将完全融入穹隅，以至于两个结构成为一个结构，通过去除穹隆环形平面的4个部分来制造方形（*e*和*f*）。在实例（*f*）中，浅穹隆与穹隅的区别完全消失了。

穹隅的替代品可以是穿过方形一角的拱券，人们称之为突角拱（*g*）。可能是单一拱券或者一系列拱券逐渐向下沿着穹隅的线条延伸至边角处的设计。在其最简单的形制中，突角拱构成了八边形，正方形的每一边上都有4个拱。这一设计反过来支撑了一个八边形的鼓形座（*h*）——在这一实例中拱形窗起拱于穹顶内。突角拱是另一种拜占庭式设计手法。八边形能够毫不费力地融入环形穹顶（*i*）或者以板片的形式（*j*）延伸至穹顶中心点。如果八拱开口在穹顶内部起拱（*k*），拱形结构能够在每部分中逐渐削减直到最终在中间会合。这种特殊的形式得名伞形穹顶。

这些形状以及其他形状都应用于简单穹顶的设计中。有很多不同的变体与组合。比如，伞形穹顶可能会有12条边，每边带有拱形窗起拱于穹顶中并且（从穹顶内部观察）为凹面结构。这一设计可置于带有穹隅的正方形拱券结构的圆形开口（*l*）上。

TYPES OF DOME

In the history of Classical architecture, domed structures have taken many different forms influenced by the shape of the supporting structure, the means of admitting light, the method of construction and the desire for visual effect.

The circular plan of the dome sits comfortably on a circular supporting structure called a drum (*a*). This arrangement has been used frequently since antiquity. When the shape of the underlying walls is square, however, the relationship between the square and circular plans becomes a significant feature of the design.

The dome can just be placed on the square structure and the geometric difference clearly expressed (*b*), but this arrangement is not often used as the weight of the dome is only supported by the walls at four points. To overcome this problem the dome can gradually merge into the square. This type of construction is generally associated with arches in each side of a square (*c*). The triangular space at each corner, between the curves of the arches and the edge of the dome, can be curved down to the springing of the arch and is known as a pendentive. Pendentives were particularly favoured by Byzantine architects. A pendentive structure does not have to merge directly into the circumference of the dome but can support an intermediate drum (*d*). If the dome is less than a full hemisphere it will eventually merge so completely with the pendentives that the two forms become a single dome, made square by removing four segments from the circular plan of the dome, (*e*) and (*f*). In example (*f*) the distinction between the shallow dome and the pendentives has completely disappeared.

A substitute for pendentives is an arch across the corner of a square. This is known as a squinch (*g*). It can be a single arch or a series of arches diminishing downwards towards the corner, following the line of a pendentive. In their simplest form, squinch arches form an octagon with the four arches on the sides of the square. This in turn can support an octagonal drum (*h*)—in this example with arched windows rising into the dome. Squinch arches were another Byzantine device, but domes on octagonal drums are also to be found on early Roman domed structures. The eight sides of the octagon can quite effortlessly merge into a circular dome (*i*) or extend to the central point of the dome as flat leaves (*j*). If eight arched openings are placed to rise into the dome (*k*) the arched forms can be gradually diminished in each segment until they meet in the centre. This distinctive shape gives the name umbrella dome.

These shapes and others have been used in the design of simple domes. There are many varieties and combinations. An umbrella dome can, for example, have twelve sides each with an arched window rising into the dome and (viewed from the inside) concave segments. This can sit on a circular opening in a square, arched structure with pendentives (*l*).

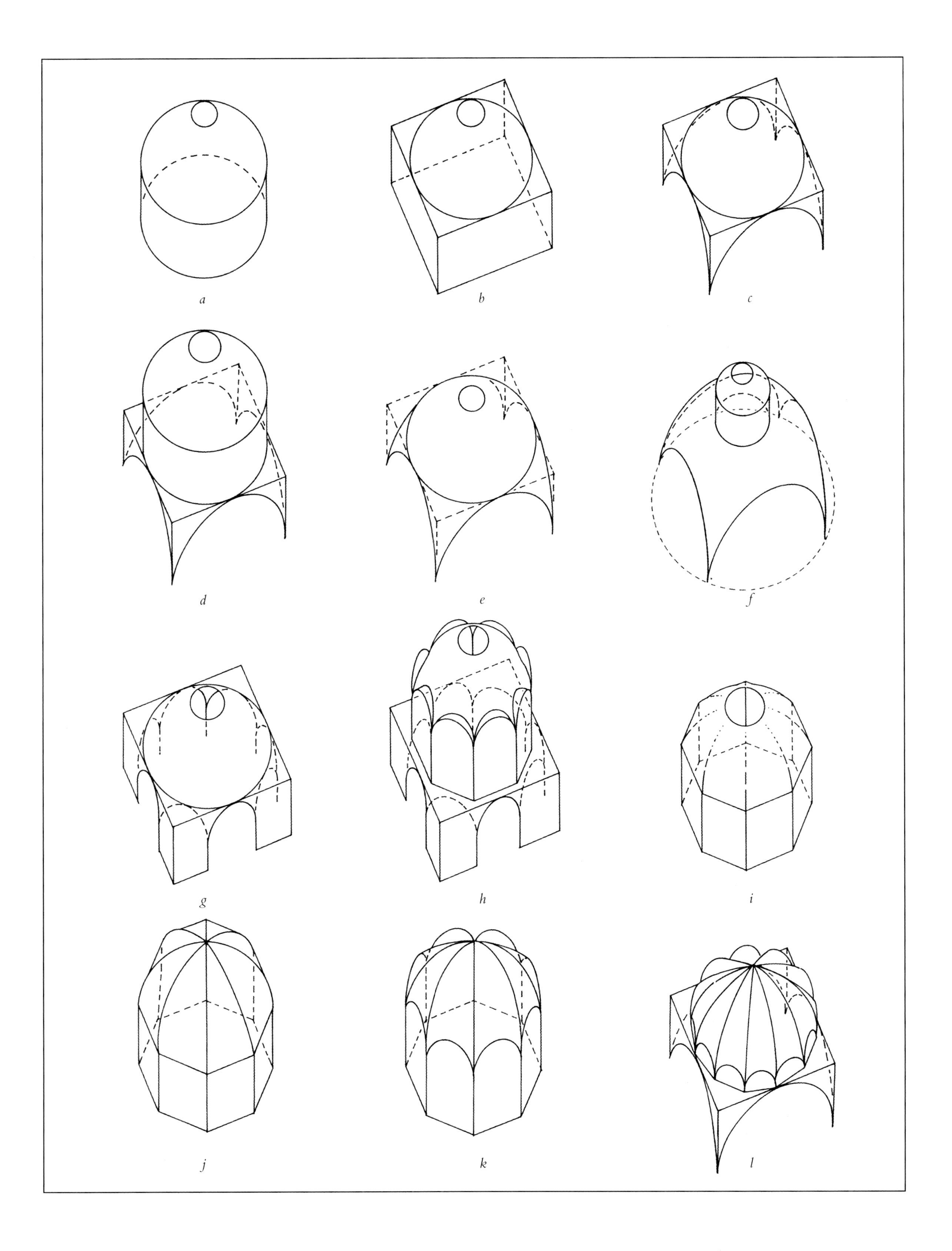
a
b
c
d
e
f
g
h
i
j
k
l

复合穹顶

罗马建筑师从很早就开始探索水泥结构的可能性，设计了多个穹顶体量组合的建筑物。建于90年的罗马帕拉丁山皇家宫殿（*a*）包含一个带有穹顶八边形房间，八边中的7个边带有一系列壁龛。其中4个壁龛为带有半个穹顶的半圆形，其余3个为长方形。这些长方形的壁龛当中还有更深的凹陷结构。两个为长方形，一个为半圆形。位于罗马城外的哈德良皇帝别墅，建于2世纪早期，包含几个穹顶，并且尝试了复杂的几何形制。这些实例中有黄金广场（*b*）。这一建筑内部设计有蜿蜒的柱廊和彼此相连的壁龛，人们认为这些壁龛由四周环绕着较小半穹顶的大型穹顶结构群所覆盖。

意大利中部城市托迪城外的抚慰圣母教堂（*d*）建于1508年，是一座集中于中央穹顶空间的对称结构。周围半穹顶结构与具有这一时期穹顶建筑特色的中央体量之间有着简单几何关系。

17世纪巴洛克建筑师设计穹顶，其复杂与新颖，达到了前所未有的水平。1634年，弗朗西斯科·博罗米尼（1599—1667年），在一系列的革新建筑中，首先设计了罗马的四泉圣卡洛教堂（*e*）。采用十字形平面，延长建筑物从门口至神坛的距离并覆盖以椭圆形为基底的纹案设计。博罗米尼将平面上所有传统上独立的元素全部融入一个空间，在这一空间中每一种元素都沿着一系列的弧线延伸至下一种纹案当中。平面中椭圆形结构由下部穹隅设定，穹隅中则有椭圆形半穹顶穿透并且支撑着更高的椭圆形穹顶，中央有小型穹隆顶塔。

这些非凡结构的建造大师是建筑师与僧侣瓜里诺·瓜里尼，他的设计遍布欧洲罗马天主教堂。他的一部分设计作品在意大利城市都灵。他于1668年设计的圣劳伦佐教堂（*g*）当中，沉重的穹顶位于有8排向内弯曲的柱子构成的八边形底座上，内部集中设计了4个大型穹隅，起拱于外露的16条交错的拱肋，包含8扇大型椭圆形窗户，其间点缀着小型开口。他于1666年设计的圣绍德礼拜堂（*f*）令人叹为观止。为容纳3个入口而建起3个大拱支撑穹顶，6个拱券以其尖端撑起一系列的浅拱，每个拱券中有两扇窗户并呈波浪状向上延伸至中央穹隆顶塔处。在建于1673年的无玷圣母教堂（*c*）中，两个相对简单的圆形穹顶，每一个带有四对扁平拱肋和三扇窗户，在第四面上与中央扁平八边形穹顶融合，创造出多变的内部构造，既不完整，也非割裂。

COMPLEX DOMES

From an early date Roman architects, exploiting the possibilities of concrete construction, designed buildings incorporating several domed volumes. Remains of the Imperial Palace on the Palatine hill in Rome (*a*) from about AD 90 include an octagonal domed room with a series of recesses on seven of the eight sides. Four of these recesses are semicircular with half-domes and three are rectangular. The rectangular recesses have further recesses within them, two rectangular and one semicircular. The villa of the Emperor Hadrian outside Rome was constructed in the early second century and contains several domes and buildings which experiment with complicated geometric forms. One of these is the Piazza d'Oro (*b*). This building has an interior of winding colonnades and interconnecting niches thought to have been covered by a large dome surrounded by a cluster of smaller half-domes.

S. Maria della Consolazione, outside the central Italian city of Todi (*d*), was designed in 1508 and is a symmetrical form focusing on a central domed space. The surrounding half-domes have a simple geometric relationship with the central volume typical of domed buildings of this period.

Baroque architects of the seventeenth century took the design of domes to levels of unprecedented complexity and novelty. In 1634 Francesco Borromini (1599-1667), in the first of a series of revolutionary buildings, designed the tiny S. Carlo alle Quattro Fontane in Rome (*e*). By taking a cross-shaped plan, lengthening it from the door to the altar and overlaying a pattern based on the geometry of an ellipse, Borromini merged all the conventionally separate elements of the plan into the one space where each part flows into the next in a series of curves. The elliptical structure of the plan is defined by the lower pendentive dome which is penetrated by oval half-domes and supports a further tall oval dome with a small domed lantern at the centre.

The master of these extraordinary structures was the architect and monk Guarini, who designed throughout Roman Catholic Europe. A number of his buildings are in the Italian city of Turin. In his church of S. Lorenzo (*g*) of 1668 the dome sits heavily over an octagon formed by eight inward-curving rows of columns and, gathering inwards on four large pendentives, rises on an exposed framework of sixteen interlacing ribs containing eight large oval windows interspersed with smaller openings. His Chapel of the Holy Shroud (*f*) of 1666 is even more extraordinary. To accommodate three entrances, three large arches rise to support a dome where six arches support from their apex a series of shallow arches containing two small windows each and forming a pattern which ripples upwards to a central lantern. In the church of the Immacolata Concezione (*c*) of 1673 two relatively simple circular domes, each with four pairs of flat ribs and three windows, are brought together on their fourth sides to merge with a central flattened octagonal dome, creating a restless interior that is neither whole nor divided.

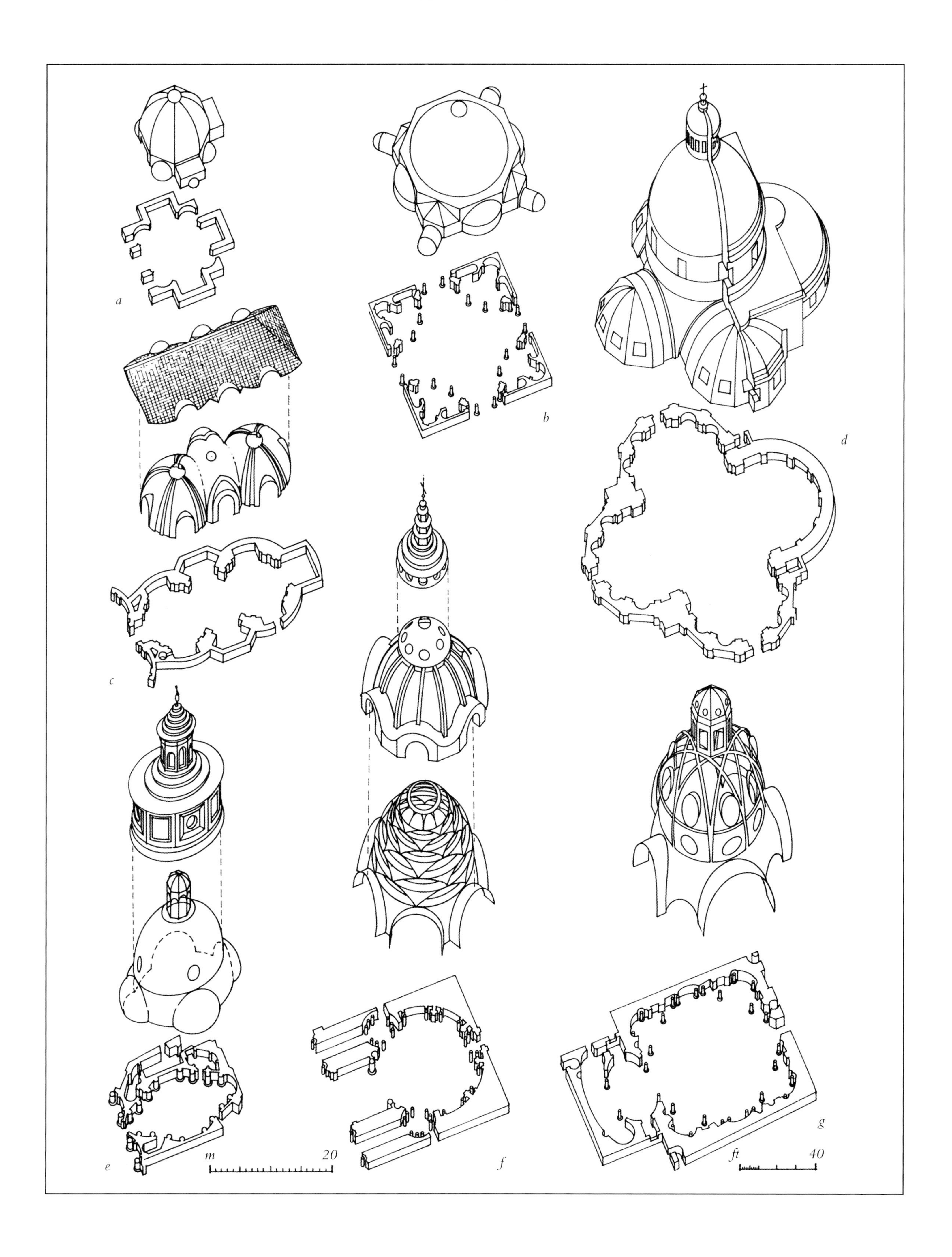
a
b
d
c
e
m
20
f
g
ft
40

假穹顶

布鲁内列斯基在弗洛伦塞修建的穹顶（参见P174），拥有砖石内部结构和独立的外部结构保证其不受天气影响。这与古代及拜占庭时代的穹顶截然不同，那一时代的结构外部坚固，但符合哥特式的构造传统。尽管许多穹顶建筑是坚固的，双穹顶的结构在文艺复兴时期仍然得以延续。这些双穹顶的内部与外部表面仅由狭小的空间分割。然而，在17世纪，实现了内部与外部结构创造出不同的视觉效果的可能性，首先出现在威尼斯，随后更加富有戏剧效果地出现在北部国家。

由儒勒·哈杜安·孟萨尔（1599—1667年）于1680年设计的恩瓦立德新教堂（*a*）在其所在广场与巴黎都是地标建筑。这一显赫的公众地位赋予建筑师以特殊的外部设计要求。穹顶与顶塔在视觉上与内部设计没有关系。外部穹顶由实木和铅建造，置于两个穹顶之上，这两个穹顶可以从教堂内部看到。内部穹顶拥有更大中央开洞，即眼洞窗，使得中间穹顶拥有坚固的外观，装饰有绘画并且通过窗子从内部可见，置于外部上层鼓形座上。

圣保罗大教堂（*b*）于1675年由克里斯多弗·雷恩设计，作为他在1666年伦敦大火后，该城巴洛克式设计重建工程的中心建筑。同样，这座建筑的外部穹顶与内部穹顶没有视觉关联，但是修建了砖石圆锥结构支撑沉重的石质穹隆顶塔，这一锥形结构穿透一系列隐藏的圆形开口，使得在外面鼓形座上部的窗户能够间接地将光线投射进圆锥体量中，其上部可通过一个眼洞窗看到光滑的、带有彩绘并且是灰泥砖石材质的内部穹顶。外部穹顶由木材与铅建造。

建筑师雅克斯·热尔曼·苏弗洛（1713—1780年），于1755年设计圣热纳维埃芙教堂（*c*）——后改称先贤祠，当时人们对古代以及哥特式建筑充满热情，巴洛克式设计开始衰退。尽管如此，他的设计，包含了创新的巴洛克式三穹顶设计。3个穹顶全部是坚固的结构，并且成排纤细的柱子取代了通常用于支撑巴洛克式穹顶的沉重支墩，这些支墩曾被证明是不适合结构重量的。其内部穹顶拥有古代内部穹顶的简单设计。

即使是法国新古典主义革新建筑师克劳德·勒杜，在18世纪后期德国的黑森卡塞尔公国图书馆（*d*）设计中也求助于三穹顶结构。然而勒杜设计的眼洞窗景观史无前例，放眼望去，不是虚假的天空或者彩绘装饰，而是一层附加的、不能进入的新古典主义内部设计。

FALSE DOMES

Brunelleschi's dome in Florence (see page 174) had an inner structure of brick and a separate outer structure to keep out the weather. This was quite unlike the domes of antiquity and Byzantium, where the structure was solid to the exterior, but was in the tradition of Gothic construction. Although many domed structures were solid, the construction of double domes continued during the Renaissance. The exterior and interior surfaces of these double domes were only separated by a narrow space. In the seventeenth century, however, the visual possibilities of different interiors and exteriors were realised, first in Venice and then, more dramatically, in the northern nations.

The Dome des Invalides (*a*) was designed by Jules Hardouin Mansart (1599-1667) in 1680 to act as a landmark in its own square and in the city of Paris. This public prominence gave the architect special design requirements for the exterior. The dome and its lantern have no visual relationship to the appearance of the interior. The external dome is constructed of timber and lead and sits above a further two domes which can be seen from within the church. The inner dome has a large central opening, or oculus, giving a view of the solid middle dome, which is decorated with paintings and lit by windows invisible from the interior and located in the external upper drum.

St Paul's Cathedral (*b*) was designed in 1675 by Christopher Wren as a centerpiece to his Baroque replanning of the City of London, following the Great Fire of 1666. Again, this external dome has no visual relationship to the interior dome, but a brick cone was constructed to support the heavy stone lantern. This cone is pierced by a series of concealed circular openings which allow windows in the external upper drum indirectly to light the conical volume, the upper part of which is visible through an oculus in the smooth, painted, plaster-and-brick inner dome. The outer dome is built of timber and lead.

When the architect Jacques-Germain Soufflot (1713-80) designed the church of Ste-Geneviève (*c*)—later to become the Panthéon—in 1755, Baroque design was starting to decline in the face of increased interest in ancient and Gothic architecture. His design, nonetheless, included the Baroque invention of the triple dome. All three domes were of solid construction and the usual heavy supporting piers of Baroque domes were replaced with rows of slender columns, which proved inadequate for the weight of the structure. The inner dome had the simple design of the interior of ancient domes.

Even the revolutionary French Neo-Classical architect Claude Ledoux had recourse to the triple dome in his design for a library in the principality of Hesse-Kassel in Germany (*d*) in the late-eighteenth century. However, Ledoux offers for the first time a view through the oculus, not of a false sky or painted design, but of an additional, inaccessible, Neo-Classical interior.

a
b
m
15
c
d
50
ft

半穹顶，法官席，半圆形后殿和壁龛

罗马建筑师首先设计了半穹顶，目的是为了给房间或室外空间提供一个视觉焦点，或在较小的尺度上为厚重的罗马水泥墙与支墩增加一些特点。大型半穹顶，根据其位置与功能又可称为半圆形屋顶、法官席、半圆形后殿或半圆形室外座椅，小型的通常称为壁龛。

大型罗马浴场通常包括几个半圆形屋顶，面向内部中央大厅，或朝向外部花园或者游泳池。在建于216年的罗马卡拉卡拉浴场（*b*）中，有两个纪念碑式的开敞谈话间向外面向冷水池，又称冷水浴室，位于一个同等大小的拱券两侧。每一个这样的开敞谈话间包括中央的一个拱形开间，两侧设有较小的半穹顶壁龛，中间可能有雕像。这些既创造出视觉特征，又将其与主要空间分割开来，用于集会或者座席。

地方官、行政官、皇帝或者任何重要人物的座位都置于巴西利卡中央凹陷处的明显位置。这一凹陷处通常为半圆形并且为了凸显其执法功能，称其为法官席。这一空间的重要性既在于其象征意义，又在于其实用性，皇帝的塑像通常安放在这个位置来体现皇家权威。北非城市大莱普提斯（*a*）中巴西利卡的法官席，建于3世纪早期，通过位于中央的两根巨大的柱子来展现对权威的强调。

法官席的联想意义与巴西利卡的设计一同延续给后来的教堂建筑，并演变为半圆形后殿。半圆形后殿是罗马、罗马式以及哥特教堂的主要特征，象征着上帝的存在。起初，简单的半圆形后殿就像法官的席位一样被置于教堂中殿末端，并且用马赛克或者彩绘的半穹顶加以强化，位于主入口的对面。随着圣母以及圣徒们信仰的发展，附属半圆形后殿或礼拜堂攒聚在中央半圆形后殿周围。

较小的法官席可见于民居中。这些与半圆形室外座椅具有相似的功能，定义了一个附属于主要空间的区域。这些在18世纪后期开始流行，例如，在餐厅里设置凹陷处放置送餐桌（*d*），两个偏离中心的大门中间的一个入口大厅（*c*），或者柱子作为屏障分割图书馆，凹陷处设置书架（*e*）。

壁龛是包括雕塑的较小建筑形态，为一个走廊的末端即远景提供焦点，抑或是为一面朴素的墙壁增加情调。壁龛一词源于意大利语贝壳一词，用贝壳装饰半穹顶，如（*f*）和（*i*），十分常见。也有很多其他形式的装饰，或者源于其较大的对应物（*h*），或者从较小型的（*g*）演变出来，以适合壁龛的大小。

HALF-DOMES, TRIBUNES, APSES AND NICHES

The half-dome was first developed by Roman architects to provide a focal point in a room or external space and, at a small scale, as a feature within the bulk of thick Roman concrete walls and piers. A large half-dome can be called a semi-dome, tribune, apse, or exedra, according to its position and function; smaller ones are generally referred to as niches.

Large Roman baths often contain several semi-domes facing internally into the central hall and externally on to gardens or pools. Two monumental exedrae face outwards into the cold pool, or frigidarium, on either side of an arch of equal size in the Baths of Caracalla in Rome (*b*) of AD 216. Each one of these exedrae contains an arched opening in the centre flanked by two smaller half-domed niches, which probably contained statues. These acted both as visual features and areas separated from the main space for assembly or seating.

The seat of the magistrate, governor, emperor, or any person of importance was distinguished by a recess at the focal point of a basilica. This recess was often of a semicircular form and, in recognition of its judicial function, is called a tribune. The significance of the space became as symbolic as practical, and statues of emperors were placed in tribunes to represent imperial authority. The tribune of the basilica in the city of Leptis Magna in North Africa (*a*), built in the early third century AD, has a display of columns that reinforces the authority of the centre with two giant columns.

The associations of the tribune were passed on, together with the design of the basilica, to church architecture, where it became the apse. Apses were major features in the design of late Roman, Romanesque, and Gothic churches, symbolising the presence of God. At first, a single apse was placed like the seat of the judge at the end of the nave and emphasised by mosaics or paintings in the half-dome and the location of the main entrance at the opposite end. As the cult of the Virgin and the Saints developed, subsidiary apses or chapels gathered around the central apse.

Smaller versions of tribunes were included in domestic buildings. These served a similar function to exedrae by defining a space subordinate to the principal space. These became popular in the late eighteenth century and could form a recess for a serving-table in a dining-room (*d*), an entrance lobby for two off-centre doors (*c*) or a recess for shelving in a library separated by a screen of columns (*e*).

Niches are invariably smaller features containing statues, providing a focus at the end of a passageway or vista, or adding interest to a plain wall. The word niche derives from the Italian for shell, and decorations of shells in the half-dome, (*f*) and (*i*), are common. There are many other forms of decoration, either derived from their larger counterparts (*h*) or of a small scale (*g*) to suit the size of the niche.

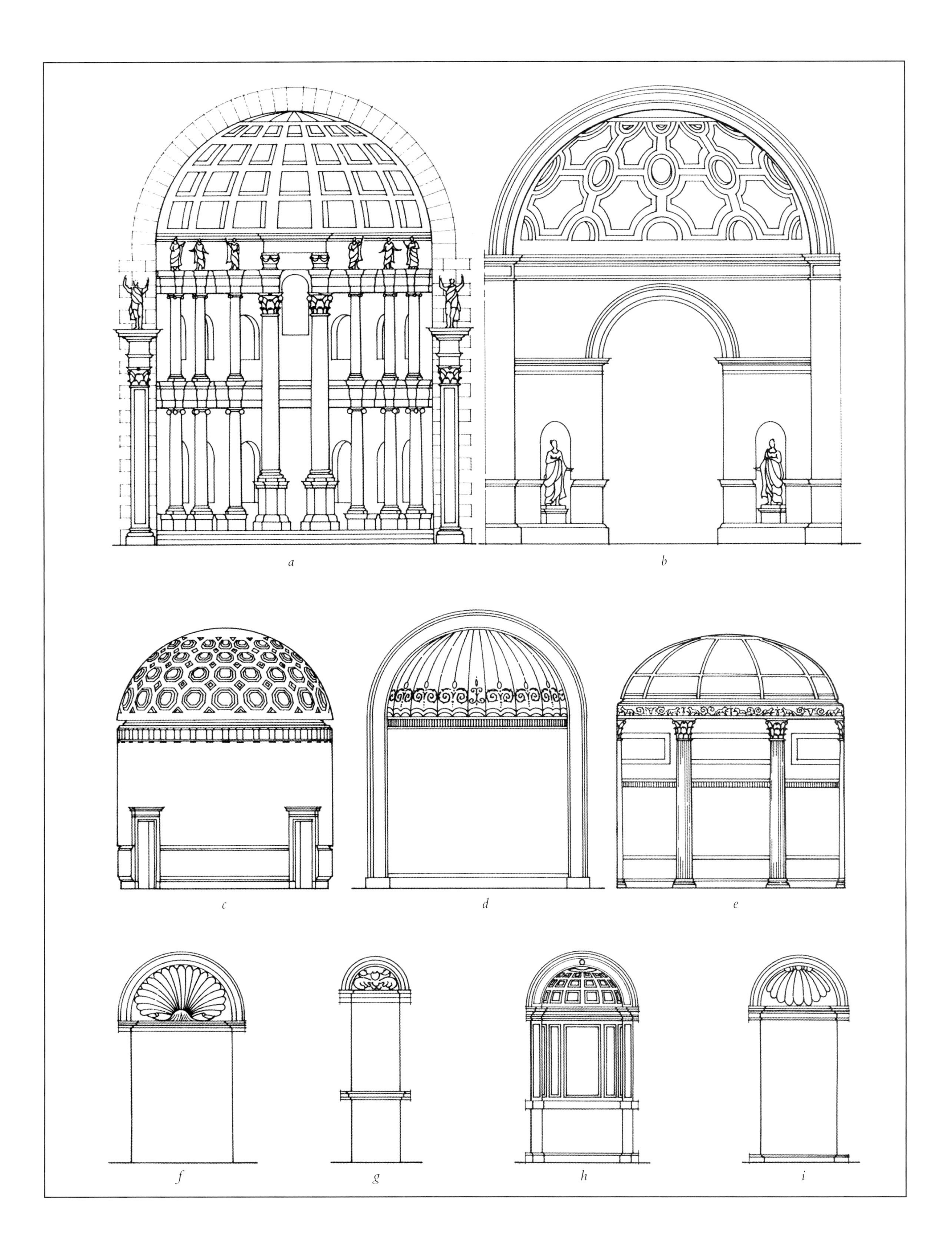
a
b
c
d
e
f
g
h
i

穹顶内部

穹顶底部，即拱腹的装饰有其独特的传统。绘画装饰、马赛克和浅灰泥浮雕用于无数穹顶表面的装饰。这些装饰的本质与变化导致找到一种令人满意的分类方式，并且这里展示的只有高浮雕。

罗马装饰纹案，由于穹顶与半穹顶起源于罗马，因而为后来几个世纪的设计确立了标准。建于2世纪的罗马万神殿（*a*）所使用的简单花格镶板，基本上保存完好，并且可能比其他形式的装饰更具影响力。水平和垂直的拱肋向上或者环绕着穹顶彼此交叉，在拱腹表面创造出一种类似方形的刻痕，即花格镶板。在万神殿附有四种向内阶梯，置于每个花格镶板内部，阶梯的表面角度朝向穹顶的中央。每个花格镶板中央都可能设计圆花饰。随着垂直线条向上聚拢至穹顶顶部，水平线条的间距也逐渐缩小，以创造出比例相同、大小却递减的花格镶板。这种设计增加了穹顶内部自然的戏剧性效果并且成为大部分花格镶板体系的设计原则。

其他罗马花格镶板通常是这种罗马万神殿更加复杂的版本。花格镶板也可是六边形或者长方形的连锁纹案，这样的装饰可见于4世纪早期的罗马马克森提乌斯巴西利卡（*b*）中。也可以创造其他几何变体。18世纪巴黎圣日内维耶修道院（*d*）的穹顶设计简化成相似的圆形与方形纹案。4世纪早期的罗马维纳斯神庙（*c*）的眼洞窗设计有对角线条，从相对的方向穿过穹顶表面，创造出文艺复兴时期建筑经常使用的螺旋纹案。

在表面弯曲拱肋（在罗马穹顶与花格镶板中的确存在，但是隐藏的）的手法在文艺复兴时期引入——大概是由于哥特式与拜占庭式建筑的影响。早期的拱肋实例是简单的，并且确定了穹顶分割刻面的扁平区块（*e*和*k*）。但是，在巴洛克建筑师追求提升穹顶垂直效果（*g*）的过程中变得更加夸张。在罗马穹顶和花格镶板中可增加拱肋（*h*和*j*），赋予其更多的垂直特征，并且省略了水平与垂直区块的交替使用，变得更加简单。正方形的花格镶板设计（*i*）同样赋予其巧妙的垂直强调效果。

尽管穹顶起源于罗马，但其地位如此稳固以至于在19世纪希腊复兴的设计中无法被忽略。菲利普·哈德维克（1792—1870年）在其伦敦尤斯顿车站购票大厅的穹顶（*f*）设计中创造了拱肋圆花饰，而约翰·索恩将带有中央圆花饰的花格镶板与沟槽置于原始希腊柱式之上的设计，用于其伦敦城外的达利奇学院美术馆大厦（*l*）中。

INSIDE THE DOME

The decoration of the underside, the soffit, of the dome has its own tradition. Painted decoration, mosaic, and shallow plaster in relief have adorned innumerable domed surfaces. The nature, and variety of these make them impossible to categorise satisfactorily and only decoration in the form of deep relief is shown here.

Roman decorative patterns have, due to the Roman origin of the dome and half-dome, set a standard over the following centuries. The simple coffering of the second-century dome of the Pantheon in Rome (*a*) has survived virtually intact and has probably been more influential than any other form of ornament. Horizontal and vertical ribs running up and around the dome intersect to create almost square indentations, or coffers, in the surfaces of the soffit. On the Pantheon there are four additional inward steps set inside each coffer and the faces of the steps are angled towards the centre point of the dome. A decorative rosette may also have sat in the middle of each coffer. As the vertical lines converge towards the top of the dome, so the distance between the horizontal lines diminishes to create coffers of similar proportions and decreasing size. This effect increases the natural drama of a domed interior and is the principle behind most coffering systems.

Other Roman coffering is generally a more complex version of this Pantheon type. The coffers can become an interlocking pattern of hexagons and rectangles, which can be seen in the Basilica of Maxentius in Rome (*b*) of the early fourth century. Many other geometric variations can be created. The eighteenth-century dome of the Abbaye-Sainte-Geneviève in Paris (*d*) reduces a similar pattern to circles and squares. The design of the apses of the Temple of Venus and Rome, in Rome (*c*) from the early fourth century has lines diagonally across the surface of the dome in opposite directions, creating a spiral pattern often used in Renaissance buildings.

Ribs arching over the surface (existing, but concealed, in Roman domes and coffers) were introduced in the Renaissance—probably due to the influence of Gothic and Byzantine architecture. On early examples the ribs are simple and define the flat segments of faceted domes, (*e*) and (*k*), but became more exaggerated where Baroque architects sought to enhance the vertical impression of their domes (*g*). Ribs can also be added to Roman coffering patterns, (*h*) and (*j*), to give them a more vertical character, and the omission of alternate horizontal and vertical divisions in a simple, squared, coffered design (*i*) also gives a subtle vertical emphasis.

In spite of its Roman origin the dome was too well established to be dismissed in Greek Revival designs in the nineteenth century. Philip Hardwick (1792-1870) created a ribbed rosette design for his dome over the booking-hall in Euston Station in London (*f*) while Soane mixed a central rosette with coffering and fluting to sit over primitive Greek columns in his Mausoleum at Dulwich Picture Gallery outside London (*l*).

a
b
c
d
e
f
g
h
i
j
k
l

11. 粗琢

起源

粗琢是指除了接缝外保留石面粗糙且未完工的一种石雕工艺。垂直与水平接缝处为光滑的沟槽，石造工艺表面却是粗糙的。可用砖块与灰泥达成同样的效果。这个词暗示着粗糙与不精细的建造工艺，但是粗琢却成为古典建筑装饰中最流行、最精细的一种手法。

在古代，石墙表面的石块留下粗糙的凿工或者在连接处向外凸出是十分常见的现象，并且在希腊城墙（*b*）上是十分实用的结构。也有一些罗马人建筑的石墙，为了刻意造成视觉效果，而在连接处设计浅沟槽。

1世纪中期，克劳狄皇帝（公元前10—公元54年）在位时将粗琢作为一种美学手段而广泛采用，成为这种工艺得以应用的最初例证。一系列非凡的设计，例如克劳狄神庙（*a*），就可以追溯至这个时代。这是一座采用同样夸张手法的建筑，将光滑完整与粗糙未完工石材相结合。众所周知，克劳狄热衷于文物收藏，很有可能是这种对远古时代的想象力，激发出一种前所未有的建造氛围。不论这项工程背后隐藏的目的如何，这种情形在古代未曾再次出现过。

中世纪的意大利城镇演变出另一种独立的传统。城市要塞或宫殿，例如14世纪意大利中部城市蒙特普齐亚诺的市政厅（*d*），刻意使用巨大且粗糙的石块修建而成，用以表达力量与坚不可摧，因而象征领主的高贵地位。文艺复兴时期的建筑师致力于效仿古代的同时，不得不在同样的城市暴力背景下进行设计并且采用与其中世纪先驱者们同样的、用以表达力量与地位的传统。大概是在克劳狄遗留下来的建筑的影响之下，创造出了更加规范的粗琢体系，这一特殊的建筑类型独树一帜并且模仿石头在自然环境中受到侵蚀的样态（*c*），这样就与未经加工的石块具有同样显著的特征。

尽管这一工艺在古代昙花一现，粗琢的应用体系是文艺复兴时期对中世纪传统的延续。例如，同样位于蒙特普齐亚诺的切尔维尼宫（*e*）的建筑，由安东尼奥·达·桑迦洛于1520年设计，在已有的哥特式祖先的基础上增加了古典和谐之美。

11. RUSTICATION

ORIGINS

Rustication is stonework that is left rough and unfinished except where it fits together at the joints. The even joints form vertical and horizontal grooves in the uneven face of the stonework. A similar effect can be achieved with brick and stucco. The word implies construction that is crude and unsophisticated, but rustication has become one of the most widespread and subtle forms of decoration in Classical architecture.

Stone walls where the face of the blocks show the marks of coarse chisel work or bulge outwards between the joints are quite common in antiquity and can be found in utilitarian structures such as Greek city walls (*b*). There are also some Roman buildings where the joints in masonry are set in shallow channels for deliberate visual effect.

The first evidence of rough stone rustication being used extensively as an aesthetic device is on buildings constructed in the reign of the Emperor Claudius (10 BC-AD 54) in the middle of the first century AD. A series of remarkable designs, such as the Temple of the Deified Claudius in Rome (*a*), date from this period and have the same exaggerated mixture of smooth finished and very rough unfinished stone. It is known that Claudius was a keen antiquarian and it is quite possible that this unprecedented architectural phenomenon arose out of a desire to evoke an imaginary antique primitiveness. Whatever the intention behind this work, it was not repeated in this form in antiquity.

In medieval Italian towns another unconnected tradition developed. Urban fortresses or palaces, such as the fourteenth-century Palazzo Comunale in the central Italian town of Montepulciano (*d*), were built of deliberately massive and rough blocks of stone to express their strength and impregnability, and consequently the nobility of the occupier. Renaissance architects, while striving to emulate antiquity, had to design for the same climate of urban violence and with the same traditional expressions of strength and status as their medieval precursors. Perhaps influenced by the remnants of Claudius' buildings, a more orderly system of rusticated decoration was created, distinct types became recognised and the imitation of stone eroded by weather (*c*) became as pronounced as that of unfinished stone.

In spite of its brief flowering in antiquity, the systematic use of rustication is a Renaissance extension of a medieval tradition. Buildings such as the Palazzo Cervini, also in Montepulciano (*e*), designed by da Sangallo the Elder in 1520, add Classical harmony to the established forms of their Gothic forebears.

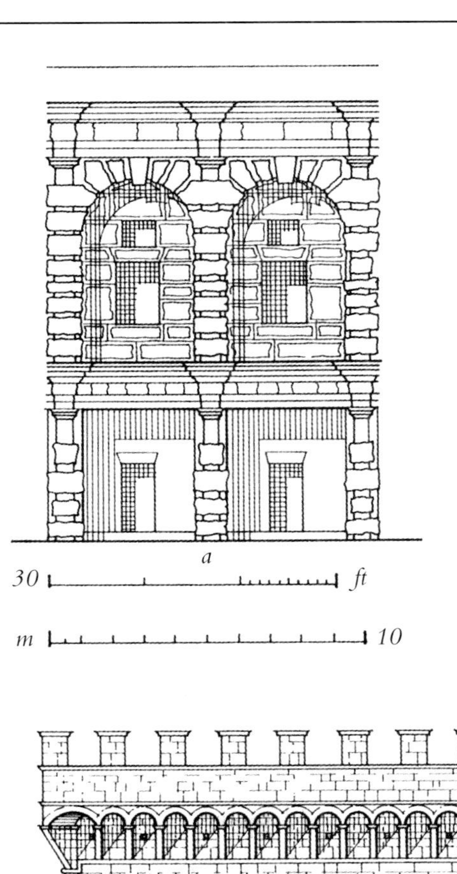

a

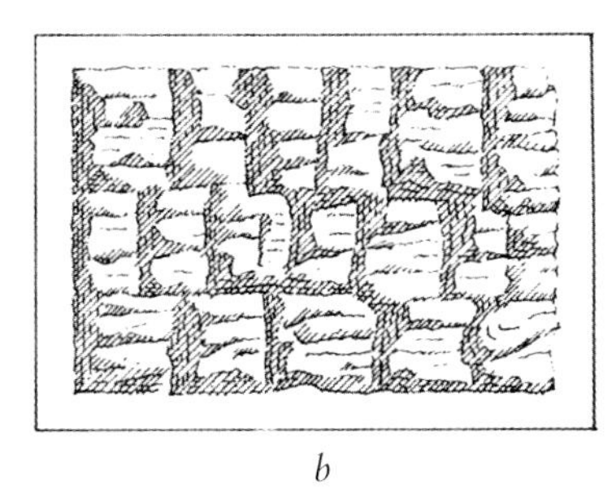

b

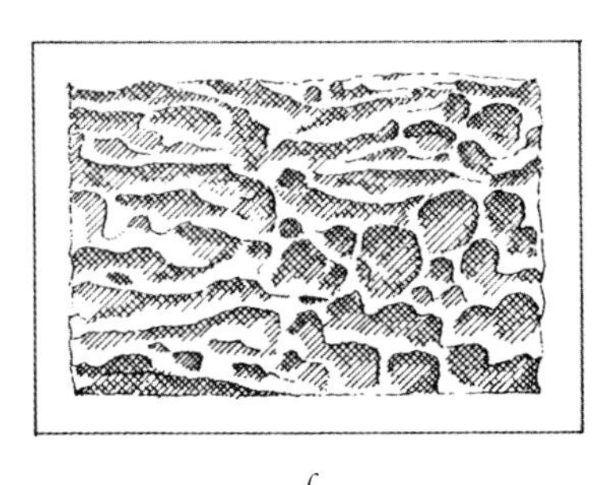

c

d

e

粗琢的种类

"粗琢"一词在过去一直局限于"粗糙的石块"之意，但后来逐渐演变出许多石造砖或灰泥工艺，连接处呈凹陷状或者每块石头的表面是单独表现出来的。只有水平连接处是凹陷的，有时砖块的轨迹呈后缩状，粗琢通常指镶边。

许多不同种类的粗琢得以演变出来。它可能局限于石块之间凹陷的连接部分（*a*），这种光滑的粗琢通常用石膏或者灰泥，甚至是木料进行再造。连接处的变化（*m*～*r*），可以大大地改变粗琢的外观。石块的分割能够通过将凸出表面（*b*）的边角变圆而加以强化，尽管这一细部设计并不常见。凸出的垫石粗琢（*c*）更加普遍并且经常带有如图所示的连接。石造工艺的表面有不同的处理方式。网状粗琢（*d*）覆盖着网纹图案，虫蚀状粗琢石面上（*e*）装饰有一系列扭曲的形状，象征虫洞，可以切割成不同深度和复杂程度。这两种粗琢通常相互混淆。岩石表面的装饰处理（*f*）通常十分流行，并且是未经切割的石料最直接的表达方式。岩石表面粗糙工艺可以通过雕凿使其变成柔和、啄刻的表面（*g*）或者成为更加粗糙、穿凿的表面（*h*）。最不同寻常的是一种磨砂，或者霜冻表面（*i*），用于巴洛克式建筑，切割成类似山洞中石灰石、钟乳石的样态。矫饰主义建筑师发明了棱柱形或又称菱尖形石材（*j*），在建筑中产生了一种非常有气势的纹案。这种纹案可通过凹陷连接（*k*）加以强化或者重复这一设计来修饰（*l*）。

可以使用不同种类的凹陷连接。简单直线形连接（*m*）与倒角连接（*n*）都可以与各个种类的粗琢一同使用，然而半圆角连接（*o*）就不那么常见。两层阶梯状连接（*p*～*r*），可强调光滑表面的粗琢效果，在区分粗糙表面的粗琢连接与粗质表面时非常实用。

粗琢通常是混合的。轨迹高度可能不同（*s*），石块大小也多变，以便创造出不规则的纹案（*t*）。连接处也是多变的，例如交替使用直角和倒角（*u*），粗琢的石块可与加工过的光滑石材混合使用，创造出带状、环绕和边角纹理（*v*），不同的粗琢纹理也可以混合使用（*w*）。粗琢的水平轨迹通常是平行的，但是有少数建筑的轨迹是随意的（*x*）。

展示出的实例绝不是各个种类的综合集成。也有大量来自于建筑师和石匠的对粗琢的单独解读。混合种类与细部设计能够在墙面上创造出许多不同纹理和效果。

TYPES OF RUSTICATION

The word rustication has in the past been limited to rough stone blocks but has come to mean any form of stonework, brickwork, or stucco where the joints are recessed or the face of each stone is separately expressed. Where only the horizontal joints of stonework are recessed, or occasional courses of bricks are set back, rustication is sometimes referred to as banding.

A large number of different types of rustication have evolved. It can be limited to recessing the joints between the stones (*a*). This smooth rustication is often reproduced in plaster or stucco, or even wood. Variations in the joints, (*m*) to (*r*), can significantly alter its appearance. The separation between the stones can be exaggerated by rounding the corners of the projecting faces (*b*), although this detail is unusual. Bulging, cushioned rustication (*c*) is much more common and frequently has even joints as illustrated. The face of the stonework can have many different types of finish. Reticulated rustication (*d*) is covered with a net-like pattern, while vermiculated stones (*e*) have a series of contorted forms which resemble worm-casts and can be cut to varying degrees of depth and complexity. These two types of rustication are often mistaken for each other. A rock-faced finish (*f*) has always been popular and is the most straightforward representation of uncut stone. The coarseness of rock-faced work can be diminished by chiseling to give a gentle, pecked finish (*g*) or a harsher, punched face (*h*). Most extraordinary of all is the frosted, or congelated, finish (*i*), used on Baroque buildings and cut to imitate the petrified drops of limestone found in caves. Mannerist architects developed a prismatic, or diamond-pointed, stone (*j*) which produces a very powerful pattern on a building. It can be emphasised by recessing the joints (*k*) or embellished by repeating the design (*l*).

Different types of recessed joints can be used. Simple straight (*m*) and chamfered (*n*) joints are found associated with all types of rustication while the half-round joint (*o*) is less common. Joints with two steps, (*p*) to (*r*), can strengthen the effect of smooth-faced rustication distinct from the coarse surface.

Rustication is often mixed. Courses can be of different heights (*s*) and the sizes of the blocks can be varied to produce an irregular pattern (*t*). Joints can be varied, for example by alternating straight and chamfered joints (*u*). Rusticated blocks can be mixed with smooth, dressed stone to create textured bands, surrounds or edges (*v*) and different textures of rustication can be mixed (*w*). The horizontal courses of rustication are usually level, but a few buildings have random coursing (*x*).

The illustrated examples are by no means a comprehensive collection of types. There has been a great deal of individual interpretation of rustication, both by architects and stonemasons. Mixing types and details can produce many different textures and effects on the surfaces of walls.

a
b
c
d
e
f
g
h
i
j
k
l
m
n
o
p
q
r
s
t
u
v
w
x

粗琢柱式

尽管最早的一些粗琢为科林斯式，但是当文艺复兴理论家将柱式排序后，根据柱式的纤细程度，确立了托斯卡纳与多立克柱式最常应用于底层建筑的地位。同样，一些理论家扩充了罗马作家维特鲁威的神秘柱式人体特征，建议在更加刚毅的柱式，例如托斯卡纳和多立克等柱式上采用粗琢工艺这样表现粗壮的细部设计最为恰当。最终，文艺复兴对粗琢工艺的发展与柱式的理解同步进行，演变成这样的思想，托斯卡纳与多立克柱式最适合采用粗琢工艺并且这些柱式通常应用于粗琢的底层建筑结构中。因此，托斯卡纳与多立克柱式比其他柱式更常应用粗琢工艺。

然而，所有的柱式都能够进行粗琢，但是在细部设计上没有一贯的区分。一种柱式上的特色能够用于另一种柱式，并且这些都能够采用不同的形式。

柱身上规则的圆形石块（*a*）赋予柱式整洁的外观并且能够保持这些更加精美的细部设计的完整性。这些带饰能够展现更多柱身的真实形态。维尼奥拉于1550年在罗马修建的朱利亚别墅（*c*），作为基础的柱身与嵌入式粗琢拱券的拱墩成一直线。随着粗琢石块之间的空间逐步扩大而展现更多的柱式，如（*k*）、（*f*）、（*d*）和（*i*）。伦敦的萨默赛特府邸拱门（*d*），由威廉·钱伯于1776年设计，拱墩之上拱券的柱子并没有采用粗琢工艺。

粗琢石块能够进行规则或者不规则摆放（*g*和*l*），与1551年塞里奥非凡的粗质石料设计（*e*）有所不同，直到18世纪后期，使用简洁、带有垂直沟槽的方形石块（*b*）。最初的重型粗琢应用于1世纪早期科林斯式马吉奥莱大门（*j*），在科林斯柱头上使用了一系列类似未经切割的石块。在另一个塞里奥的设计中，柱子捆束在墙上（*f*），好似被墙面绑缚，如图（*h*）所示，方形石块将两根柱子连接在一起。

有时拱券穿过额枋置于柱子之上。萨默赛特府邸（*d*）多立克柱式上层的楔形石，与雕带上的三陇板协调一致，然而塞里奥通过使用简单的多立克细部设计（*f*）来避免这种情况。尽管檐口常常免于粗琢结构的干扰，偶尔这一设计也能应用于整个檐部（*i*）。

扁平拱上的粗琢楔形石可以置于柱头的水平区域内，正如在佛罗伦萨的皮蒂府邸花园立面（*k*），由巴尔托洛梅奥·阿曼那提于1560年设计，正是这样的例子。它们也可以切入额枋（*e*）或者额枋能够通过增加楔形石成为扁平拱券。

RUSTICATED ORDERS

Although some of the earliest rustication is Corinthian, when Renaissance theorists fixed the Orders in a sequence, according to the slenderness of columns, they established Tuscan and Doric as the Orders most often used for lower storeys of buildings. The same theorists enlarged upon the Roman author Vitruvius' mythological human characteristics for the Orders, suggesting that robust details such as rustication would be most appropriate for the masculine Orders, Tuscan, and Doric. The Renaissance development of rustication was, consequently, contemporary with developments in the understanding of the Orders that led both to the idea that Tuscan and Doric were most suitable for rustication and to the location of these Orders in the often rusticated lower storeys of buildings. Tuscan and Doric rustication is, therefore, much more common than rustication in the other Orders.

All the Orders can, however, be rusticated, but there is no consistent difference in the details. Features seen on one Order can be applied to another and those can take many different forms.

Regular circular blocks on the columns shaft (*a*) give an orderly appearance and leave the finer details such as the capitals and entablature intact. These bands can be set out to reveal more of the true column shaft. Vignola's doorway on the Villa Giulia in Rome (*c*) of 1550 has the underlying column shaft exposed in line with the imposts of the inset rusticated arch. The spaces between the rusticated blocks can be increased progressively to expose more column, (*k*), (*f*), (*d*), and (*i*). An archway at Somerset House in London (*d*), designed by Chambers in 1776, has no rustication on the columns above the impost of the arch.

Rusticated blocks can be regularly or unevenly spaced, (*g*) and (*l*), and can vary from the coarse stones of Serlio's extraordinary design (*e*) of 1551, to the late-eighteenth century use of neat, square blocks with vertical joints (*b*). One of the first uses of heavy rustication, in the Corinthian Porta Maggiore (*j*) of the early first century AD, has stones like a series of uncut blocks for Corinthian capitals. In another detail by Serlio, the columns are strapped into the wall (*f*) as if captured by the stones, while illustration (*h*) shows how square blocks can bind two columns together.

Arches set into the Order at times push through into the entablature. The upper voussoirs of the Doric arch at Somerset House (*d*) are very carefully coordinated with the triglyphs in the frieze, while Serlio avoids this by using simplified Doric details (*f*). Although the cornice often escapes the intrusion of rustication, at times it can occupy the whole entablature (*i*).

Rusticated voussoirs from flat arches can be contained within the horizontal zone of the column capital, as on the garden elevation of the Palazzo Pitti in Florence (*k*), by Ammanati in 1560. They can also be cut into the entablature (*e*) or the entablature can, by the addition of voussoirs, become the flat arch itself.

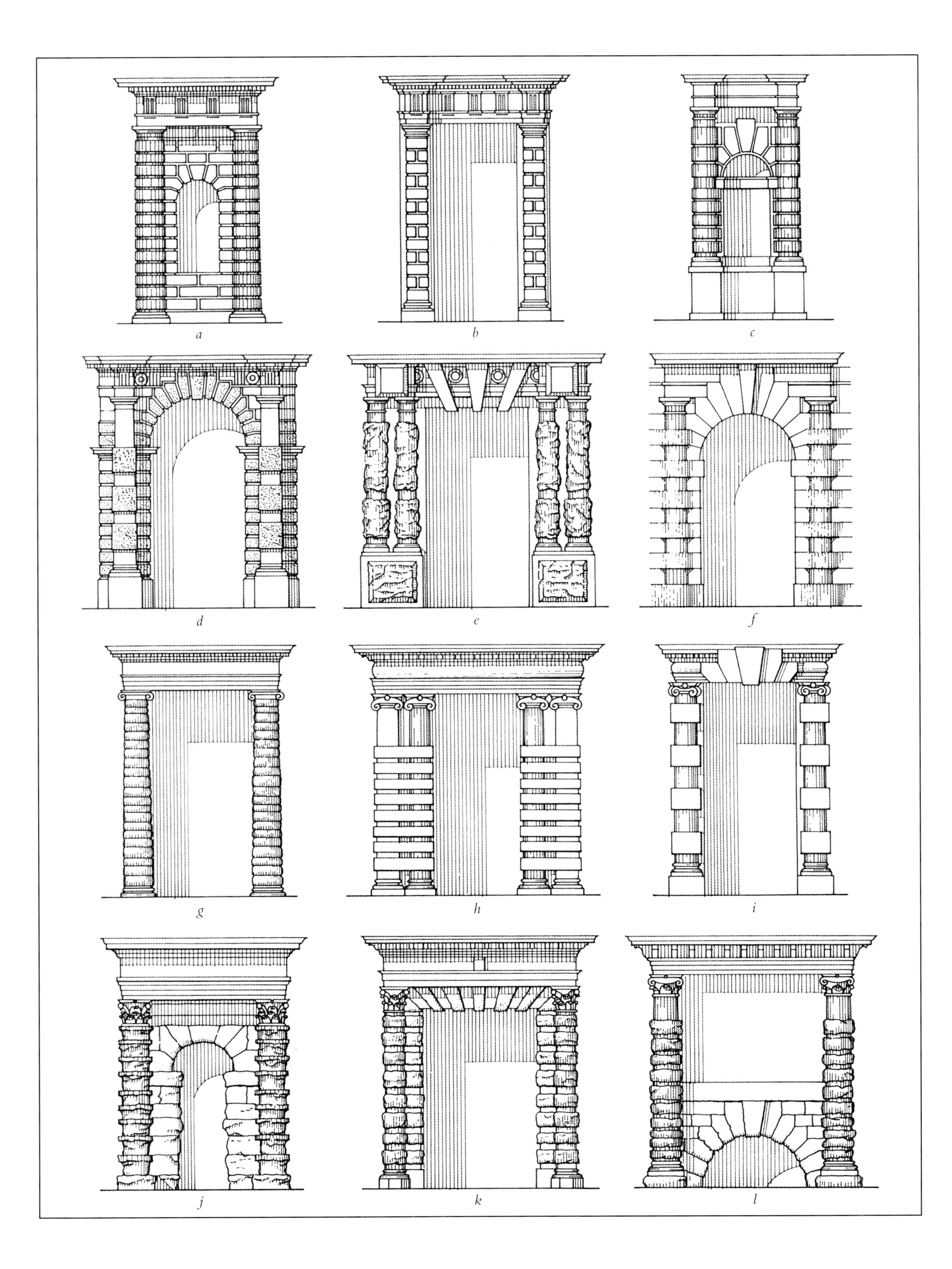
a
b
c
d
e
f
g
h
i
j
k
l

粗琢墙壁与开口

在16世纪的一系列粗琢柱式出现以前，便已经建立了在粗琢墙壁上开口的诸多方法。由于粗琢工艺实质上增强了正常用于墙壁建筑中石头的效果，因而简易开口中的细部设计也是对传统石头工艺的强化。

形成平拱的系列拱石可以被提升到粗琢墙壁中适当接缝的水平面上（*d*）。有一种结构更加有趣，结合了拱石和若干石层形成砌块，使拱石转向外侧连接到墙壁（*a*）。墙壁上的粗琢结构可以从外部逐渐嵌入平拱的拱石之中，具体做法是降低一块（*c*）或多块（*b*）拱石，形成一系列向上的阶梯，类似圆拱中的拱石布局。示例（*b*）出自罗马的海岸宫，因有两块而不是一块拱顶石而与众不同。

圆拱中也可以进行相似的改动，主要由拱石与墙壁上粗琢结构水平线的交接方式所决定。拱形高度上有超过三层的粗琢结构，拱石在与每条水平层相交时逐渐向上，在石头上形成尖拱的大致轮廓（*n*和*o*）。这种样式的特点是每个交叉点都带有接缝（*n*），或者保持连续的线条延展至水平层上（*o*），带有镶边的情况下往往采用后者的形式。将两种排列方式结合起来，可以形成L形石头结构（*m*）。在限制阶梯式辐射状拱石的高度时，可以使拱石停止在拱形上方接缝处的水平层上（*p*）。

通过分离式的粗琢缘饰可以将平拱和圆拱的粗琢结构从墙壁中隔离出来（*e*和*i*）。也可以将缘饰缩减成石头框架，不再表明开口的结构意义（*h*和*l*）。

16世纪时在源于古代的矩形开口和缘饰的设计中都包括粗琢工艺。将砌块布置在额枋周围，并且作为拱石穿透部分檐部（*f*），或者像1550年维尼奥拉在罗马设计的朱利亚别墅（*g*）一样，将粗琢平拱和古典缘饰融合在一起。两者融合的方法众多并不断演化，小乔治・唐斯（1741—1825年）1770年在伦敦设计的新门监狱（*j*）中，拱形内部的粗琢结构实际上已经完全取代了缘饰。意大利文艺复兴宫殿结合了矩形和拱形粗琢的开口。约1550年，帕拉迪奥在意大利北部维琴察设计蒂恩宫（*k*），在粗琢样式内部设计中轻松结合了这两种形式，唐斯则试图在拱石中匹配拱形的曲线，因此逐渐失了去与水平墙壁的联系。

RUSTICATED WALLS AND OPENINGS

Before the creation of a series of rusticated Orders in the sixteenth century, a number of methods of forming openings in rusticated walls had become established. As rustication is essentially an enhancement of the stone normally used for the construction of walls, the details of simple openings are an enhancement of the traditional method of constructing stonework.

A series of voussoirs forming a flat arch can be brought up to the level of an appropriate joint in a rusticated wall (*d*). A more interesting arrangement is the integration of the voussoirs with several courses of stone to form blocks that turn integration of the voussiors outwards into the wall (*a*). The pattern of the rustication on the wall can cut into the voussoirs of the flat arch progressively from the outside by lowering one (*c*) or more (*b*) of them to create a series of upward steps that resemble the arrangement of voussoirs for round arches. Example (*b*) from the Palazzo Costa in Rome is unusual as it has two keystones, instead of one.

Similar variations in round arches are dictated by the way the voussoirs meet the horizontal lines of the wall rustication. With more than three courses of rustication in the height of the arch the voussiors will step progressively upwards as they meet each horizontal course to form the rough outline of a pointed arch in the stone, (*n*) and (*o*). This pattern can be marked with a joint at each intersection (*n*) or, particularly with banding, can run in unbroken lines into the horizontal courses (*o*). These two arrangements can be combined to form L-shaped stones (*m*). The height of the stepped, radiating vousoirs can be limited by stopping them in a horizontal line at a joint above the arch (*p*).

The rustication of both flat and round arches can be isolated from the wall by forming separate rusticated surrounds, (*e*) and (*i*). The surrounds can also be reduced to stone frames that contain no indication of the structural means of forming the opening, (*h*) and (*l*).

In the sixteenth century, designs for rectangular openings and surrounds derived from antiquity were rusticated. Blocks were placed around the architrave and penetrated part of the entablature as voussoirs (*f*) or, as on the window in Vignola's Villa Giulia in Rome (*g*) of 1550, a rusticated flat arch and Classical surround were fused together. Different varieties of this union of rustication with Classical detail have evolved and in George Dance the Younger's (1741-1825) Newgate Prison in London (*j*) of 1770 the rustication has virtually consumed the surround, which is contained within an arch. This combination of rectangular and arched rusticated openings was established in Italian Renaissance palaces. Palladio's arch from the Palazzo Thiene in Vicenza in northern Italy (*k*) of about 1550 combines the two forms effortlessly within the pattern of the rustication, while Dance attempts to match the curve of the arch in the voussoirs and so gradually loses their relationship with the horizontal wall.

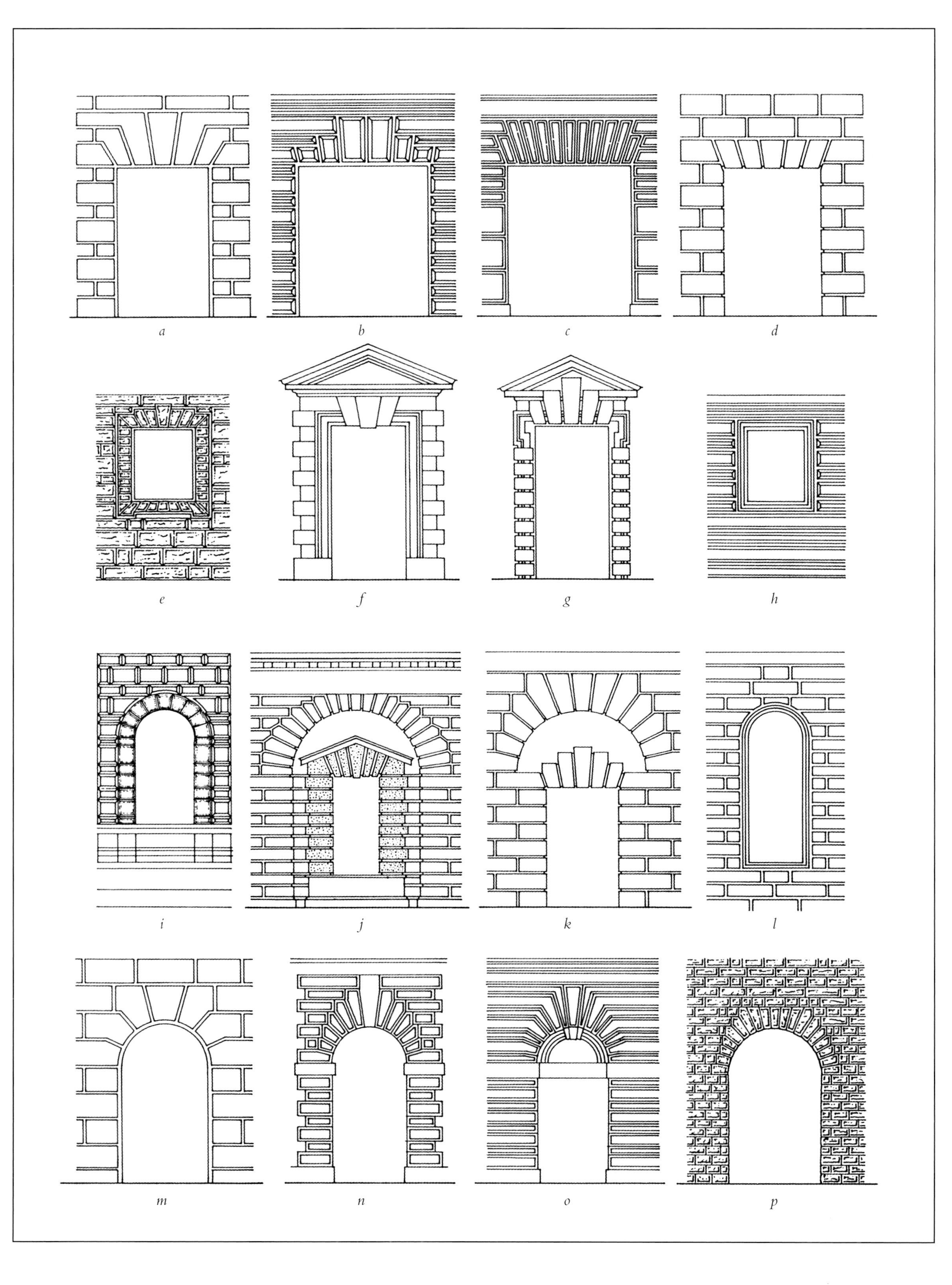

a
b
c
d
e
f
g
h
i
j
k
l
m
n
o
p

粗琢建筑

粗琢工艺在传统上一直用于建筑的低矮楼层，示例参见P29、P33、P37、P39、P43、P45、P59、P81、P89、P97、P101、P131、P135、P139和P143。文艺复兴粗琢工艺使用在宫殿建筑中，牢牢地奠定了这种传统的起源。此类宫殿的一楼通常只包括马厩或服务间，精致的细部设计开始于上层，即主层。粗琢工艺的粗制表面不仅表达了一楼低贱的功能，也显示出建筑的防御特点，此类建筑中的低矮楼层既坚固又表达了生人勿进的意思。粗琢结构也可以构成类似岩石的底座或形成精美结构的底层。设计延续了希腊尤其是罗马的风格，将建筑建在带有粗糙表面饰或凹槽接缝的石头底座上。

粗琢结构可以成为建筑设计中的决定性要素。城市宫殿中粗琢的完整表面装饰是佛罗伦萨文艺复兴建筑中最早的特征之一。随着文艺复兴传播到意大利的其他地方，这种表面装饰也应用到相似的建筑中。1482年由比亚焦·罗塞蒂（1447—1516年）在意大利北部城市费拉拉建造的钻石宫（*a*）之所以得名，便是由于在建筑表面镶嵌锋利的钻石尖形粗琢，因其独特的粗琢形式成为意大利北部设计系列之一。

1537年由雅各布·桑索维诺（1486—1570年）设计的威尼斯造币厂（*d*）原来仅有两层，与同一建筑师设计的圣马可广场相邻，但风格迥异。由于建筑物需要容纳该城的金银储备而必须坚固，因此桑索维诺随即选择粗琢工艺表达这种意图，首次使用带有环形粗琢砌块的立柱，在接下来的几个世纪中成为经久不衰的设计方式。

英国巴洛克风格建筑师尼古拉斯·霍克斯莫尔频繁使用带状粗琢，增加建筑的厚重和坚实感。1712年他为剑桥大学国王学院设计的院长府邸（*b*）虽然没有最终实施，但是表现出在建筑中大量应用带状粗琢的设计意图，在粗琢环带中只简洁地体现出柱头和柱础，装饰也仅限于狮面和两个涡卷形的半山花。

粗琢工艺也可以有选择性地进行使用。督政官官邸（*c*）是法国东部贝桑松附近制盐厂的焦点，由富有创新精神的建筑师克劳德·勒杜在1775年设计。大型门廊中的粗琢结构带有显著的方形砌块，门廊主控整个建筑，成为该综合企业中原始粗琢结构的系列之一。与之相比，威廉姆·塔克（死于1805年）1784年在英国金斯林设计的监狱（*e*）中带有的粗琢结构更加拘谨，有效地使用了粗琢转角，即外墙角和其他小面积区域，使本来未加装饰的外立面有所不同。

RUSTICATED BUILDINGS

Rustications has traditionally been used on the lower floors of buildings and examples have been illustrated on page 29, 33, 37, 39, 43, 45, 59, 81, 89, 97, 101, 131, 135, 139, and 143. The origin of Renaissance rustication in the design of palaces firmly established this tradition. These palaces usually had only their stables and service rooms on the ground floor and the fine detail began on the upper floor, or piano nobile. The coarse surface of rustication expressed not only the humble function of this level but also the defensive character of buildings that needed robust and forbidding lower floors. Rustication can also form a rock-like base for the Orders or provide the first level of a design that becomes progressively finer as it rises. This follows Greek and, in particular, Roman practice where some buildings were constructed on bases of stone with a rough finish or channelled joints.

Rustication can be a dominant element in the design of buildings. A total surface decoration of rustication for urban palaces was one of the earliest features of Renaissance architecture in Florence. The spread of the Renaissance to other parts of Italy led to the introduction of this surface on similar buildings. In the north Italian town of Ferrara the Palazzo dei Diamanti (*a*) of 1482, by Biagio Rossetti (1447-1516), is so called because of the encrustation of sharp, diamond-pointed rustication. It is one of a series of north Italian designs with this very distinctive form of rustication.

The Zecca, or Mint, in Venice (*d*), designed in 1537 by the Florentine exile Jacopo Sansovino (1486-1570), was originally only two storeys. It is one of a series of adjacent but different buildings by the same architect in and near the Piazza S. Marco. As it was to house the bullion reserves of the city it had to be robust and Sansovino chose to express this with rustication which included the first use of columns with the circular rusticated blocks that would become so popular in the following centuries.

The English Baroque architect Hawksmoor frequently used banded rustication to add a feeling of weight and solidity to his buildings. His unexecuted design for the Provost's house at King's College, Cambridge (*b*) of 1712 is an essay in the dramatic application of banded rustication. Column capitals and bases make only brief appearances from within their girdle of rustication, and decoration is limited to lions' masks and two scrolled half-pediments.

Rustication can also be used selectively. The Director's house is the focal point at the saltworks near Besançon in eastern France (*c*) and was designed by the revolutionary architect Ledoux in 1775. The large portico rusticated with prominent square blocks dominates the building and is one of a series of original rusticated compositions in this manufacturing complex. The rusticated features on the gaol in the English town of King's Lynn (*e*), by William Tuck (*d*.1805) in 1784, are restrained by comparison and show the effective use of rusticated corners, or quoins, and other small areas to give some distinction to an otherwise unadorned façade.

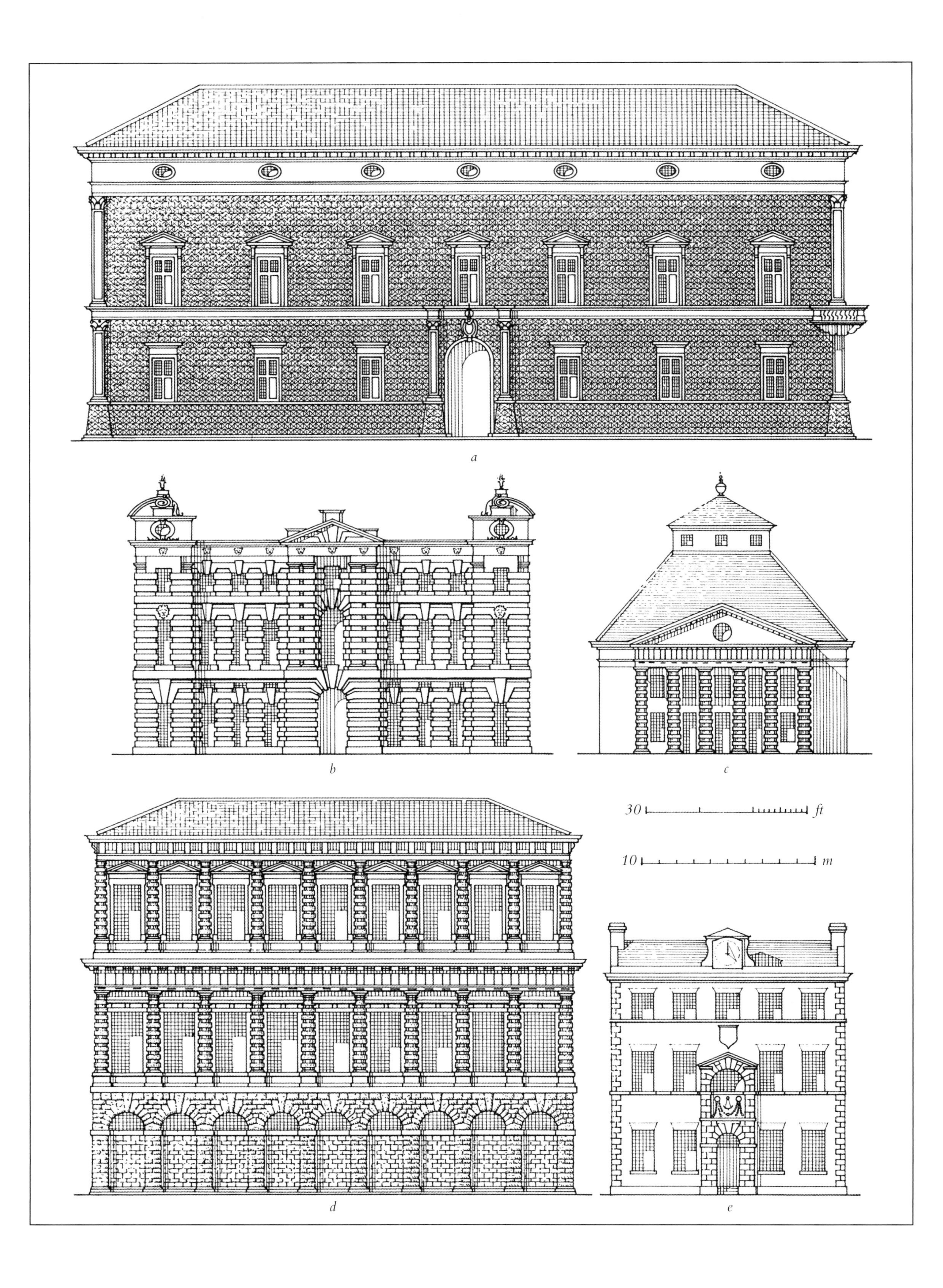
a
b
c
30 ft
10 m
d
e

12. 门与门廊

建筑与入口

入口是建筑设计最为重要的元素之一。入口划分了外部与内部、公共与私人、暴露在外与遮蔽在内的空间——简而言之，入口标记了建筑外壳所能表述的任何东西。几乎所有类型和时期的建筑都特别强调门的设计，门口的交叉地点常带有象征甚至宗教意义。

若干世纪以来，人们普遍接受和理解将入口设置在外立面中心处的做法。依据此传统，入口的设计可以极具重要性，也可以不加强调，仅仅位置居中。

意大利北部维琴察附近的基耶里凯蒂别墅（*a*）1554年由帕拉迪奥设计，门廊很大，与身后的别墅成比例，巨大的门廊出入口成为该设计中唯一重要的元素。布拉曼特约1512年在罗马设计并在后来遭到毁坏的卡普里尼宫即拉斐尔府邸（*b*）中的门廊除了其中心位置以及缺少店主长凳的特征以外，出入口不包含任何其他细部设计。

具有附加意义的入口装饰设计可以采取与建筑设计相关的诸多不同形式。罗伯特·利明奇（活跃于1607—1628年）1609年在设计伦敦附近哈特菲尔德宅邸的正面入口（*c*）时，将其装饰得富丽堂皇，增强了装饰的细部设计和高度。两个世纪之后，许多建筑设计与伦敦郊外里士满的房屋（*d*）类似，细部设计稀疏，只剩下古典门缘。

入口并不总是位于外立面的正中位置。由于宽度不够，或者特殊的功能需求，可以将门设置在一侧。这种设计在小房屋中十分常见。在伦敦19世纪早期的房屋（*g*）中，门与相邻的窗子装饰相同，使整个设计产生平衡感。埃德温·勒琴斯1922年在位于伦敦的银行建筑（*e*）中使用相似的手法，但是用隐蔽的壁龛与门匹配，并且使用大窗子强调中心位置。1925年，西格德·莱维伦茨（1885—1975年）在斯德哥尔摩设计的复活礼拜堂（*f*），即停尸房，运用更加极端的方法，将部分分离的入口门廊牢牢设置在一端。

12. DOORS AND PORCHES

BUILDINGS AND ENTRANCES

One of the most important elements in the design of a building is the entrance. It marks the division between the exterior and the interior, the public and the private, the exposed and the sheltered—everything, in short, that the enclosure of a building represents. In architecture of almost all types and periods particular attention had been paid to the door, and the crossing of the threshold can have symbolic and even religious significance.

The location of the entrance centrally on the façade has for centuries been universally accepted and understood. In response to this convention the design of the doorway can assume supreme importance or receive no emphasis except its central position.

The Villa Chiericati near Vicenza in northern Italy (*a*) was designed by Palladio in 1554 with a portico so large in relation to the size of the villa behind that it acts as a massive entrance porch, and the only significant element in the design. The Palazzo Caprini, or House of Raphael, in Rome (*b*), designed by Bramante in about 1512 and since destroyed, includes no detail specific to the entrance except its central location and the lack of shopkeepers' benches.

The decoration of the entrance for additional significance can take many different forms in relation to the design of the building. Robert Lyminge's (*fl.*1607-28) entrance front to Hartfield House near London (*c*) of 1609 is richly decorated and marked by an increase of decorative detail and height. Two centuries later there were many buildings like the house at Richmond outside London (*d*) where the detail is so sparse that the Classical door-surround is virtually the only decoration.

It is not always possible to locate the entrance centrally on the façade. A narrow width or particular functional requirements can necessitate a door to one side. This is quite common on small houses. In an early-nineteenth-century house in London (*g*) the door and its adjacent window have been given the same decoration to balance the design. Lutyens used a similar device in 1922 for his bank building, also in London (*e*), but matched the door with a blind niche and used a large window for a central emphasis. The Chapel of the Resurrection, a mortuary, by Sigurd Lewerentz (1885-1975) in Stockholm (*f*) in 1925, is much more radical, with a fractionally detached entrance portico sited firmly at one end.

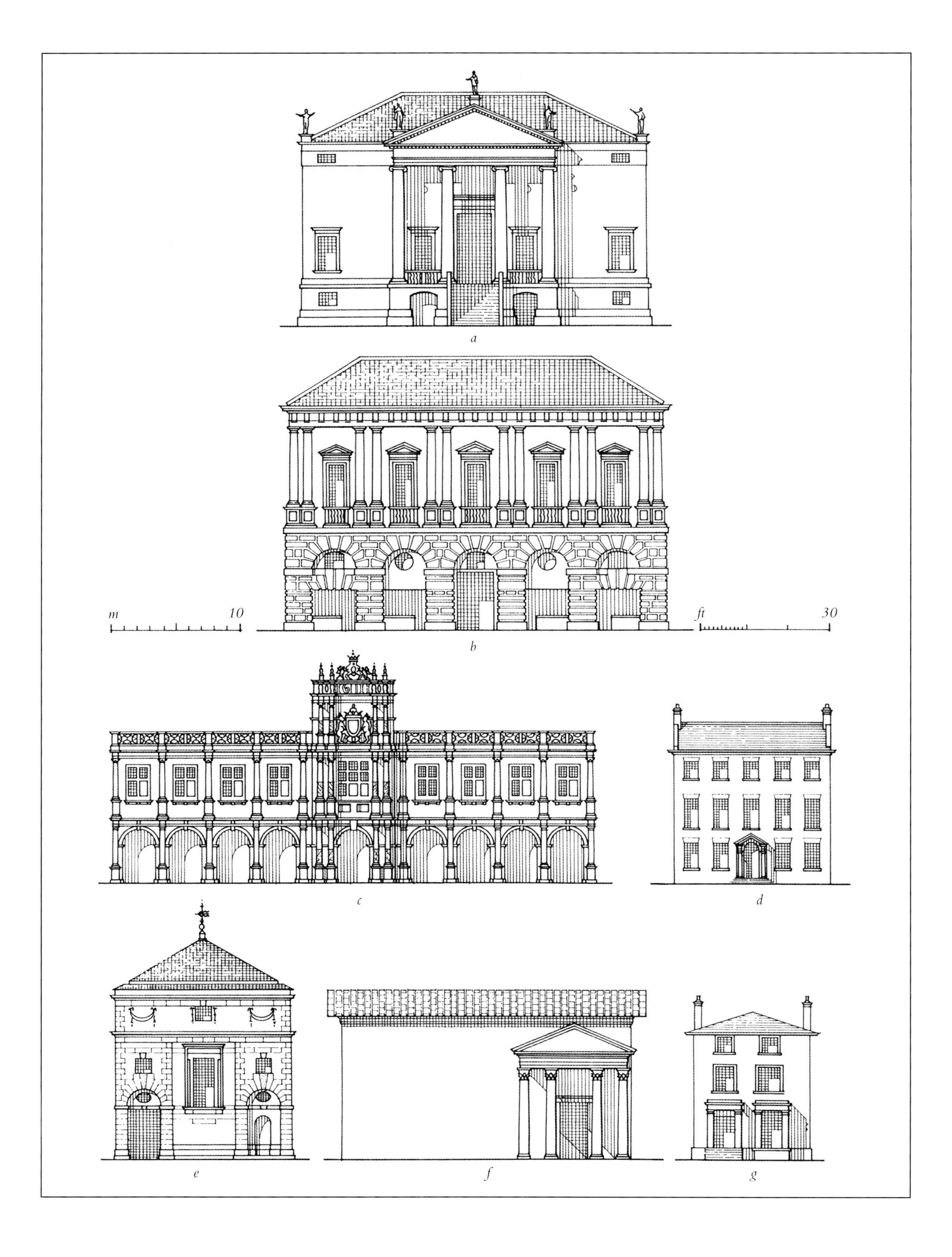
a
m
10
ft
30
b
c
d
e
f
g

古代的门

早在古代便确立了古典门的设计原则。门的细部设计记录了早期出入口建筑的历史，神庙中门的设计表达出进入神殿的象征意义。

门的横梁即过梁必须有足够强的力量来承受上部墙体的重量，并将其转移到开口的侧面。木材具有合适的弹力能有效地承担这一任务，因此成为最早期的过梁材料。但只有早期的石头过梁留存下来，保留了矩形形态和更加适合天然木料形态的结构比例。大多数古典门应用垂直比例，以此限制狭窄横梁的长度，当然也考虑到了人类体型的需要。如果门比较宽，如100年罗马图拉真市场的商店门面（*b*），则需要设置拱形，以减轻对狭长过梁的压力。

建于公元前438年的雅典万神殿中的门（*c*）只包括柱和横梁，过梁两端凸出在墙壁上，边缘用简单的线脚进行强调。和众多古代的门一样，宽度逐渐减小，保存了使开口变窄以减小过梁宽度的传统特征。同期出自意大利中部的伊特鲁里亚风格的出入口（*a*）具有相同的设计，但夸大了细部设计风格。

公元前421年雅典厄里希翁神殿内的门（*d*）更加华丽。石头框架与建筑中爱奥尼亚柱式额枋的装饰方式相同，额枋上方和四周环绕另一种装饰带，与柱式的雕带相对应，上方由两个托架即托脚支撑经改造之后的檐口。木门自身未能留存下来，而是根据雕刻和绘画进行了重建。诸如此类的出入口装饰原则源于建筑物中柱式的细部设计，此传统已经受到广泛认可。柱式和门的额枋都起源于木头横梁，成为大多数古典门中不可分离的一部分。雕带通常省略，尤其在其作为垂直特征的时候更是如此，还可以选择檐口是否带有托架或雕带。

罗马万神殿中的大型出入口（*e*）带有原始的青铜门，是从2世纪保存下来的杰出建筑。大开口的额枋、雕带和檐口根据传统模式建造，但并未减小顶部的宽度。在开口内部，一对多立克壁柱略高于门，支撑自身的额枋即内部过梁。在内部额枋的上方设计了带框的大窗子即楣窗。

DOORS OF ANTIQUITY

The principles of design for Classical doors were established in antiquity. The history of the construction of early doorways is recorded in the detail of the doors they framed, and, in the case of temple doors, their design expresses the symbolic significance of entering into the sanctuary of the god.

The beam, or lintel, of a door must have sufficient strength to take the weight of wall above and transfer it to the sides of the opening. As wood has the right kind of elastic strength to perform this task efficiently, it is likely that the earliest lintels were timber. Only early stone lintels have survived and they retain the rectangular form and proportions that are more suited to the natural form of wood. The need to limit the length of the narrow beam and, of course, the shape of the human figure have established vertical proportions for most Classical doors. Where doors were wider, as in the shop fronts of Trajans's Markets in Rome (*b*) of AD 100, an arch to relieve the weight on the slender lintel was required during construction.

The door of the Parthenon in Athens (*c*) of 438 BC has only the posts and beam expressed, with the ends of the lintel projecting into the wall. The edges of these features are emphasised with a simple moulding. In common with many doors in antiquity, the door diminishes in width and this traditional feature may be the survival of a narrowing of openings to reduce the width of the lintel. An Etruscan doorway from central Italy (*a*) of a similar date has the same design with the details stylistically exaggerated.

One of the doors of the Erecntheion in Athens (*d*) of 421 BC is more heavily embellished. The stone frame is decorated in the same way as the architrave of the Ionic Order of the building. Above and around the architrave is another decorated band, which corresponds to the frieze of the Order, and above this is a modified cornice supported on two brackets, or consoles. The wooden doors themselves have not survived but are reconstructed from the evidence of sculpture and paintings. The decorative principles of doorways such as this, derived from the details of the Order of the building, have become established conventions. The architrave of the Order and the door share a common origin in the wooden beam and are an inseparable part of most Classical doors. The frieze, particularly as a vertical feature, is often omitted and the cornice, with or without brackets or the frieze, is also optional.

The huge doorway of the Pantheon in Rome (*e*), with its original bronze doors, is a remarkable survival from the second century AD. The architrave, frieze, and cornice of the large opening are to the conventional pattern but without a reduction in width at the top. Inside this opening are a pair of Doric pilasters, a little higher than the doors, that support their own architrave, or inner lintel. Above this inner architrave there is a large window, or fanlight, in its own frame.

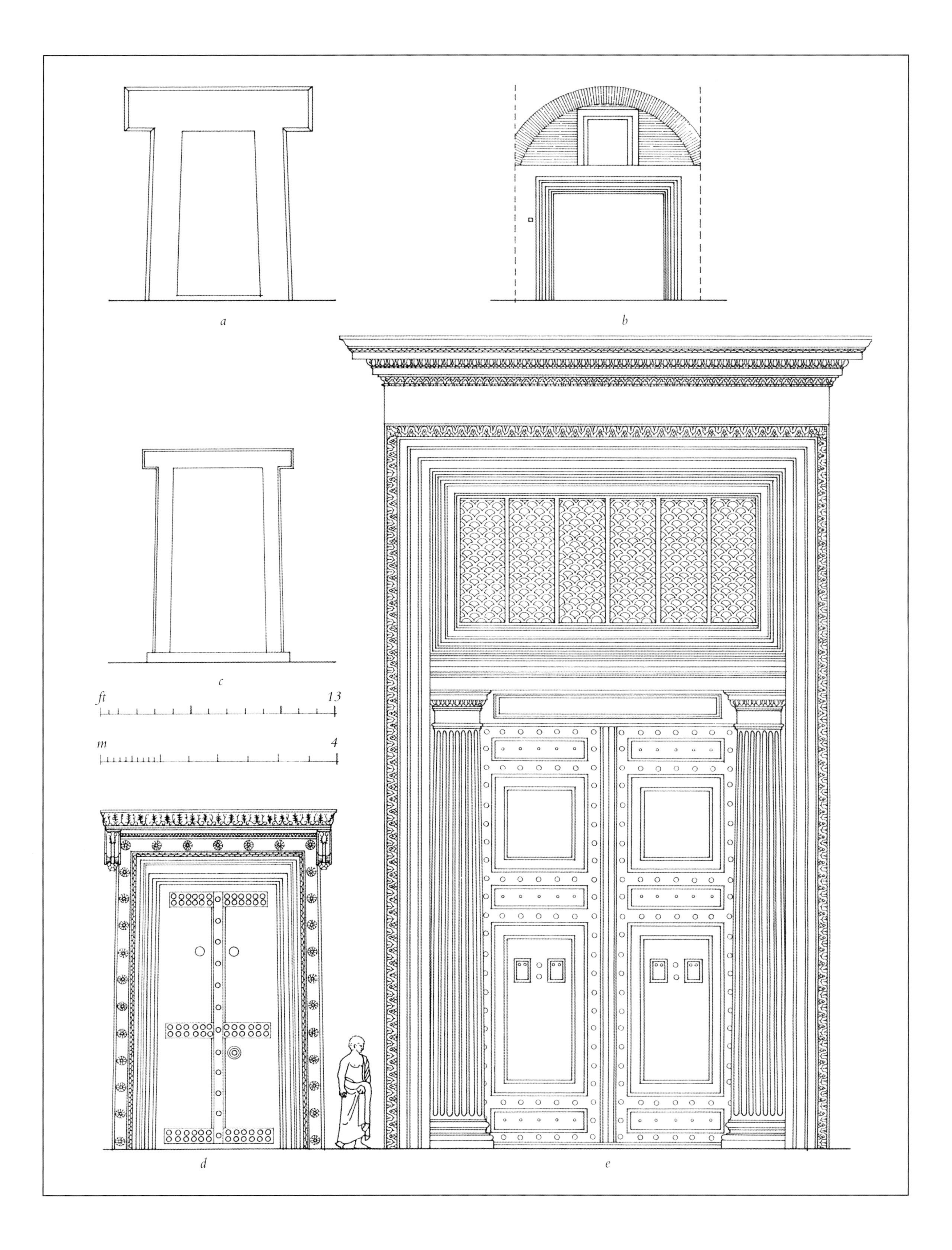
a
b
c
ft
13
m
4
d
e

古代门廊

古典神庙的设计以古希腊的简单原始居所为原型，在古希腊和罗马世界中的小神庙中保留下来，通常包含一个单间即内殿，内殿正对门廊，即内殿入口，门廊由入口前方的立柱支撑。约公元前500年建于古希腊圣城特尔斐的雅典宝库（*a*）便保留了这一布局。

在温暖的气候中，露天门廊成为居所的重要组成部分。在古希腊和罗马房屋发展的过程中，门廊逐渐扩展进而形成完整的庭院柱廊，又称作列柱廊，隐藏在封闭围墙之后。居所中的门廊不再具有公共意义，只有在亚洲的一些古希腊风格的房屋中，部分柱廊结构向上凸起成为主接待室的门廊。

门廊依旧是古希腊神庙的重要组成部分。希腊神庙的内部并不用于接待信众，而是供奉着神像及其宝藏。礼拜和祭祀活动在外面的神坛举行，普通信众在门廊朝拜神像，便是他们与神像最近的距离。厄里希翁神殿（参见P9）在同一屋檐下设置了若干宗教场所，建立了3个门廊代表神庙中的多种供奉。

神庙形式也用于大门即山门，信众经此进入，每条中心通道都经过一处门廊。公共喷泉处建有小门廊，保护其不受天气的影响，大厅的入口正面通常建有露天的柱廊。

当古希腊风格的建筑师在希腊神庙建筑形式中改造这些简单的建筑以适应更加复杂用途的时候，门廊、立柱以及山花得以确立重要的意义，并被视为独立的建筑特征，设置在不同建筑的入口处凸显其地位。入口处添加了形似小型神庙正面的门廊结构，例如约公元前40年建于雅典的风之塔（*c*）成为现在仅存的包括水钟、承重日晷仪以及风向标在内的古代钟塔建筑。

罗马建筑师使用与希腊相同风格的神庙门廊。罗马的万神殿（参见P13）坐落在庭院内部，其圆形平面被庭院围墙和主宰入口正面的传统门廊所遮蔽，现在已被发掘，引人注目。罗马建筑广泛使用小型门廊和门缘装饰，强化门的重要性。公元前1世纪意大利庞贝古城的壁画（*b*）表现出装饰华丽的门缘和栅栏门。甚至连2世纪罗马港口奥斯蒂亚的一处寒酸的仓库建筑（*d*）都带有混合柱式门缘，在雕带中镶嵌石饰板。

PORCHES OF ANTIQUITY

The Classical temple originated in the simple primitive dwelling of the Greeks and the design survives in the small temples of the Greek and Roman world which have a single room, or naos, faced with a porch, or pronaos, supported on columns in front of the entrance. The Treasury of the Athenians in the sacred Greek city of Delphi (*a*) from about 500 BC retains this layout.

In warm climates open porches are an important part of living accommodation. In the development of Greek and Roman houses the porch became extended to form complete colonnades around courtyards, known as peristyles, which were hidden behind enclosing walls. The public significance of the porch in dwellings was lost, except in some Greek houses in Asia where a raised section of the colonnade created a porch for the principal reception room.

Porches remained an essential part of the Greek temple. The interiors of Greek temples were not entered by worshippers but housed the image of the god and his treasures. Worship and sacrifice took place at the altar outside. The closest the ordinary worshipper could come to the sacred image was the porch. When the Erechtheion (see page 9) was constructed to house several religious sites under one roof, three porches were erected to represent the multiple dedications of the building.

The temple form was also used for the gate, or propylon, which gave access to sacred enclosures; the roadway in the centre passed through a porch at each end. Public fountains were protected from the weather by small porches and public halls and were often constructed with an open colonnade on their entrance façade.

When Hellenistic architects adapted these simple buildings in the Greek temple form to more complex uses, the porch, its columns, and its pediment had an established significance and could be taken as an isolated feature and added to the entrance of different buildings to give them additional status. Porches that looked like miniature temple fronts were added to doorways such as the two entrances to the Tower of the Winds in Athens (*c*) of about 40 BC, the only surviving clock tower from antiquity, containing a water-clock and supporting sundials and a weather-vane.

Roman architects used the temple porch in the same way as the Greeks. The Pantheon in Rome (see page 13) sat inside a courtyard and its circular shape, now dramatically revealed, was disguised by the courtyard walls and the conventional temple porch that dominates the entrance façade. The use of small porches and door-surrounds to give additional importance to doors was a widespread Roman practice. A complete doorway is realistically illustrated in a wall painting in the Italian city of Pompeii (*b*) of the first century BC, which shows a richly decorated door-surround and doors with grilles. Even a humble second-century-AD warehouse in Ostia, the port of Rome (*d*), has its own Composite door-surround, executed in brick with a stone plaque in the frieze.

a
b
m 6
m 1
ft 20
ft 3
c
d

门与柱式

门缘中的柱式既可以与主宰建筑或独立楼层比例的柱式直接相关，也可以不与之相关甚至构成对立。尽管门缘或许缺乏支柱或其他特征，但其比例仍与柱式相关。然而，当结构中既包括柱也包括檐部的时候，有多种方法将柱式应用到门中。

在示例（*a*）中，门独立于其周围的多立克柱式的细部设计。门与缘饰之间的间隙并不均等，但是门顶和柱头底部在一条直线上。爱奥尼亚柱式的示例（*d*）显示出两者相似的关系，但是门额枋与柱式的细部设计不同。门额枋的顶部与柱头底部成一条直线，如果门是科林斯柱式（*g*）的话，那么要在两额枋之间保留较大的空间以便作垂花装饰。混合柱式却极力避免这样的构造，在示例（*l*）中，门顶与柱头底部成直线，并由较大的门额枋填充上方的部分空间。

在周柱与门额枋的交界处，檐部额枋和门额枋之间形成明显的视觉冲突。当两者结构相同时，如多立克柱式的示例（*b*），可用匾额分隔。在科林斯柱式示例（*h*）中，不仅门额枋不同，而且檐部的雕带和额枋被限制在柱身的垂直延长线内，在两者之间留出了装饰的空间。在另一科林斯柱式示例（*j*）中，门额枋也有变化，但仅在檐部省略了巴洛克曲线额枋的下面部分，在余下狭窄的空间进行装饰。爱奥尼亚柱式的示例（*e*）则解决了这一问题，长凹槽仅延伸到门边的上部。在混合柱式的门（*k*）中，则用拱门和拱顶石做出分隔。

门顶也可以抬高到檐部下边处，使门额枋与檐部额枋成直线。这样的话，立柱与门垂直额枋之间的关系会出现问题。多立克柱式的门（*c*和*b*）相似，表现出利用垂直额枋将支柱减半的一般技巧，并进一步强调用门额枋上的突耳和上方特殊的雕带细部设计进行分隔。

文艺复兴后期衍生出一种避免额枋之间冲突的更为极端的方法。柱头紧贴在雕带的下方，额枋被支柱减半后沿柱头伸展，如爱奥尼亚柱式示例（*f*）和科林斯柱式示例（*i*）。尽管这种设计改变了柱身和檐部的比例关系，却成功应对了门额枋和柱式之间的关系，因此一经问世便广为使用。

DOORS AND THE ORDERS

The Order of the door-surround may relate directly to the Order that governs the proportions of a building or of an individual floor. It may also be a contrast. The proportions of a door-surround will relate to its Order although it may lack columns or other features. Variations in the application of the Orders to doors are, however, most readily understood when both columns and entablature are included.

In example (*a*) the door is independent of the Doric details that surround it. The margin between the door and the surround is uneven, but the door-head aligns with the bottom of the capital. A similar relationship is to be seen in the Ionic example (*d*) but the architrave details of the door differ from those of the Order. The top of the door architrave aligns with the bottom of the capital and when a similar relationship is applied to a Corinthian door (*g*) the space between the two architraves becomes large enough to be decorated with swags. This is avoided in the Composite example (*l*) where the door-head aligns with the bottom of the capital and the space above is partly filled with the large door architrave.

Where the surrounding Order comes into contract with the architrave of the door, a visual conflict can be created between the architrave of the entablature and the door architrave. Where they are identical, as in the Doric example (*b*), a tablet helps to separate them. In the Corinthian example (*h*) not only is the door architrave different but the frieze and architrave on the entablature are limited to vertical extensions of the columns leaving a decorated space between. In the other Corinthian example (*j*) the door architrave is also varied, but only the lower section of the curved Baroque architrave is omitted on the entablature and the remaining narrow space is decorated. In the Ionic example (*e*) the problem is solved by bringing a fluted architrave only up the sides of the door. On the Composite door (*k*) the separation is maintained with an arch and keystone.

It is also possible to bring the door-head up to the underside of the entablature so that the door architrave will be in line with the entablature architrave. This creates a problem in the relationship between the columns and the vertical architraves of the door. The Doric door (*c*), like (*b*), shows the common technique of halving the columns with the vertical architraves, and further emphasises the separation with ears on the door architrave and special frieze details above.

In the later Renaissance a more radical method of avoiding any conflict between the architraves was evolved. A column capital sits immediately below its frieze and the architrave, halved with the column, lies alongside the capital. This can be seen on the Ionic example (*f*) and the Corinthian door (*i*). Although it altered the normal proportional relationship between the columns of the Order and the entablature, this design so successfully dealt with the relationship between the architraves of the door and the Order that since its development it has been widely used.

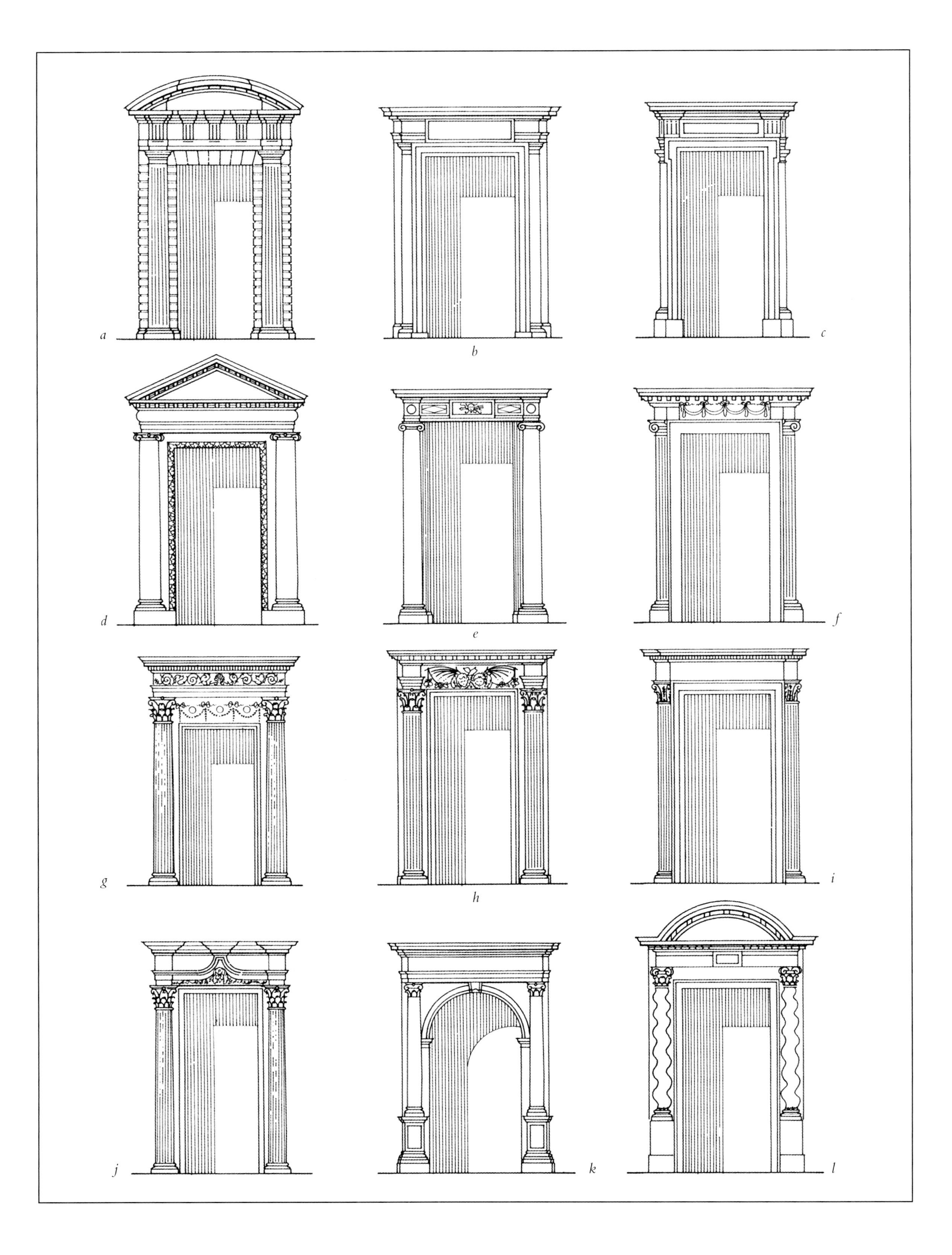

a
b
c
d
e
f
g
h
i
j
k
l

门与托架

在门缘上使用托架的古典传统从早期古希腊神庙一直沿用至今。托架最先用来支撑门上方凸出的檐口，并在后来的大多数门缘中保持不变的位置。

埃尔姆斯和查尔斯·罗伯特·科克雷尔1839年在英格兰利物浦建造的圣乔治厅门缘（*b*）再现了门周围线脚外侧的托架在古希腊时期的原始位置。设计直接源于雅典厄里希翁神殿中由想象而诠释的门。托架通常置于沿门额枋而建的特殊狭窄壁柱之上。罗马文艺复兴的门（*a*）和英国18世纪的门（*c*）体现出这一细部设计特征，但建筑中托架的位置不同，文艺复兴门（*a*）的托架位于开口的上方，仅支撑檐口的上半部分，因此减少了檐口向侧面的凸出。英国门（*c*）的托架在开口的正下方向外凸出，依靠更大的侧面凸出来支撑整个檐口。

托架可以直接安置在与门等高的整个壁柱上。高托架将替代檐部的下部，并将檐口抬至一定高度用窗子填充托架之间的空间。此设计可见于18世纪早期（*f*）和19世纪早期（*d*）的门中。16世纪意大利建筑师塞里奥的设计（*e*）也将檐口抬高到托架的高水平位置，但是将高檐口置于完整的门缘上方，门缘本身自带较小的檐口结构。

托架可以设置成与门顶同高，并支撑部分雕带和檐口（*k*）。如果门侧面的空间受限，它甚至能够放置在上方位于额枋内（*j*），以便限制檐口的凸出。米开朗琪罗在1524年设计佛罗伦萨劳伦森图书馆（*l*）的时候，甚至在门额枋内部使用简化的托架，以便产生为沿额枋顶部放置的檐口雏形提供支撑的效果。

巴洛克建筑中额枋上的突耳与托架具有不确定的关系。两者可以形成直接且独立的关系（*g*）。在1576年由贾科莫·德拉·波尔塔（1532—1602年）设计的罗马大学的门（*h*）中，附加的装饰托架紧靠在墙壁上，而到了下一世纪中期，已经简化成门上的简单轮廓，例如弗朗西斯科·博罗米尼在罗马建造的布法罗赌场（*i*）中所体现出的门的托架设计。尽管轮廓以托架为原型，但已经被博罗米尼及其追随者转变成为门和窗子额枋的上半部分，平实地再现出突耳。

DOORS AND BRACKETS

The use of brackets on door-surrounds has been a consistent Classical tradition since they were introduced in early Greek temples. Brackets first supported the projecting cornice over the door and this has remained their position on most subsequent door-surrounds.

The original Greek position of the bracket outside the mouldings around the door is reproduced in the door-surround of St George's Hall in Liverpool, England (*b*), by Elmes and C.R. Cockerell in 1839. This design is taken directly from the door of the Erechtheion in Athens, although the door is an imaginative interpretation. The bracket is more often placed on a special narrow pilaster alongside the door architrave. A Renaissance door from Rome (*a*) and an eighteenth-century English door (*c*) show this detail. The position of the bracket, however, differs on the two examples. On the Renaissance door (*a*) the bracket is above the opening and supports the upper part of the cornice only, thereby reducing the projection of the cornice to the sides. On the English door (*c*) the brackets project well below the opening and support the full cornice with a correspondingly greater projection to the side.

Brackets can be placed directly on to full pilasters the same height as the door. A tall bracket will then replace the lower sections of the entablature and raise the cornice to a height that will allow a window to fill the space between the brackets. This can be seen on doors of the early eighteenth (*f*) and early nineteenth (*d*) centuries. An unusual design (*e*), by the sixteenth-century Italian architect Serlio, also raises the cornice to a high level on brackets, but places the high cornice above a complete door-surround with its own smaller cornice.

The bracket can be set at the same height as the door-head and support a section of the frieze as well as the cornice (*k*). Where the space at the side of the door is restricted, it can even sit above and inside the architrave (*j*) in order to limit the projection of the cornice. Michelangelo, in a design for the Laurentian Library in Florence (*l*) of 1524, even took a simplified bracket within the door architrave in order to give the impression of support for a rudimentary cornice placed along the top of the architrave.

In Baroque architecture the ears on the architrave developed an ambiguous relationship with the bracket. Ears on an architrave can have a straightforward and independent relationship with the bracket (*g*). Extra decorative brackets set flat against the wall, as on the door of the university of La Sapienza in Rome (*h*), by Giacomo della Porta (1532-1602) in 1576, had, by the middle of the next century, been reduced to a simple profile on doors, such as those designed for the Casino del Bufalo in Rome (*i*) by Borromini. This profile, although by origin a bracket, was transferred to the upper part of the architraves of doors and windows by Borromini and his followers to create a design that was a very literal representation of ears.

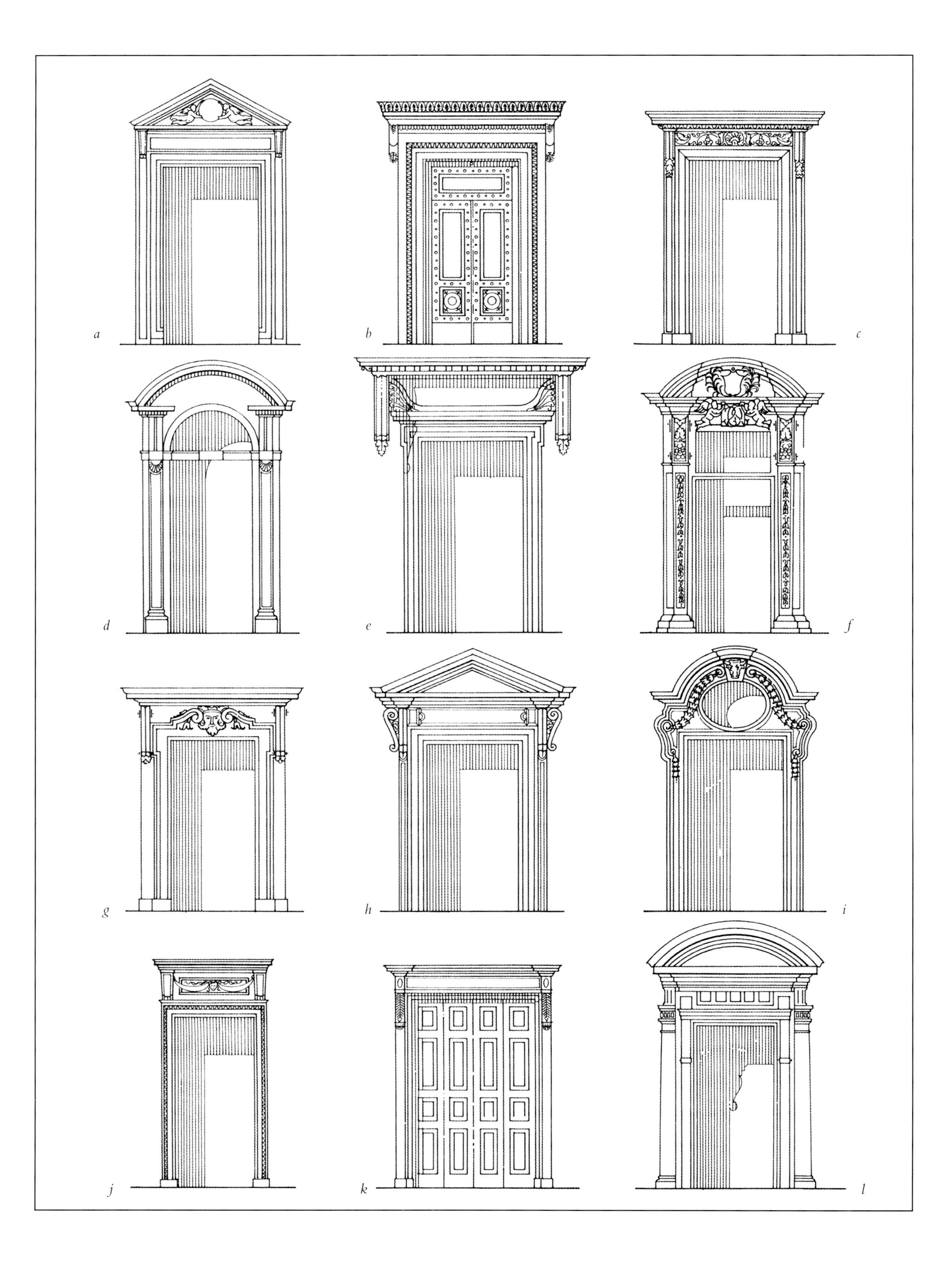
a
b
c
d
e
f
g
h
i
j
k
l

门缘1

罗马的文艺复兴建筑示例（*a*）诠释了无柱的完整古典门缘的标准形式。完整的檐部和山花从门头的位置开始延伸。额枋向下勾勒出门开口的框架，在底部转向内部形成底座。或者省略山花，只保留简单的檐部，额枋向下延伸至砌块，或终止于地面而无回路。此类门缘可以使用任意柱式比例和装饰，采用多变的细部设计，频繁出现在所有古典时期。

示例（*b*和*c*）展示了文艺复兴时期的其他一些门缘。示例（*b*）取自意大利北部的热那亚，内外额枋勾勒出门上方雕刻镶板的框架。外部额枋和上方的檐口十分狭窄，且檐部省略了雕带。内部额枋较宽，开口的上角包括小型托架。示例（*c*）取自意大利北部的博洛尼亚，也省略了雕带且额枋中包括简朴的内部框架，拱形檐口搭配涡卷形装饰，独立于水平檐口的上方。

简单的门缘也有多变的样式。额枋顶部可以延伸，可以添加凸出即鼓突的雕带，也可以中断山花结构（*d*）。当难以容纳檐口的凸出部分时，可以将雕带缩窄成为涡卷形装饰（*e*）以减小檐口的宽度。1728年詹姆斯·吉布斯（1682—1754年）在英国萨里郡设计萨德布鲁克公园（*f*），便将雕带和檐口缩减至曲面镶板，实现减小山花宽度的目的。

简单的文艺复兴门缘（*g*）展示了如何在标准形式中容纳拱形，此例中使用粗琢工艺。未经装饰的拱形可以置于简朴的额枋内部，但贝尔尼尼于1658年设计的罗马奎琳岗圣安德鲁教堂（*h*）却特别注意侧门的设计，带有阶梯状拱形和冠顶涡卷形装饰。粗琢结构可独立构成门缘，示例（*i*）1901年由埃德温·勒琴斯设计，取自英国的霍姆伍德，展示了平拱内部的粗琢横梁。

标准的经典样式也衍生出诸多的变化。约翰·福尔斯顿（1772—1841年）1823年在英国普利茅斯的圣安德鲁礼拜堂中设计了希腊复兴风格的门（*j*），虽然源于希腊传统，但设计异常大胆。同样令人惊艳的还包括取自1714年伦敦圣乔治东由霍克斯莫尔设计的侧门（*k*），用粗琢的拱顶石打穿极度简化的标准门缘，并包括椭圆形窗和极简的内部门缘。1831年柏林建筑专科学校（*l*）出自申克尔，装饰性框架和古代希腊檐口结合在一起，使人联想到文艺复兴时期。

DOOR-SURROUNDS 1

The standard form of a full Classical door-surround without column is illustrated with a Renaissance example from Rome (*a*). From the head of the door there is a full entablature and pediment. The architrave turns down to frame the door-opening and at the bottom turns inwards to form a base. The pediment can be omitted to leave a simple entablature and the architrave can come down to a block, or end at the floor, without a return. This type of door-surround can have the decoration and proportions of any of the Orders with varied degrees of detail and has been used repeatedly in all Classical periods.

Other Renaissance door-surrounds are illustrated in examples (*b*) and (*c*). Example (*b*), from Genoa in northern Italy, shows an outer and inner architrave which frame a sculpted panel over the door. The outer architrave and the cornice above are narrow and the frieze has been omitted from the entablature. The inner architrave is wider and the upper corners of the opening contain small brackets. In example (*c*), from Bologna in northern Italy, the frieze is also omitted and the architrave contains a plain inner frame. An arched cornice, finished with scrolled ornaments, sits independently above the horizontal cornice.

The simple door-surround can be varied too. The top of the architrave can be extended, a bulging, or pulvinated, frieze can be added and the pediment can be broken (*d*). When the projection of the cornice is difficult to accommodate, the width of the cornice can be reduced by narrowing the frieze in the form of a scroll (*e*). A Baroque door-surround from Sudbrook Park in Surrey in England (*f*), by James Gibbs (1682-1754) in 1728, shows the width of the pediment decreased by the reduction of the frieze and cornice to a curved panel.

A simple Renaissance door-surround (*g*) shows how the standard form can accommodate an arch and, in this case, rustication. An undecorated arch can sit within a plain architrave, but Bernini has given his side-door of S. Andrea al Quirinale in Rome (*h*) of 1658 greater interest with a stepped arch and crowning scrolls. Rustication can itself form the door-surround and the example (*i*), by Lutyens in 1901, from Homewood in England, has a rusticated beam suspended inside a plain arch.

There are many departures from the standard Classical type. The bold Greek Revival door from St Andrew's Chapel in Plymouth, England (*j*), by John Foulston (1772-1841) in 1823, is revolutionary in spite of its derivation from Greek sources. Equally dramatic is Hawksmoor's side-door from St George-in-the-East in London (*k*) of 1714, where a savagely simplified standard door-surround is punctured with a rusticated keystone and contains an oval window and severely simplified inner door-surround. Schinkel's weighty design, from the Bauschule in Berlin (*l*) of 1831, combines a decorative frame reminiscent of the Renaissance with an ancient Greek cornice.

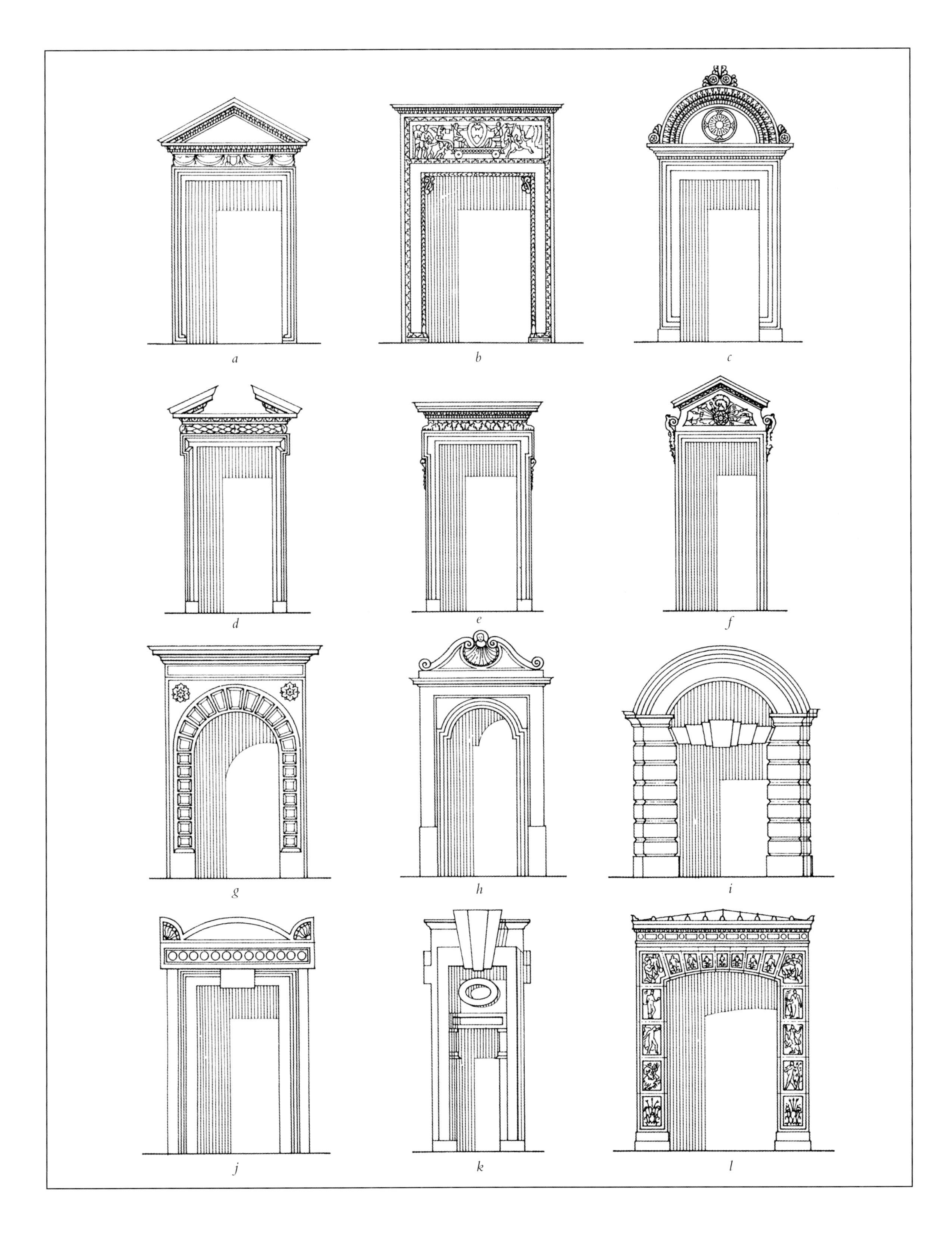
a
b
c
d
e
f
g
h
i
j
k
l

门缘2

门缘通常嵌入拱形开口之中，且门的上方带有半圆形窗子。这也成为18世纪末和19世纪初的一般性特征，半圆形窗子这一体现出创造力的设计也得名扇形窗。

早期的范例包括帕拉迪奥1548年在威尼斯附近的博雅纳别墅中所做的大胆设计（*b*）。内部拱形置于极简的檐口形式之上，一列小圆形窗环绕四周，形成独具特色的室内光线样式。带有内部装饰的拱门构造简易，与之相比柱式的细部设计更为完整（*a*）。门的两侧可以添加两扇窗，根据情况调整柱式的比例，以减少立柱和檐部的宽度和高度，给玻璃留出更大的空间（*d*）。通过限制门周至额枋的细部设计，可以制造出略为粗糙但相似的效果，示例（*c*）在外部拱形中添加了装饰性的石膏扇形，将扇形窗限制在门上方的内部拱形中。

巴洛克和洛可可门缘中的柱式经过改造之后，能够实现丰富的装饰效果。檐部的额枋通常在中心向上提起，在门上方添加小型雕刻特征（*e*）。山花也常常改变形态，省略部分上、下结构（*e*、*f*和*h*），并使用涡卷形装饰或多条曲线来调整轮廓。例如一扇17世纪后期的英国门（*f*）使用山花在非常狭小的空间内制造出复杂的装饰方案。檐部简化为柱体上方的柱顶石，山花被相对的曲线分割开，且山花内部包含双弧线拱形，用雕刻装饰填充。德国慕尼黑的门缘（*h*）将多数装饰置于檐口的上方，檐口则以椭圆形窗子为中心，然而在巴西圣若昂-德尔雷伊，18世纪门（*i*）中的额枋从檐部延伸出来，支撑圣母像的华丽框架。在建于1705年的英国布伦海姆宫（*g*）中，由霍克斯莫尔所设计的纪念碑上的巴洛克风格门缘相对比较素净，与位于平拱开口上方且带有简单涡卷形装饰的大型多立克柱式形成鲜明的对比。

19世纪的门缘也能够实现对柱式的改造，但目的仅限于创造更加一致性的效果。某设计样书中所列举出的多立克门（*k*）源自意大利乡村建筑，属于古典样式，并将山花简化为上部线脚。德国建筑师申克尔为柏林一所音乐学校而做的设计中包括一扇门（*j*），置于外部柱式内，柱式混合了多立克和科林斯柱式的细部设计并带有希腊风格的内部门缘。路易斯·沙利文（1856—1924年）1894年在纽约州水牛城保险大楼内设计的门（*l*）结合了文艺复兴设计和用赤陶土制成的新奇几何形装饰。

DOOR-SURROUNDS 2

Door-surrounds are often fitted into arched openings with a semicircular window over the door. This was a common feature in the late eighteenth and early nineteenth centuries and the inventive designs of the semicircular windows gave rise to the name fanlights.

An early example is a bold design by Palladio from the Villa Pojana near Venice (*b*) of 1548. An inner arch sits above the simplest representation of a cornice and a row of small circular windows rotates around it, giving a distinctive pattern of light to the interior. The details of the Order can be more complete (*a*) contrasting the simple shape of the arch with the details inside. Two more windows can be added on either side of the door and the proportions of the Order are sometimes modified to reduce the width and height of the columns and entablature and give more space for glass (*d*). A similar but less elaborate effect is created by limiting the details around the door to the architrave, and in example (*c*) a decorative plaster fan has been added to the outer arch, restricting the fanlight to the inner arch over the door.

The Orders on Baroque and Rococo door-surrounds can be modified to achieve a rich decorative effect. The architrave in the entablature is often lifted in the centre and a small carved feature added above the door (*e*). Pediments are frequently varied by omitting sections from above and below, (*e*), (*f*), and (*h*), and by using scrolls or multiple curves to modify the profile. A late-seventeenth-century English door (*f*) has used the pediment to create a complex decorative scheme in a very small space. The entablature is reduced to dosserets above the columns, the pediment is split with opposing curves and the space inside the pediment contains a double-curved arch, spilling over with carved decoration. A door-surround from Munich in Germany (*h*) has most of the decoration above the cornice concentrated around an oval window, while the architrave on the late-eighteenth-century door of São João del Rei in Brazil (*i*) explodes through the entablature to support an ornate frame for a statue of the Virgin. Hawksmoor's monumental Baroque door-surround at Blenheim Palace in England (*g*) of 1705 is sober by comparison, contrasting the large Doric Order with the simple scrolls above the plain arched opening.

Nineteenth-century door-surrounds could also include modifications to the Orders, but these were often introduced to create a greater feeling of solidity. The Doric door (*k*) from a pattern-book of designs is typical and includes a pediment reduced to its upper mouldings, derived from Italian rural buildings. The German architect Schinkel's design for a music academy in Berlin included a door (*j*) inside an outer Order of mixed Doric and Corinthian details with an inner door-surround of a Greek type. A door by Louis Sullivan (1856-1924) on the Guaranty Building in Buffalo, New York State (*l*) of 1894 combines a Renaissance design with novel geometric decoration executed in terracotta.

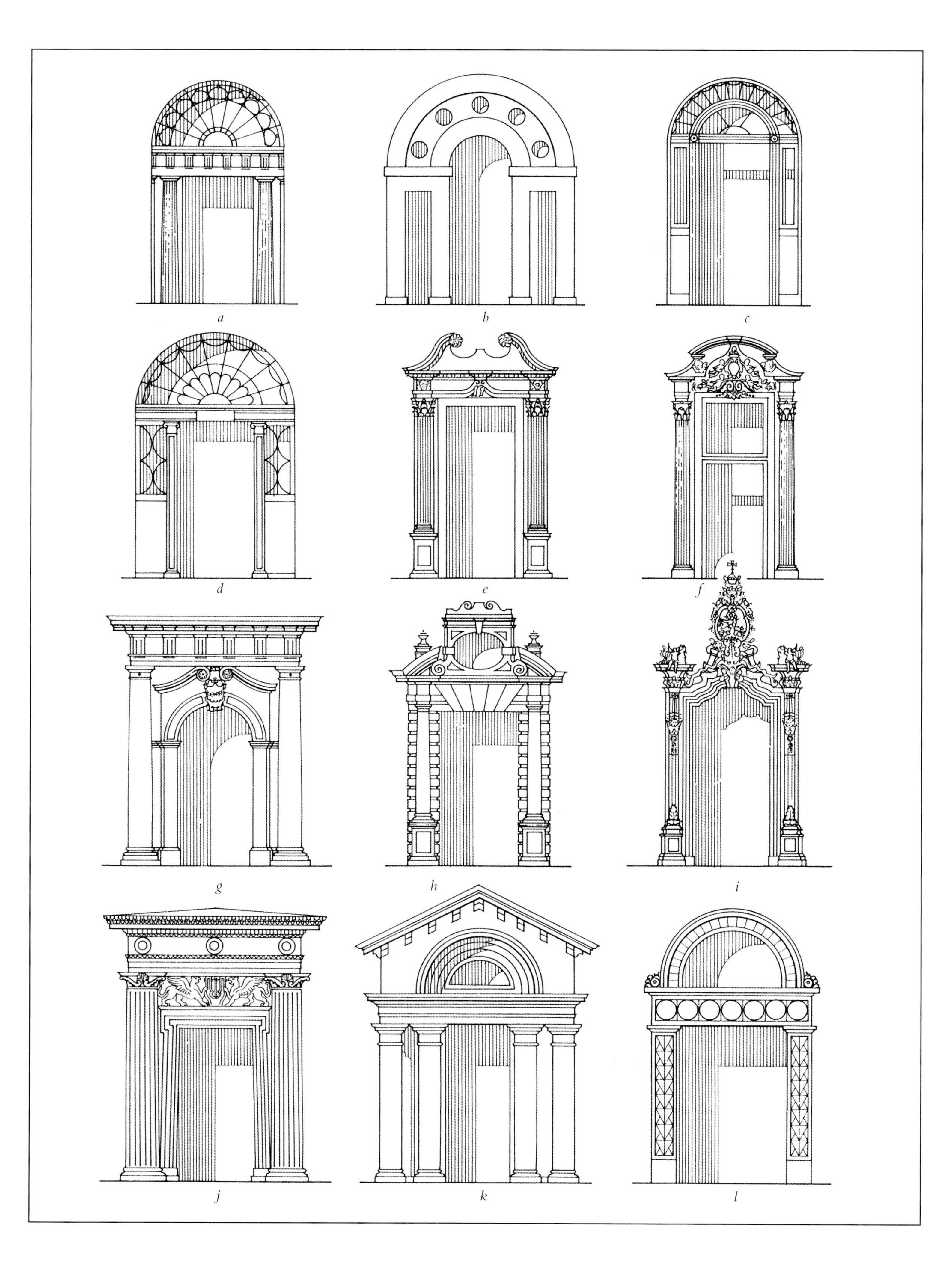

a
b
c
d
e
f
g
h
i
j
k
l

双层高门缘

在不同建筑物中，门上方窗子的设计、屋檐或山花的细部设计通常各不相同，使建筑的外立面发生变化，并且更加强调入口或中心的位置。在一些设计中，门上方的建筑部分包含在独特且分离的门缘设计之中。此类设计的高度从一层至多层不等，甚至能够与建筑等高，但在实际中多为两层，很少超过三层。

三层门缘是16世纪北欧建筑的一般特征，展示了新发现的古典柱式。1590年威尔士博普雷城堡（*a*）包括逐渐变小的上升系列柱式，但依靠柱脚的变化或多柱脚制造出两个同等高度的上部水平面。1674年牛津大学圣凯瑟琳学院的大门入口处（*c*）的设计更加拘谨，在大科林斯柱式内部包含两扇上层窗户。下面的入口成为多立克柱础的一部分，支撑上方的主要支柱。该设计因彻底改变了在垂直方向上缩小柱式的正常做法而闻名。雅各布·普兰陶尔（1660—1726年）1712年在克洛斯特斯的圣弗洛里安教堂的入口处（*b*），应用德国巴洛克设计充分展示了装饰雕塑以及在三层结构中逐渐消失的多曲线设计。

维尼奥拉1550年在罗马朱利亚别墅入口（*e*）的设计要素可以延伸至整栋建筑，但由于细部设计和装饰的不同而使它成为外立面上的独立要素。巴洛克建筑大多使用更多的装饰衔接门框和上方的窗子。1701年英国索尔兹伯里蒙柏森官邸（*g*）中的门、窗仅仅通过门上方墙壁上的稍微凸起和窗上增强的细部设计相联系。在贾科莫·莱尼（1686—1746年）1723年在伦敦设计的阿盖尔官邸（*i*）中，栏杆扶手和瓶与上方的窗子重叠在一起，创造出更为统一的结构。在阿尔多·安德烈阿尼（1887—1971年）1924年为米兰菲迪亚宫设计的门（*f*）上，三个截然不同的细部设计在纵向彼此连接，使用石头和个性化特征而区别于石膏外立面的其余部分。

双层高度的入口不必包括附加柱式和特征，但可以进行个性化设计。埃德温·勒琴斯1904年为伦敦乡村生活建筑而设计的门（*d*）提升了单一山花柱式的高度，包括门上方山花装饰内的窗子。彼得·斯伯瑞思（1772—1831年）1811年在德国维尔茨堡设计的士兵营房（*h*）十分出色，门缘中包含置于粗琢拱门之上的柱廊，形成视觉上的独立感，将要素隔离出来并支配整个外立面。

DOUBLE-HEIGHT DOOR-SURROUNDS

The design of the window above the door and the detail of the eaves or pediment are often varied to give some rhythm to the façade and to give additional emphasis to the entrance or centre. In some designs the section of the building above the door is included in the design of a distinct and separate door-surround. The height of such a design could be any number of storeys, to the full height of the building, but in practice they are rarely greater than three and more often two storeys high.

Three-storey door-surrounds were a common feature in northern European buildings of the sixteenth century and displayed the newly discovered Classical Orders. The porch of Beaupre Castle in Wales (*a*) of 1590 has an ascending series of Orders diminishing in size, but manipulated with varying or multiple pedestals to create two upper levels of equal height. A gateway to St Catherine's College, Oxford (*c*) of 1674 is more restrained, containing two upper-storey windows within one large Corinthian Order. The gateway below becomes part of a Doric base for the principal columns above. The design is notable for the reversal of the normal vertical reduction in the size of the Orders. The German Baroque design of the gateway to St Florian in Kloster (*b*), by Jakob Prandtauer (1660-1726) in 1712, is an extravagant display of figured sculpture and multiple curves diminishing over three storeys.

The elements in the design of the entrance to the Villa Giulia in Rome (*e*), by Vignola in 1550, could be extended over the whole building, but a difference of detail and decoration makes it an independent element in the façade. An increase in decoration was often used on Baroque buildings to link the door-surround with the window above. On Mompesson House in Salisbury, England (*g*) of 1701 the door and window are only just brought together by a slight projection in the wall above the door and by the enhanced detail on the window. On Argyll House in London (*i*), by Giacomo Leoni (1686-1746) in 1723, a balustrade and urns overlap with the window above, creating a more unified composition. A door for the Palazzo Fidia in Milan (*f*), by Aldo Andreani (1887-1971) in 1924, has three quite different details just touching one another in a vertical sequence which is distinguished from the rest of the stucco façade by the use of stone and by the individuality of the features.

Double-height entrances do not have to be formed from additive assemblies of Orders or features but can be individual designs. Lutyens's door for the Country Life building in London (*d*) of 1904 raises the height of a single pedimented Order to contain the window above the door in the high tympanum. The remarkable guards' barracks in Würzburg, Germany (*h*), by Peter Spreeth (1772-1831) in 1811, has a door-surround with a colonnaded gallery over a heavily rusticated arch, giving it a visual independence which both isolates the element and allows it to dominate the façade.

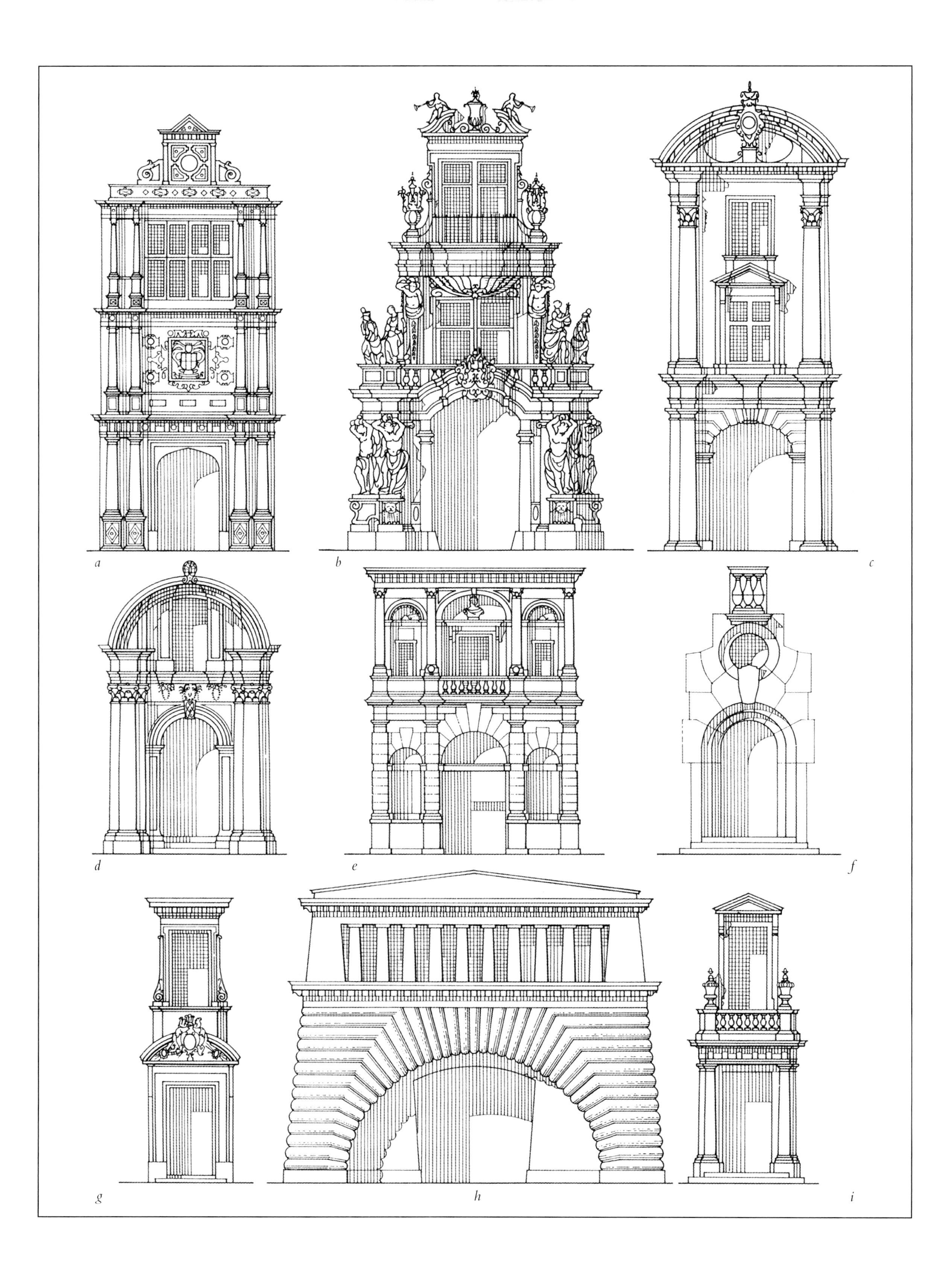
a
b
c
d
e
f
g
h
i

门廊与入口

门廊的实用优势保证了在建筑中的持续使用。门廊位于门的前方，成为入口处的设计，与门缘具有相同的重要性。前文中展示的许多门缘细部设计都可以添加独立支柱或托架而成为门廊。

规模大到支配整个外立面且一般高于单层的门廊或入口，产生于古代神庙式立面建筑转变成为不同建筑形式的过程之中。虽然低矮的连续柱廊自4世纪首次出现之后一直用于教堂正立面，但仅在文艺复兴时期与多神教神庙不再直接相关联的时候，才增加了带有山花的大型柱式。连续柱廊被称作门廊，并且在英语中该词逐渐转变成为独立的门廊之意。

将神庙立面式的门廊应用到住宅建筑中的做法几乎完全要归功于16世纪意大利建筑师帕拉迪奥及其名著《建筑四书》。帕拉迪奥错误地认为罗马人重视他们的住宅建筑就如同他们注重神庙一样，因此会将神庙立面式的门廊应用在住宅之中。尽管依据这样一个错误的假设，高门廊在住宅中仍然流行起来。神庙立面式门廊通常带有山花，偶尔省略，可以仅包括一种高柱式，如威廉姆·维克菲尔德1700年在约克郡邓肯公园中的设计（*a*），或者包括上升柱式和阳台，如1553年帕拉迪奥本人在威尼斯附近的科尔纳罗别墅（*b*）。阳台也可以沿大柱式从中途向上修建，这种同样由帕拉迪奥引入住宅建筑的细部设计在美国十分流行。

建于支柱上的小型门廊可采取多种形式。比较普遍的设计是在两个或多个支柱中简单凸出檐部或山花（*g*），并且根据柱式的全部变体有所变化。通常包括两个壁柱，檐部与墙壁在此处相接，并且可以在侧面重复这一设计（*c*）以此扩大结构。门廊可以由侧面墙壁部分围绕，支柱置于墙壁之间（*d*），采用小型希腊神庙门廊的样式。或者门廊有半圆形（*e*）、半椭圆形或弓形平面。

如果有托架支撑部分或全部凸出的檐部，那么门廊可以不带独立式支柱。托架可参考P205所阐述的任何位置。为了限制凸出的天棚的重量，通常仅仅凸出檐口，托架位于雕带中（*f*）。天棚采用平铺或山花的形状，抑或其他形式，诸如英国18世纪早期的装饰性半圆形出檐（*h*）。

PORCHES AND PORTICOS

The practical advantages of porches have ensured their continuous use. Located in front of the door, the design of the porch becomes the design of the entrance and shares all the significance of the door-surround. Many of the door-surround details shown on previous pages can, by the addition of free-standing columns or brackets, become porches.

Porches, or porticos, large enough to dominate the façade and generally more than a single storey in height, derive from the transfer in antiquity of the temple front to buildings of a different form. Although low continuous colonnades had been used for the front of churches since their first construction in the fourth century, the addition of giant Order columns with a pediment could only be introduced in the Renaissance when the direct association with pagan temples had ceased to be significant. Continuous colonnades are called porticos and the word seems to have been transferred in the English language to isolated porches.

The use of the temple front portico for houses is almost entirely due to the influence of the sixteenth-century Italian architect Palladio, and the popularity of his *Four Books of Architecture*. Palladio held the erroneous idea that the Romans gave as much importance to their houses as to their temples and would, consequently, have used temple front porticos for their houses. Although based on an incorrect assumption, the application of a high portico to houses has become widespread. Temple front porticos, often with but occasionally without pediments, can have one giant Order, as on William Wakefield's Duncombe Park in Yorkshire (*a*) of 1700, or can have ascending Orders and a balcony, as on Palladio's own Villa Cornaro near Venice (*b*) of 1553. Balconies can also be included halfway up giant Order columns and this detail, also introduced to house fronts by Palladio, is popular in the United States.

Smaller porches sitting on columns can take many forms. The simple projection of an entablature or pediment on two or more columns (*g*) is universal and subject to all the variations available in the Orders. There are usually two pilasters where the entablature meets the wall and these can be repeated at the sides (*c*) to enlarge the composition. The porch can be partially enclosed by side-walls and the columns can sit between the walls (*d*) in the form of a small Greek temple porch. Alternatively, the porch can have a half-circular (*e*), half-elliptical, or segmental plan.

A porch can be made without free-standing columns by projecting a part or all of the entablature on brackets. The brackets can take any of the positions illustrated on page 205. To limit the weight of the projecting canopy it is often restricted to a projecting cornice with the brackets in the frieze (*f*). Canopies are flat or pedimented or can take other forms such as the English early-eighteenth-century decorative semicircular hood (*h*).

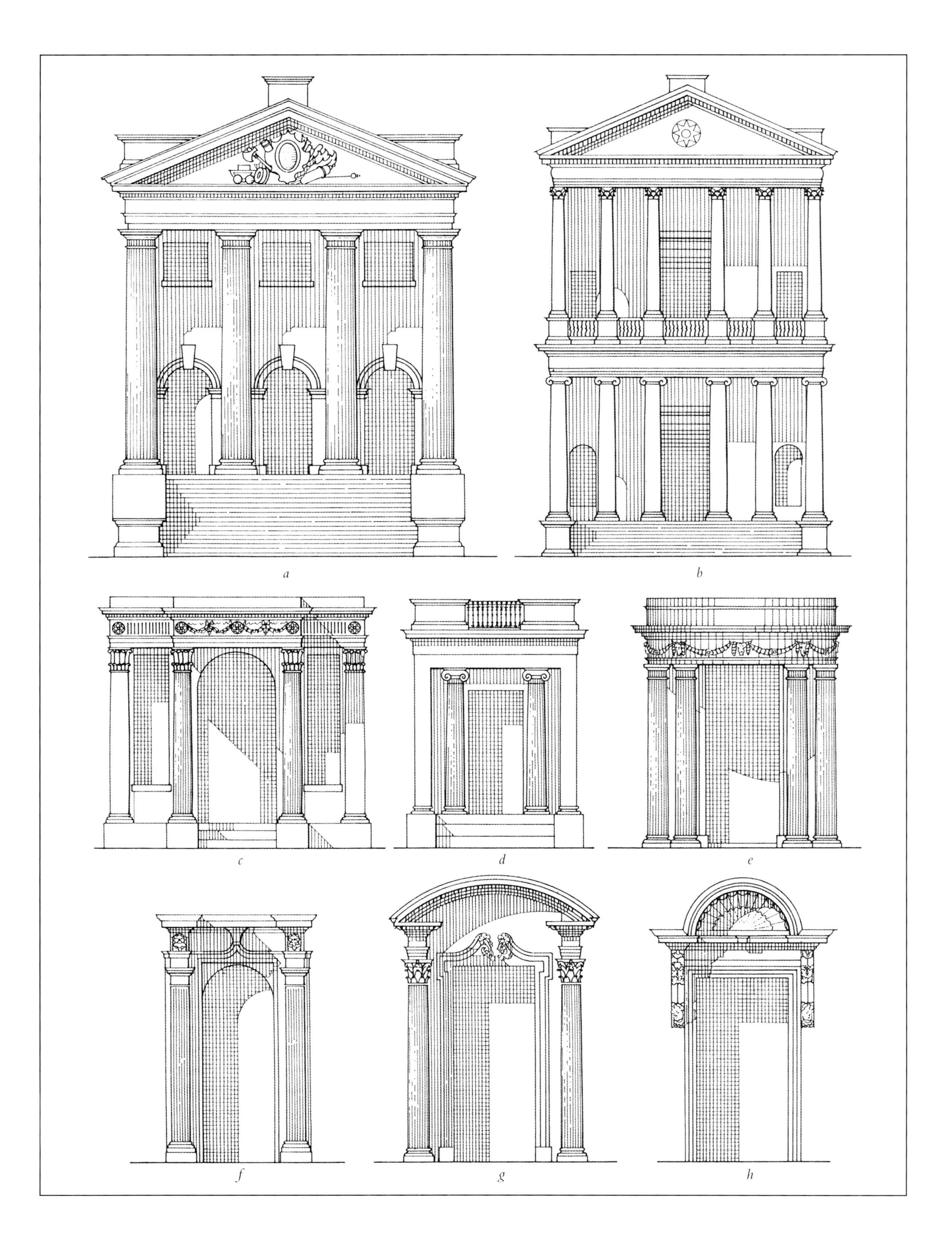
a
b
c
d
e
f
g
h

多立克门缘

应用柱式能够设计简单的门缘。该多立克门缘基于维尼奥拉1563年发布的两版柱式，但此处进行了改造以适应门的大小，门缘不含支柱。经改造后可以创造出前面阐述的五种版式，或者通过考虑柱式、山花和其他门例的细部设计，使用众多其他的方法进行改造。

该设计中的门缘可以用于室内和室外，用石头、石膏、木材或其他任何适合的材料都可以构建。如果用在室外，且建筑材料不耐腐蚀，则需要用防水材料保护檐口的凸出部分；如果是平的，则要将顶部设置成向外下垂。防水材料通常是某种金属薄片，必须仔细布置在檐口的顶部边缘，以便即使不悬于波纹线脚的顶端之上，也能保护建筑不受风吹雨水的侵蚀。

维尼奥拉的多立克柱式的版式之一在檐口安置飞檐托块（*a*），另一种带有檐下齿饰（*b*）。这两种版式都可以带有三陇板，但是在齿饰檐口中被省略了，用另一种设计替代，与三陇板和飞檐托块的间距不成比例。如果使用三陇板和飞檐托块的常用间距，那么在图中三陇板的数量下，门高与宽的比例是2.6：1，增加一个后则变为1.7：1。

门高通常是标准样式或者由其他因素所决定，假设已经确定门高，采用门高作为模数，这种做法与柱式支柱高度作为模数的做法相同。额枋应为半个模数，并环绕门的开口。如果在额枋底部添加一个砌块，也应为半个模数，以便与柱础同高，并且决定墙裙或护壁板的高度。

如果包括三陇板并且将其置于所示的位置，额枋应当有突耳，以便为额枋上的螺钉，即圆锥饰留出空间。圆锥饰应当为半圆形，但是如果门比较小，这样的细部设计便难以实现，可在材料的水平处开凹槽嵌入以实现相似的效果。门缘中包括三陇板，特别是在飞檐托块内小螺钉的间距很小的情况下，会构成十分精巧的细部设计。托块上的螺钉可以省略，但一般不省略三陇板下方的螺钉。

爱奥尼亚、科林斯和混合柱式门缘可参照其他柱式的细部设计完成。可以考虑添加壁柱，P203中的门缘和柱式展示出一些可以实现的方法。

A DORIC DOOR-SURROUND

A simple door-surround can be designed by the application of the Orders. This Doric surround is based on two versions of the Order published by Vignola in 1563, but modified here to suit the small size of a door. This is a door-surround without columns. It can be altered to create the five versions illustrated or, by reference to details of the Orders, the pediment and examples of other doors, can be modified in a number of other ways.

A door-surround of this design could be used either internally or externally and could be constructed of stone, plaster, timber, or any other suitable material. If it is used externally and constructed of perishable materials the projection of the cornice will have to be protected with a waterproof material, and, if it is flat, the top will have to be laid to fall outwards. The waterproof material, usually a sheet metal of some form, must be dressed carefully over the top edge of the cornice to provide protection against windblown rain without overhanging the top cyma moulding.

Vignola's Doric Orders include a version with mutules in the cornice (*a*) and a version with dentils (*b*). Both versions can have triglyphs, but these are omitted in the dentil cornice to illustrate an alternative which will not be tied proportionally to the spacing of the triglyphs and mutules. If the usual spacing of the triglyphs and mutules is used, the ratio of height to width of the door will have to be 2.6 : 1 with the number of triglyphs shown, or 1.7 : 1 if another is added.

It is assumed that the height of the door has already been determined, as this is commonly a standard or dictated by other factors. This height is taken as the height of the column of the Order to obtain a module. The architrave will then be one half of that module and will surround the door-opening. If a block is to be added to the base of the architrave this could also be half-module to equal the height of a column base and to determine the height of the skirting, or baseboard.

If triglyphs are included and located as shown, then the architrave should have ears in order to allow space for the pegs, or guttae, which project into the architrave. The guttae should be half-round, but with small doors this can create a difficult detail. It is possible to notch a flat section of material to achieve a similar effect. The inclusion of triglyphs and, in particular, mutules with their closely spaced pegs will give very fine detail to a door-surround. It is possible to omit the pegs on mutules, but the pegs below the triflyphs are usually one of the last details to go.

By reference to the details of the other Orders, Ionic, Corinthian, and Composite door-surrounds can be made. Pilasters can be added and page 203, on door-surrounds and the Orders, shows some of the ways this can be done.

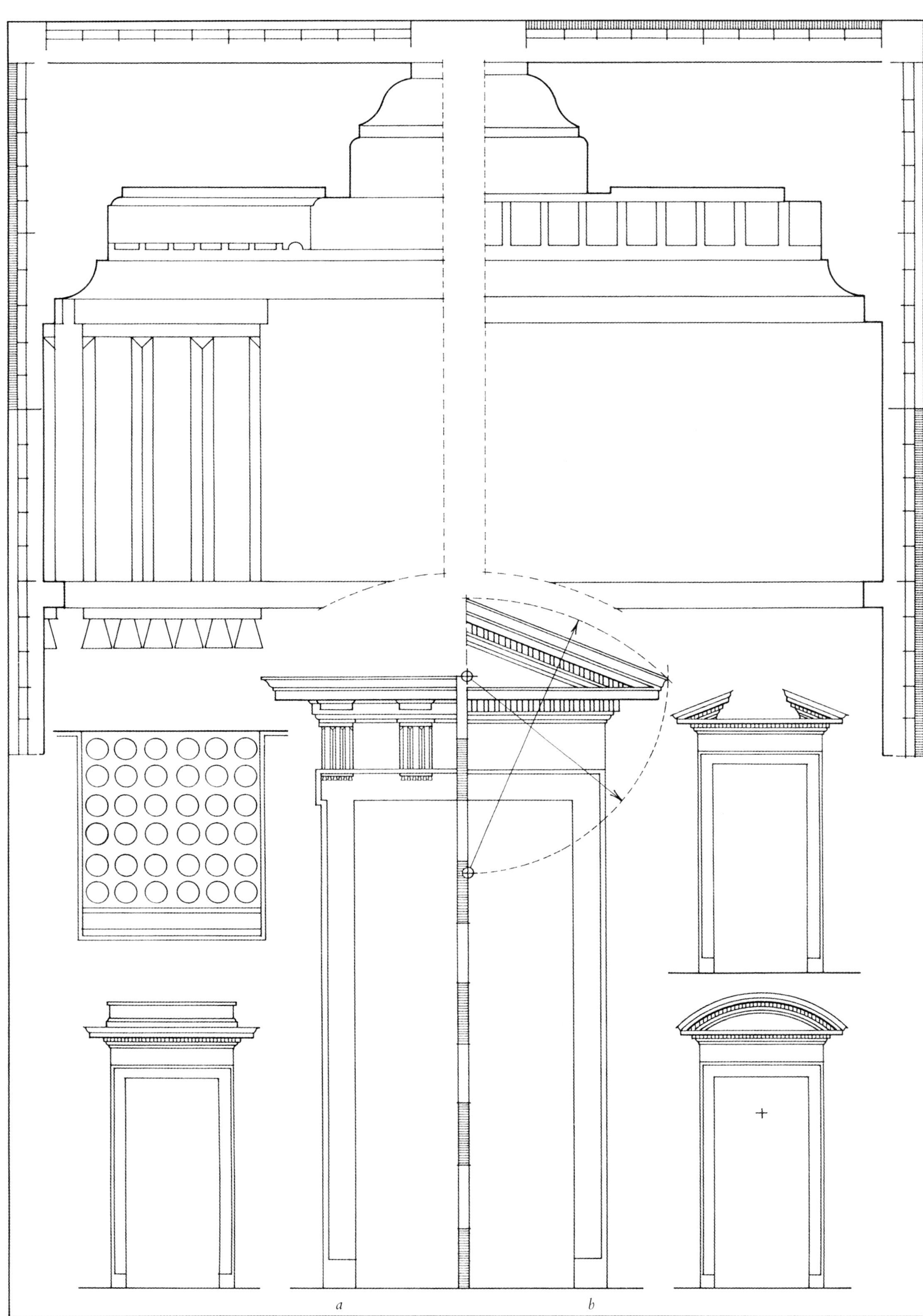

a
b

多立克门廊

柱式最直接的应用之一是设计包含两个独立式支柱在内的简单门廊，发展或改造该设计原则能够创造出许多简单多变的结构。门廊的大视图展示了一种带有山花（*f*）和矮墙块（*e*）的设计版式。上方的小视图展示了矮墙块上方的山花（*a*）、部分山花（*c*）、简单檐部的顶端缓缓倾斜到前方（*d*）、一对支柱间隔半个支柱的宽度，以及略高的矮墙块（*b*）。大视图也展现了完整的圆形支柱，有凹槽（*g*）和无凹槽（*i*），还有方柱（*h*）。图中靠墙的壁柱为方形，也可采用半圆形，此处只列出了少数几种能够实现的设计。

门廊可用石头、木材或任何合适的材料制成，也可在粗糙的砖或混凝土外面覆盖精美的石膏而成。当有山花或平屋顶时，可以让水自然落到门廊周围的地面上，当门廊很大的时候，可以添加檐槽。位于最上部的波纹线脚最先便是作为檐槽而使用的。如果添加阁楼形成低矮挡墙，雨水将在后面的屋顶收集起来。为了避免管子影响美观，在不会严重结霜的地区可以使其穿过垂直的支柱向下延伸。

该设计源自14世纪早期罗马戴克里洗浴室的多立克柱式，但现在经历了简化能够适应小型门廊。这种柱式的特点是缺少檐口托块，因此檐口凸出部分被减小到一定的深度，与其他柱式更为协调。所示版式带有三陇板（*e*）和省略三陇板（*f*）。如果带有三陇板，那么必须计算门廊的宽度，给予满意的间隔。支柱没有柱础，但如图所示，用支柱下面的低台阶将支柱抬高离开铺面层。

在建筑正面，门廊支柱间隔四个模数。一般认为这是最大化的间隔，除特殊考虑外不应该使用更大的间隔。如果需要特别大的门廊，那么最好额外增加支柱。从支柱返回到壁柱的两个模数的间隔可以增加至四个。支柱或靠墙壁柱终止了建筑中的檐部结构，既实用又产生使人满意的视觉效果。

其余的细部设计直接源于柱式，且建造方法应当依据地方建筑的实际情况而确定。不同的多立克、爱奥尼亚、科林斯和混合柱式门廊也可根据相似的原则进行设计，并参照柱式的细部设计。

A DORIC PORCH

The design of a simple porch with two free-standing columns is in one of the most straightforward applications of the Orders. The principles for setting out this design can be expanded or modified to create many simple variants. The large view of the porch shows a version with a pediment (*f*) and with a small attic block (*e*). The small views above show a pediment over an attic block (*a*), a segmental pediment (*c*), a simple entablature with a top sloping gently to the front (*d*) and an example with paired columns half a column width apart and a high attic block (*b*). The large views also show a full circular column, fluted (*g*) and unfluted (*i*), and a square column (*h*). The pilasters against the wall are shown square but could be half-round. These are only a few of the alternatives possible.

These porches could be made out of stone or timber or any suitable material, or be finished with a coat of fine plaster over a rough brick, or concrete core. When there is a pediment or a flat roof the water can be left to fall on to the ground around the porch or, particularly if the porch is large, a gutter can be included. The uppermost cyma moulding is by its origin a gutter and could be used for this purpose. When an attic block is added this will form a parapet and rain-water will collect in the roof behind. To avoid unsightly pipes these can be taken vertically down through a column in areas that do not suffer severe frosts.

A Doric Order from the early-fourth-century Baths of Diocletian in Rome is used as the basis of this design, although it has been simplified to suit a small porch. This Order is notable for its lack of mutules and the projection of the cornice is, consequently, reduced to a depth more in keeping with the other Orders. A version with triglyphs (*e*) is shown together with a version where they have been omitted (*f*). If triglyphs are included then the width of the porch will have to be calculated to give them a satisfactory spacing. The column has no base, but a low step below the column is shown to lift them off the paving.

The porch has columns spaced four modules apart at the front. This is sometimes considered to be a maximum and a greater spacing should not be used without careful consideration. If a significantly wider porch is required, it is better to include additional columns. The spacing of two modules from the columns back to the pilasters on the wall could be increased to four. Columns or pilasters against the wall provide a satisfactory practical and visual method of stopping the entablature against the building.

The remaining details are derived directly from the Order and the method of construction would be according to local building practice. Different Doric and Ionic, Corinthian and Composite porches can be designed according to similar principles and by reference to the details of the Order.

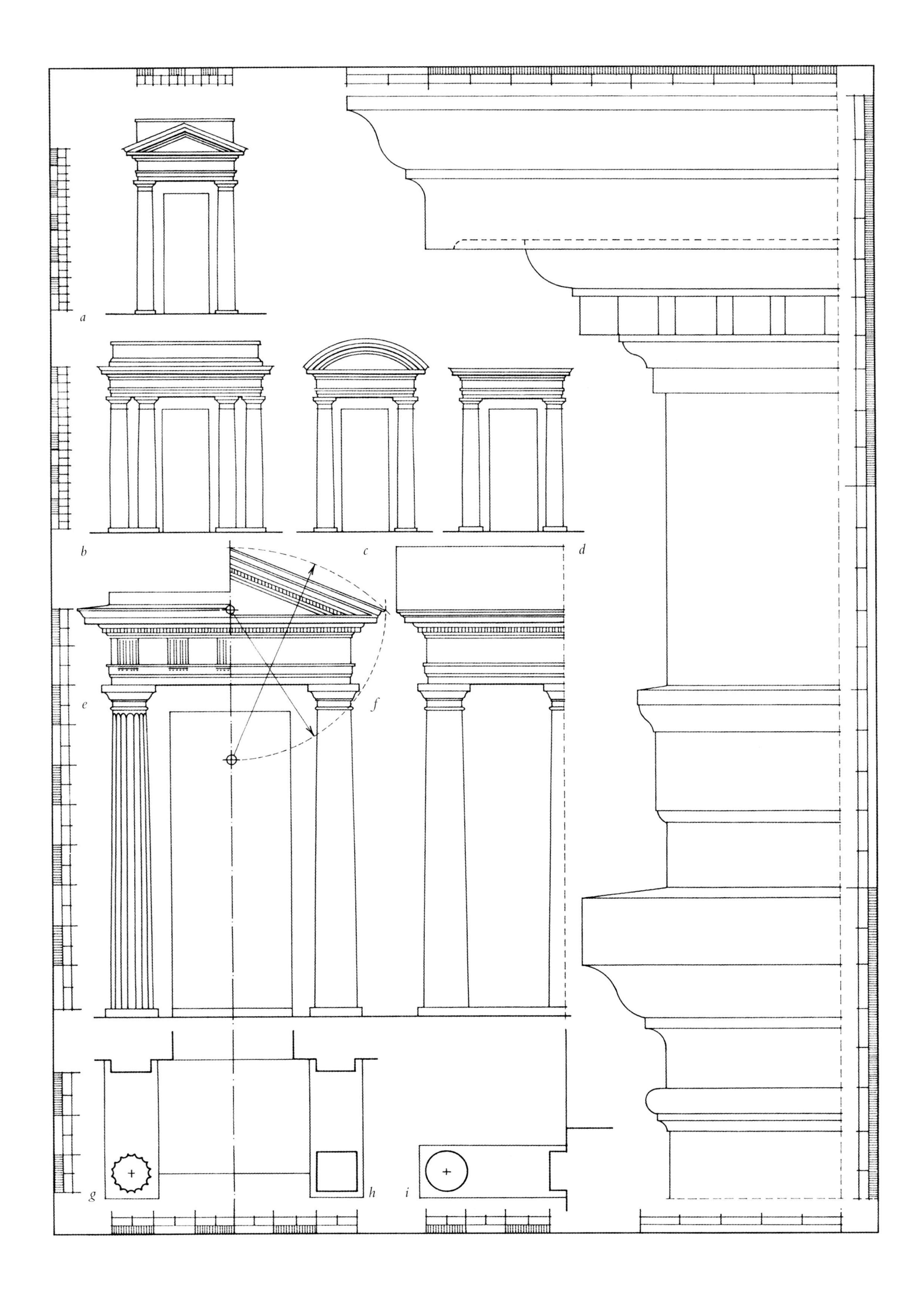
a
b
c
d
e
f
g
h
i

13. 窗

窗子与设计

在设计窗子的细部和位置时，必须结合建筑的内部功能和外观。窗子的设计通常直接表达出建筑的布局和房间的用途。

1458年建于意大利佛罗伦萨的碧提宫（*a*）包括粗琢的拱形开口。建筑并未强调门，上层窗子成为外立面上精美的细部设计。大窗子确定了居所的位置，而下方的小开口则指示出佣人房的所在。

1620年盖拉尔多·希尔瓦尼（1579—1675年）同在佛罗伦萨建造戴恩女士宫（*b*），包括简朴的墙壁及其上方无关紧要的檐口。各层不同的精巧窗缘装饰体现出设计效果。罗伯特·史密森1590年设计的哈德维克庄园（*c*）位于英格兰的德比郡，因在巨大的石框中展示极其炫耀的昂贵玻璃而闻名于世。外部除了塔楼上的顶饰和低屋顶上的栏杆之外几乎没有任何其他细部设计。

这些建筑都可以通过窗子的大小或装饰而识别出主楼层的位置。最主要的楼层通常在一楼之上，意大利语中称作*piano nobile*（主楼层）。图例中的建筑结构共享各楼层中窗间的水平装饰带，还可以将此细部设计倒置。17世纪末期建于英格兰萨塞克斯的佩特沃斯庄园（*d*）在房屋的两翼中将上下窗子连接到一起，形成强烈的垂直特征。

外立面上的窗子彼此间可以不具备任何关联。1511年，位于彼得库夫的城堡（*e*）落成，成为波兰首批古典建筑之一，建筑师疑为本尼迪克·列特（1454—1534年），或为Rejt，Ried von Piesting，其装饰窗框中不含连接线脚或带状装饰。该布局常见于新古典风格建筑师的简朴设计。约翰·索恩1786年为英格兰萨福克郡的布伦德斯顿庄园所进行的设计（*f*）仅仅由未经装饰的窗子开口抽象地组成。弗兰克·劳埃德·赖特1891年在美国芝加哥建造的查恩里别墅（*g*）也体现相同的设计理念，只用低楼层的石头结构将上方的两个外部窗子连接到一起，且在中央隆起将前门和两个大窗子包含在内，还形成底座支撑多立克柱式覆盖下的阳台。

13. WINDOWS

WINDOWS AND DESIGN

In the detail and location of windows, the internal function and the internal and external appearance of the building must be combined. The arrangement of the windows is often a direct expression of the layout of the building and the uses of the rooms.

The Palace Pitti in Florence in Italy (*a*), by an unknown architect in 1458, has an arrangement of arched openings in rustication. No emphasis is given to the door and the upper-floor windows provide the fine detail on the façade. These large windows identify the living accommodation while the small openings below light the service rooms.

The Palace Covoni-Daneo, also in Florence (*b*), designed by Gherardo Silvani (1579-1675) in 1620, has plain walls and above them an insignificant cornice. The design depends for its effect on elaborate window-surrounds, different on each floor. Hardwick Hall in Derbyshire, England (*c*), designed by Smythson in 1590, is remarkable for its ostentatious display of expensive glass in huge stone-framed windows. There is little other detail on the exterior except the crests on the towers and the balustrade on the lower roof.

These buildings all share the identification of principal floors by the size or decoration of windows. The most important floor is often that above the ground floor, known by its Italian name piano nobile. The buildings illustrated also share the horizontal bands of decoration that join the windows on each floor. This detail can be reversed. On the wings of Petworth House in Sussex, England (*d*), by an unknown architect in the late seventeenth century, the windows have been linked above and below to give a strong vertical accent.

Windows can be placed on a façade without any direct connection between one another. The decorated window-surrounds on one of Poland's first Classical buildings, the castle at Piotrków (*e*) of 1511, probably by Benedikt Ried (or Rejt, Ried von Piesting) (1454-1534), have no connecting mouldings or bands. This arrangement is more often found on the severe designs of Neo-Classical architects. Soane's design for Blundeston House in Suffolk, England (*f*) of 1786 is an abstract composition of little more than unadorned window-openings. A similar concept lies behind Frank Lloyd Wright's Charnley House in Chicago in the United States (*g*), built in 1891, except that the stone lower floor joins the two outer windows above and rises in the centre to contain the front door two large windows and give a base to the Doric covered balcony.

a
b
c
d
m
20
60
ft
e
f
g

古代窗子

早在古代便确立了古典窗子的设计原则，也保存下来大量的范例，形成文艺复兴风格。在早期的希腊和后来的罗马出现了矩形窗，与门采用相同的形式，且布局和细部变化也应用相同的模式。

公元前5世纪后期，雅典厄里希翁神殿内的窗子（*a*）与现代的门在形式和设计方面相同。底部开口比顶部开口略宽，窗台由类似门槛的石头砌块构成。窗子由平整的石头装饰带环绕，外部边缘用线脚带装饰，与建筑中爱奥尼亚柱式的额枋顶部相对应。石带的凸出部分或额枋在开口的顶部指示出横梁的位置形成突耳。

公元前1世纪位于罗马郊外蒂沃利的希贝尔神庙内的内部窗框（*c*）拥有更加精致的额枋，以略小的比例与建筑中科林斯柱式的完整额枋以及相连的门相匹配。窗顶的突耳与窗台处相匹配，底部开口略宽，顶部檐口无雕带。

在2世纪的罗马万神殿中（*b*），日光透过窗子照在长廊内，窗子上的连续线脚比较密集地围绕在内部。窗子呈矩形包含雕带和檐口。由于依据原来的隔板和栅栏进行重建使大量证据保存了下来而备受瞩目。门上方窗栅栏的石箍采用相同的设计（参见P199），属于常见的类型。这些栅栏即格栅用木料、石头和金属材料制成，用来填充开口，在保证安全的同时也透射进光线。此外，还安装了百叶窗封闭窗子。环形神庙内的罗马浅浮雕展示了柱间格栅（*d*），英格兰还发现了小型铁质窗栅栏（*g*）。

1世纪开始广泛使用玻璃窗，虽然大部分证据已经遗失，但已知玻璃与铅和灰泥石膏固定在一起，置于木材或石头框架中。在罗马圣莎宾娜教堂的中殿（*h*）中，考古发现了5世纪带有石栅栏的大型窗扇，并进行了修复重建。栅栏包括半透明的雪花石膏，有时也用这种石膏替代玻璃。这一设计在早期的开口栅栏中安装了玻璃，是大型罗马和哥特风格窗子的祖先。

更大的玻璃窗格也被制造出来，1世纪带玻璃的墙壁结构在意大利南部赫库兰尼姆古城中的镶嵌细工中庭别墅（*e*）中保存下来。在窗子中使用相似的大块玻璃或雪花石膏，诸如罗马后期的窗子示例（*f*）。

WINDOWS OF ANTIQUITY

The principles of Classical window design were established in antiquity, and a sufficient number survived to inform the Renaissance. Early Greek and later Roman rectangular windows took the same form as doors and the disposition and variety of details follow the same pattern.

A window from the Erechtheion in Athens (*a*) from the late fifth century BC is the same shape and design as a contemporary door. The opening is wider at the bottom than the top and the window-sill is a block of stone similar to the threshold of a door. The window is surrounded by a band of flat stone with a moulded strip at its outer edge, corresponding to the top part of the architrave of the Ionic Order of the building. At the head of the opening the position of the beam is indicated by a projection of the stone band, or architrave, to form ears.

The internal window-surround from the first-century-BC Temple of the Sibyl at Tivoli outside Rome (*c*) has a more elaborate architrave which matches at a smaller scale both the full architrave of the Corinthian Order of the building and the adjacent door. The ears at the head of the window are matched at the sill. The opening is slightly wider at the bottom, and at the top there is a cornice but no frieze.

Windows lighting an interior gallery in the Pantheon in Rome (*b*) of the second century AD sit on a continuous moulding which encircles the interior at a high level. These windows are rectangular and include a frieze as well as a cornice. They are of particular interest as sufficient evidence survived of their original divisions and grilles to restore them. The stone hoops are of the same design as the window grille above the door (*see* page 199) and were a common type. These grilles, or transennae, in timber, stone, and metal were used to fill openings for security while admitting light. Shutters were also fitted to seal windows. A Roman relief carving shows transennae between columns in a circular temple (*d*) and a small iron window grille (*g*) has been discovered in England.

In the first century AD the use of window glass became widespread. Much of the evidence for this has been lost, but it is known that glass was fixed together with lead and plaster as well as sitting in timber or stone frames. Large fifth-century windows with stone grilles were discovered and restored over the nave in the church of S. Sabina in Rome (*h*). These grilles contained translucent alabaster which was sometimes used as a substitute for glass. These elaborate designs are a glazed version of early open grilles and are the ancestors of the great Romanesque and Gothic windows.

Larger panes of glass were also made and framework of a glazed wall from the first century AD has survived in the House of the Mosaic Atrium in Herculaneum, southern Italy (*e*). Similar large sheets of glass or alabaster were used in windows such as the late Roman example (*f*).

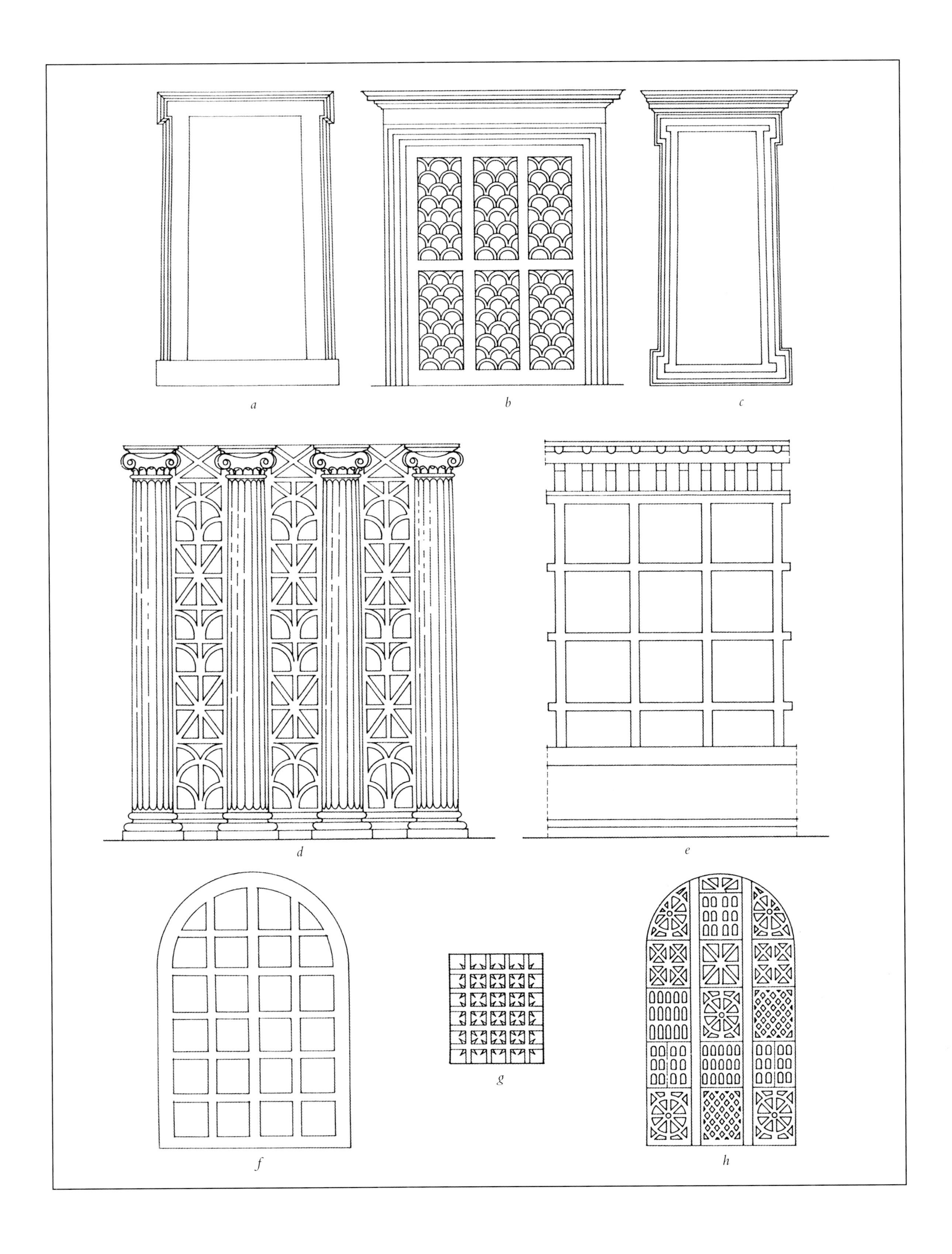

a
b
c
d
e
f
g
h

窗子的语汇

自古代开始，窗子类型始终与门的设计相对应。示例（*i*）至（*xiv*）显示了支柱、山花、栏杆桩、额枋和托架的诸多结合方式，可以用在所有矩形和部分圆顶窗中。其中大多数也出现在门的设计中，图示见其他页。窗子开口的主要细部设计集中在窗子下方，其独有的特征将窗子连接至地面或至外立面上更低的水平带上。素板（*vi*）或栏杆柱小镶板（*i*和*xiv*）构成窗下的底座。也可以在窗台下方布置形状各异的托架，支撑沉重的装饰细部。全托架（*ii*和*xiii*）以及小托架（*iii*和*xii*）适用于不同尺寸的窗框。巴洛克和矫饰主义细部设计（*ix*）将托架平贴在窗下凸出墙体中的镶板上，既能够进行装饰也可以保留成为简单的矩形窗台。

在开口内安装玻璃的方法很多，仅受玻璃窗尺寸的限制。小玻璃窗嵌入不同模式的古代或仿古代的格栅内（*xv*和*xxix*）。在诸多不同的设计中频繁使用铅来结合小块的玻璃（*xx*、*xxi*、*xxiii*、*xxiv*、*xxxvi*、*xl*和*xli*）。通常用石栅栏（*xxi*）或木栅栏（*xx*）也就是直棂，分割大面积的铅窗，当18世纪出现更大的玻璃时，用木制直棂或栅栏支撑整个玻璃窗。起先木栅栏规模很大（*xix*），但后来逐渐变小，19世纪时对样式进行了改造进而容纳更大的玻璃窗（*xvii*）。19世纪平板玻璃的发明使窗子不再需要格栅或直棂，同时继续有选择性地使用小玻璃窗（*xxxvi*）。大幅玻璃也应用在设计大胆而不需要精致小幅玻璃的建筑物中（*xxxvii*）。

诸多方法可以对环形或圆头窗进行分隔。分隔线可随着圆形或椭圆形线布置，从中心向外呈辐射状（*xxii*、*xxiii*、*xxiv*和*xxviii*）。分隔线可以呈矩形，独立于环形（*xxxi*和*xxxiii*）。戴克里先窗又称浴场式窗，是一种特殊的窗子形式，更常用于大型半圆形窗，在纵向分成三个不均等的部分（*xxix*和*xxx*）。门上方的小型半圆形窗又称扇形窗，可以使用变化多端的装饰设计进行分隔（*xxv*～*xxvii*）。

通常，古典窗子高长于宽。通常情况下宽度由两块或三块以上的窗子并排放置而构成。为此，P161中由帕拉迪奥设计的窗子一直非常流行。在其他由三扇窗子组合成的样式中，通常会出现更大的中心开口（*xxxvii*、*xxxviii*、*xxxix*），而两扇和四扇窗子之间的分隔线是相等的（*xxxiv*、*xxxv*、*xxxvi*、*xl*、*xli*）。

THE VOCABULARY OF THE WINDOW

Window types have developed parallel to the design of doors since antiquity. Examples (*i*) to (*xiv*) show many of the combinations of column, pediment, baluster, architrave, and bracket that can be used with all rectangular and some round-headed windows. Most of these are also found on doors and are illustrated on other pages. The principal details exclusive to window-openings are the features that are added beneath to link them to the ground or to a lower horizontal band on a facade. Plain panels (*vi*) or small panels of balusters (*i*) and (*xiv*), can give a base from which a window can rise. Brackets of various forms can also be placed below sills to support heavy decorative details. Full brackets, (*ii*) and (*xiii*), and small brackets, (*iii*) and (*xii*), will be appropriate to different sizes of window-surround. A Baroque and Mannerist detail (*ix*) flattens the bracket to a panel of projecting wall below the window, which can be decorated or left as a simple rectangle known as an apron.

The arrangement of glazing within the opening has taken numerous forms, constrained only by the limitations of glass manufacture restricting the size of pane available. Small panes of glass have been contained in grilles of various patterns in, or in imitation of, antiquity, (*xv*) and (*xxix*). Lead has frequently been used to bind together small pieces of glass in many different designs, (*xx*), (*xxi*), (*xxiii*), (*xxiv*), (*xxxvi*), (*xl*), and (*xli*). Stone (*xxi*) or timber (*xx*) bars, or mullions, often divided up large areas of leaded windows and, when glass came in larger sizes in the eighteenth century, wooden mullions or bars supported whole panes. Timber bars were originally broad (*xix*) but later became very narrow, and patterns changed to accommodate larger panes in the nineteenth century (*xvii*). The invention of plate glass in the nineteenth century made bars or mullions redundant, although the traditional use of small panes of glass continued as a choice not governed by necessity (*xxxvi*). Large sheets of glass were also used on buildings where bold designs did not need the delicate scale of small panes (*xxxvii*).

Circular or round-headed windows can be divided in a number of ways. The divisions can follow the line of the circle or ellipse, (*xxii*), (*xxiii*), (*xxiv*), and (*xxviii*), and can radiate out from the centre. Divisions can form rectangles independent of the circle, (*xxxi*) and (*xxxiii*). A particular form, more often used for large semicircular windows, is the Diocletian, or thermal, window, (*xxix*) and (*xxx*), divided vertically into three uneven parts. Small semicircular windows, or fanlights, placed over doors have been divided with a great variety of decorative designs, (*xxv*) to (*xxvii*).

Classical windows tend to be taller than their width. Width is usually achieved by placing two, or more often three, windows alongside one another. The Palladian window illustrated on page 161 has been very popular for this purpose. On other triple windows a larger centre opening is normal, (*xxxvii*), (*xxxviii*), (*xxxix*), while divisions of two and four, (*xxxiv*), (*xxxv*), (*xxxvi*), (*xl*), (*xli*), are equal.

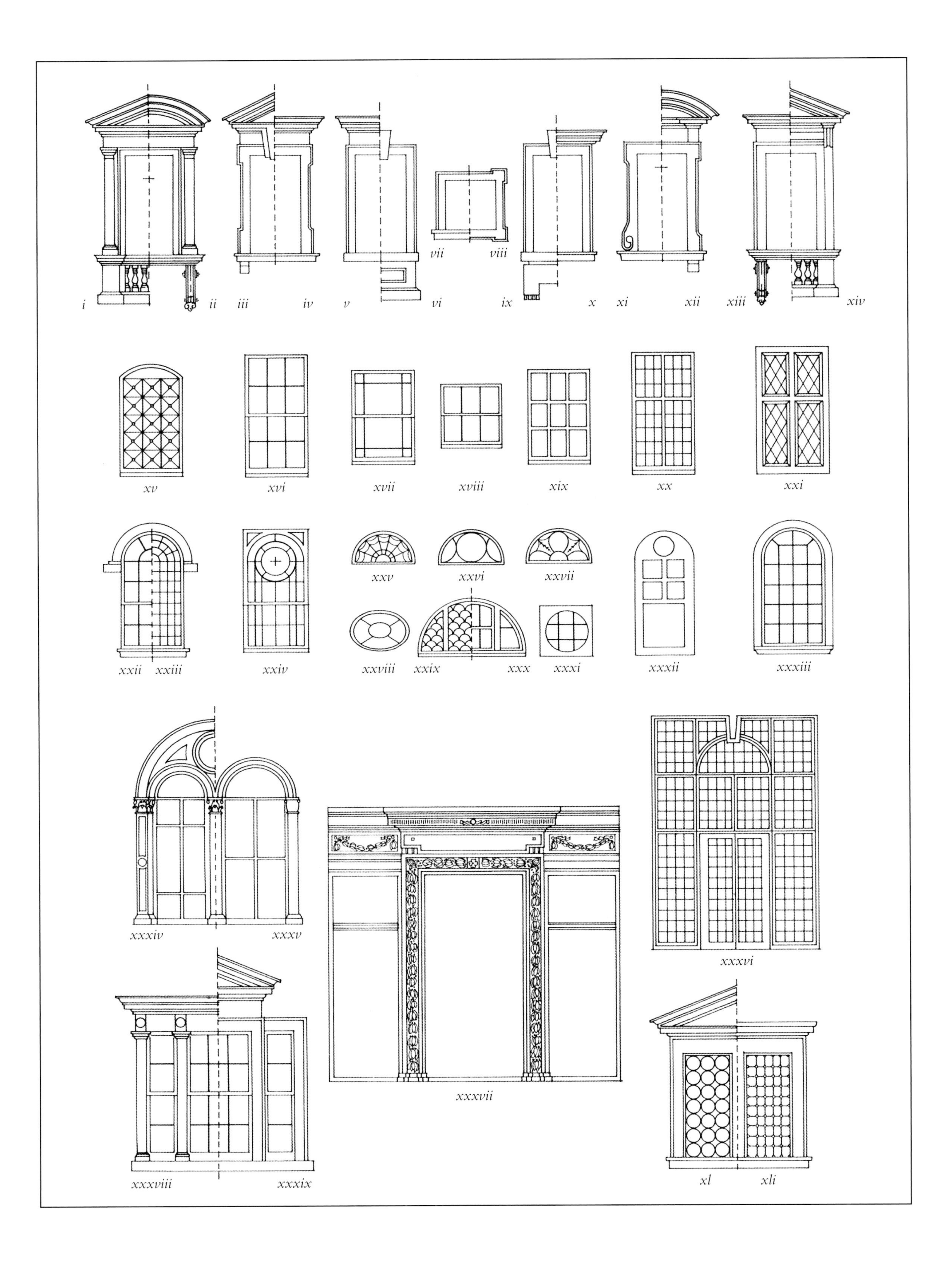
i
ii
iii
iv
v
vi
vii
viii
ix
x
xi
xii
xiii
xiv
xv
xvi
xvii
xviii
xix
xx
xxi
xxii
xxiii
xxiv
xxv
xxvi
xxvii
xxviii
xxix
xxx
xxxi
xxxii
xxxiii
xxxiv
xxxv
xxxvi
xxxvii
xxxviii
xxxix
xl
xli

凸窗和凸肚窗

中世纪后期的大型家庭建筑内通常包含凸肚窗，即从上层的外立面中向外凸出的窗子。中世纪后期的房屋，尤其在英格兰，一楼也会有凸出的大窗子，称凸窗，为主房间或主厅采光。尽管原始设计中的凸肚窗或许是出于防御目的而设立，但逐渐变得豪华奢侈，包括座位空间和大面积的玻璃。

英国文艺复兴建筑保留了中世纪建筑物的这一特征，由于玻璃的大量使用体现出主人的社会地位，因而在16世纪后期和17世纪早期更加受到重视。1605年汉普郡布莱姆希尔庄园的三层入口塔楼上有一个大半圆形凸肚窗（*a*），作为主要的特征悬在门的上方。塔斯福德郡的伍滕小屋（*b*）大约建于1607年，建筑师疑为史密森，尽管后来对房屋进行了改建，但仍然完整地展示了全高度的凸窗。入口经过方形开间，外展的三面开间延伸到每一个侧面上，两个半圆形凸窗即凸肚窗嵌在边墙上。这些开间也可以在低矮挡墙的上方向外凸出而形成塔楼。凸出的窗扇不仅增加了玻璃的使用，使房间更加奢华，也突出了外立面的纵向特征，使人联想到中世纪的塔楼。

18世纪早期，此类窗子已经过时，但到世纪中期却在英国广泛复兴，此后也一直应用在众多大大小小的房屋之中。

早期的复兴，例如在埃塞克斯的两处建筑之中，萨弗伦瓦尔登的单层凸窗（*g*）和科尔切斯特的双层凸窗（*h*），外立面由连接到建筑物上的多个窗子构成（*g*），或者由房屋的部分正面墙壁组成（*h*），凸窗成为外立面的简单外延。后来的18世纪，曲线形式更加流行。在大房屋中，房间的整个墙壁和部分外立面都进行了曲线设计，在偏小的房屋中，建筑更加关注简单的凸肚窗。白金汉郡马洛的房屋（*c*）中唯一的特征就是将简单的凸肚窗置于支柱上方形成门厅。多塞特郡内韦茅斯的凸肚窗（*d*）包含完整的上升系列壁柱，邻街表层上偏矮的窗子退回到独立支柱之后。在伦敦一处医生的住宅（*e*）中，窗扇的设计与之相似，壁柱已经简化成单层凸肚窗，由此抵消了门所产生的不对称感。

这些窗子的便利性确保其在19世纪英格兰的大量小房屋中保存下来（*i*），一直很受欢迎，并且当英国重燃对17世纪建筑的兴趣时受到特别的追捧，1886年工艺美术建筑师欧内斯特·牛顿（1856—1922年）在伦敦郊区的银行建筑中所设计的凸肚窗（*f*）便成为此说法有力的佐证。

BOW, BAY, AND ORIEL WINDOWS

Large domestic buildings in the later Middle Ages often had an oriel window, a window that projected from the façade of an upper floor. Late medieval houses, particularly in England, could also have a large projecting window on the ground floor, known as a bay window, which usually lit the principal room or hall. Although the oriel window was probably defensive in origin, both of these window types came to have a luxurious character with seating space and large areas of glass.

English Renaissance architecture retained this feature of earlier medieval buildings, which was further emphasised in the late sixteenth and early seventeenth centuries due to the status attached to the lavish use of glass. The three-storey entrance tower of Bramshill House, Hampshire(*a*) of 1605 has a large semicircular oriel window suspended as a principal feature over the door. Wootton Lodge in Staffordshire (*b*), probably by Smythson in about 1607, although subsequently altered still displays a complete collection of full-height projecting windows. The entrance is through a square bay with splayed three-sided bays to each side and two semicircular bay windows, or bow windows, on the side-walls. These bays may have projected above the parapet to create turrets. The projecting windows not only increased the use of glass, giving a luxurious character to the rooms, but added strong vertical features to the façade reminiscent of medieval towers.

In the early eighteenth century these windows became unfashionable, but by the middle of the century they were extensively revived in Britain and since then have been used on many large and small houses.

Early revivals, such as the single-storey bays in Saffron Walden (*g*) and the two-storey bays in Colchester (*h*), both in Essex, tend to be simple extensions of the façade made up either of multiple windows attached to the building as in (*g*) or, as in (*h*), of part of the front wall of the house. Later in the eighteenth century curved forms became more popular. On large houses whole walls of rooms and sections of façades were curved and on smaller houses simple bow windows provided architectural interest. A house in Marlow, Buckinghamshire (*c*) has no features except simple bow windows, one of which sits on columns to provide a porch. A bow window in Weymouth, Dorset (*d*) has a complete ascending series of pilasters and at street level the lower window is recessed behind free-standing columns. A similar window with pilasters is reduced to a single-storey oriel on a doctor's house in London (*e*) counteracting the loss of symmetry created by the door.

The convenience of these windows ensured their survival among the large number of small houses constructed in England in the nineteenth century (*i*). They remained popular and enjoyed particular favour with the renewed interest in English seventeenth-century architecture, seen in the oriel windows of a bank building in a London suburb (*f*) of 1886, by the Arts and Crafts architect Ernest Newton(1856-1922).

a
15 ft
5 m
b
c
d
e
f
g
h
i

14. 托架

涡卷形托架

涡卷形托架又称飞檐托饰或托臂（*a*），是古典建筑所特有的一种装饰特征。形式十分独特，但其起源和含义不明。

公元前421年雅典厄里希翁神殿北部门廊中门的上部檐口（*c*）由托架支撑，全图参见P199。其涡卷或螺旋十分复杂，并且和该建筑中的爱奥尼亚柱头拥有相同的多拱肋。低位的涡卷形向外展开以支撑托架，而上部涡卷形在正面又带有自身的小螺旋形，内部是倒置的叶丛状装饰。这些细部设计是此托架及其复制品独有的特征，也建立了未来涡卷形托架（*a*）的主要特征。两个涡卷形转向相对的方向，正面分成两部分：一个是从上部的涡卷形中抽出的苞芽装饰；另一个是从外部折入低位涡卷形的莨菪叶形装饰。

在罗马，涡卷形托架被应用到其他建筑细部设计中。罗马科林斯柱式包括一系列涡卷形托架，水平放置在檐口的上部。托架细部设计与希腊风格的设计有所不同。公元前36年，罗马阿波罗神庙中的托架（*b*）包含两个涡卷形装饰，中心有花的图案，底边的大型莨菪叶装饰自大涡卷形的正面中心抽芽而出。

涡卷形托架经改造也可以作拱门的拱顶石，如约82年罗马的提图斯凯旋门（*d*）和其他一些凯旋门。细部越重要则装饰水平越高。涡卷形上添加了一行行的珠子和凹槽线脚，用天平装饰填充外部表面。莨菪叶装饰的上方是具有象征意义的人像，并用缠绕的花蕾、叶子和花朵填充后面粗劣的缺口。

后来，托架在许多不同的情况中充当支架，外形轮廓可以改变以便适应特定的结构，例如支撑门上方的浅檐口（*e*）。巴洛克设计师赋予该形式更多的变化，打断了螺旋形之间的线条（*f*和*g*），有时省略除必要的细部以外的所有部分，将托架简化成装饰框架（*f*）。

14. BRACKETS

SCROLLED BRACKETS

The scrolled bracket, also known as the modillion or console (*a*), is a decorative feature unique to Classical architecture. The form is very specific, but its origin and meaning are unknown.

An early example supports the upper cornice of the door of the north porch of the Erechtheion in Athens (*c*) of 421 BC, shown in full on page 199. The scrolls, or volutes, are complex and have the same multiple ribs as the Ionic capitals of the building. The lower scroll opens out as a back to the bracket while the face of the upper scroll has its own small volutes, out of which grows an inverted anthemion. These details are individual to this bracket and its copies, but in other respects it establishes the principal feature scrolled brackets (*a*). It has two scrolls turning in opposite directions, the outer face is divided into two parts, a bud sprouts from the upper scroll and an acanthus leaf from the outward turn in the lower scroll.

In Rome, scrolled brackets were applied to other architectural details. The Roman Corinthian Order included a series of scrolled brackets set horizontally in the upper part of the cornice. The bracket details differ from the Greek design. The bracket from the Temple of Apollo in Rome (*b*) of 36 BC has two scrolls, both of which have flowers in the centre, and what is now the underside has a large acanthus leaf sprouting from the centre of the face of the larger scroll.

The scrolled bracket was also adapted to act as keystone for arches on the Arch of Titus in Rome (*d*) of about AD 82 and other triumphal arches. The level of decoration has increased in keeping with the importance of the detail. Rows of beaded and fluted mouldings have been added to the scrolls and the outer face is filled with a scale decoration. An acanthus leaf supports a symbolic figure, and the awkward gap behind is filled with a twisting series of buds, leaves, and flowers.

The bracket has subsequently been used as a support in many different circumstances and the profile can be modified to suit particular configurations, such as the support of shallow cornices over doors (*e*). Baroque designers added further variations to the form, breaking the line between the volutes, (*f*) and (*g*), and at times carving away all but the essential details, reducing the bracket to a decorative skeleton (*f*).

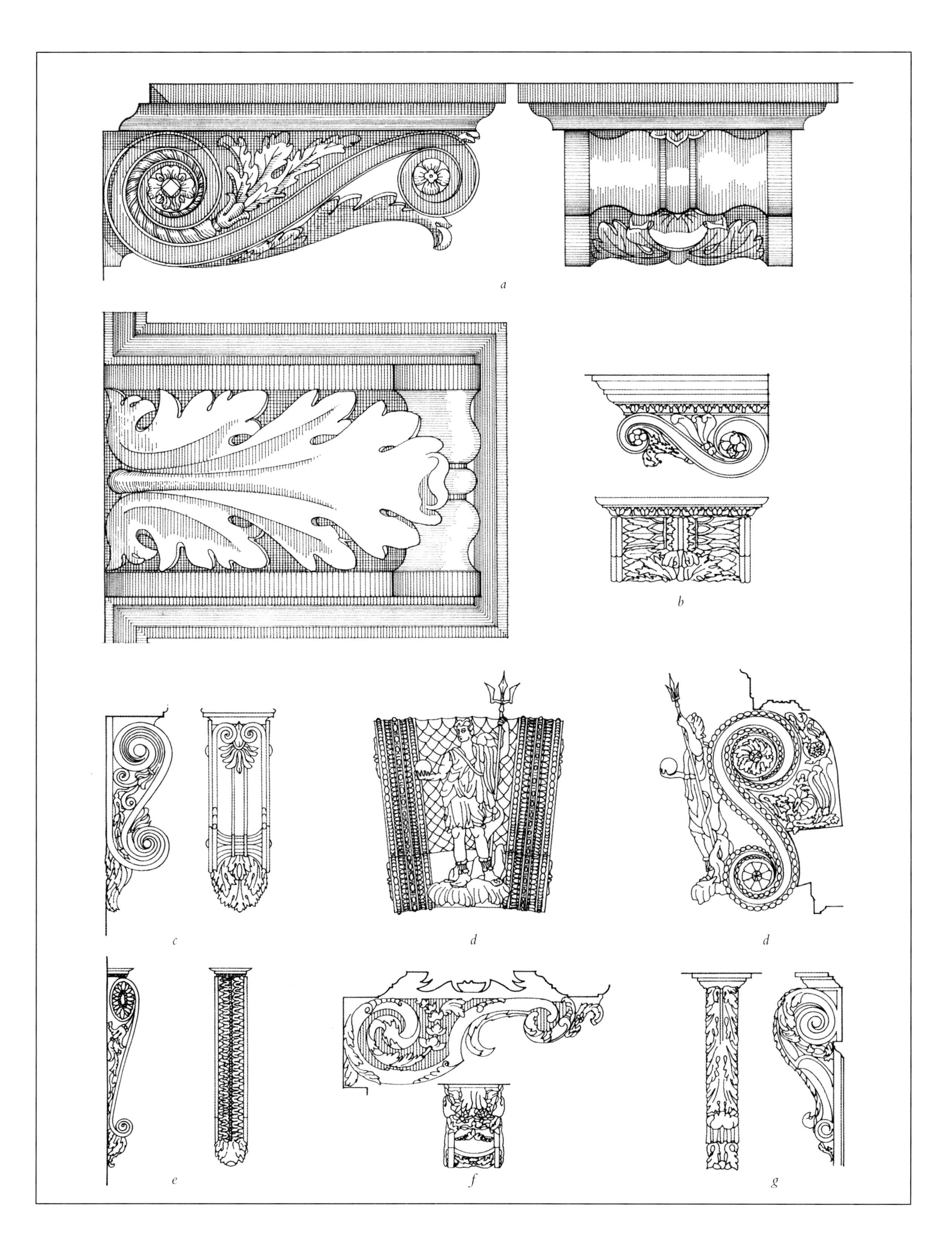

a
b
c
d
d
e
f
g

其他托架

从柱式中发展出一系列能够替换涡卷形的结构。希腊多立克建筑中的石头飞檐托块（*a*）从墙壁中凸出来以支撑悬着的挑檐滴水板和波纹线脚，代表了斜屋顶上椽子的末端。此时的飞檐托块充当托架，但在希腊建筑中却保留了倾斜的形式。罗马建筑师通常将其减小至水平模式，嵌入挑檐滴水板凸出部分的下方。公元前174年雅典的宙斯奥林匹斯神庙（*b*）是最早期的纪念碑式科林斯神庙，由罗马建筑师设计，似乎采用简朴的托架支撑挑檐滴水板。此类托架在后期的建筑中被涡卷形托架所取代，看似融合了爱奥尼亚齿饰和多立克飞檐托块，并且可能代表着凸出的横梁这一现今已经失传的结构传统。包括维尼奥拉在内的文艺复兴建筑师进一步调整了多立克飞檐托块（*c*），不再与椽子相似，而是像宙斯奥林匹斯神庙中那样成为水平的横梁末端，但仍然保留了多立克圆锥饰的特点。

挑檐滴水板的石头凸出部分丧失了木材结构的所有优势，还产生出结构问题。直到后来的希腊或罗马时期才出现能够支撑挑檐滴水板的深托架，此时原始的木材柱式已经成为历史。80年在罗马竞技场中，混合柱式上层的檐口（*d*）带有支撑天棚的桅杆，反波纹线脚托架为挑檐滴水板提供了额外的支撑。由于罗马竞技场保留了下来，因此这种实用的细部设计也成为普遍的形式。维尼奥拉将其转变成涡卷形装饰，放置在标准的科林斯托架下（*e*），塞里奥将反波纹线脚轮廓和多立克三陇板结合在一起，接在多立克檐口的飞檐托块上（*f*）。

上述所有托架已经脱离檐口单独使用，而且又进行了改造。法国文艺复兴托架（*g*）将涡卷形放在三陇板上，并在19世纪德国建筑师申克尔设计的三陇板托架的后面布置相似的结构（*h*）。米开朗琪罗在1546年罗马法尔内赛宫中上部楼层的窗子上（*i*）创新性地将凸出的飞檐托块托架下降到雕带之中。法国18世纪早期的托架（*j*）最终失去了与柱式传统的一切联系。

在拱形靠墙的边界处需要一类托架或梁托，在筒形穹顶之下则不需要。出于这一目的，在中世纪的建筑物中产生出一种特殊的柱头托架，例如12世纪法国努瓦永的天主教堂（*k*）。文艺复兴时期对此形式进行了改造，最初的设计是对古典柱头的自由诠释，例如1467年出自意大利北部帕维亚的陶土托臂的柱头（*l*），但很快进行了改造（*m*），更加响应罗马柱式。

OTHER BRACKETS

A series of alternatives to the scrolled bracket has evolved from the Orders. The mutule in Greek Doric buildings (*a*) was a stone representation of the end of the rafters from the pitched roof as they projected over the wall to support the overhanging corona and cyma. In this respect the mutule is a bracket, but on Greek buildings it retained its sloping form. Roman architects often reduced this to a horizontal pattern cut into the underside of the projection of the corona. In the cornice of the earliest monumental Corinthian temple, the Temple of Zeus Olympius in Athens (*b*) of 174 BC, designed by a Roman architect, there are plain brackets seemingly supporting the corona. These brackets, on later examples substituted for scrolled brackets, seem to be a mixture of Ionic dentils and Doric mutules and perhaps represent projecting beams from a structural tradition now lost. Renaissance architects such as Vignola further modified the Doric mutule (*c*) so that it no longer resembled rafters and became a horizontal beam-end, like that of the Temple of Zeus Olympius, while retaining the characteristic Doric guttae.

The projection of the corona in stone loses all the structural advantages it had in timber and creates instead a structural problem. It is significant that the introduction of deep brackets capable of supporting the corona did not occur until the later Hellenistic, or Roman, period when the timber origins of the Orders had become historic. The cornice of the upper Composite level of the Colosseum in Rome (*d*) of AD 80 held masts to carry awnings and the corona was given the extra support of cyma reversa brackets. The survival of the Colosseum made this useful detail common. Vignola turned it into a scroll and set it below a standard Corinthian bracket (*e*) and Serlio combined the cyma reversa profile with a Doric triglyph and joined it to a mulute in a Doric cornice (*f*).

All these brackets have been used on their own without their cornices. Further modifications have also been made. A French Renaissance bracket (*g*) shows a scroll put on to a triglyph and a similar combination lies behind the nineteenth-century German architect Schinkel's triglyph bracket (*h*). On the upper-floor windows of the Palazzo Farnese in Rome (*i*) of 1546 Michelangelo introduced projecting mutule brackets that dropped down into the frieze. An early-eighteenth-century bracket from France (*j*) finally loses virtually all association with the traditions of the Orders.

The termination of an arch against a wall created the need for a bracket of a type that was unnecessary below barrel vaults. A special column capital bracket, or corbel, had been used for this purpose in medieval buildings, such as the twelfth-century cathedral at Noyon in France (*k*). This form was adapted in the Renaissance and at first the designs, such as the terracotta corbel capital from Pavia in northern Italy (*l*) of 1467, were free interpretations of Classical capitals, but they were soon adapted (*m*) to correspond more closely to the Roman Orders.

a
b
c
d
e
f
g
h
i
j
k
l
m

涡卷形托架：变化

涡卷形托架已经成为如此独具特色的古典形式以至于经改造后用在大量不同的实际应用和装饰主题之中。

涡卷形的原始形状并不总是适合于支撑大型或沉重的凸出结构，因此在设计中需要使用成对的涡卷形托架。在文艺复兴早期，这些结构常常围绕在方形或矩形块（*a*）四周，形成更大的凸出结构，并且体现出复杂的设计。示例（*c*）所展示的英国巴洛克门廊托架也结合了两个涡卷形托架，但此例改造了单独的托架以适应所需的形状。将小天使的头像嵌入低位托架上大螺旋形的下方，延展的莨苕叶形装饰将其连接到更为传统的上部涡卷形托架上。示例（*b*）是文艺复兴托架，用深矩形托架支撑沉重的石头凸出结构，托架表面的标准设计已经极度简化，几乎没有任何的曲线装饰。托架是传统的中世纪设计样式，但被装饰赋予了古典特征。

自从最初涡卷形托架经改造在罗马的凯旋门中用作拱顶石以来，便通常作此用途。最初的涡卷形拱顶石经扭曲方便负载人像并适合拱门的深线脚（参见P227），而后期的示例通常更加简单。传统的涡卷形有序地置于拱形的檐部以及拱门墙侧装饰线条的下方（*d*），并且可在外侧表面进行装饰。如果拱顶石进一步后退至拱门墙侧装饰线条之中，那么可省略涡卷形的侧面，只在前表面进行装饰（*e*）。此外，在涡卷形拱顶石中添加天平装饰已经成为传统。

涡卷形托架的叶饰是传统样式的莨苕叶装饰，改造后可以适应特殊的装饰主题。示例（*f*）是罗马托架，带有橡子和橡树叶，从外表面上常见的莨苕叶装饰中抽芽而出。出自叙利亚帕尔米亚的另一罗马托架（*g*）则包含狮子的头像。

动植物的组合形式是古典装饰的独特之处。将动物的足或爪子添加到涡卷形中（*h*），形成桌腿或其他支撑，在古代十分常见，矫饰主义建筑师热情复兴了这一设计，并进一步在设计中添加具有象征意义的人像（*i*）。在巴洛克风格中，托架在侧面支撑窗子和火炉等开口的边框，示例（*j*）带有经雕刻的承梁木，由简单的阶梯引导到涡卷形中，涡卷形产生于框架之中，且自身也成为框架的一部分（*k*）。用类似的方法还可以结合涡卷形和支柱，两者通常结合到一起，在开口一侧成为含义模糊的壁柱（*l*和*m*）。巴洛克风格的爱奥尼亚涡卷形壁柱（*m*）包括垂花雕饰。

SCROLLED BRACKETS: VARIATIONS

The scrolled bracket has become such a characteristic Classical form that it has been adapted to suit a large number of different practical applications and decorative themes.

The original shape of the scroll is not always suitable for supporting large or heavy projections. This has given rise to the use of pairs of scrolled brackets combined in one design. In the early Renaissance these were often grouped around a square or rectangular block (*a*) to give a greater projection and create a complex design. Example (*c*) shows an English Baroque porch bracket, which also combines two scrolled brackets, though the individual brackets have in this case been modified to adapt to the required shape. A cherub's head is lodged below the large volute of the lower bracket and an extended acanthus leaf links it to the more conventional upper scrolled bracket. Example (*b*) shows a Renaissance bracket on which the standard design has been reduced to a barely curved decorative face on a deep triangular bracket supporting a heavy masonry projection. The shape of the bracket is a traditional medieval design, but the decoration has given it a Classical character.

Scrolled brackets have often been used as keystones since their first adaptation for this purpose in Roman triumphal arches. These first scrolled keystones were contorted to carry figures and to fit into the deep mouldings of the arch (see page 227). Later examples are often more simple. A conventional scroll can sit neatly below an entablature and the archivolt mouldings of an arch (*d*) can be decorated on the outer face. If the keystone is to be recessed further into the archivolt the sides of the scrolls can be omitted and the decoration limited to the front face (*e*). Scale decoration has become a traditional addition to scrolled keystones.

The decorative foliage of the scrolled bracket is traditionally stylised acanthus. This can be varied to suit a particular decorative theme. Example (*f*) is a Roman bracket with acorns and oak leaves sprouting within usual acanthus leaf on the outer face. Another Roman bracket, from Palmyra in Syria (*g*), incorporates the head of a lion.

This combination of plant and animal forms is a peculiarity of Classical decoration. The addition of an animal's foot or paw to a scroll (*h*), to form a table leg or other support, was common in antiquity and was enthusiastically revived and expanded to include symbolic representations of the human figure (*i*) by Mannerist architects. The Baroque use of the bracket as a side-support for frames to openings such as windows and fireplaces, shown here (*j*) with carved tassels, led by a simple step to scrolls that grew out of and became a part of the frames themselves (*k*). A similar transformation led to the combination of scrolls and columns, usually joined together as ambiguous pilasters on either side of an opening, (*l*) and (*m*). The Baroque Ionic scrolled pilaster (*m*) includes a drapery festoon.

a
b
c
d
e
f
g
h
i
j
k
l
m

简化的涡卷形托架

涡卷形托架是正式建筑装饰中独特的核心内容。它们结合了涡卷形与科林斯柱式中的莨菪装饰，其中涡卷形类似于爱奥尼亚柱式复杂柱头结构中的螺旋装饰。托架作为装饰特征，能够负载建筑装饰，并且展现出雕刻家的艺术水平。但在一个建筑主题中包括如此丰富的装饰特征却并不总是相宜或经济的。与古典设计的大多数方面相同，完备的装饰特征所具有的特殊形式和建筑要素的既定传统使得即使减少细部设计的数量和规模，也能够保留要素的辨识特征。涡卷形托架成为一种传播最广并且独具特色的古典建筑特征，进而衍生出一系列简化的细部设计水平。

作为飞檐托饰，涡卷形托架被包括在科林斯檐口的上层部分。如果科林斯柱式相对较小，便不可能包括完整的装饰范畴。示例（*b*）、（*i*）和（*j*）取自科林斯檐口，无论削减了多少细部设计，只要程度恰当，便可以应用到适当比例的飞檐托饰之中。

涡卷形自身用来定义边缘的狭窄拱肋可以省略，并且整个涡卷形可以向内卷曲，涡卷形要么由沟槽要么由涡卷形向内盘卷时外面凸出的部分所定义。示例（*a*）、（*e*）、（*f*）和（*g*）中的涡卷形由沟槽定义，而（*b*）和（*h*）则依赖能够在正面两侧看到的凸出部分。

托架正面的细部设计受与之相似的简化过程所影响。（*a*）和（*b*）的中心珠子可以省略，而在前表面形成连续的曲线轮廓，如（*d*）和（*h*）。表面也可以是平的，并有简单的凹槽嵌入其中（*g*），或不含任何轮廓及装饰，如（*f*）、（*i*）和（*j*）。

随着轮廓的不断简化，叶形装饰也一直简化至完全消失。在托架正面的叶形装饰起先失去部分细部设计（*a*），随后失去了全部（*b*），最终被一起省略。省略了从较大的涡卷形中生长出的特色苞芽或将其简化为轮廓（*c*）。可以重新使用经过完全雕刻的孤立叶片，以便使托架不至于极度简化。此技巧是古典装饰中独特的一个方面，以实用的方式显示细部设计，进而展示出要素的全部装饰。低位的莨菪叶形装饰单独进行完全雕刻（*c*），而（*f*）和（*h*）中的苞芽在细部设计的省略程度和形式的简化方式上都进行了对比。

最后，形态自身也能够进行简化。用矩形细部设计替代小涡卷形（*f*、*g*和*j*），且将托架简化至轮廓的情况也很常见。但是省略涡卷下部的情况（*h*）却并不多见。

SCROLLED BRACKETS SIMPLIFIED

Scrolled brackets are a unique concentration of formalised architectural decoration. They combine scrolls similar to the volute of the Ionic capital with a complex enriched profile and the acanthus of the Corinthian Order. As a decorative feature they can be heavy with architectural ornament and a display of the sculptor's art. It is not, however, always appropriate or economic to include this wealth of decoration in an architectural scheme. In common with most aspects of Classical design, the specific form of the fully developed feature and the established tradition of the architectural element make it possible to reduce the quantity and scale of the detail while maintaining the identity of the element. The scrolled bracket is one of the most widespread and individual Classical architectural features and a series of levels of reduction in detail has evolved.

As the modillion, the scrolled bracket is included in the upper part of the Corinthian cornice. If the Corinthian Order is relatively small it is not possible to include the full range of decoration. Examples (*b*), (*i*), and (*j*) are from Corinthian cornices although any appropriate reduction of detail could be applied to a suitably proportioned modillion.

The scrolls themselves can have the narrow ribs that define their edges omitted and the whole body of the scroll can curl inwards. The scroll is then defined either by a groove or by the outward projection of the scroll as it coils to the centre. Examples (*a*), (*e*), (*f*), and (*g*) have scrolls defined by grooves, while (*b*) and (*h*) rely on the projection that can be seen on either side of the face.

The detail on the face of the bracket is subject to a similar process of simplification. The central bead seen on (*a*) and (*b*) can be omitted to give a continuous curved profile across the face as on (*d*) and (*h*). The face can also be flat and have simple flutes cut into it (*g*) or be left without any profile or decoration, (*f*), (*i*), and (*j*).

As the complexity of the profiles diminishes, so the leaf decoration simplifies until it disappears. The acanthus on the face of the bracket at first loses some (*a*) then all (*b*) of its detail and finally is omitted altogether. The characteristic sprouting buds that emerge from the larger scroll are either omitted or reduced to an outline (*c*). Fully carved isolated leaves can also be reintroduced to give added richness to an otherwise severely simplified bracket. This technique, a particular aspect of Classical decoration, allows detail to be introduced economically to suggest the full decoration of the element. The lower acanthus leaf alone is fully carved on (*c*) and the sprouting buds in the scrolls on (*f*) and (*h*) contrast not only with the omission of detail but also with a simplification of form.

Finally the form itself can be simplified. The substitution of the smaller scroll with a rectangular detail, (*f*), (*g*), and (*j*), and the reduction of the bracket to no more than its profile are common. The total omission of the lower scroll (*h*) is unusual.

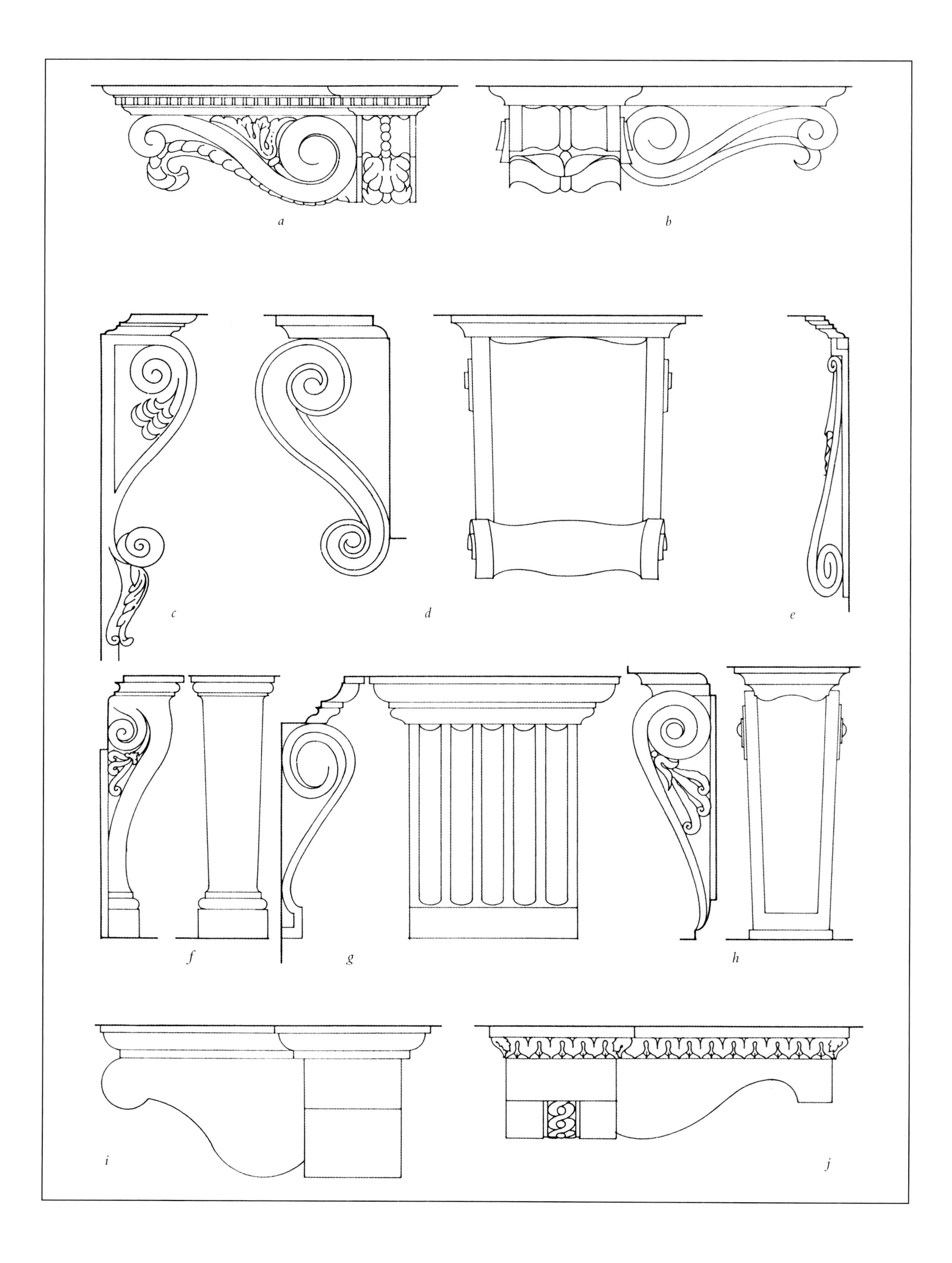
a
b
c
d
e
f
g
h
i
j

设计涡卷形托架

涡卷形托架可以用多种不同的方法进行设定，产生不同的形状，此处阐述一种方法。涡卷形的设计技巧与在爱奥尼亚柱头的涡卷形中所使用的方法相似，还可以用来替代P87中的结构。

托架高是8个测量单位，不包括任何毗邻的线脚，长是21个单位。设定大涡卷形（*a*）时，从外角引两条对角线，在对角线交会处从涡卷形的起点并且朝远离外角的方向画一个正方形，其对角线等于1个单位，也是圆花饰中孔的周长。在方形内部再画一个正方形，与第一个正方形的侧边中心点相交。用相同的方法在第二个方形内画另一个正方形。在所示的柱式中将第二个正方形的边向外延伸，把圆规放在点1，从涡卷形的外面画一个圆弧，1*a*到2*a*。然后把圆规放在点2，从2*a*画圆弧到3*a*，以此类推至4*a*。从点1*b*（在下一段中确定其位置）继续这个过程，在各点处由内部正方形切分第二个正方形内的对角线直至线消失在直径为3个单位的圆形圆花饰之中。用相同的方法画小涡卷形（*b*），但单位的尺寸减半。3.5个小单位的圆花饰按比例计算尺寸更大，线条也消失得更快。

为了连接涡卷形（*d*），从小涡卷形中圆花饰的内边向下画一条垂直的线，直到大涡卷形下方1个单位处的*A*点。将这个单位分成3份，使*B*点在*A*点上方2/3个单位处。把大涡卷形上的点1垂直向上延伸至涡卷形上方两个单位处的*C*点。将*A*点和*C*点连接起来，并且从大涡卷形的外面画一个以*C*点为中心的弧，再从小涡卷形的内侧画一个以*B*点为中心的弧，在*AC*线上交会。从小涡卷形的外侧再画一个以*B*点为中心的弧至*AC*线，继续该线并且以*C*点为中心画另一个弧使该线继续下去直至返回到大涡卷形。在垂直线上此弧停止的地方，开始画大开口涡卷形的内边。使用相同的中心作为该涡卷形的外部边缘，从1*b*处将内侧线转向内，直至消失在圆花饰中。

可以添加莨菪叶和其他装饰，以填充涡卷形外侧以及涡卷形之间的空隙，大叶片从大涡卷形下方向外卷曲至小涡卷形的中心。该叶片覆盖了托架外表面上部分复杂的曲线和中心楞条（*c*）。

简朴的断开式托架（*e*）和（*f*）以涡卷形托架为基础，可以在简单的曲线周围构造。

SETTING OUT A SCROLLED BRACKET

The scrolled bracket can be set out in a number of different ways to produce varied shapes and one method is illustrated here. The technique for forming the scrolls is similar to that used for the Ionic capital and is an alternative to the system shown on page 87.

The height of the bracket, excluding any adjacent mouldings, is taken as eight units of measurement and the length as twenty-one. To set out the large scroll (*a*) take two diagonals from the outside corners and at their intersection, away from the corners and from the start of the scroll, form a square with diagonals equal to one unit. These diagonals are the circumference of the eye of the rosette. Inside this square, from another square touching the centres of the sides of the first square and, inside this square, form another in the same way. Extend the sides of the second square outwards in the Order shown and, setting the compass on point 1, form an arc, from the outside of the scroll, 1*a*, to 2*a*. Then set the compass at point 2 and from 2*a* form an arc to 3*a* and so on to 4*a*. Continue the process from point 1*b* (the position determined in the following paragraph) with points where the inner square cuts the diagonals of the second square until the line disappears into a circular rosette three units in diameter. With the same method, but units of half the size, draw the small scroll (*b*). The rosette, at three and a half of the smaller units, will be proportionately larger, and the line will disappear sooner.

To connect the scrolls (*d*) drop a line vertically from the inside edge of the rosette of the small scroll to a point *A* one unit below the large scroll. Divide that unit into three and make a point *B* two thirds of a unit above point *A*. Extend point 1 in the large scroll vertically upwards to a point *C* two units above the scroll. Join points *A* and *C* and draw an arc centred on point *C* from the outside of the large scroll and another centred on point *B* from the inside of small scroll to meet at line *AC*. Draw a further arc centred on point *B* from the outside of the small scroll to meet at line *AC* and continue this line by drawing another arc, centred on point *C* back to the large scroll. Where this arc ends on a vertical line the inner edge of the large open scroll starts. Using the same centres as the outer edge of this scroll, turn the inside line inwards from 1*b* onwards until it disappears into the rosette.

Acanthus leaves and other decorations may be added to fill the spaces at the outside of the scrolls and in the gap between the scrolls. A large leaf can curl out from below the large scroll to the centre of the outside face of the bracket (*c*).

Plain cut-out brackets, (*e*) and (*f*), based on the scrolled bracket, can be formed around simple curves.

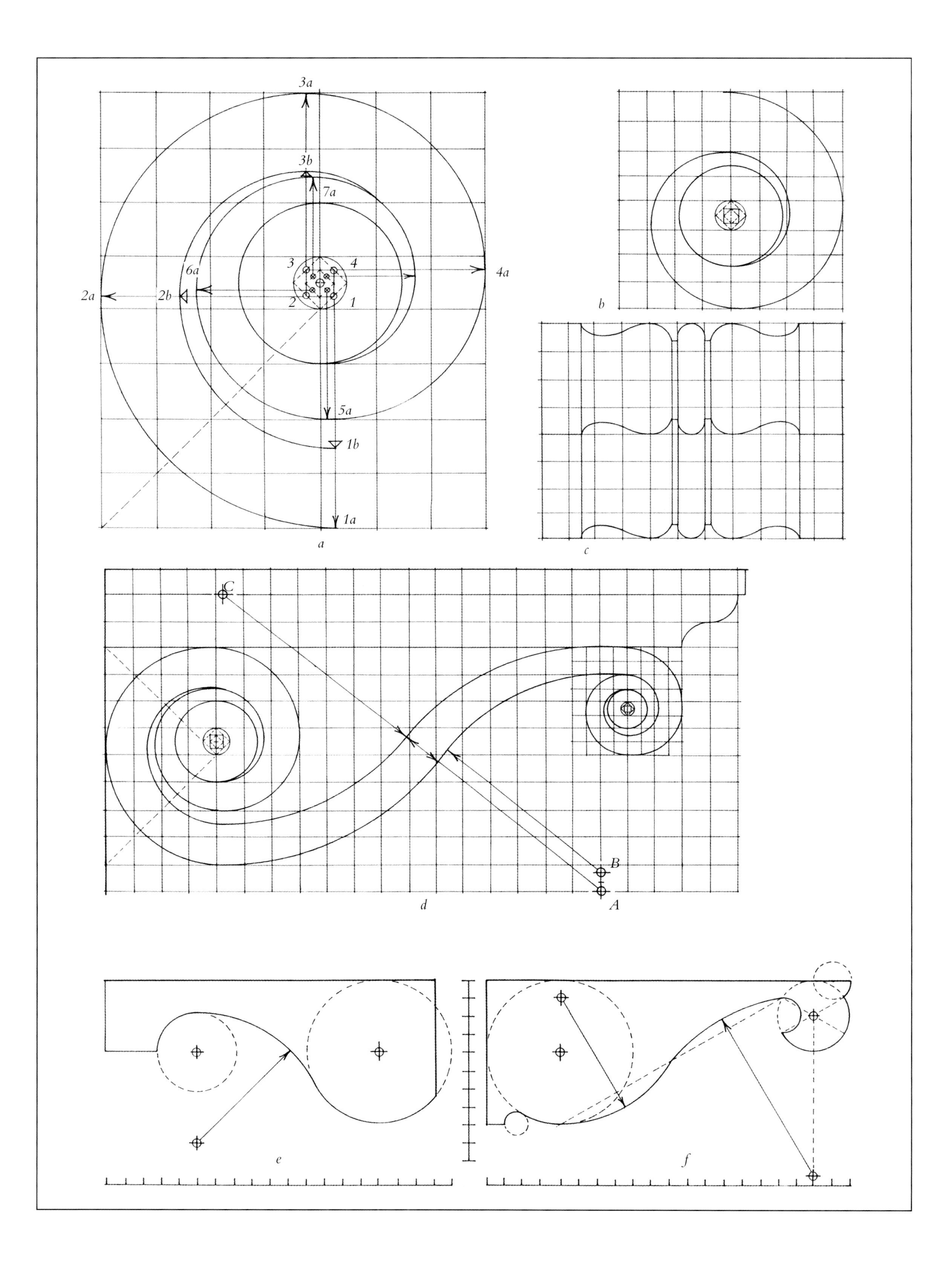

3a
3b
7a
3
4
6a
4a
2a
2b
2
1
5a
1b
1a
a
b
c
C
B
A
d
e
f

15. 壁炉

起源

尽管有迹象表明壁炉存在于罗马北部省份，实例却很少见。大多数罗马建筑由单独房间内的火盆供暖，或者带有火炕式供暖系统，即穿过中空的地板沿墙壁的暖气管向上传输热量和烟。我们所熟知的壁炉首创于中世纪。

中世纪壁炉最先建于11世纪和12世纪，取代大礼堂中心那类传统且令人头疼的开放式壁炉。为了建造壁炉，开放式的炉子被移到房间的一边，排风罩高耸在上方，将烟排到户外。这些火炉属于奢侈的建筑，通常仅限于私人房间，而且只有上层社会府邸和城堡的巨幅墙壁才适合建造烟囱。英国阿宾顿修道院的一处13世纪火炉（*a*）展示了遍布欧洲的典型布局，其中的细部设计明显源自该时代的建筑。两根支柱支撑托架，托架在高悬的排风罩下方负载着横梁，排风罩向上逐渐变细，延伸到后面墙壁上的烟囱之中。

早期文艺复兴壁炉只在原有设计的基础之上添加了经典的细部设计。威尼斯公爵宫的15世纪火炉（*b*）完全复制了中世纪的模式。支柱已经变成独具特色的古典栏杆柱，其中涡卷形托架负载着横梁，装饰后的横梁成为纵深的古典檐部，同时又支撑着高悬的排风罩。标准装饰十分精美，体现出壁炉一贯的重要地位。

到了16世纪，意大利建筑师已经意识到壁炉以及用来替代排风罩的凸出式壁炉腔所具备的创造性潜力。此时，壁炉变成了巨大的建筑结构，通常主导整个房间（*c*和*d*）。开口可以采用和门相同的设计原则，使用柱子或额枋支撑檐部或山花。与门不同的是，檐部通常带有长托架，保留了中世纪凸出式排风罩的印记。

18世纪时壁炉的使用更加广泛，几乎安置在每一个房间内部，因而不宜保留奢侈的壁炉设计。设计变得比较内敛（*e*和*f*），尤其是小房间内的壁炉也变小了。英国等国家使用煤炭替代木料，也是造成壁炉变小的一个原因。尽管如此，却保留了在前一个世纪确定下来的壁炉设计原则，并延续了与众不同且过于丰富的细部设计。

15. THE FIREPLACE

ORIGINS

Although there is evidence of fireplaces from the northern Roman provinces, their incidence is rare. Most Rome buildings were heated with braziers brought into individual rooms or with hypocaust floors which drew the heat and fumes from fires through hollow floors and up flues in the walls. The fireplace as we know it is a medieval invention.

Medieval fireplaces were first built in the eleventh and twelfth centuries to replace the traditional and troublesome open hearth in the centre of the great hall. To make the fireplace, the open hearth was moved to the side of the room and a tall hood was built over it to carry the smoke to the outside. Often restricted to private rooms, these fires were luxuries, and the massive walls of aristocratic houses and castles were needed to contain the chimneys. A thirteenth-century fireplace at Abingdon Abbey in England (*a*) shows a typical arrangement found throughout Europe. The details are clearly derived from the architecture of the period. Two columns support brackets carrying a beam below an overhanging hood which tapers upwards to a chimney in the wall behind.

Early Renaissance fireplaces merely added Classical detail to the established design. The fifteenth-century fireplace from the Doges' Palace in Venice (*b*) repeats the medieval pattern exactly. The columns have become distinctive Classical balusters with scrolled brackets carrying a beam, decorated as a deep Classical entablature, which supports an overhanging hood. The standard of decoration is very fine, reflecting the continuing high status of fireplaces.

By the sixteenth century, Italian architects had realised the creative potential of the fireplace and of the projecting chimney-breast that replaced the hood. Fireplaces now became huge architectural compositions, often dominating the room, (*c*) and (*d*). The openings could be designed according to the same principles as doors, with columns or architraves carrying entablatures or pediments. Unlike doors, the entablature was often carried on long brackets which retained the impression of the medieval projecting hood.

As the use of the fireplace became more widespread in the eighteenth century and they were introduced to almost every room, this kind of extravagance could not be justified. Designs came to be more restrained, (*e*) and (*f*), and, particularly in smaller rooms, less massive. In countries such as England, the replacement of wood with coal was a further reason for a reduction in size. The principles of fireplace design established in the previous centuries none the less remained, and the application of unusual and lavish detail continued.

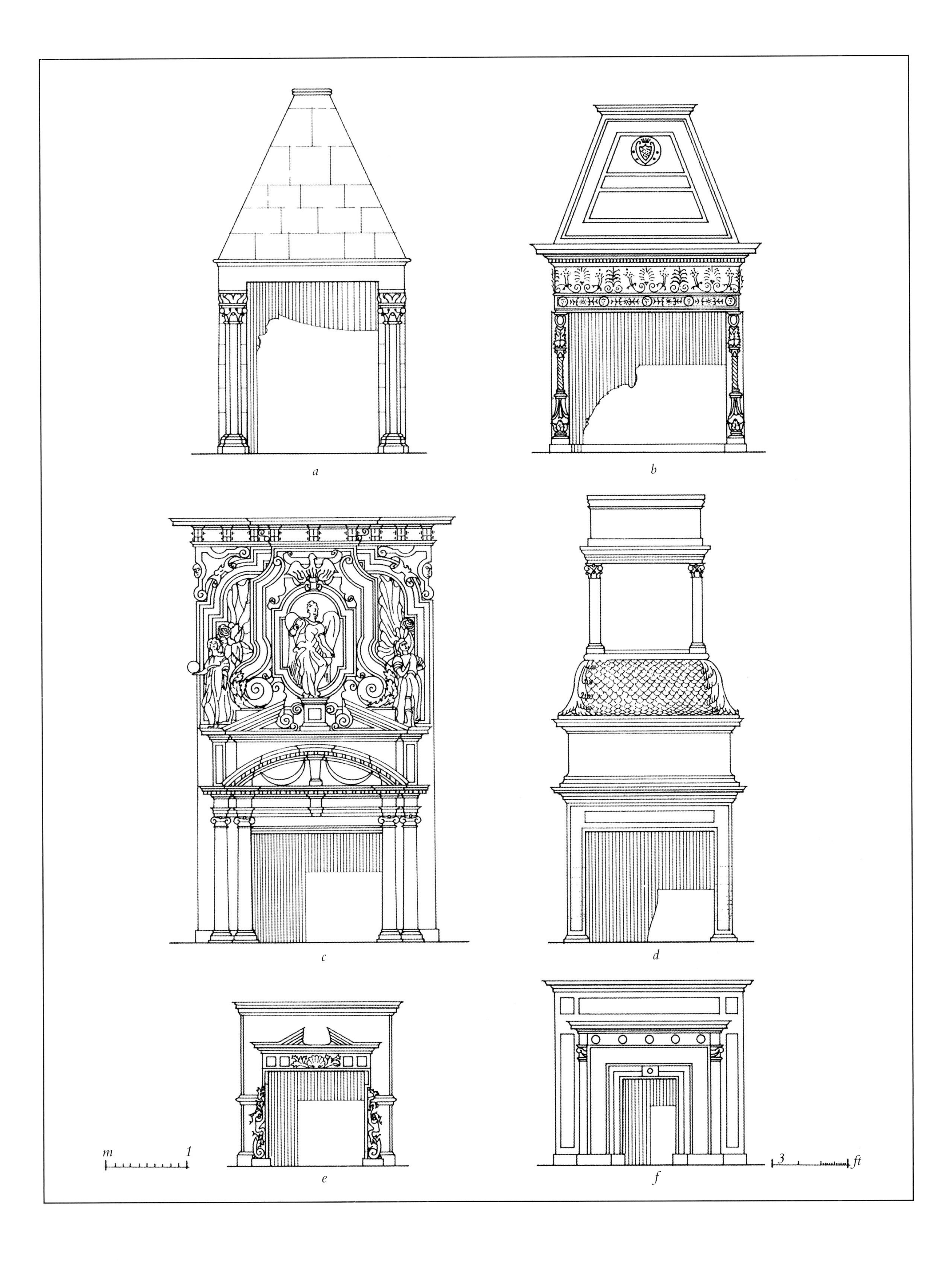
a
b
c
d
m 1
e
f
3 ft

壁炉与柱式

柱式壁炉的设计与门的设计有诸多相同的原则。壁炉与门的功能虽然不同，但正如门是建筑外立面的主要特征一样，开放式壁炉作为房间中唯一的热源，自然成为室内的焦点。壁炉的设计也反映出这种重要性，重要房间内的壁炉通常具有原创性且细部设计程度较高，这一传统赋予壁炉超出自身功能的重要性。尽管近来中央加热系统的发展使壁炉的存在失去了必要性，然而壁炉仍然继续使用下去，只是一般局限在主房间内，保存自然火源的美学效果，因而成为房间的焦点。

诸多壁炉设计中的创造性特征之一体现在利用想象力所表达出的柱式。意大利建筑师塞里奥在1547年发表的著作《建筑五书》中表达了一系列带有柱式的壁炉设计。多立克壁炉（*a*）表面上只包括简单的一对立柱和檐部，由狮子足部上的涡卷形装饰支撑，涡卷形装饰的正面经切割体现出多立克支柱的细部设计。他设计的爱奥尼亚壁炉（*c*）在狮足上用女性半身像代替支柱，表达爱奥尼亚柱式传奇的女性起源。

柱式可以通过更加传统的方式来表现，下面的例子全部来自18世纪的英国。简朴的多立克设计（*b*）简化了檐部，但保留了常规比例，檐部向前穿透立柱，并且在中心处体现出变化。爱奥尼亚壁炉的檐部（*d*）也向前穿透支柱，但省略了额枋，留出更大的装饰空间，在下方添加更多的线脚，与柱头成一条直线。此类设计中的重点是中心的镶板，镶板成为科利斯壁炉（*e*）中的决定要素。檐部也向前穿透支柱，但是将额枋和雕带限制在支柱上方的砌块内，因而保留了比例。柱式可以不包括支柱，但是檐口细部设计表达模糊不清会引发困惑。此例中檐口的飞檐托饰能够辨识科林斯设计（*g*），其中使用更多的涡卷形托架支撑檐口。雕带上充满丰富的莨菪叶形装饰，额枋则被削减至小型线脚。

16世纪和17世纪，火炉常常被设计成与整个房间等高的两层（*f*）甚至三层结构。此类设计在后续的几百年里逐渐减少，却并未有一刻完全消失。随着19世纪时早期文艺复兴风格再度流行起来，这类壁炉也随之复兴。如果设计包含若干层结构，那么柱式便如同传统顺序那样逐一向上排列。

FIREPLACES AND THE ORDERS

The design of fireplaces with the Orders follows many of the same principles as the design of doors. The function is quite different, but when the only source of heat in a room was the open fire it was natural for the fireplace to become the focal point of the interior in the same way as a door is a principal feature on a façade. Fireplace designs reflect this significance and are often original and highly detailed when located in important rooms. This tradition has given the fireplace a prominence that goes beyond its function. Although the recent development of heating from a central source has made the fireplace unnecessary, the use of fireplaces has nevertheless continued but is generally limited to the principal rooms in houses to retain the aesthetic effect of a natural fire and to give a focus to a room.

One aspect of the inventive character of many fireplace designs has been the imaginative expression of the character of the Orders. The Italian architect Serlio published a series of fireplace designs representing the Orders in his *Five Books of Architecture* in 1547. One of the Doric fireplaces (*a*), while apparently a simple pair of columns and entablature, is supported on scrolls on lion's feet with the face of the scroll cut to represent the details of a Doric column. One of his Ionic fireplaces (*c*) has female busts on lion's feet in place of columns, to represent the legendary female origin of the Ionic Order.

The Orders can be represented more conventionally and the following examples are all English from the eighteenth century. A plain Doric design (*b*) has a simplified but regularly proportioned entablature which breaks forward over the columns and in the centre to give some variety. The entablature on the Ionic fireplace (*d*) also breaks forward over the columns, but the architrave has been omitted altogether, giving a larger space for decoration. A further moulding has been added below in line with the column capitals. The central panel that features on both of these designs becomes the dominant element in the Corinthian fireplace (*e*). The entablature also breaks forward over the columns, but the proportions have been retained by limiting the architrave and frieze to blocks over the columns. The Orders can be represented without columns, but the ambiguities of cornice details can cause some confusion. The Corinthian design (*g*) can be identified by the modillions in the cornice, which is supported on further scrolled brackets. The frieze is filled with a rich acanthus pattern and the architrave has been reduced to a small moulding.

In the sixteenth and seventeenth centuries, fireplaces were frequently designed to the full height of the room and in two (*f*) or even three levels. This type of design became less frequent in later centuries but never died out altogether. It was revived as a part of the nineteenth-century interest in the early Renaissance. When the designs are of several stages the Orders can be placed in their conventional sequence one above the other.

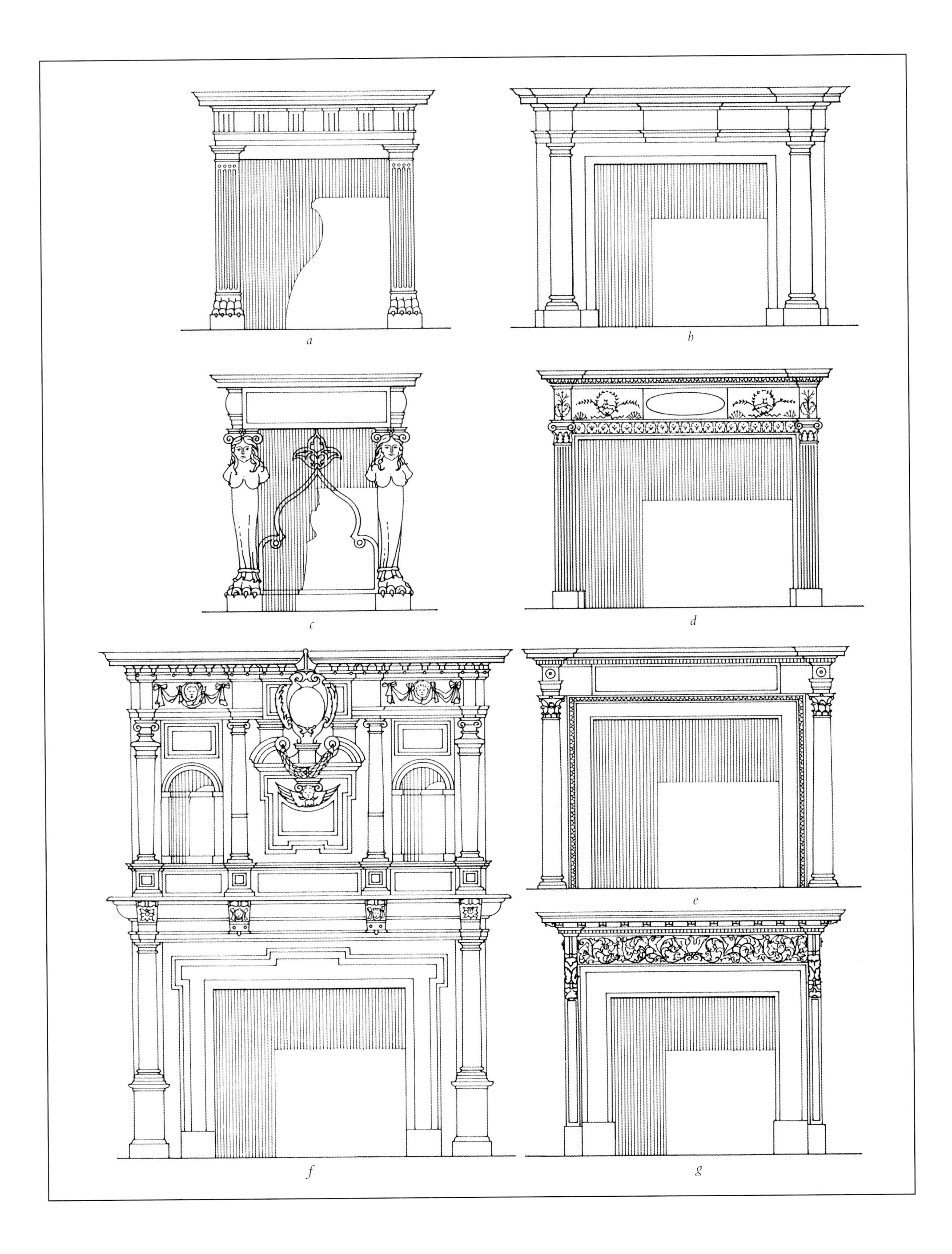
a
b
c
d
e
f
g

壁炉：变化与创新

壁炉可以仅仅是简单的额枋，对室内设计贡献甚微，但它也常常是房间内的主要建筑特征，并根据具体情况完成设计。

巴洛克后期的设计（*a*）发表在巴蒂·朗利（1696—1751年）1740年于英国出版的《设计集锦》一书中，设计虽然简单，但其中的双弧线檐口、圆雕饰浮雕人像和垂花饰足以使其与众不同。大约建于1750年的洛可可壁炉（*b*）也属于英国样式，并成为普遍的类型。壁炉从侧面向外倾斜成一个角度，此例中采用延长的涡卷形装饰，并在其中添加葡萄叶和葡萄装饰。檐口成为隔板，并简化为狭窄的线脚，也在侧面倾斜并支撑样式奇特的涡卷形装饰。装饰结构汇聚在中心的不对称旋涡处。这种形式的火炉非常流行，也可以进行更为拘谨的设计，或者在其上方放置匹配的镜框或画框。

涡卷形托架一直是火炉常见的特征之一。托架可以像在门缘中一样支撑檐口，也可以仅仅作为装饰。壁炉侧面的大型托架（*c*）是该设计中唯一的雕塑特征。托架的位置从开口处退后，从壁炉独立的中心部分中分离出来，暗示其装饰功能。在示例（*d*）中，两对涡卷形托架被安置在突耳周围和简易额枋的侧面。用狭长的托架装饰侧面，而上方的短托架托举起理论上的檐口并形成大型的雕带镶板。

涡卷形也出现在由科林·坎贝尔（1676—1729年）在18世纪早期设计完成的双层高度的英国壁炉中（*f*），但并非设计中最重要的元素。双层结构中从理论上的柱础到檐口的高度相同，但由于完全省略了雕带和额枋，并为檐口添加了山花，致使上部框架主宰了整个结构。尽管初看时并不明显，但壁柱上部的涡卷形装饰或许暗示其设计基础与爱奥尼亚柱式更为相像，而区别于多立克柱式。

18世纪末期的壁炉（*e*）和19世纪壁炉（*g*）的比例截然相反。早期的设计（*e*）包括细长、方形且带镶板的壁柱，没有柱头，壁柱支撑檐部。除了檐口的线脚，其余装饰浮雕都是后加的，使之成为一种流行且生产成本不高的壁炉。后期由查尔斯·罗伯特·科克雷尔1824年进行的设计（*g*）是19世纪普遍样式的早期示例。宽且平的支墩貌似柱子，檐口的突出部分成为柱头。这些支墩框住了里面更小的柱子，小柱子用更加传统的比例关系支撑雕带镶板和同一檐口。

THE FIREPLACE: VARIATIONS AND INVENTIONS

The fireplace can be no more than a simple architrave and play only a minor contributory role in the interior. It is, however, often the major architectural feature in a room, and is designed accordingly.

The late Baroque design (*a*), published in England by Batty Langley (1696-1751) in his *Treasury of Designs* of 1740, is very simple, but is given some distinction by the double-curved cornice, the medallion relief portrait and swags. The Rococo fireplace (*b*) is also English, from about 1750, but is a universal type. The sides are canted outwards at an angle and in this case are decorated with elongated scrolls enriched with vine leaves and grapes. The cornice which forms the shelf is reduced to one narrow moulding which also cants at the sides and is supported on further fanciful scrolls. The decoration meets at the centre in an asymmetrical swirl. This form of fireplace, which could have more restrained decoration or a matching frame for a mirror or picture above, was very popular.

Scrolled brackets have always been a popular feature for fireplaces. Brackets can provide support for the cornice in the same way as a door-surround or be added as solely decorative devices. The large brackets on the side of the fireplace (*c*) are the only sculptural feature on the design. Their position set back from the opening indicates their decorative role by separating them from the self-contained central part of the fireplace. In example (*d*), two pairs of scrolled brackets are fitted around the ears and sides of a simple architrave. Thin brackets decorate the sides while short brackets above hold up a notional cornice and create a large frieze panel.

Scrolls also appear on a double-height English fireplace (*f*) designed by Colin Campbell (1676-1729) in the early eighteenth century. They are not, however, the most significant elements in the design. The height of the two levels from the notional column base to the cornice is the same, but the total omission of the frieze and architrave and the addition of a pediment to the cornice make the upper frame dominate the composition. Although it is not immediately apparent, the scrolls above the upper pilasters suggest that this design may be based on an Ionic Order over a Doric Order.

The late-eighteenth-century fireplace (*e*) and the nineteenth-century fireplace (*g*) are of contrasting proportions. The earlier design (*e*) has slender, square, panelled pilasters without capitals, which carry a full entablature. With the exception of the cornice mouldings, all the decorative reliefs are applied, making this a popular and inexpensive fireplace to manufacture. The later design (*g*), by C.R. Cockerell in 1824 is an early example of a widespread nineteenth-century pattern. Wide flat piers give the appearance of columns with the projecting sections of the cornice acting as capitals. These piers frame smaller inner columns which support a frieze panel and the same cornice with a more conventional proportional relationship.

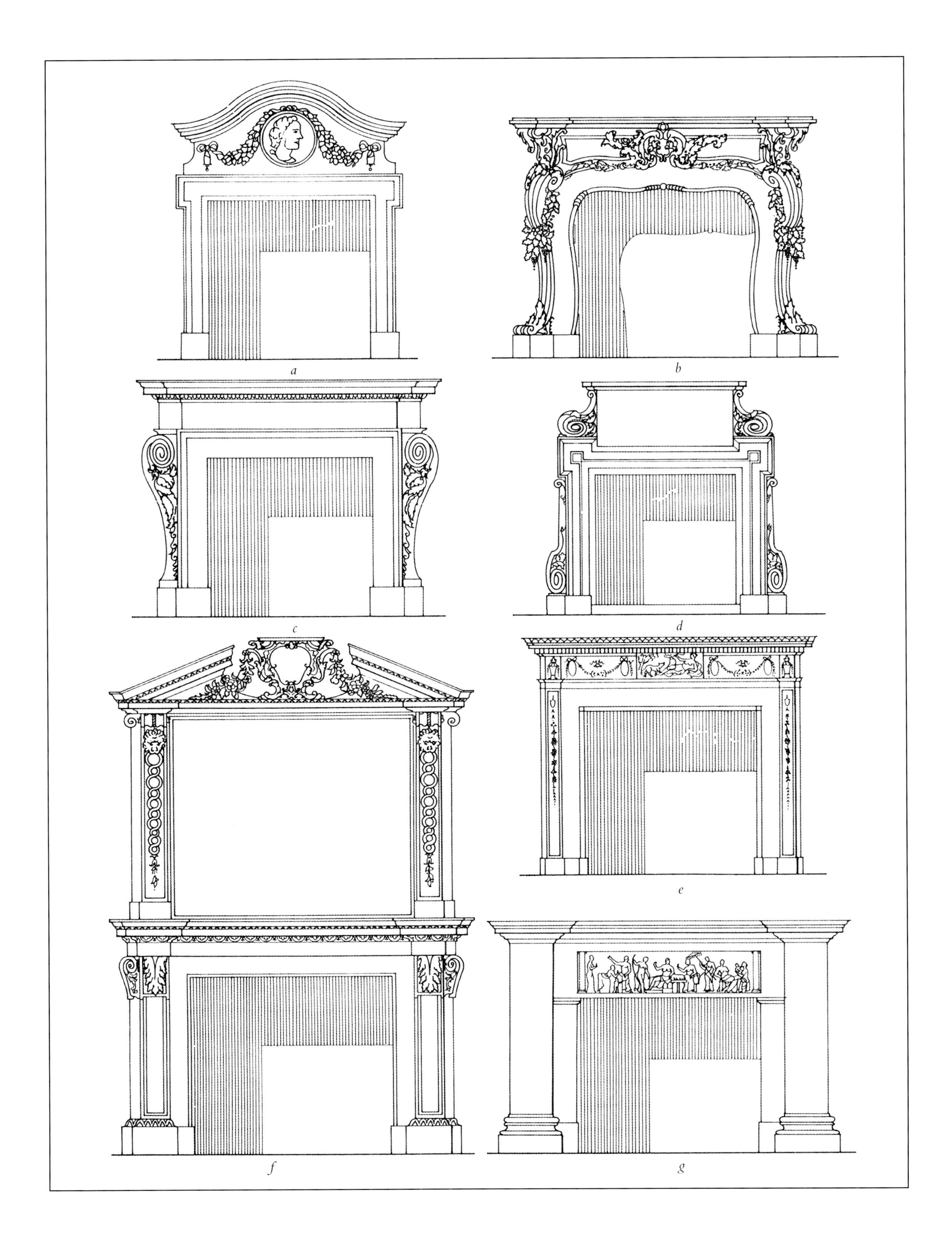
a
b
c
d
e
f
g

壁炉与雕像

壁炉设计中丰富的装饰细部设计也包括具有象征意义的雕像，壁炉的重要性使其特别适合展示雕像。壁炉与室内设计的其他方面相同，通常包括富有表现力和非传统样式的细部设计，而这些特征并不出现在同一建筑的公众外立面上。

法国建筑师皮埃尔・科罗（活跃于17世纪30年代）17世纪早期发表的双层壁炉（*a*）使用开口和缘饰作为底座展示全尺寸雕像，而雕像支配了整个建筑结构。尽管设计表现出参照了上升的柱式序列，柱式却并不能直接辨认出。壁炉周边有两个涡卷形装饰支撑蓄须的男性头像，上层结构更高并紧贴在简朴的托斯卡纳或多立克柱式檐口下方，将女性人像置于科林斯柱式檐口之下。在25年后由让・勒・保特利（1618—1682年）发表的法国壁炉设计（*b*）中，则看不到与柱式的联系。开口边缘无关紧要，实际上两个侧面托架已经隐蔽在橡树叶垂花饰末端之后。壁炉仅仅成为底座，支撑两个悬垂的人像和燃烧着的炉腔。

上一世纪的英国壁炉（*c*）在双层结构上的布局更加传统，但是用荷兰方式不能够完整地传递古典的细部设计，也因此赋予设计独特的个性。荷兰矫饰主义柱状人像，即赫尔墨斯像具有象征意义，显示出稚拙的活力。下层柱式计划为多立克式，上层人像上方的涡卷形表现出爱奥尼亚上层柱式。然而，人像和柱式之间并无联系，男、女像在两层中均交替出现。两个上部镶板中包括小规模的假透视图。

雕塑壁炉通常使用赫尔墨斯像作支座。18世纪早期的英国设计（*d*）带有出自《伊索寓言》故事中的浅浮雕镶板，每个侧面的上部都包括两个蓄须的赫尔墨斯像，据推测雕像代表哲学家。檐口和额枋保持传统关系，且顶部恰好置于雕带之中，但檐口保持了托斯卡纳柱式的简洁风格，而额枋则是爱奥尼亚或科林斯式。全像又称女像柱，也经常用作支座，18世纪法国壁炉（*e*）带有完整的檐部，由两座神像支撑，雕刻的是优雅的女性半人半羊农牧之神。

在19世纪，壁炉普遍的矩形开口发生改变，完整的拱形开口越来越常见。1818年的早期示例（*f*）明显出自罗马凯旋门，包括两个胜利女神的浅浮雕。其他的细部设计极为简洁，科林斯支柱失去了叶子装饰，未表现檐部，只剩下檐口线脚。

FIREPLACES AND SCULPTURE

The rich detail that is often incorporated in fireplace designs can include figurative sculpture. The importance of many fireplaces makes them particularly suitable for the display of sculpture. In common with other aspects of interior design, fireplaces often include expressive and unconventional details that are not found on the more public face of the same building.

A double-height fireplace (*a*) of the early seventeenth century, published by the French architect Pierre Collot (*fl.* 1630s), uses the opening and its surround as a base for a display of full-length figures that dominates the composition. No Orders are directly identifiable although there appears to be some reference to an ascending sequence. The fire-surround is supported by two scrolls with bearded male heads immediately below a plain Tuscan or Doric cornice and the taller upper level contains female figures below a cornice that could be Corinthian. Another French fireplace design (*b*), published some twenty-five years later by Jean Le Pautre (1618-82), has no recognisable association with an Order. The surround to the opening is insignificant and two flanking brackets are virtually hidden behind the ends of oak-leaf swags. The fireplace is no more than a plinth for the two draped figures and flaming urn.

An English fireplace (*c*) from the previous century has a more conventional arrangement on two levels, but the incomplete transmission of Classical detail by way of Holland has given the design a characteristic individuality. Figurative sculpture is contained within Dutch Mannerist columnar figures, or herms, and has been executed with a naïve vigour. The lower Order is intended to be Doric and the scrolls above the upper figures indicate an Ionic upper Order. There is, however, no association between the figures and the Orders and at both levels male and female sculptures alternate. The two upper panels contain small false perspectives.

The use of herms as supports has remained a popular theme for sculptural fireplaces. An early-eighteenth-century English design (*d*) has two bearded herms, probably representing philosophers, on each side of a relief panel with a scene from *Aesop's fables*. The relationship between the cornice and the architrave is conventional and places the heads precisely in the frieze, but the cornice has a Tuscan simplicity while the architrave is Ionic or Corinthian. Full figures, or caryatids, are also frequently used as supports and the eighteenth-century French fireplace (*e*) has a full entablature carried on two elegant, female, half-goat and half-human fauns.

In the nineteenth century the almost universal convention of rectangular openings for fireplaces was changed and fully arched openings became common. An early example (*f*) from 1818 includes relief sculpture of two figures of Victory, clearly derived from Roman triumphal arches. The other details are heavily simplified; the Corinthian columns have been stripped of their leaves and there is no reference to an entablature except for the suggestion of a cornice moulding.

a
b
c
d
e
f

多立克壁炉

为了展示壁炉中柱式的应用，此处绘制了一款简单得多立克式设计。图中所示的是一种传统的柱式应用，然而之前的样例已经表明，有诸多方式将古典细部应用于壁炉设计。

假设设计的对象是一个相对较小的壁炉开口，柱式的细部设计也做出了相应的调整。由于隔板的下层会减小飞檐托块的冲击，且小尺寸的线脚也使其并不显眼，因此已经从檐口省略了飞檐托块。方案之一（*b*）包括三陇板，但另一方案（*a*）省略三陇板，并在支柱上方的雕带中包含无装饰或有装饰的圆花饰。如果火炉较大或者需要更加精细的设计，则可以增加装饰力度。P73和P75的多立克柱式插图展现出适当的形式。

柱式壁炉设计的传统之一是将柱式的细部设计置于壁炉开口四周的独立框架周围。框架可以采用不同的比例，图中为方形。框架自身为石头缘饰，带有圆凸形外部线脚装饰。壁炉的其他部分可以选用任何适当的材料，但与开口直接相连的位置应当采用石头或其他防火材料。

将开口和框架的高度分成8.5份，形成模数。由于支柱的柱础相对比较小，用半个模数的小支座支撑，可以布置适当高度的墙裙或护壁板。支座样式多变，所示支柱不含圆柱收分线且作为壁柱高度为8个模数。其他比例依据此高度分割而成，但如果宽度变化并使用三陇板，那么就必须计算面积以适合间距。

为了使壁柱的凸出线脚独立于开口周围的框架，需要间隔半个模数。由于这种小尺寸的柱式显得比较狭窄，因此可以在外部重复这样的间距，以便在凸出壁柱的后面形成较宽的支墩（*a*和*d*）。图中所示的凸出的壁柱伸向前方，为由檐口所产生的隔板提供附加的支撑。如果该檐口不向前开辟出路径，那么壁柱的凸出部分应当限制在半个模数，参见檐口底面的向上视图（*e*）和截面图（*c*）。

壁柱自身呈方形，正面无装饰（*b*），且嵌有镶板（*a*和*d*）。可以使用半圆形支柱，带有凹槽装饰和圆柱收分线。示例（*a*）不带三陇板，其中的雕带可以用许多不同的方式进行装饰，而三陇板（*b*）可以装饰圆花饰或其他样式。

A DORIC FIREPLACE

To demonstrate the application of the Orders to fireplaces a simple Doric design is illustrated. This shows a conventional use of the Orders, although previous examples will have demonstrated that there are a bewildering number of ways of applying Classical details to fireplaces.

It has been assumed that the design is for a relatively small fire-opening and the details of the Order have been adapted accordingly. Mutules have been omitted from the cornice as the low level of the shelf would reduce their impact and the small size of the mouldings could make them insignificant. One alternative (*b*) includes triglyphs, but the other (*a*) omits them and has a plain or decorated rosette in the frieze above the column. The level of decoration can be increased if the fireplace is larger or a more intricate design is required. The appropriate forms are shown in the illustrations of the Doric Order on pages 73 and 75.

One of the traditions of fireplace design with the Orders places the details of the Order around an independent frame surrounding the fire-opening. The shape of the frame is shown as a square although it could be of different proportions. The frame itself is shown as a stone surround with an ovolo outer moulding. The other parts of the fireplace could be of any suitable material, but that part immediately adjacent to the opening should be of stone or another fire-resistant material.

The height of the opening and its frame is divided into eight and a half parts to give the module. As the column base would be relatively small it is raised on a small pedestal of half a module to allow for a skirting, or baseboard, of a satisfactory height. This pedestal could be varied, but the column, shown without entasis and as a pilaster, is eight modules high. The remaining proportions follow from this division of the height, but if the width varies and triglyphs are used then the dimension will have to be calculated to fit their spacing.

To allow the projecting mouldings of the pilaster to stand free of the frame around the opening they are spaced half a module away. As the Orders can appear thin at this small size it is possible to repeat this spacing on the outside to form a broad pier behind a projecting pilaster, (*a*) and (*d*). The projecting pilaster can, as shown, stand forward to provide additional support to the shelf created by the cornice. If this cornice is not to break forward then the pilaster projection should be limited to half a module and this can be seen on the upward view of the underside of the cornice (*e*), and the section (*c*).

The pilaster itself is shown as square and with a plain (*b*) and a panelled face, (*a*) and (*d*). The column could be half-round and have fluting and entasis. The frieze in the example without triglyphs (*a*) could be decorated in many different ways and the triglyphs (*b*) could have rosettes or other suitable decoration between them.

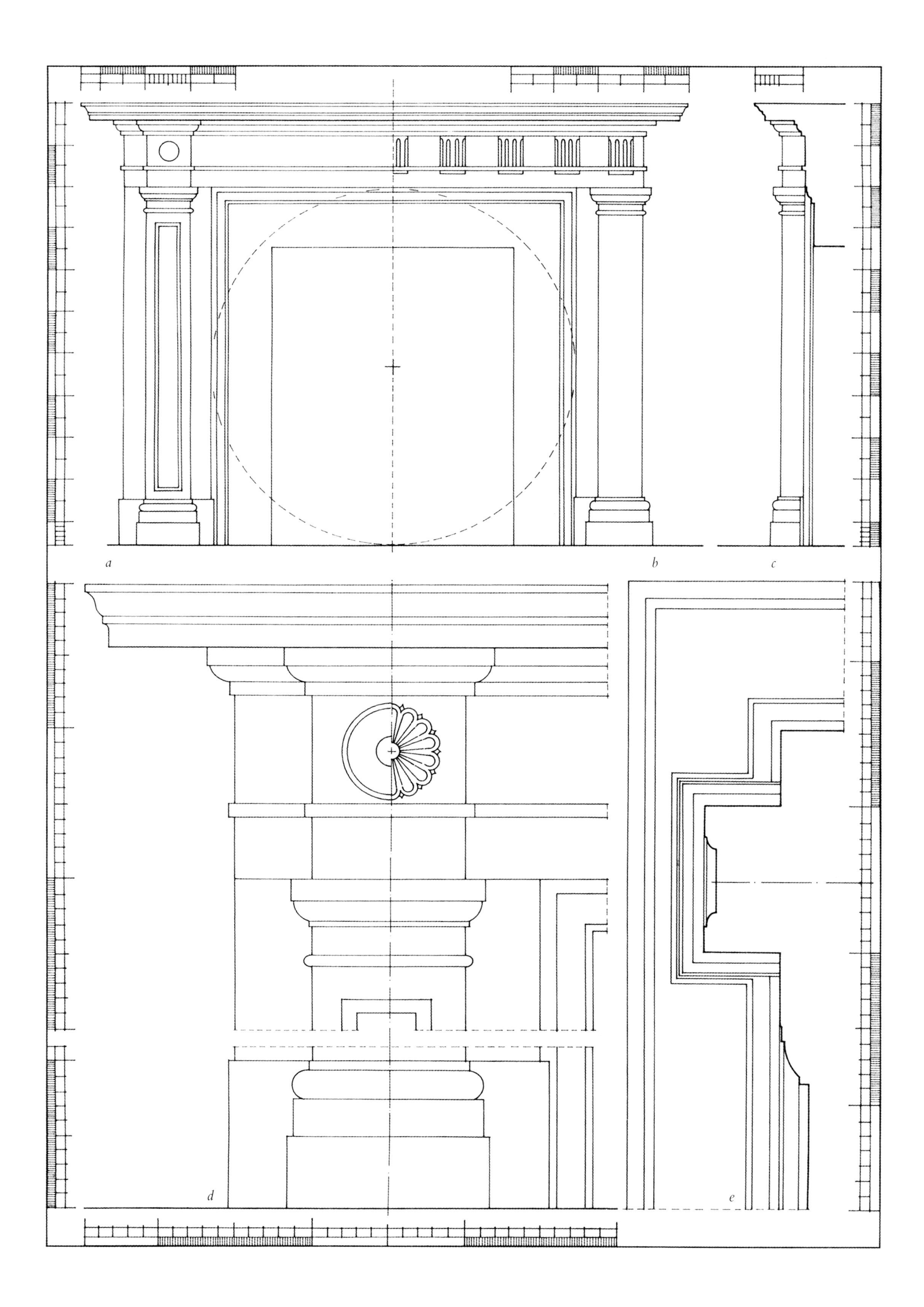
a
b
c
d
e

科林斯壁炉

当柱式应用到小型壁炉中的时候，与壁炉的宽度相比高度较矮，便显得过度纤细。可以将多立克和托斯卡纳支柱的比例设置得足够宽以克服这一问题，但是在许多旧式火炉中，双柱、附加的支墩或柱式经过严格的比例调整，可以用上方的细部设计来平衡侧面。额枋和门一样可以成为主要框架，檐部的其余细部设计可加诸在上方。该做法会使火炉侧面更加纤细，当省略支柱时可以更加满意地调节柱式比例的变化。这种将柱式应用到壁炉中的方法在过去应用得比较频繁，此处用科林斯柱式进行说明。

假设一个带有方形开口的小壁炉，其比例可以进行调整。与前一页的示例不同，该壁炉直接将柱式细部设计应用在开口的侧面，结果使额枋直接毗邻壁炉侧面。如果壁炉由石头或防火材料制成，这种做法没有问题，但如果壁炉由易燃材料建造，那么额枋需要建成防火样式。有完整细部设计的额枋（*b*和*e*）可以用防火材料建造，或者进行简化仅包括简单的石板（*a*）。

此示例中将开口高度分成5份，形成细部设计的模数，这是柱式传统比例的一半。通常最多可以在柱高1/3处分界，以便增加细部设计，当额枋单独构成壁炉且省略檐部的其余部分时尤为如此。如果用简单的石板做额枋，那么其比例通常根据雕带和檐口而加大。

由于假设的壁炉尺寸比较小，图中所示的额枋仅带有两个分界线。分界线可以增加到3个，这种情况更加常见，并进行装饰以提供更多的细部设计。突耳带有石板额枋（*a*），可添加到传统或扩大了的额枋之中。雕带保持简朴样式，亦可进行装饰（*a*），壁炉通常包括中部檐板（*b*），保持简朴或进行装饰。

檐口包括齿饰和飞檐托饰，之间的圆凸形线脚装饰上可以添加卵镖饰装饰线脚。飞檐托饰类型简洁，但可以进行任何程度的细部设计。一种设计方法（*d*）采用檐口底面视图两倍大小的尺寸，并在其中显示出间距（*c*）。檐口通常经简化后模糊了柱式的类型，此处出于展示科林斯壁炉的目的，并未显示那些简化形式（*e*）。

A CORINTHIAN FIREPLACE

When the Orders are applied to a small fire-opening, the low height in comparison with the width required for a fire produces columns that can appear excessively slender. The proportions of Doric and Tuscan columns can be sufficiently wide to overcome this problem, but many fireplaces of the past have relied on double columns, additional piers, or Orders with severely modified proportions to balance the sides with detail above. In the same manner as a door, the architrave can be the principal frame and the other details of the entablature can be added above. This would produce even more slender sides for the fireplace but, when the columns are omitted, variations to the proportions of the Order can be more satisfactorily accommodated. This method of applying the Orders to the fireplace has been used frequently in the past and is illustrated here with the Corinthian Order.

A small fireplace with a square opening has been assumed although these proportions can be varied. This type of fireplace, unlike the example on the previous page, applies the details of the Order directly to the sides of the opening. The architrave, consequently, directly abuts the sides of the fire. If the fireplace is of stone or a fire-resistant material this presents no problems, but if the fireplace is of an inflammable material the architrave will need to be fire-resistant. The architrave with all its details, (*b*) and (*e*), can be created in fire-resistant material or can be simplified (*a*) to allow for simple slabs of stone.

This example has taken the height of the opening and divided it by five to arrive at a module for the details. This is one half of the conventional proportions of the Order. A division of up to one third of the column height has often been used to increase the size of the details, particularly if the architrave alone is to be the fireplace and the rest of the entablature is to be omitted. If simple stone slabs are used for the architrave, its proportions are often increased relative to the frieze and cornice.

The architrave is shown with only two divisions, due to the assumed small size of the fireplace. These could be increased to the more usual three and decorated to provide more detail. Ears are shown with a stone slab architrave (*a*) but these could be added to a conventional architrave or increased in size. The frieze can be quite plain or decorated (*a*) and central tablets (*b*) are often included on fireplaces and either left plain or decorated.

The cornice includes dentils and modillions and the ovolo between them could have egg and dart moldings added. The modillions are of a simple type but could be to any appropriate degree of detail. A method of setting them up is shown (*d*) at twice the size of the view of the underside of the cornice (*c*) which shows their spacing. Cornices are often simplified to a degree that obscures the identity of the Order, but for the purposes of illustrating a Corinthian fireplace no such reduction is shown (*e*).

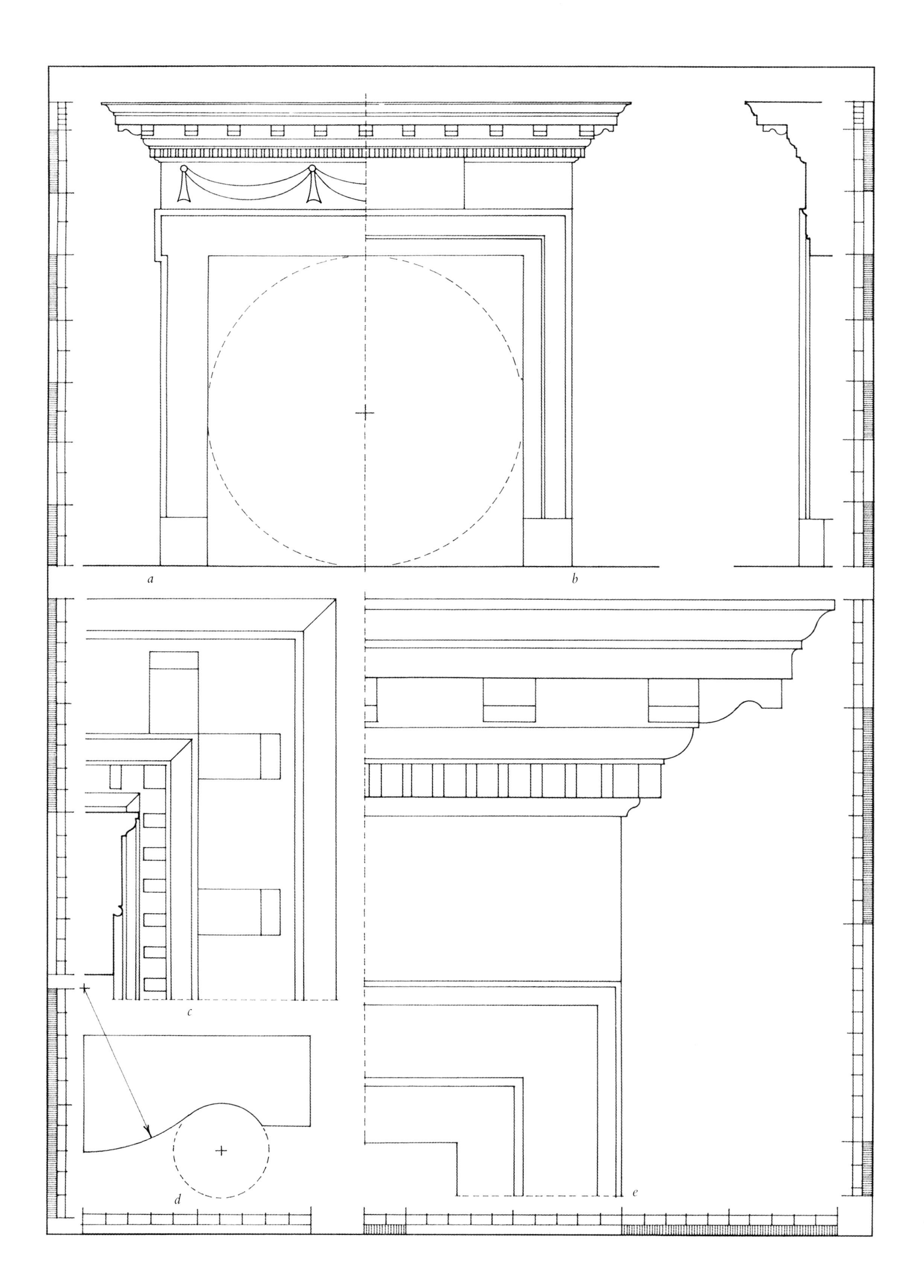
a
b
c
d
e

16. 栏杆和护栏

古代和文艺复兴时期

在古代，位于高处通道或开口边缘的阳台和低矮挡墙由简朴的墙壁或围栏构成，它们置身于支柱之间，作为独立的屏障直接毗连支柱的侧面（*a*和*b*）。围栏通常采用格状结构，称为格子细工，也可以用石料（*a*）再现，或设计成更加精致的青铜结构（*c*）。罗马绘画中出现许多栏杆的图案，但因为可能是由木材或青铜建造而没能保存下来。简朴的阳台墙壁可以用浅浮雕进行装饰（*b*）。

中世纪衍生出一系列独特的阳台设计，开放式阳台产生透明感。在威尼斯，运河沿岸建筑中的阳台成为建筑的一个主要特征，有两种不同的类型，一种是由小立柱构成的护栏（*e*），不带装饰或在扶手下方带有拱形；另一种是墙壁，在上面穿透了哥特风格的样式（*f*）。

在文艺复兴早期，哥特式设计经改造后吸收了古典细部设计。成排的小型柱式构成扶手（*i*），镂空围屏上的哥特式窗花格在浅浮雕低点原本应该出现的位置被古典风格的基本图案（*j*）所替代。

与此同时，出现了一种新型的护栏，现在普遍称作栏杆。文艺复兴建筑师的灵感似乎源于大量保留下来的罗马仪式中所用的大型蜡烛台（*g*）的遗迹。

最早期的文艺复兴扶手栏杆（*h*）在中心有两处凸起，会合在圆剖面小线脚或凹弧边饰线脚处。尽管该蜡烛台类型比较罕见，但是这种形式的罗马装饰柱仍然有据可证，罗马浅浮雕中便有如此图案（*d*）。后期的形式（*k*）明显源自一种常见的罗马蜡烛台（*g*），据说最先出现在16世纪早期，大量存在于罗马教皇的收藏品中。栏杆的这些新形式与基座结合在一起成为中间的支墩，自出现以来已经成为古典样式中必不可少的一个要素。

16. BALUSTRADES AND RAILINGS

ANTIQUITY AND THE RENAISSANCE

In antiquity, balconies and parapets on the edge of high-level walkways or openings were either plain walls or fences. Where these ran between columns, they were constructed as independent barriers directly abutting the sides of the columns, (*a*) and (*b*). These fences were often of a lattice construction and called transennae. They could be reproduced in stone (*a*) or be more elaborately designed in bronze (*c*). There are many illustrations of railings in Roman paintings, but as they were probably made of wood or bronze they have not survived. Plain balcony walls could be decorated with relief sculpture (*b*).

In the Middle Ages a specific range of designs for balconies evolved which were open and gave some impression of transparency. Two distinct types developed which can both be seen in Venice, where balconies were a feature of buildings which overlook the canals. One was a railing of small columns (*e*), either plain or with arches below the handrail, and the other was a wall with pierced Gothic patterns (*f*).

In the early Renaissance these Gothic designs were modified to incorporate Classical details. Rows of miniature Orders were designed to form rails (*i*) and Gothic tracery on pierced screens was replaced by Classical motifs (*j*) pierced where the low points of a relief would have occurred.

At the same time a new form of railing, now known as balustrading, was invented. The form seems to have been suggested to Renaissance architects by the remains of large Roman ceremonial candlesticks (*g*), a number of which have survived.

The earliest Renaissance balustrading (*h*) has two swellings in the centre which meet at an astragal or scotia moulding. There is evidence of Roman decorative columns of this form although it is a relatively rare type of candlestick; illustration (*d*) is from a Roman relief. A later form (*k*), said to have been first used in the early sixteenth century, is clearly derived from a common type of Roman candlestick (*g*), a number of which are in the Pope's collection in Rome. Both these new forms of balustrade were combined with column pedestals acting as intermediate piers and, since their invention, these have become an essential element in the Classical pattern.

a
b
c
d
e
f
g
h
i
j
k

设计栏杆柱

文艺复兴时期的两种栏杆柱都由小支柱构成，用特色凸起代替了圆柱收分线。其比例、间距和细部设计可以进行很大程度的改造。尽管个别作者表示每种柱式都对应自身独特的栏杆柱类型，但是并未在栏杆柱的细部设计和比例与不同柱式之间建立一致公认的关联。栏杆柱通常设置在支墩之间，支墩即是柱式的基座或等同于基座。因此栏杆柱的高度由相关柱式的基座决定，一般来说，柱式细部设计越丰富、越细长，那么扶手的细部设计也越多越纤细。所有栏杆柱都可以选择使用圆形或者方形的形状，但是方形栏杆柱的体积显得更大。

示例中每个栏杆柱类型都采用细部设计的平均水平和比例，出自18世纪的标准栏杆柱范例。

在双栏杆柱中（*a*），给定的高度被分成10个测量单位，每个栏杆柱宽两个单位，彼此间距1个单位。半条栏杆柱紧靠在墩身上。顶部和底部单位决定柱头和柱础上砌块的大小，这些砌块在平面上呈方形。每个凸起中大球形部分的直径由中心上下的两个单位所决定。栏杆柱中心位置的一个单位决定柱身或扶手的最小宽度。中心的球形由拱形连接到柱身上，拱形以从中心向外两个单位处为中心，垂直方向向上或向下两个半单位。两个球形的交叉点被对称的凹弧边饰所遮盖，边饰的直径是1/4个单位，用方形饰线连接到凸起处。砌块上下的柱头和柱础近似于多立克柱头和古雅典柱础，起始于1/3个单位处，高一个单位。

栏杆柱的比例可以更宽并且更为狭长，高度达到12个或更多个单位。尤其是可以用中心的圆剖面小线脚或装饰扩展中心凹弧边饰周围的细部设计，然而间距和规划的原则在本质上保持不变。

将单栏杆柱（*b*）的给定高度分成8个单位。水平布局、柱身宽度、柱头以及上下端的砌块都与双栏杆柱相同。凸起的部分由以柱础上方3个单位处为中心的球形构成，并由一个辐射形状连接到柱身，辐射形的中心距相邻栏杆柱中心线向下3个单位。古雅典柱础包括柱脚圆盘线脚和凹弧边饰，分别高1/3和2/3个单位。圆盘线脚上部及其饰线高1/3个单位。

该栏杆柱的设计比例可以加宽或者缩窄，常常通过增加柱础的高度或赋予栏杆柱基座的方式来提升高度。

SETTING OUT BALUSTERS

The two types of Renaissance baluster are small columns with their characteristic bulges replacing the entasis. Their proportion, spacing, and details can vary considerably and, although individual authors have suggested particular types for each Order, no universally accepted link between the detail and proportion of the baluster and different Orders has been established. Balusters commonly sit between piers which are, or are identical to, the pedestals of the Orders. The height of the balusters is, consequently, given by the dado of the relevant Order and, generally, the more detailed and slender the Order the more detailed and slender the baluster. All balusters can be round as well as square, but square balusters give the appearance of greater bulk.

An example of each type of baluster of average detail and proportion is illustrated, derived from standard eighteenth-century examples.

The double baluster (*a*) has the given height divided into ten units of measurement and each baluster will be two units wide and spaced one unit apart. There will be a half-baluster against the dado. The top and bottom units will give the dimensions of the blocks at the capital and base, which will be square in plan. The two units above and below the centre will give the diameter of the spheres which act as the larger part of each swelling. One unit located centrally on the baluster gives the shaft, or minimum width of the baluster. The spheres at the centre are joined to the shaft with an arc centred two units out from the centre and two and a half units up or down. The intersection of the two spheres is masked by a symmetrical scotia one quarter of a unit in diameter and joined to the swelling with square fillets. The capital and base above and below the blocks approximate to a Doric capital and an Attic base. Each is one unit high and set out on one-third divisions of a unit.

This baluster can be based on broader and more slender proportions of twelve units high or more, and the detail around the central scotia in particular can be expanded with a central astragal or decoration. The principles of the spacing and setting out will, however, remain essentially the same.

The single baluster (*b*) has the given height divided into eight units. The horizontal setting out, shaft width, capital and top and bottom blocks will be the same as for the double baluster. The swelling is made by a sphere centred three units above the base and this is joined to the shaft with a radius centred three units down on the centre line of the adjacent baluster. The Attic base has a torus and scotia one third and two thirds of a unit high respectively. The upper torus and its fillets are one third of a unit high.

This baluster can also have broader or narrower proportions. Extra height is often given by increasing the height of the base or giving the baluster its own pedestal.

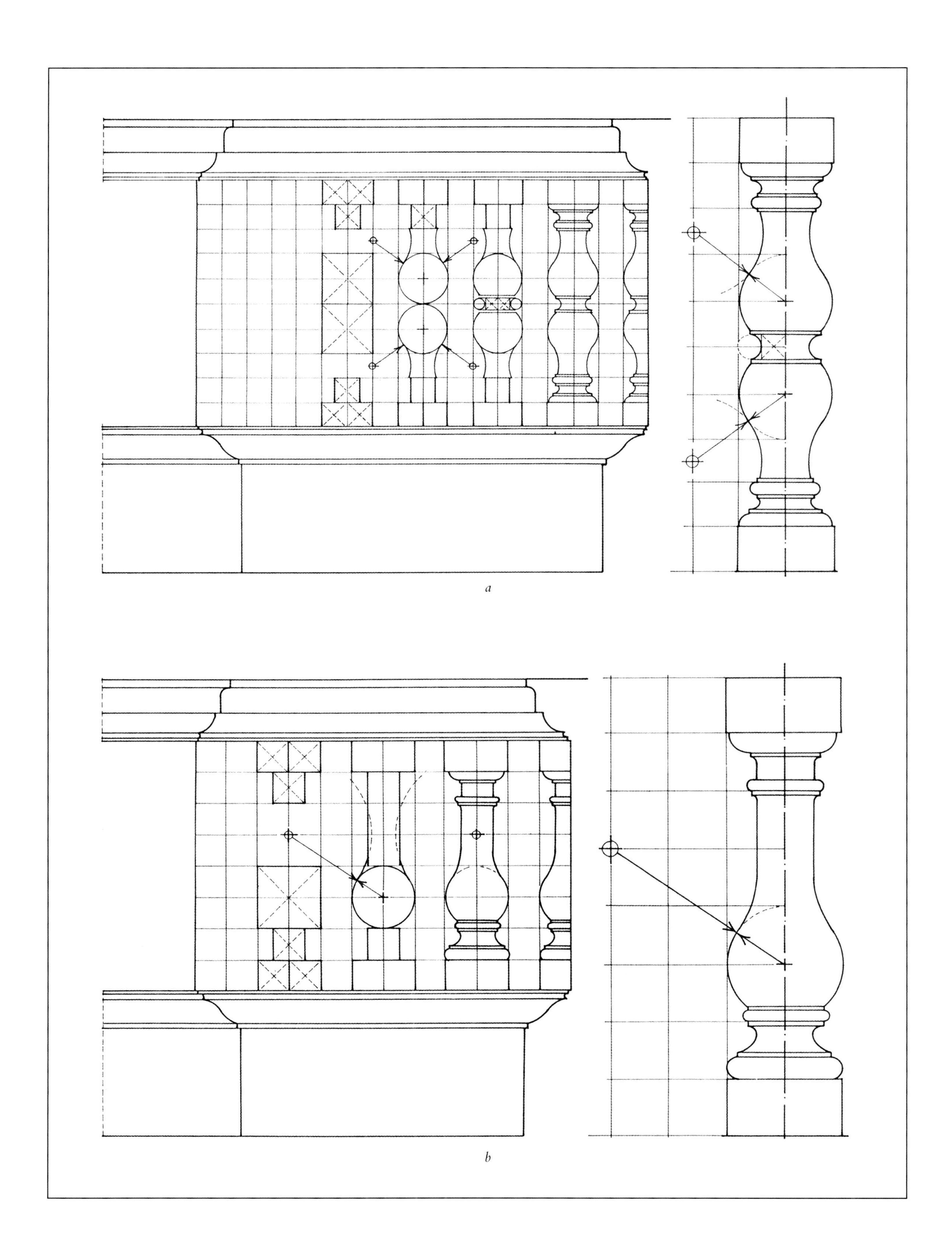
a
b

栏杆柱的变化

自栏杆柱出现以后，其设计一直不断发生变化，直至18世纪才充分确立设计传统，并由作者发布标准的版式，但不久之后考古兴趣揭示出栏杆柱的现代性，文艺复兴和巴洛克建筑风格的复兴也刺激产生了更多的创新性设计。

尽管文艺复兴时期栏杆柱的细部设计不尽相同，却拥有一致的普遍形式（*a*～*i*）。双栏杆柱（*a*）、（*b*）和（*c*）只在比例和两部分栏杆柱交会处的装饰程度上做出细微的改动。双栏杆柱出现得更早，却通常在同一建筑的不同部分中与单栏杆柱结合到一起。示例（*c*）和（*i*）都由布拉曼特建造，并出自罗马梵蒂冈的同一建筑。同样，不同的单栏杆柱也应用在同一建筑的不同部分，示例（*d*）和（*h*）同出自罗马附近由维尼奥拉设计的卡普拉罗拉别墅。这些栏杆柱的比例不同，但主要的变化体现在柱头，以及特别是底座上的细部设计中。维尼奥拉在罗马的朱利亚别墅中设计的栏杆柱（*e*）不带柱头，但同在罗马由巴尔达萨雷·托马索·贝鲁奇（1481—1537年）设计的兰特宫（*f*）和小安东尼奥·达·桑迦洛（1484—1546年）的法尼斯宫（*g*）中，栏杆柱柱础和柱头都包括更多的细部设计。

巴洛克时期打断了这种形式上的一致性，1705年范布勒在英国布伦海姆宫中设计的栏杆柱（*j*）夸大并打断了凸起部分中的连续曲线。这种巴洛克类型的变化已经获得认可，并出现在19世纪（*k*）和20世纪（*l*）的建筑中。在同一时期还出现了将凸起部分倒置的栏杆柱。1630年巴尔达萨雷·隆盖纳（1596—1682年）在威尼斯的安康圣母教堂中设计了复杂的栏杆柱（*m*），它们在平面上呈方形。倒置形式与被打断的变化形式相似，复兴于19世纪（*n*）。其他奇异的版式体现在巴洛克和19世纪的建筑物中。18世纪早期出自威尼斯佩萨罗宫的方形粗琢栏杆柱（*r*）包括交替的细蜡烛装饰，模糊了与单双栏杆柱的关系。另一方面，栏杆柱与相关柱式支柱的关系在19世纪示例（*q*）中更加明确，单栏杆柱专指爱奥尼亚柱式。

栏杆柱的文艺复兴起源对新古典风格的建筑师来说产生了一个问题。尽管人们认为栏杆柱由于混合了现代风格因而不再拥有纯正的血统，但是栏杆柱的使用还是得到了广泛的认可因此得以保留。查尔斯·罗伯特·科克雷尔将栏杆柱改造成瓶子的形状，1839年在牛津的阿什莫林博物馆中采用颠倒的凸起部分（*o*）。稍早时期的解决方法是用古代特有的叶形装饰替代栏杆柱（*p*），但由于使用希腊复兴风格的小支柱（*s*）和小拱形（*t*）而在无意中靠近了哥特式风格。真正的新古典风格中的栏杆柱（*u*）源自古代的格子细工栅栏。

BALUSTER VARIATIONS

Since the introduction of balusters there has been a considerable variety in design. It was not until the eighteenth century that the tradition had become sufficiently well established for authors to publish standard versions, but before long an archaeological interest in antiquity had revealed the modernity of the baluster and the revival of Renaissance and Baroque architecture stimulated more inventive designs.

Although Renaissance balusters differ in detail, the general form is consistent, (*a*) to (*i*). The double baluster, (*a*), (*b*), and (*c*), has only minor variations in proportion and in the degree of decoration where the two parts of the baluster meet. Although the double baluster is an earlier form, it is often combined on different parts of the same building with the single baluster. Examples (*c*) and (*i*) are both by Bramante and from the same building in the Vatican in Rome. Equally, different single balusters were used on different parts of the same building - examples (*d*) and (*h*) are both on the Villa at Caprarola near Rome by Vignola. The proportions of these balusters differ, but the major variations are the details at the capital and, in particular, the base. Balusters from the Villa Giulia in Rome (*e*) by Vignola have no capital, while there is more detail on both the base and capital on balusters from the Palazzo Lante (*f*) by Baldassare Tommaso Peruzzi (1481-1537) and the Palazzo Farnese (*g*) by Antonio da Sangallo the Younger (1484-1546), also in Rome.

This consistency of form was interrupted in the Baroque period; the balusters on Blenheim Palace in England (*j*), by Vanbrugh in 1705, have the even curve of the swelling exaggerated and broken. This Baroque type became an established variant and is found on nineteenth-century (*k*) and twentieth-century (*l*) buildings. In the same period, balusters were introduced that reversed the direction of the swelling. The complex examples from Baldassare Longhena's (1596-1682) S. Maria della Salute in Venice (*m*) of 1630 are square in plan. This reserved form, like the broken variant, was revived in the nineteenth century (*n*). Other fantastic versions belong to Baroque and nineteenth-century buildings. The square rusticated balusters from the early-eighteenth-century Palazzo Pesaro in Venice (*r*) have an alternating taper which makes their relationship to double and single balusters ambiguous. The relationship between the baluster and the column of the relevant Order, on the other hand, is made more explicit in a nineteenth-century example (*q*) which makes a single baluster specifically Ionic.

The Renaissance origin of balusters created a problem for Neo-Classical architects. The use of balusters had become so well established that they could not be abandoned in spite of, what was regarded as, an impure modern pedigree. C.R. Cockerell adapted the baluster to a vase shape with a reverse swelling on the Ashmolean Museum in Oxford (*o*) in 1839. An earlier solution was to substitute a specific leaf form from antiquity for a baluster (*p*), while the use of small Greek Revival columns (*s*) and small arches (*t*) unwittingly came closer to Gothic examples. The most authentic Neo-Classical baluster (*u*) was a version of a transcenna fence from antiquity.

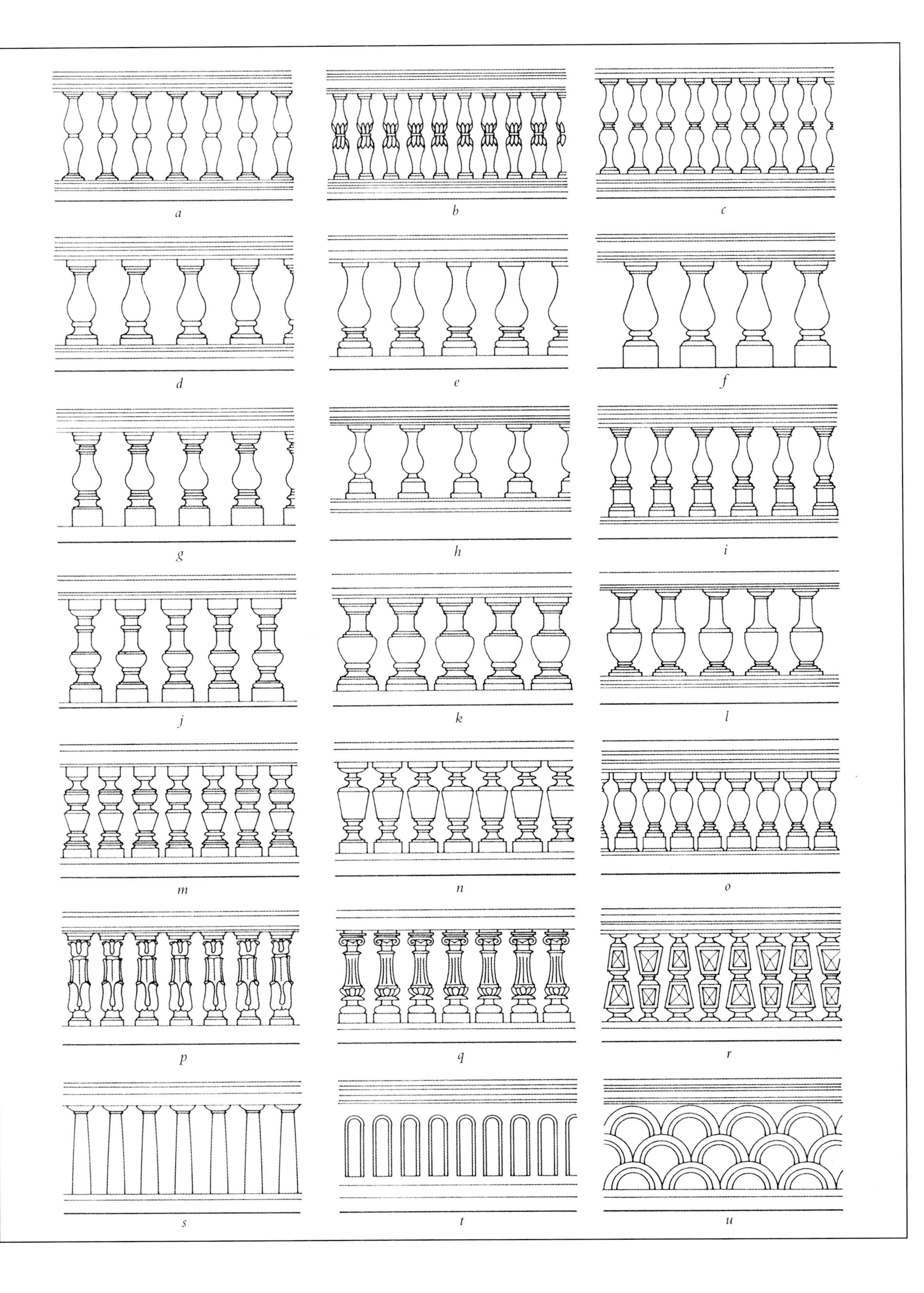
a
b
c
d
e
f
g
h
i
j
k
l
m
n
o
p
q
r
s
t
u

镂空墙壁

栏杆柱在15世纪和16世纪的意大利建筑中成为矮护墙的标准方法，并且在几个世纪中一直受到意大利风格的影响，此时北欧的建筑师们发展出在屏风墙上穿孔构成矮护墙的传统。

尽管法国文艺复兴建筑开始之初直接接触了意大利的风格，但仍保留了自身的民族特征。最具影响力的本土古典风格建筑师之一当属菲利贝·德·洛梅，他保存下来为数不多的作品设计中包括大量的镂空屏风护墙。示例（*a*）出自1547年阿内的一处门房，由一系列交错并且可以单独识别的古典涡卷形构成。在法国，相似的设计一直持续到下一个世纪，但设计中的古典起源不再那么显著（*b*）。

在新教盛行的北方，古典建筑大多通过低地国家的建筑出版物进行传播，在这些国家中，意大利的矫饰主义著作经诠释创造出一种奇特的建筑装饰主题，由于明显类似于切割皮革而被称作带状装饰。镂空屏风护墙依赖剪影，使其尤其适合在建筑物的天际线上进行带状装饰和自由设计，如16世纪德国巴登-巴登的城堡设计（*c*）具有该时期北方建筑的特征。或许是由于与低地国家的政治联系，西班牙在17世纪时也出现了类似的矮护墙（*d*），此类设计19世纪时在英国复兴（*l*）。

镂空屏风墙的设计传统16世纪时在英国得以延续。巴洛克风格精致的自然设计（*e*）和（*f*）出现在复杂精细的内部阳台和楼梯之中，由技艺无比高超的木雕家完成。

18世纪时，巴洛克建筑的后期包括不同的栏杆柱和其他形式的矮护墙。在欧洲，巴洛克楼梯设计带有相对素净的屏风墙，用传统的扭索饰交错样式（*g*）或与之相反用夸张的旋涡状石雕进行分割。诸如1713年由卢卡斯·冯·希尔德布兰特（1668—1745年）在维也纳的道尼宫中的屏风墙（*h*）。被称作扶手之家的意大利还出现了巴洛克（*i*）和洛可可（*j*）风格的镂空屏风墙。

新古典风格的建筑师意识到镂空屏风墙和围栏是古代世界唯一的保护性护栏形式，便重新兴起对镂空屏风墙的兴趣，其形式更加婉约，大量设计带有古代示例的影子，诸如1863年戈特弗里德·森佩尔（1803—1879年）为瑞士温特图尔市政厅的楼梯设计（*k*）。

PIERCED WALLS

While the baluster became the normal method of forming a parapet in Italian architecture of the fifteenth and sixteenth centuries, and in countries under Italian influence, north European architects developed the traditional of parapets formed by pierced screens.

Although French Renaissance architecture began with direct contact with Italy, it retained an individual national identity. One of the most influential native Classical architects was Philibert de l'Orme. His few surviving works include a number of pierced screen parapets designs. An example from the gatehouse at Anet (*a*) of 1547 is made up from an interlacing series of Classical scrolls which remain individually recognisable. Similar designs, but with a less obvious Classical derivation (*b*), continued to be produced in France into the following century.

In the Protestant north, much of the spread of Classical architecture was though publications from the Low Countries where Italian Mannerist books had been reinterpreted to create a fanciful architectural decorative theme known as strapwork, due to its apparent similarity to cut leather. The reliance on silhouette in pierced screen parapets made them particularly well suited to strapwork decoration and free designs on the skyline of buildings, such as on the sixteenth-century castle at Baden-Baden in Germany (*c*), are characteristic of northern architecture of this period. Similar parapets can also be found in Spain (*d*) in the seventeenth century, perhaps through a political association with the Low Counties. These designs were revived in the nineteenth century in Britain (*l*).

The traditional of the pierced screen continued in Britain in the sixteenth century. Rich naturalistic designs of the national Baroque style, (*e*) and (*f*), were executed on intricate internal balconies and stairs by woodcarvers whose skills have seldom been surpassed.

Later developments in Baroque architecture in the eighteenth century included variants of the baluster and other forms of parapet. In middle Europe, Baroque stairs were designed with relatively sober screen walls, cut with a form of the traditional guilloche interlacing pattern (*g*) or, by contrast, with a dramatic swirl of sculpted stonework, such as the screen in the Daun Palace in Vienna (*h*), by Johann Lukas von Hildebrandt (1668-1745) in 1713. In Italy, the home of the baluster, Baroque (*i*) and Rococo (*j*) pierced screens were also introduced.

The realisation by Neo-Classical architects that screen walls and fences were the only form of protective railing in the ancient world led to renewed but more restrained interest in pierced screens and a number of designs, such as Gottfried Semper's (1803-79) stair for the Winterthur Town Hall in Switzerland (*k*) of 1863, were loosely based on examples from antiquity.

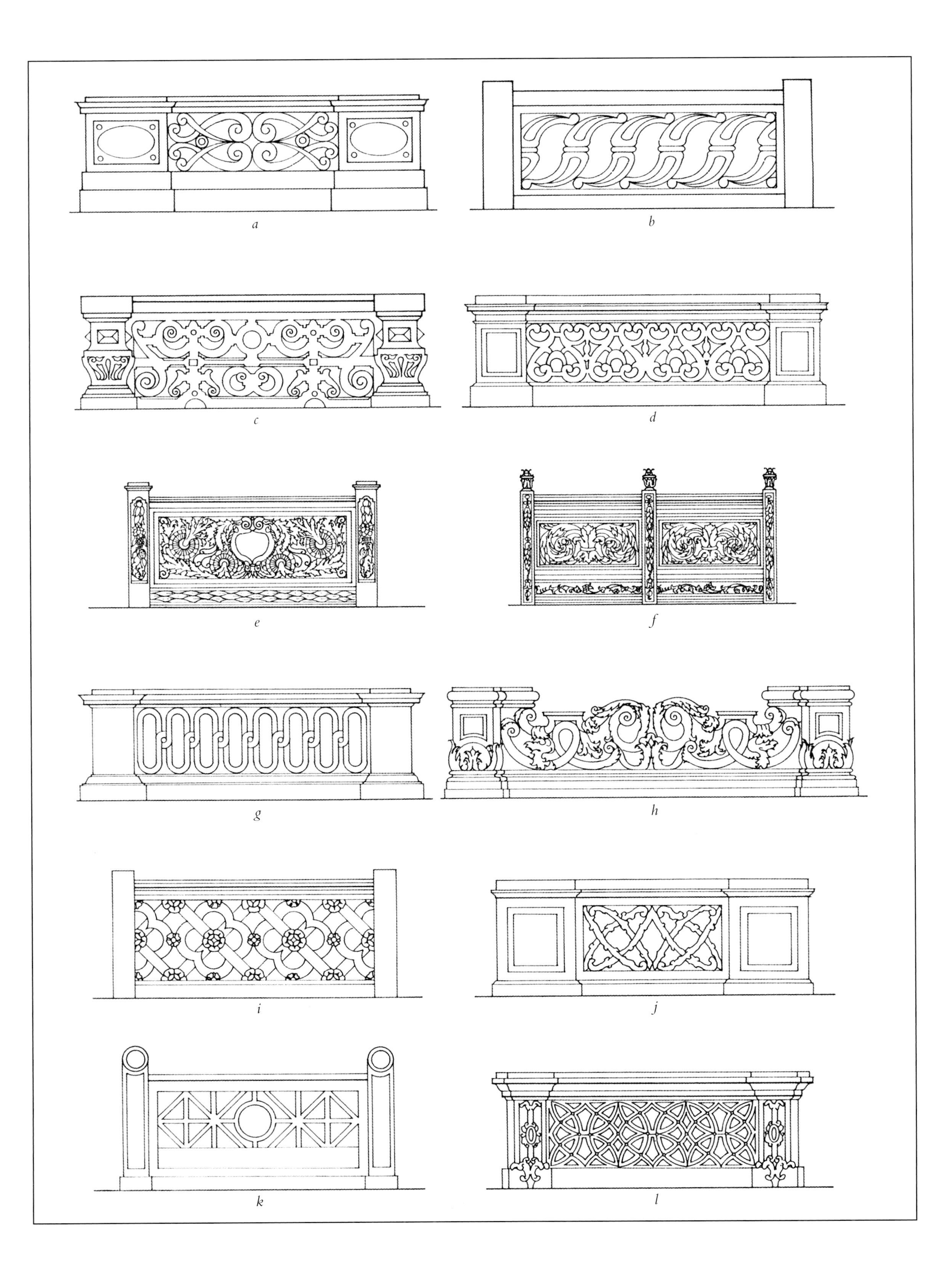
a
b
c
d
e
f
g
h
i
j
k
l

楼梯护栏

楼梯护栏是一种栏杆柱，与楼梯扶手和楼梯边缘经过改造之后适应了轻便的结构和大多数室内小型楼梯的外形。改造主要涉及减小宽度以便形成通常由木材建造的狭窄护栏。此外，其他细部设计也进行了类似的调整。

调整了早期木楼梯的比例和细部设计，使其成为近似倾斜的石头栏杆柱（*a*）。踏步板位于充当基座柱基的大木板或楼梯斜梁之后，且沉重的方柱即楼梯端柱充当位于楼梯段端点上的完整基座。扶手也很大并采用檐口的制作形式。18世纪时在英国出现了另一种楼梯类型（*b*），源自牢固的传统石阶布局。省略了闭合的楼梯斜梁，并露出踏步板的两端。为了使这种设计方便用木料实现，装饰性托架通常掩盖住重叠的踏步板。同时护栏与扶手变得更窄，而楼梯端柱变成扩展的护栏而非支墩，一般采用小型多立克立柱的形式。尽管沉重的早期设计样式在19世纪得以复兴，但带有闭合式楼梯斜梁的楼梯还是越来越少见了。

从16世纪到18世纪，楼梯护栏朝着越来越纤细的方向发展。示例（*c*）、（*d*）和（*e*）都出自17世纪，（*f*）出自16世纪末期。其比例与石头栏杆柱相似，但由木头制成，并严格遵循两种栏杆柱的样式标准。示例（*h*）也出自17世纪后期，其中的栏杆柱过于纤细不能用石头建造。示例（*g*）出自18世纪早期，与前面的一些例子相似，但是包括更加精细的巴洛克特征。随着护栏在18世纪后期变得越来越窄（*i*和*j*），便更加难以表达两种栏杆柱的形式。

18世纪早期开始使用一种新型护栏，它的出现或许是对更加纤细的流行趋势所做出的反应。新型护栏（*k*～*r*）将多立克支柱置于小栏杆柱的上方。该设计使比例更加纤细，却避免了结构变形后无法从垂直方向上识别栏杆柱的问题。起初，支柱自身采用传统比例，如1709年的示例（*k*）。但随着护栏变得更加纤细，支柱宽度也变得更窄。巴洛克风格的变体（*l*）出自1720年，支柱包括倒置的渐窄结构，且支柱和栏杆柱也颠倒过来。支柱上可以开辟凹槽（*n*），或进行扭曲（*o*和*p*），栏杆柱设计或精巧（*m*和*p*）或简单（*q*和*r*）。一般来说，随着18世纪的到来，细部设计和栏杆柱宽度一直在缩减直至达到木护栏能够承受并且不会危险易坏的程度，如（*r*）。

STAIR RAILS

Stair rails are a form of baluster which, together with the handrail and the edge of the stair, has been adapted to suit the lightweight construction and appearance of most smaller-scale internal stairs. The principal adaptation has been a reduction in width to form a narrow rail often turned out of timber. The other details have been similarly modified.

Early timber stairs were proportioned and detailed approximate to a sloping stone balustrade (*a*). The treads were concealed behind a large timber board, or string, acting as the base of the pedestal, and heavy square posts, or newels, acted as full pedestals at each end of the flight. The handrail was also large and took the form of the pedestal cornice. In eighteenth century Britain another type of stair (*b*) came into use, derived from a traditional arrangement of solid stone steps. The closed string was omitted and the ends of the treads were exposed. To ease the construction of this design in timber, decorative brackets often concealed the overlap of the treads. At the same time the rails and handrail became narrower, while the newel-posts became enlarged rails rather than piers, generally taking the form of small Doric columns. Closed-string stairs became less common, although the heavy early designs were revived in the nineteenth century.

From the sixteenth to the eighteenth century the general tendency was for stair rails to become progressively more slender. Examples (*c*), (*d*), and (*e*) are all from the seventeenth century and (*f*) is from the late sixteenth century. Their proportions are similar to stone balusters, but they are turned out of wood and follow precisely the pattern of the two types of baluster. Example (*h*) is also from the late seventeenth century and has a baluster form but is too slender for stone. Example (*g*) from the first years of the eighteenth century is similar but has a more intricate Baroque character. As rails became narrower in the later eighteenth century, (*i*) and (*j*), the two baluster forms became more difficult to represent.

In the early eighteenth century a new type of rail came into use, probably as a response to the fashion for greater slenderness. This new rail, (*k*) to (*r*), had a Doric column located over a small baluster. This allowed more slender proportions but avoided the contortion that would have made the baluster form virtually unrecognisable. At first the column itself, as in example (*k*) from 1709, was of conventional proportions, but as the rail became more slender the column was reduced in width. There were Baroque variations such as example (*l*) from 1720 where the column has a reverse taper, and the column and baluster are also reversed. The columns can be fluted, (*n*), or twisted, as in (*o*) and (*p*); the baluster can be elaborate, (*m*) to (*p*), or simple, as in (*q*) and (*r*). Generally, as the eighteenth century progressed, the detail and baluster width was reduced until, as with (*r*), the rail was as thin as the timber would allow without becoming dangerously brittle.

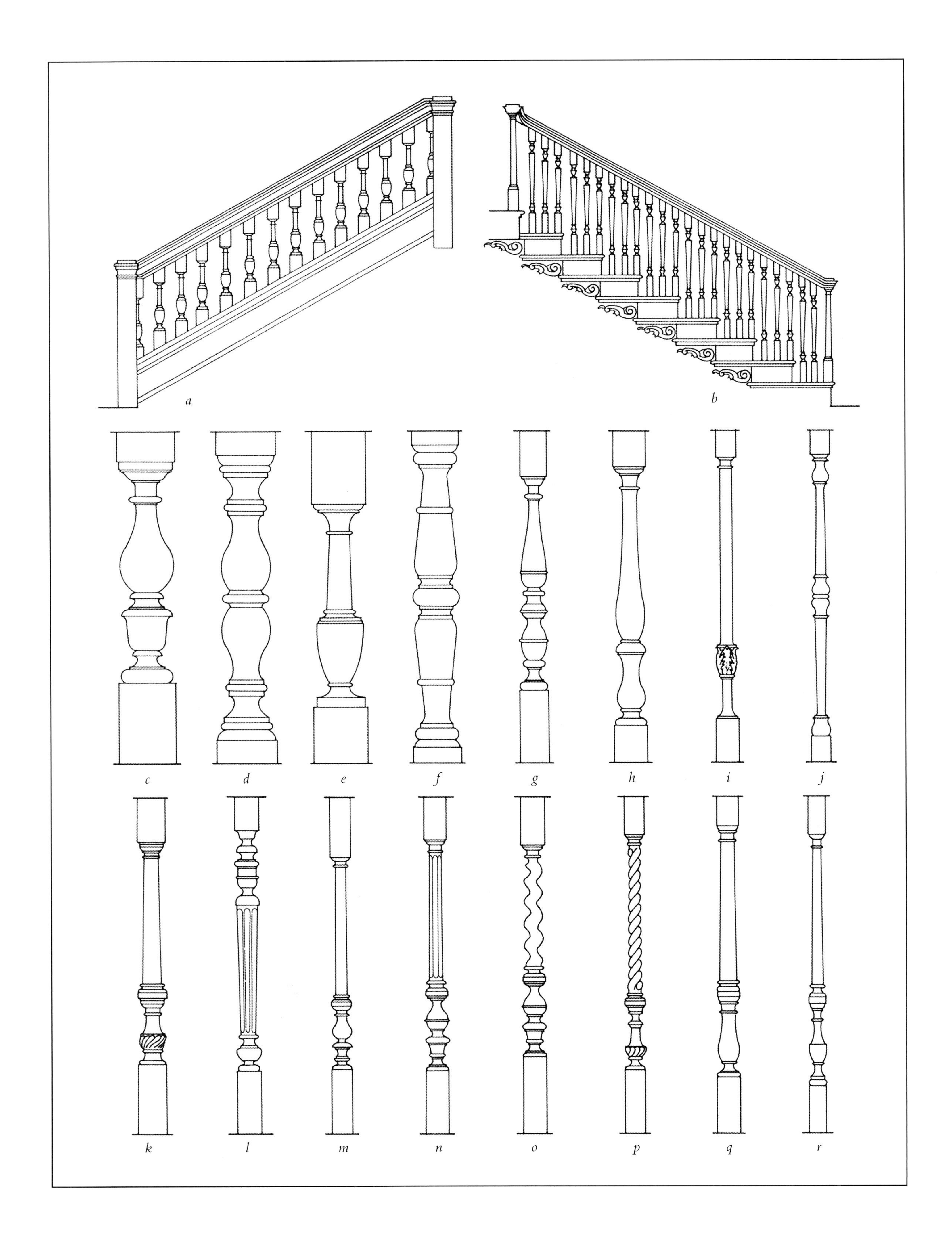
a
b
c
d
e
f
g
h
i
j
k
l
m
n
o
p
q
r

金属护栏与屏风墙

从文艺复兴到19世纪，金属护栏和屏风墙一般由锻铁或铸铁制成，在古代多为青铜所制。钢铁从19世纪开始使用，20世纪时已经使用了大量不同种类的金属和合金。所选图示全部为锻铁或铸铁，但设计大多来源于锻铁。18世纪末和19世纪的铸铁通常大量仿造手工制造并且更加高级的锻铁工艺。如今这两种材料都不常用了，但钢仍然可以用与锻铁相同的方式进行手工锻造。

金属栏杆的设计样式没有穷尽。这种材料在加热和锤制、溶解并导入铸模的过程中，似乎能够形成无限可能的形状。使用金属时，唯一重要的局限性是铁锈和大多数金属的重量。不过这两个问题可以通过使用现在的轻质金属和不锈合金与涂层的方法完美地解决。

传统上用来制造护栏的金属在力量和重量的方面都适合做出精致且纤细的设计，与18世纪末和19世纪初更加狭窄的护栏和栏杆柱趋势一致，如（*b*）、（*c*）、（*d*）、（*f*）和（*g*）。将细薄的矩形炙热锻铁材料用手工弯曲并塑造成形的过程也建立了由曲线、螺旋状和小型平展叶子形状所组成的设计传统，如（*i*）、（*j*）、（*k*）和（*l*）。铸铁也沿用了这一传统，比较容易脆裂，不过在浇筑液体的过程中可以在设计表面实现更加复杂的样式，如（*o*）、（*p*）、（*q*）、（*r*）和（*s*）。

金属结构不像木结构那样包含接缝，复杂的纤细金属护栏可以由不需要弯曲的直段组成，如（*c*）和（*d*）。简单的曲线设计几乎不带有锻造工艺，如（*a*）、（*b*）、（*f*）和（*g*），建造起来十分经济划算。精致华丽的巴洛克护栏和支柱（*i*）、（*k*）、（*l*）和（*m*）通常在选定的区域镀金，需要高超的技能来实现，且热金属具有韧性，能够进行连续的设计，使其连接到重复式的护栏（*i*）或形成长长的独立镶板（*k*）。18世纪后期的样式通常更加朴实并包括大量出自古代的基本图形，如叶丛状装饰（*j*和*o*）和里尔琴（*h*）。精巧且贵重的金属护栏或支柱（*e*）继续用锻铁制造，更加倚重雕塑质量而非材料性质，但可以大量使用更加经济的铸铁模具（*n*、*q*和*r*）制造出相似的效果。

还有大量其他的设计，包括从带有矛尖的简单圆、方形锻铁护栏到精致的锻铁镶板。这些示例在类型、时期和材料等方面受到限制而有不足，却展示出英国在处于工业财富和制铁工业巅峰时期的设计传统。

METAL RAILS AND SCREENS

From the Renaissance to the nineteenth century, metal rails and screens have generally been of wrought or cast iron. In antiquity they were often bronze. Steel was used from the nineteenth century and in the twentieth century a number of different metals and alloys have been used. The patterns illustrated are all either wrought or cast iron, but the designs mostly derive from wrought iron. Cast iron in the late eighteenth and the nineteenth century was often a mass-produced reproduction of hand-worked and higher status wrought iron. Neither material is used frequently today although steel can be hand-forged in the same way as wrought iron.

There are endless different railing designs that can be created in metal. The material can take on seemingly limitless shapes when heated and beaten, or melted down and poured into moulds. The only significant limitation to the use of metals has been the rusting of steel and the weight of most metals. Both of these problems can be overcome with the lightweight metals and rust-resistant alloys and coatings now available.

The strength and the weight of the metals traditionally used for rails has made many of the designs fine and slender, in keeping with the general tendency towards narrower rails and balusters, (*b*), (*c*), (*d*), (*f*), and (*g*), in late eighteenth and early nineteenth century. The process of bending and shaping hot wrought iron by hand from thin rectangular lengths of the material has established a tradition of designs made up of curves, spirals, and small flat leaf forms, (*i*), (*j*), (*k*), and (*l*). Cast iron, while following this tradition, was more brittle, but the liquid process of moulding allowed more complex patterns, (*o*), (*p*), (*q*), (*r*), and (*s*), in the surface of the design.

Without the joints that limit the possibilities of timber construction, slender metal railings of complex design can be made of straight pieces, (*c*) and (*d*), which require no bending. With very little forging work, simple curved design, (*a*), (*b*), (*f*), and (*g*), can be created economically. Rich Baroque rails and posts, (*i*), (*k*), (*l*), and (*m*), often with gilding on selected areas, require considerable skill in manufacture and the flexibility of the hot metal allows continuous designs to be produced which either link a repeated rail (*i*) or form long individual panels (*k*). Late-eighteenth-century patterns are often more sparse and include a number of motifs taken from antiquity, such as anthemions, (*j*) and (*o*), and lyres (*h*). Elaborate and costly metal rail or posts (*e*), which rely more on their sculptural qualities than the nature of the material, continued to be produced in wrought iron, but similar effects could be produced more economically in large numbers from moulds in cast iron, (*n*), (*q*), and (*r*).

There are numerous other designs, from simple round or square cast-iron railings with spear tops to elaborate wrought panels. There examples are limited in type, period, and material but illustrate the British tradition which developed at a time when the nation's industrial wealth and iron manufacture were at their peak.

a
b
c
d
e
f
g
h
i
j
k
l
m
n
o
p
q
r
s

倒置的护栏

此处将倒置护栏中的两条直径应用到不同类型的楼梯或阳台的设计之中，设计也可以在整个高度上保持正方形而不必倒置。

比例依据护栏的高度而定，为了达到较宽栏杆柱的最大成品宽度，需要将高度分成12份，窄一些的护栏需要分成20份。两种类型护栏的中心部分都采用最大宽度的一半，并且要确定所使用的木材或其他材料在这条直径上起到必要的强调作用，狭窄样式尤为如此。在所有示例中，将顶部和底部的砌块留作方形，且偏下的砌块要更高一些。设计中可以削减砌块上的凸角以避免后期受到损坏。

双栏杆柱的柱础和柱头（*a*）各高1个单位，分别开始于1/3和1/6个单元分界处。两处凸起置于柱头和柱础之间的中心位置，并交叉于相关单位一半的地方。拱形的中心形成凸起，距此交叉点向上或向下1个单位，从护栏中心向外1个单位。凸起返回到中心的直径，曲线中心分别距柱头和柱础的上下末端8个单位。

单扶手柱头（*d*）与（*a*）的设定方式相同。柱础高2.5个单位，开始于3/4个单位分界处。凸起的拱形中心距柱础向上1个单位，距护栏中心向外1.5个单位。凸起返回到中心，曲线中心距柱头向下2个单位，距护栏中心向外8个单位。

狭窄栏杆柱护栏的柱头和柱础（*b*）高1个单位，分别开始于1/3个单位和1/6个单位处。由于护栏宽度有限，凸起被限制在两个高1.5个单位的小双弧线中，曲线各自基于直径为1个单位的球形。其余从柱础和柱头至中心的柱身部分向外逐渐变细。

狭窄栏杆柱（*c*）的柱头与（*b*）的设定相同。小栏杆柱高4个单位，其柱础和柱头分别高1个单位。凸起部分的拱形中心距柱础向上1个单位，距中心向外1个单位。凸起返回到柱头，其中的曲线以柱头一半处为中心，距护栏中心向外9个单位。小支柱的柱础高0.5个单位，偏低的矩形线脚留成方形，且未倒置。支柱在向上延伸直至柱头内径的过程中逐渐变细。

TURNED RAILS

Two diameters of turned rail, which are applicable to different types of stair or balcony design, are illustrated here. Both designs could be square throughout their height rather than turned.

The proportions are taken from the height of the rail. To establish the maximum finished width of the wider rails divide the height by twelve, and for the narrower rails divide the height by twenty. Both types have a continuous central core one half of the maximum width and, particularly with the narrow examples, it is important to establish that the timber or other material used can take the necessary stresses at this diameter. In all cases the top and bottom blocks are left square and the lower block is higher. These blocks can have their projecting corners cut back to avoid later damage.

The base and capital of the double baluster (*a*) are each one unit high and set out on divisions of one third and one six of the unit. The two swellings are placed centrally between the capital and base and their intersection falls halfway across the relevant unit. The centres of the arcs that create the swellings are one unit up or down from this intersection and one unit out from the centre of the rail. The swellings return to the diameter of the core with curves centered eight units from the lower and upper extremity of the capital and base respectively.

The capital of the single baluster (*d*) is set out in the same way as (*a*). The base is two and a half units high and set out on divisions of one half and one quarter of the unit. The center of the arc for the swelling is one unit up from the base and one and a half units out from the centre of the rail. The swelling returns to the core with a curve centered two units down from the capital and eight units out from the centre of the rail.

The capital and base of the narrow baluster rail (*b*) are each one unit high and set out on division of one third and one sixth of a unit. Due to the limited width of the rail, the swelling is limited to two small double curves one and a half units high, each based on a sphere one unit in diameter. The remaining shafts taper out to the centre from the base and capital.

The narrow column-on-baluster rail (*c*) has the capital set out in the same way as (*b*). The small baluster is four units high and has a base and capital each one unit high. The centre of the arc for the swelling is one unit up from the base and out from the centre. The swelling returns to the capital with a curve centered halfway up the capital and nine units out from the center of the rail. The small column base is one half of a unit high and its rectangular lower moulding is left square and unturned. The column tapers up to the core diameter at its capital.

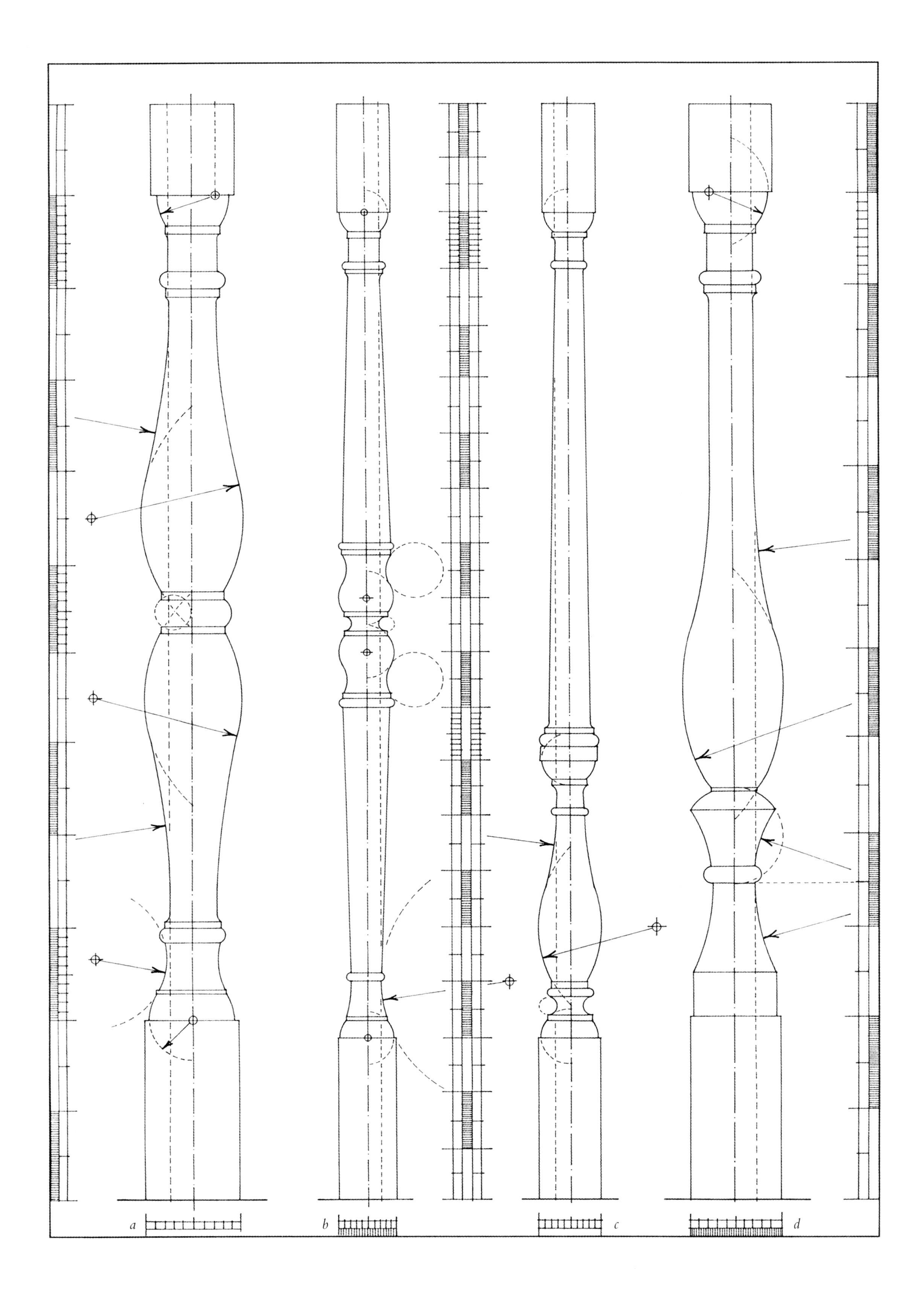
a
b
c
d

17. 天际线

天际线设计

在古代和近几个世纪，古典建筑在屋顶应用了大量的设计特征，突出整个建筑物或其中的某些部分。

希腊神庙的屋顶带有向外凸出的固定特征。屋脊的瓦片包含陶土脊饰，用彩色叶片设计进行装饰。平台又称作饰座，位于屋脊和屋檐的山花上，负载雕塑装饰物，包括大型神像或神话怪兽以及颜色亮丽的2米宽圆盘或青铜三脚架。在对约建于公元前460年的希腊奥林匹亚宙斯神殿（*a*）进行重建时，充分体现出这一效果。

罗马神庙的设计受到其祖先伊特鲁里亚人在装饰屋顶时的影响，从沿着山花成排的角状物到沿着屋脊成群的大型陶土塑像。后期的罗马神庙在希腊神庙的框架中延续了此传统，如141年古罗马广场中的安莱尼和福斯蒂娜神庙（*b*）。

后来的文艺复兴建筑师显然受到古代遗迹、钱币和塑像中此类特征的影响，如雅各布·德安东尼奥·桑索维诺（1486—1570年）1536年在威尼斯设计的著名的圣马可图书馆（*c*），在低矮挡墙上放置了方尖塔和塑像。

然而，巴洛克建筑才最终充分利用了天际线。建筑师常向上拉伸穹顶以显示垂直特征，如博罗米尼1642年在罗马的圣伊芙教堂（*d*）中，在建筑物顶端覆盖具有象征意义的螺旋藤蔓和花冠，并在递降层面布置有火焰的瓮、球和纹章。英国建筑师范布勒1705年在布伦海姆宫中（*e*）使用一系列金球、瓮和塑像，突出如门一般建筑要素的轮廓。1666年伦敦大火烧毁了众多哥特式塔尖，后来用一系列显著的古典塔尖来替代，如雷恩设计的圣布莱德教堂（*f*）创造性地使用古典细部设计产生富有影响力的垂直地标。

巴洛克传统由建筑师一路传承下去，查尔斯·卡尼尔（1825—1898年）便是其中之一，他1861年在巴黎歌剧院（*g*）的设计中，用高耸的塑像群装饰各端。诸如此类的建筑物影响了20世纪巴洛克风格的复兴。

17. THE SKYLINE

DESIGN FOR THE SKYLINE

In antiquity and in more recent centuries Classical architects have used a number of features at roof level to accentuate buildings or parts of buildings.

The Greek temple roof had certain specific projecting features. The tiles along the ridge of the roof could have clay crests decorated with coloured leaf designs. On the pediment at the ridge and eaves, platforms were created, known as acroteria, that carried sculptural ornaments ranging from large statues of gods or mythological creatures to brightly colored two-meter-wide discs or bronze tripods. The effect can be seen on a reconstruction of the Temple of Zeus at Olympia in Greece (*a*) of about 460 BC.

The design of Roman temples was influenced by the roof decorations of their ancestors, the Etruscans, which ranged from rows of horns along the pediment to groups of large clay figures along the ridge. Later Roman temples, such as the Temple of Antoninus and Faustina in the Forum in Rome (*b*) of AD 141, continued this tradition within the framework of the Greek temple.

Later Renaissance architects, like Jacopo d'Antonio Sansovino (1486-1570) who placed obelisks and figures on the parapet of his prominent Library of St Mark in Venice (*c*) in 1536, were perhaps influenced by evidence of these features on ruins and coins and sculpture from antiquity.

It was, however, Baroque architects who exploited the skyline to the full. The dome, always a vertical feature, was stretched upwards by architects such as Borromini who, in the church of S. Ivo in Rome (*d*) of 1642, surmounted the building with a symbolic spiral ramp and open crown and placed flaming urns, balls, and heraldic devices at descending levels. The English architect Vanbrugh used a series of golden balls, urns, and figures to punctuate the profile of important elements, such as gates, on Blenheim Palace (*e*) in 1705. In London the loss of a series of Gothic spires in the Great Fire of 1666 brought about their replacement with a remarkable series of Classical spires, such as St Bride's (*f*), designed by Wren, which made an inventive use of Classical details to create powerful vertical landmarks.

The Baroque tradition was continued by architects like Charles Garnier (1825-98), who decorated each end of his Opéra in Paris (*g*) in 1861 with tall sculptural groups. Buildings such as these influenced the twentieth-century Baroque revival.

a
b
c
d
e
f
g

球、瓮和松球

在建筑屋顶使用尖顶饰而非塑像作为特征的设计传统可以追溯到古代，能够从现存的建筑物中找到佐证。但是由于此类装饰物非常易碎而仅有极少数量保留了下来，主要通过绘画和其他表现形式来证明。文艺复兴早期延续了拜占庭和哥特式风格的做法，将球和相似的细部设计应用到斜屋顶和穹顶的顶端形成底座，在上方安置十字架表示上帝对世界的统治。文艺复兴后期，球和瓮成为建筑屋顶和低矮挡墙中常见的附加装饰，并且无论规模大小都成为古典装饰最熟悉的要素之一。

瓮从古代大量保存下来，并受到文艺复兴风格的追捧而被收集起来。示例（*a*～*i*）取自罗马教皇的收藏。这些容器的质量反映出它们在古代的地位。古希腊艺术留存下来的最精美的作品包括陶制的瓶和瓮，这些物品的大量生产不仅满足国内市场，同时也用于出口。它们用来盛放死者的骨灰，之后葬在地下。此外，由于价格低廉且原料耐用，再加上其他一些原因而得以大量保留下来。金属容器地位更高，通常被熔为金银块。示例（*a*～*i*）都由精致的理石雕刻，确保其保留下来，但其形状、细部设计和装饰如陶制瓶一样通常源自金属。

这些形状被用在后来的古典主义建筑中，有时含有特殊的意义。瓮表示死亡，尤其当部分被遮盖时这种含义更加明显。顶端带有火焰的瓮（*j*）曾经代表上帝对人类的爱，但是随着时间的流逝，火焰仅充当标准的装饰。（*a*）、（*c*）、（*f*）和（*h*）形状的瓮常见于新古典主义建筑，而（*j*）和（*k*）属于巴洛克样式。最后的示例（*l*）样式奇特，中心包括粗琢砌块，与下方的粗琢建筑相关联。此细部设计常常出现在球形装饰中。

松球塞（*k*）和（*l*）可以支配整个瓶（*p*）。独立使用松球作为尖顶饰（*m*）属于常见的古典装饰特征，源自著名的大型青铜球，该球中世纪发掘于罗马万神殿，被误认为是其尖顶饰。

球形是最为常见的尖顶饰，一般坐落在带有凹弧边饰线脚的底座上（*o*）。这一简单的特征经历了大规模的改造，常添加粗琢砌块（*n*），或将球压扁至椭圆形，或支撑在叶片形状之上（*q*），又或产生类似老式榴弹所发出的火焰（*s*）。球形和涡卷形进行不同形式的组合，示例（*t*）出自18世纪早期，（*r*）出自20世纪早期。

BALLS, URNS, AND PINE-CONES

The use of finials rather than sculptures as features on the roofs of buildings can be traced to antiquity through a few surviving buildings but, as such fragile ornaments rarely survive, principally through paintings and other representations. In the early Renaissance, balls and similar details were applied to the apex of pitched roofs and domes following Byzantine and Gothic practice, often as a base for a cross signifying God's rule over the world. In the late Renaissance, balls and urns became common additions to the roofs and parapets of buildings and have become, at both a large and small scale, one of the most familiar elements of Classical decoration.

A large number of urns survived from antiquity and were admired and collected in the Renaissance. Examples (*a*) to (*i*) are from the Pope's collection in Rome. The quality of these vessels reflects their prestige in antiquity. Some of the finest surviving works of early Greek art are the pottery vases and urns that were mass produced not only for the domestic market but for export. Their use as containers for the ashes of the dead and consequent burial, together with the low value and durability of the raw material, are the principal reasons for their survival in large numbers. The higher status metal vessels have, however, more often been melted down as bullion. Examples (*a*) to (*i*) are all carved out of fine marble, which has ensured their survival, but the shape, detail, and decoration often derives, as with pottery vases, from metal originals.

All of these shapes have been used on later Classical buildings and at times have specific meaning. An urn, particularly if partially covered, can signify death. An urn with flames at the top (*j*) once represented charity, but in time the flames came to be no more than a standard decorative stopper. Urns the shape of (*a*), (*c*), (*f*), and (*h*) are often found on Neo-Classical buildings, while (*j*) and (*k*) are Baroque forms. The last example (*l*) is unusual and includes a block of rustication in the centre to relate to a rusticated building below. This detail is more common on balls.

The pine-cone stopper of (*k*) and (*l*) can grow to dominate the vase (*p*). The use of pine-cones on their own as finials (*m*) is a frequent Classical decorative feature and derives from a famous large bronze pine-cone unearthed near the Pantheon in Rome in the Middle Ages and erroneously thought to be its finial.

The ball is the most common decorative finial, usually sitting on a base with a scotia moulding (*o*). This simple feature is subject to a surprising number of variations. Rusticated blocks are often added (*n*) the ball can be flattened to an oval, be supported on leaves (*q*), or issue forth flames (*s*) like an old-fashioned grenade. Different combinations of balls and scrolls can be made: example (*t*) is from the early eighteenth century and example (*r*) from the early twentieth century.

a
b
c
d
c
f
g
h
i
j
k
l
m
n
o
p
q
r
s
t

尖顶饰、方尖塔和战利品装饰

特殊物品经设计可以作为建筑物屋顶和低矮挡墙的装饰特征。这些尖顶饰有时源自瓶或蜡烛台，有时属于原创设计。古代最著名的例子位于雅典李西克拉特唱诗班纪念碑的屋顶上（*a*），可追溯至公元前4世纪后期，用精致的组合装饰作为纪念碑中主要展示品——青铜三脚架的底座。布拉曼特16世纪早期在罗马欧雷奥设计的圣乔瓦尼礼拜堂（*b*）中包含同样复杂的叶形和球形设计。同时期还有两处蜡烛台尖顶饰，其一出自罗马的一处墓地（*c*），由安德里亚·桑索维诺（约1467—1529年）设计，另一出自西班牙的圣地亚哥德孔波斯特拉（*h*）。简朴的支柱和球（*k*）与栏杆和球（*l*）相似，但由于简洁而更加流行。这些特征中许多都与瓮有着含糊的关系。布鲁内列斯基在佛罗伦萨的原创性穹顶（*d*）属于最早期的设计，带有青铜叶片的巴洛克尖顶装饰（*g*）可以采用瓮的形式。1550年帕拉迪奥在维琴察基耶里凯蒂宫中（*e*）结合了瓮、蜡烛台和球形，该设计经历了一再地重复和多次的修改（*f*）。最杰出的设计出自1520年西班牙的莱昂天主教堂（*j*），其中的4个瓮形成上升序列。多尖顶饰包含代表火焰的雕塑，但灯台自身也能够成为装饰特征，如1764年苏格兰建筑师罗伯特·亚当在赛昂别墅中的设计（*i*）。

公元前1世纪埃及受罗马控制时期，方尖塔（*x*）传入罗马，这种古代针形宗教纪念碑上刻有纪念性的碑文。由于人们开始供奉埃及的神，城市中耸立起这些巨大的柱子，也引入了其他稍小一些的方尖塔来装饰神庙。方尖塔和其他埃及形式进入了古典语汇。文艺复兴时期修补并重现了倒塌的遗迹。自此，改造后各种各样的方尖塔成为许多古典建筑中的装饰性特征。增加了球和类似日晷仪中指针即圭表的尖头（*n*和*o*），且顶部可产生火焰（*p*和*q*）。通常设置在基座上，如（*n*）、（*o*）、（*r*）、（*s*）和（*t*），有时也在瓮上（*q*）。16世纪有一段时间，用球或小圆底座支撑，如（*o*）、（*p*）和（*s*），并在后来发展成曲线（*m*和*s*）。

古希腊将战败敌人的盔甲放在树干上（*v*）。罗马人还将象征征服的战利品用作庆祝胜利的一部分，把纪念品的范围扩大到旗帜、鼓和其他俘获的设备，包括战俘雕像。这些塑像群继续代表军事力量，还包括大炮或与战争相关的其他武器（*u*和*w*）。

FINIALS, OBELISKS, AND TROPHIES

Special objects can be designed to act as features on the roofs and parapets of buildings. These finials at times derive from vases or candlesticks and sometimes are original creations. The most famous example from antiquity is on the roof of the Choragic Monument of Lysicrates in Athens (*a*) dating from the late fourth century BC. This elaborate composition acted as the base for the bronze tripod that the monument was designed to display. The early-sixteenth-century chapel of S. Giovanni in Oleo in Rome (*b*), attributed to Bramante, has a similarly complex leaf and ball design. From the same period are two candlestick finials, one from a tomb in Rome (*c*) by Andrea Sansovino (*c.*1467-1529), and another from Santiago de Compostela in Spain (*h*). The plain column and ball (*k*) and baluster and ball (*l*) are similar, but their simplicity has made them more popular. Many of these features have an ambiguous relationship with urns. The earliest is from Brunelleschi's dome in Florence (*d*) and is an original invention, while a Baroque finials with bronze representations of leaves (*g*) could be an urn. A combination of urn, candlestick, and ball on Palladio's Palazzo Chiericati in Vicenza (*e*) of 1550 has been repeated and modified many times (*f*). Most extraordinary is an ascending series of four urns from León Cathedral in Spain (*j*) of 1520. While many finials have sculpted representations of flames, lamps can themselves act as features, such as example (*i*) by the Scottish architect Robert Adam at Syon House in 1764.

While Egypt fell under Roman control in the first century BC, obelisks (*x*), ancient needle-shaped religious monuments carved with commemorative inscriptions, were brought to Rome. These huge pillars were erected in the city and other smaller obelisks were imported to decorate temples as the worship of Egyptian gods became popular. The obelisk, with various other Egyptian forms, entered into the Classical vocabulary. In the Renaissance the fallen remains were repaired and erected again. Since then, the obelisk has, in a surprising number of modified shapes, become a decorative feature on many Classical buildings. A ball and spike like the needle, or gnomon, of a sundial is often added, (*n*) and (*o*), and flames can issue from the top, (*p*) and (*q*). The are often raised on pedestals, (*n*), (*o*), (*r*), (*s*), and (*t*), and occasionally on urns (*q*). At some time in the sixteenth century they were raised on ball or bun feet, (*o*), (*p*), and (*s*), and later developed curves, (*m*) and (*s*).

The ancient Greeks placed the armour of their fallen and defeated enemies on tree trunks (*v*). This symbol of conquest, or trophy, was adopted by the Romans as a part of their elaborate celebrations of victory and enlarged to incorporate flags, drums, and other captured apparatus of war including, in sculpted form, prisoners of war. These sculptural groups have continued to be used to represent military prowess, with the addition of cannon or weapons relevant to the warfare of the period, (*u*) and (*w*).

a
b
c
d
e
f
g
h
i
j
k
l
m
n
o
p
q
r
s
t
u
v
w
x

18. 自然世界

茛苕

古典装饰频繁使用自然形式，其中最常用到的当属茛苕植物。茛苕是一种土生的地中海植物，叶片很大且边缘不规则。传统的基本装饰样式取自两种茛苕植物，虾膜花和刺老鼠簕。虾膜花（*a*、*c*和*d*）与刺老鼠簕（*b*和*e*）相比，叶片更圆。

茛苕装饰最常见于科林斯和混合柱式的柱头。柱头的叶片形式各异，虾膜花（*c*）和刺老鼠簕（*e*）都应用在其中。在漫长的装饰历史中，茛苕叶被定型、简化并用多种不同的方式进行改造。中世纪继续应用这一装饰，或许连雕刻师自己对该形状的由来和植物本身都不清楚。至于为什么在古希腊时期接受这一特殊的植物，原因已经无从考据。古代世界中的大多数植物装饰都源于对神或其神秘力量的信仰，因此茛苕也不太可能产生于单纯的装饰目的。这种植物的活力和在古代葬礼仪式中的使用表达了再生或驱逐恶灵的保护性意义。

叶片和花朵已经成为大量不同装饰类型的基础，在众多样式中已经定型，致使其几乎完全模糊了与原始植物的联系。

然而，叶片或单独植物的镶板也可以进行写实设计。示例（*j*）和（*l*）属于罗马风格，而18世纪的设计最贴近生活（*k*）。利用叶子及其有凹槽纹的茎制造出连续的螺旋形叶片装饰，并且茎梗通常转变花或抽芽叶片的方向，这样产生的设计往往与自然界中的植物迥异。此模式从壁柱的柱础中垂直抽芽来装饰正面，或在水平方向无限延伸，既无起点也无终点，再或者形成独立的对称镶板。在出自罗马拱门内侧的大拱圈中（*n*），茛苕植物呈连续的螺旋形蜿蜒前进，占据了空隙。经证实这些设计既有吸引力又十分实用，在很多时期一直重复出现并历经了改造。示例（*f*）、（*i*）和（*m*）出自罗马时期，（*h*）属于15世纪，（*g*）和（*o*）出自18世纪。

18. THE NATURAL WORLD

THE ACANTHUS

Natural forms occur frequently in Classical decoration and the acanthus plant appears more frequently than any other. The acanthus is a native Mediterranean plant and has a large leaf with a broken edge. Two species have traditionally formed the basis of decorative patterns, Acanthus mollis and Acanthus spinosa. Acanthus mollis, (*a*), (*c*), and (*d*), has a more rounded leaf than Acanthus spinosa, (*b*) and (*e*).

The most familiar application of acanthus decoration is the Corinthian and Composite column capitals. The leaf form can vary considerably on these capitals and both the mollis (*c*) and spinosa (*e*) have been used. Throughout a long decorative history the acanthus leaf has been formalised, simplified, and modified in many different ways. Its use continued throughout the Middle Ages and it was probably often carved by sculptors who knew neither the derivation of the form nor the plant itself. The reason for the original adoption of this particular plant form in Greek antiquity is unknown. The decorative use of most plants in the ancient world derived from their association with a god or belief in their magical powers and for this reason the acanthus is unlikely to have been introduced for solely decorative purposes. The vigour of the plant and its use in ancient funeral rites might suggest an association with rebirth or protection against evil spirits.

The leaf and the flower have formed the basis for a large number of different types of decoration and on many patterns have become so formalised that their association with the original plant is almost totally obscured.

Nevertheless, panels of leaves or individual plants can be relatively realistic. Examples (*j*) and (*l*) are Roman, while the most true to life (*k*) is from the eighteenth century. However, the use of the leaf and its fluted stems, or caulicoli, to create a running decoration of spiraling leaves and stalks, often turning around flowers or sprouting leaves, tends to produce a design that bears little resemblance to the plant found in nature. This pattern can sprout vertically from the base of a pilaster to decorate its face, continue indefinitely horizontally, without a point of origin or completion, or form a self-contained symmetrical panel. A large Roman tympanum (*n*), from the inside of an arch, has the plant meandering in a series of spirals to fill the space. These designs have proved both attractive and useful and have been reported and modified in most periods. Examples (*f*), (*i*), and (*m*) are Roman, (*h*) is fifteenth century and (*g*) and (*o*) are from the eighteenth century.

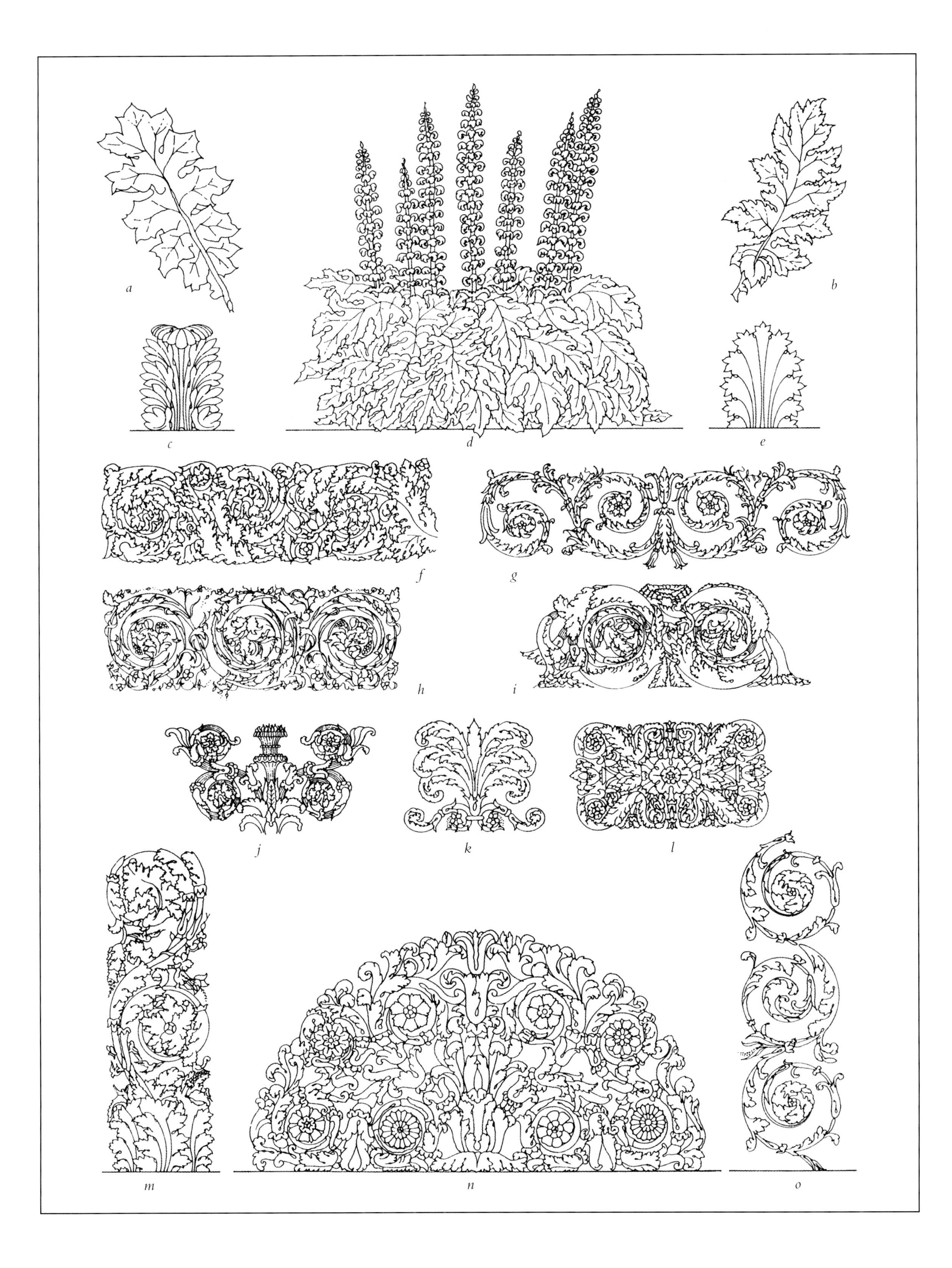
a
b
c
d
e
f
g
h
i
j
k
l
m
n
o

植物装饰

古典装饰中使用的植物种类没有限制。植物设计可以单纯出于装饰目的，也可以与物体或建筑物的用途、所有权或位置相关联。这种象征性的联系赋予装饰额外的深意和兴趣。特定的植物与神之间的关系以及植物的神秘特征都属于古代原始的精神能力的一部分。在中世纪和文艺复兴时期，这种异教徒万物有灵论转变为圣徒和植物的象征性联系以及文章学传统。此类起源下的某些植物设计已经成为传统的古典装饰。

月桂树（*d*）和橄榄树（*b*）由于叶片形状相似，并且果实都是小圆形，因而在正式设计中很难区分。橄榄树的叶子更小，在枝干上分布得也更加稀疏。月桂树在希腊神话中又称作达佛涅，因森林女神达佛涅得名，太阳神阿波罗追求达佛涅，她最终变成了月桂树以躲避阿波罗。灌木与阿波罗和缪斯女神相关联，暗示艺术能力或荣誉，尤其是在做成冠冕或花环的时候更是如此，它作为常绿植物象征永恒。橄榄花环是古代奥林匹克运动会的最高荣誉，也是女神密涅瓦的特征，和密涅瓦一样象征智慧，在异教和基督教的文化中都象征和平。

常春藤（*a*）和（*f*），与葡萄（*g*）、（*h*）和（*i*）都同生育之神和酒神巴克斯相关。常春藤作为常绿植物也代表永恒，其依附生长的特性象征忠贞。葡萄从古代（*g*）开始，经历中世纪（*h*）直到文艺复兴时期（*i*），一直长兴不衰，部分原因在于后期葡萄与救世主之血的寓言故事将其与耶稣和圣餐联系到一起。此外，葡萄也象征秋天。

橡树（*e*）象征主神宙斯，又称朱庇特，橡树叶用来制作胜利者的花环。石榴果实（*c*）起先象征普罗塞尔皮娜，她每年春天都从冥界返回人间，其多籽的特征也象征繁殖。在基督教的象征意义中，这种关联延伸至复活，但是带有坚硬外壳的种子成为贞洁和圣母玛利亚的象征，或者是教会。

一些植物设计的装饰功能胜过象征意义。一列列的水果捆在一起，如（*k*）和（*l*），通常表示富饶，不同种类的水果混合在一起代表不同的季节。18世纪末，荚或风铃草（*j*）装饰十分流行。在垂直镶板中混合不同种类的植物，如（*m*）和（*n*），有时也用瓶或动物装饰，这种做法源于1世纪的罗马设计，并在文艺复兴时期重新出现，常以浅浮雕的形式装饰正面壁柱。

PLANT DECORATION

There is no limit to the types of plant that can be used in Classical decoration. Plant designs can be used for solely decorative purposes or can have some association with the use, ownership, or location of an object or building. These symbolic connections give an added depth of meaning and interest to decoration. The relationship between specific plants and gods and the magical qualities of plants were part of the primitive mentality of antiquity. In the Middle Ages and the Renaissance this pagan animism was translated into a symbolic association of saints and plants and the conventions of heraldry. From these origins certain plant designs have become traditional Classical decorations.

The laurel (*d*) and the olive (*b*) are often indistinguishable in formal design due to the similarity of the leaf shape and their small round fruit. The leaves on the olive tend to be smaller and more sparsely distributed along the branch. The laurel, or bay, is called *daphne* in Greek and is associated with the nymph Daphne who changed into a laurel when pursued by the god Apollo. The bush is associated with Apollo and the Muses and can imply artistic prowess or honour of any kind, particularly when made into a crown or wreath. As an evergreen it can symbolise eternity. The olive wreath was the highest reward at the ancient Olympic games. It is also an attribute of the goddess Minerva and, as such, symbolises wisdom. In both pagan and Christian art, it symbolizes peace.

Both the ivy, (*a*) and (*f*), and the vine, (*g*), (*h*), and (*i*), are associated with Bacchus, the god of fertility and wine. As an evergreen, the ivy can also represent eternity and its clinging growth symbolises fidelity. The persistent popularity of the vine from antiquity (*g*), through the Middle Ages (*h*) to the Renaissance (*i*) is in part due to the later association of the plant with Christ and the Eucharist through the parable of the vine and the blood of the Saviour. The vine also symbolises autumn.

The oak (*e*) is the tree of Zeus, or Jupiter, the thunderer, rules of the gods, and its leaves are used as a victor's wreath. The fruit of the pomegranate (*c*) was originally the attribute of Proserpina, who annually returned from the underworld to earth in the spring, and the multiple seeds represented fertility. In Christian symbolism this association was transferred to resurrection, but the seeds enclosed in a hard skin became symbols of chastity and the Virgin or, as many contained by one, the Church.

Some plant designs are more decorative than symbolic. Lines of fruits tied together, (*k*) and (*l*), generally indicate abundance, and different mixtures of fruits can represent different seasons. Husks, or bellflowers (*j*), were popular in late-eighteenth-century decoration. Mixtures of plants in vertical panels, (*m*) and (*n*), sometimes with vases or animals, originate from first-century Roman designs and were revived in the Renaissance. In sculptured relief they often decorate the faces of pilasters.

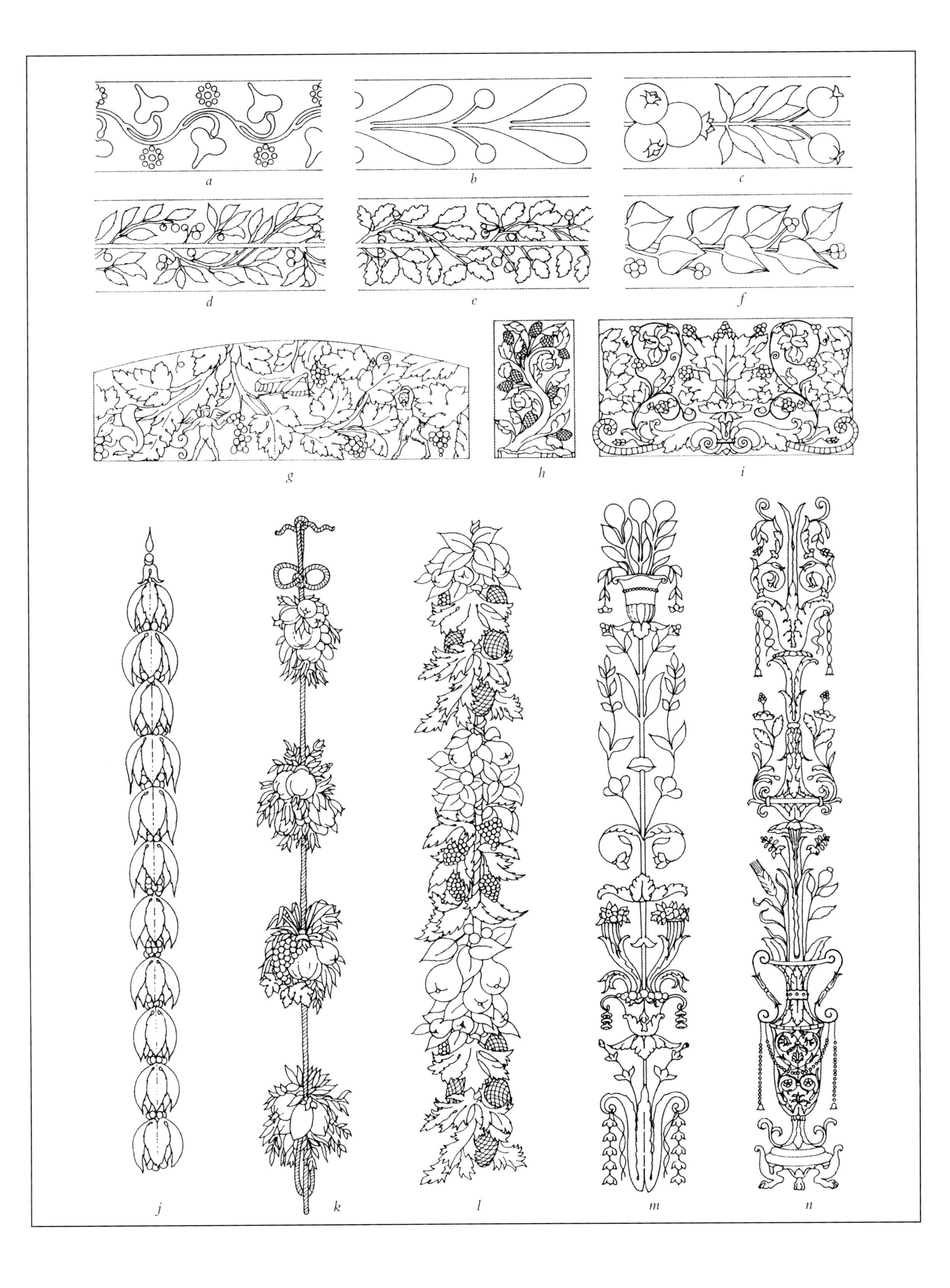
a
b
c
d
e
f
g
h
i
j
k
l
m
n

花冠与花环

花环和花冠自从在古代出现并与特殊的庆祝活动相联系之后，便成为古典建筑中经久不衰的装饰形式。尽管其起源几乎被遗忘殆尽，但是设计却保留了原始的外形和大部分含义。

由叶片制作的冠冕即花环在古代是荣誉的标志，用月桂树（*c*）、橡树（*d*）、橄榄树或其他一些合适的植物做成，并用丝带捆绑在一起。在古希腊特尔斐城举行的皮托竞技会上，胜利者被授予象征阿波罗的月桂冠冕，而奥林匹克运动会的胜者被授予橄榄冠冕。从凯旋门上带有双翼的胜利女神像手中的冠冕来看，取胜的罗马将军和皇帝得到月桂或橡树冠冕。

花环也敬献给死者，或许象征死后灵魂的胜利或者为了纪念死者在人世间的成就。古代装饰中的花环通常代表上述功用。花环装饰甚至出现在檐部的垫块即鼓突雕带中，当柱础上的圆盘线脚雕刻了用带子捆绑起来的橡树或月桂时，如（*a*）和（*b*），支柱下方也有花环作支撑。将花环用作冠冕的做法已经成为历史，但是花环与胜利、英勇和死亡的象征关系却保留了下来。尽管常绿的月桂仍然用作葬礼花环，却失去了大部分具体的象征意义，花环常常用不同植物实现简单的装饰，例如时尚的18世纪果壳装饰（*e*）。

花冠也称作垂花雕饰或垂挂装饰，集合了出自单一植物的植株、果实或叶片，用带子捆绑在一起悬挂在两侧支撑之间中心下垂。尽管与花环的布局相似，而且叶子的象征意义也与之相同，但是两者的起源不同。花冠装饰神庙，用在庆典和祭祀的场合，和祭祀工具一同安放在神庙的雕带上，悬挂在祭祀用的动物颅骨之间（*f*）或者由大型蜡烛架或小天使像支撑。在罗马圣坛（*g*）中，叶子和动物带有象征意义，更加凸显其宗教意义。代表朱庇特的橡树和天鹅组合在一起，使人联想起“朱庇特变身天鹅，强奸了勒达并使其生下两位重要的罗马战神卡斯托和普鲁克斯”的故事。在文艺复兴时期，花冠成为单纯的装饰物品，不再带有异教祭祀的象征意义。面具（*i*）、小天使像、盾形纹章、狮子头像和其他设计取代了颅骨和蜡烛架。花冠仍然是一种流行的装饰形式，文艺复兴时期之后也对设计进行了修改。示例（*h*）、（*i*）和（*j*）出自16世纪，示例（*k*）、（*l*）和（*m*）出自18世纪。

GARLANDS AND WREATHS

The wreath and the garland have persisted as decorative forms in Classical architecture since their introduction in antiquity in association with specific celebrations. While these origins have almost been forgotten, the designs retain their original appearance and much of their meaning.

The crown of leaves, or wreath, was an honorary insignia in antiquity and was made out of branches of laurel (*c*), oak (*d*), olive or some other appropriate plant and bound together with a ribbon. A laurel crown was given to the victor at the Pythian games held at Delphi in ancient Greece in honour of Apollo, while an olive crown was awarded to the victor at the Olympic Games. Victorious Roman generals and emperors received crowns of laurel or oak, and these are depicted in the hands of figures of winged Victory on triumphal arches.

Wreaths were also presented to the dead, possibly to symbolise the victory of the afterlife or to honour the earthly achievements of the deceased. Wreaths representing these functions are often found in ancient decoration. Even a cushioned, or pulvinated, frieze in an entablature could have a wreath decoration, and columns appeared to rest on wreaths when the torus moulding at the base was carved with oak or laurel tied with a ribbon, (*a*) and (*b*). The use of the wreath as a crown has long since passed, but the association with victory, prowess, and death remains. Although the evergreen laurel is still used for funeral wreaths, much of the detailed symbolism of the plant is lost and the wreath is often used as simple decoration with different plants, such as the fashionable eighteenth-century husk (*e*).

The garland, also known as a festoon or swag, is a collection of plants, fruit, or leaves from a single plant, bound together with a ribbon and suspended, drooping at the centre, from two supports. Although the arrangement is similar to the wreath and the symbolism of the leaves can be the same, the origin is different. Garlands were used for temple decoration at festivals and sacrifices and are shown on temple friezes, together with the instruments of sacrifice, suspended between the skulls of sacrificial animals (*f*) or supported on large candlesticks or by cherubs. On a Roman altar (*g*) this religious association is given additional significance with the symbolism of leaves and animals. The combination of the oak of Jupiter and swans perhaps recalls the rape of Leda, by Jupiter in the guise of a swan, which brought about the birth of Caster and Pollux, two important Roman military gods. In the Renaissance the garland became purely decorative and lost its associations with pagan sacrifice. Masks (*i*), cherubs, coats of arms, lion' heads, and other devices replaced the skulls and candlesticks. The garland has remained a popular decorative form and several variations in design have occurred since the Renaissance. Examples (*h*), (*i*), and (*j*) are from the sixteenth century and examples (*k*), (*l*), and (*m*) are from the eighteenth century.

a
b
c
d
e
f
g
h
i
j
k
l
m

叶丛状装饰、圆花饰和扇贝形装饰

诸多传统装饰模式不求精准地复制装饰原型的本质特征，而是不断地调整装饰的外形并使其抽象化，这也成为古典装饰的一个特征。尽管一些定型的设计可以追溯到用抽象的几何形状表达所有设计的时期，但后期当表达形式达到高度复杂的水平时，也展示出同样自由的组合形式。

茛苕装饰出现之前，叶丛状装饰是古典装饰中占主导地位的自然形式。叶丛状装饰属于传统形式，据说起源自忍冬科植物。该设计与棕叶饰的设计不同，后者被认为是棕榈树叶片的简化版式。棕叶饰和叶丛状装饰样式多变，尤其两者真实起源不明，因此没有必要进行区分。

这些装饰主题在古代或许具有一定的重要意义，在设计中添加内部生长出叶子或花瓣形式的螺旋形装饰，使人联想到著名的攀缘植物。叶丛状装饰中叶子是奇数的，向外或向内卷曲直至矛尖形状的中心叶片处。希腊（*a*）和一些罗马（*c*）神庙的屋顶，以及石柱或希腊墓碑（*b*和*d*），单独使用叶丛状装饰在边缘装饰瓦檐，在角落装饰雕像底座。连续的设计通常包括附带的叶片设计，叶子更显狭长，如（*e*）和（*f*），叶丛状装饰有时置于矛形状的框架内（*e*）。组合设计更加复杂（*g*），精致的石柱顶饰可以创造出集中的连续设计（*h*）。尽管这一设计从罗马时期开始逐渐失去了其统治地位，但仍然包括在罗马风格和文艺复兴风格的装饰之中，并且在18世纪后期和19世纪早期大行其道。

圆花饰包括任何环形的花卉或植物设计，或多或少带有抽象植物形状的环形装饰普遍存在于所有文化和各个时期的建筑物中。一些圆花饰清楚地表现出特殊的植物，包括叶片和花朵。此处的希腊（*j*）和罗马（*l*）示例应当是罂粟花，而文艺复兴圆花饰（*o*）很难辨识花卉的种类。普遍的形态是简易的花朵配上多叶片。示例（*i*）属希腊风格，（*m*）是罗马风格，（*n*）是文艺复兴早期，似乎全部属于菊科植物。罗马设计（*k*）出自某多立克神庙雕带上的三槽板间平面，展示橡树叶和橡树果实的辐射状设计，18世纪末的英国设计（*p*）由长叶片或羽毛组成。

扇贝形装饰属于较大的复杂设计，常带有中心圆花饰，得名于仪式中使用的大型罗马圆盘，在18世纪和19世纪早期广泛用于室内装饰。示例（*q*）、（*r*）和（*s*）全部取自此时期，在石膏上制作，设计的主题出自罗马着色石膏和马赛克装饰。

ANTHEMION, ROSETTE, AND PATERA

It is a characteristic of Classical decoration that many of the traditional patterns adapt and abstract the characteristics of growing forms rather than attempting to reproduce nature with precision. Although some of the established designs date from a period when all representation was executed with abstracted geometric shapes, later forms, produced at a time when representation had reached a high level of sophistication, show the same freedom of composition.

Before the introduction of the acanthus, the predominant natural form in Classical decoration was the anthemion. This has a conventionalized form and is said to have derived from the honeysuckle flower. A distinction is sometimes made between this design and that of the palmette, which is considered to be a simplified version of a palm leaf. The variety in palmette and anthemion design is so great that there is no useful distinction to be made, particularly as the true origin of both forms is unknown.

It is likely that these decorative themes had some significance in antiquity; the inclusion of a spiral design, from which the leaf or petal forms grow, suggests a known climbing plant. The anthemion has an odd number of leaves curling outwards or inwards to a spear-shaped central leaf. Anthemia were used singly to decorate antefixa, at the edge, and acroteria, at the corners, of Greek (*a*) and some Roman (*c*) temple roofs and the top of stele, or Greek gravestones, (*b*) and (*d*). Running patterns usually include a subsidiary narrower leaf design, (*e*) and (*f*), and the anthemion is sometimes contained in a spear-shaped frame (*e*). The composition of the pattern can become complex (*g*) and the stele crest can be elaborated to create a centralised running design (*h*). Although the dominance of this design declined from the Roman period, it was included in Romanesque and Renaissance decoration and enjoyed great popularity in the late eighteenth and early nineteenth centuries.

The rosette is any circular floral or plant design. The decoration of circular shapes with more or less abstracted plant forms is universal to almost all cultures and periods. Some rosettes are clearly representations of particular plants and include leaves and flowers. The Greek (*j*) and Roman (*l*) examples here may be poppies, while the Renaissance rosette (*o*) is hard to identify. Simple flowers with multiple leaves are common. Example (*i*) is Greek, (*m*) is Romanesque and (*n*) is early Renaissance; all seem to be some version of the daisy family. A Roman design (*k*) from a metope in the frieze of a Doric temple shows a radiant pattern of oak leaves and acorns while a late-eighteenth-century English pattern (*p*) is made up of long leaves or feathers.

Paterae are larger complex designs usually with a central rosette. The name derives from large ceremonial Roman dishes and the design became popular in the eighteenth and early nineteenth centuries for interiors. Examples (*q*), (*r*), and (*s*) are all from this date and were executed in plaster with themes developed from Roman painted, plaster, and mosaic decoration.

a
b
c
d
e
f
g
h
i
j
k
l
m
n
o
p
q
r
s

动物

古典装饰中出现了众多形态各异的动物。古代设计中的动物所具有的象征意义超出了其装饰作用。在神庙的雕带、圣坛和其他一些与动物祭祀的异教仪式相关联的物品中（*b*），经常添加公牛的颅骨（*a*）和头，以及公羊的头或颅骨（*c*）。动物的头颅一律悬挂在早期神庙的檐部，这种做法被记录在装饰之中。

动物除了体现装饰和建筑用途的关系之外，还有许多也象征单独的神或其本质特征。例如，豹子献祭给酒神，而大象表达胜利或声望。中世纪和文艺复兴时期出版的《动物寓言集》列举了此类的象征关联，并且混合了异教和基督教的观点。海豚（*j*和*n*）由海神尼普顿创造，爱与美之神维纳斯从海中诞生的一刻起便代表了四元素之一的水元素。对早期基督教徒来说，也象征信任，而通过约拿的故事则可以象征复活。同样事物而产生不同含义的相似情形现在也时有发生。鹰（*g*）是朱庇特之鸟，象征圣马可并代表美国。此处的狮子和独角兽（*o*）缠绕在莨苕雕带上，代表了大英帝国。

大多数时期的建筑中都有狮子图案，希腊神庙（*e*）中使用狮子面具作怪兽饰，并实现许多其他的装饰功能。狮子毛皮（*h*）象征希腊英雄赫拉克勒斯，可以充当简朴的装饰镶板，或者作为门或窗之类的建筑特征的缘饰。狮身人面像起源于埃及，但是很快被希腊人接受并改造（*d*），象征令人费解的智慧。狮身鹰首兽（*f*）有鹰的前肢和狮子的后肢，该兽可能出自印度，希腊人相信它守护着印度的金子。这个神话传说赋予它警惕和勇气的象征意义，后来也用来代表基督的二元本质。

将动物的特征和外形结合在一起的方法也摆脱了真实形象的限制，古代神话故事和古典装饰特征都见证了这类自由的设计。单轴设计起源于罗马，一般体现在家具的设计中。动物单腿加上头（*m*），结合处用莨苕叶装饰遮盖，或不同的头和足或人形和动物的足（*i*）合并在一起。在1645年建造的法尔科涅里府邸中，博罗米尼在外立面建造了赫尔墨斯头像柱（*k*）和带有女性半身像和猎鹰头的建筑支柱，用这一设计暗指府邸主人的家族名称（猎鹰与法尔科的英文拼写相同）。还可以将动物的某一部分变成植物。意大利文艺复兴设计（*n*）包括两个海豚的头，使人无法确定设计源自何种植物。同样，17世纪出自巴黎的唱诗班座位（*l*）将原本便已经十分复杂的狮身鹰首兽改为莨苕装饰。

ANIMALS

Animals in many different forms appear in Classical decoration. The animals that are found in the designs of antiquity generally have a significance beyond their decoration function. Ox skulls (*a*) and heads, and the heads or skulls of rams (*c*) are often added to temple friezes, alters, and other objects associated with the pagan rituals of animal sacrifice (*b*). The heads of these animals were undoubtedly hung on the entablatures of early temples and this practice has been recorded in decoration.

Beyond this relationship between the decoration and the use of buildings, many animals were associated with individual gods or had their own attributes. The leopard, for example, is the animal sacred to Bacchus, while the elephant depicts victory or fame. In the Middle Ages and the Renaissance, *bestiaries* were published which listed such symbolic connections and often mixed pagan and Christian ideas. The dolphin, (*j*) and (*n*), is the creature of Neptune and, from her birth from the sea, Venus, and represents water as one of the four elements. It was also a symbol of faith for early Christians, and, through the story of Jonah, a token of the Resurrection. Similar differences of meaning occur today. The eagle (*g*) is the bird of Jupiter, symbolises St Mark and represents the United States of America. The lion and the unicorn (*o*), here entwined in an acanthus frieze, represent Great Britain.

The lion is found in most periods and lion masks were used as gargoyles on Greek temples (*e*) and for many other decorative functions. The lion skin (*h*) is the emblem of the Greek hero Heracles and can be used as a plain decorative panel surround for a feature such as a door or a window. The sphinx was originally Egyptian but was soon adopted and modified by the Greeks (*d*) to represent obscure wisdom. The griffin (*f*) has an eagle's forequarters and the hindquarters of a lion and probably originated in India as the Greeks believed that it guarded the gold of that country. From this myth it came to symbolise watchfulness and courage and later the dual nature of Christ.

This union of the attributes and forms of creatures is one of aspect of a freedom from the constraints of literal representation founded in ancient mythology and characteristic of Classical decoration. The monopodium is of Roman origin and is generally found in the design of furniture. The single leg of an animal is joined with its head (*m*) and the junction concealed with an acanthus leaf, or different heads and feet or human forms and animal feet (*i*) are merged together. On the Palazzo Falconieri in Rome in 1645 Borromini alluded to the name of the family with a herm (*k*) on the façade that brought together an architectural column with a female bust and a falcon's head. Animals can also be partially transformed into plants. An Italian Renaissance design (*n*) reveals just enough of two dolphins' heads to make the plant origin of the design uncertain. Similarly, a seventeenth-century choir-stall from Paris (*l*) turns the already composite form of the griffin into an acanthus.

a
b
c
d
e
f
g
h
i
j
k
l
m
n
o

人形

人物形态普遍存在于古典装饰之中。寓言画通过塑造人物和动物的象征特性而传递含义，其传统手法突出了神和英雄的形象。

女像柱替代了希腊和罗马一些建筑中的支柱。最著名的当属公元前421年的厄里希翁神殿内的人像，人像位于南侧门廊，雕塑的是6个女性人像（*b*）。据说人像代表卡利伊的妇女，卡利伊是希腊的一个社会群落，由于在希波战争时的背叛行为而受到世代为奴的惩罚。阿拉特斯男像柱对应于女像柱，得名于以肩顶天的巨神阿拉特斯。巨大的男性人像被称作男像柱，常见于埃及建筑中，但是最早的希腊男像柱（*d*）出自公元前5世纪西西里的奥林匹亚克斯。男女像柱的不同形态都与流行样式有关。

异教和基督教中都包括小天使，又称作丘比特和精灵。在古代常用这些长着翅膀的婴儿表现顽皮的成人行为，尤其是在酿酒（*c*）和酒神仪式中。在文艺复兴时期，异教精灵成为基督教中的小天使，属于天使的一种，常被塑造成长着翅膀的头像（*a*）。

头像或面具出现在众多不同的情境中。在古代，上帝和神话人像（*e*、*f*和*i*）和戏剧中的面具（*l*）一样普遍。该传统一直持续到文艺复兴时期，有时不特殊所指，如（*g*）和（*h*），但通常直接取自古代示例，如（*j*）和（*k*）。在拱门的拱顶石中也常添加头像或面具，如（*h*）、（*j*）和（*k*）。这一习惯源于古代，目的或许是驱逐恶灵，面具也可以与建筑的用途或其他方面相关联。

赫尔墨斯头像柱或半身像柱向下逐渐变窄，起初用作边界柱或路标柱。赫尔墨斯是守护神并保护旅行者，因此在柱上雕刻其半身像。这些古代像柱因与酒神的生育仪式有关，而常在柱身雕刻阴茎（*m*）。古代用柱状人像作家具的支撑，底部也可以有柱脚（*n*）。罗马绘画中展示了用比例夸张的赫尔墨斯像作建筑的支座，18世纪的装饰模仿了这一设计（*o*）。文艺复兴时期，赫尔墨斯像结合了女像柱，展示大量不同的人类形态（*q*）和（*r*）。其中以矫饰主义建筑师的设计为代表，将柱式的人形比例与赫尔墨斯支柱联系到一起，如（*p*）和（*q*）。15世纪出现的倒置的锥形支柱便源自赫尔墨斯像柱。

人像中可以混合（*w*），或结合（*u*）和（*x*）动植物装饰。这一装饰传统手法源于古代并持续至中世纪。古代形式在文艺复兴时期重新出现，成为怪诞装饰主题的一部分，如（*s*）、（*t*）和（*v*）。

THE HUMAN FORM

The human figure pervades Classical decoration. The images of gods and heroes were supplemented by the convention of allegory, where meaning was conveyed by the symbolic attributes of the figure and the creatures depicted.

Caryatids replaced columns on some Greek and Roman buildings. The most famous are the six female figures (*b*) on the south porch of the Erechtheion erected in Athens in 421 BC. These are said to represent women of Caryae, a Greek community bound to slavery as punishment for a treachery in the Persian Wars. Their male counterparts are atlantes, named after the god Atlas who supported the earth. Giant male figures, called telamones, were common in Egyptian architecture, but the first known Greek atlantes (*d*) are from the fifth-century-BC Olympieum in Sicily. Caryatids and atlantes have taken different forms related to the prevailing style.

The cherub, also called putto, amorino, and genius, is both pagan and Christian. In antiquity these winged infants were frequently shown in playful versions of adult activities, particularly wine-making (*c*) and Bacchic rites. In the Renaissance, pagan genii became Christian cherubs, one of the Orders of angels, and were often depicted as winged heads (*a*).

Heads or masks appear in many different contexts. In antiquity, gods and mythological figures, (*e*), (*f*), and (*i*), were common, as were theatrical masks (*l*). This tradition continues in the Renaissance, sometimes without any specific identity, (*g*) and (*h*), but often derived directly from ancient examples, (*j*) and (*k*). Heads or masks are frequently added to the keystones of arches, (*h*), (*j*), and (*k*). The custom originated in antiquity, probably to keep out evil spirits, and the mask can also be related to the use or some other aspects of a building.

The herm, or terminus, was originally a boundary post or road marker, often tapering downwards, carved with a bust of Hermes in his role as a guardian deity and protector of travellers. These posts often had a phallus (*m*) in antiquity and were associated with Bacchic fertility ceremonies. The columnar figure was adopted as a furniture support in antiquity and could have feet (*n*) at the base. Herms of exaggerated proportions were shown as architectural supports in Roman paintings and copied in eighteenth-century decoration (*o*). In the Renaissance the herm was combined with the caryatid with varying amounts of the human form exposed, (*q*) and (*r*), and the association was made, particularly by Mannerist architects, between the human proportions of the Orders and herms columns, (*p*) and (*q*). The introduction of reverse tapering on columns in the fifteen-century derive from the herm.

Human figures can be mixed (*w*) or physically combined, (*u*) and (*x*), with plants and animals. This decorative convention originated in antiquity and was continued in the Middle Ages. The ancient forms were revived in the Renaissance as part of grotesque decorative schemes, (*s*), (*t*), and (*v*).

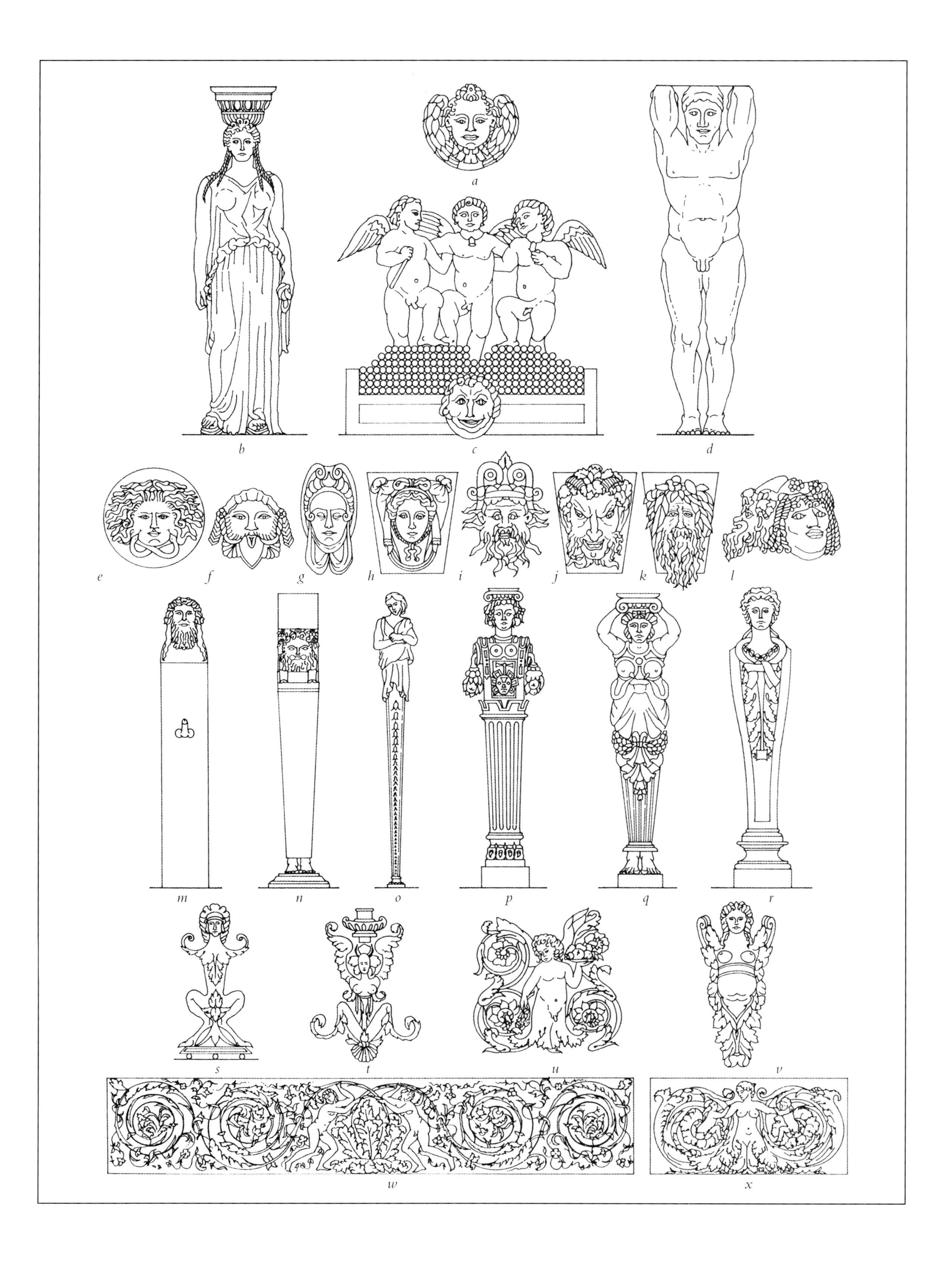
a
b
c
d
e
f
g
h
i
j
k
l
m
n
o
p
q
r
s
t
u
v
w
x

19. 几何样式

钥匙纹和波纹

古典设计中包含大量不同样式的几何装饰。在古代的浅雕刻、镶嵌细工、绘画和装饰石膏工艺中，各种各样的几何组合样式令人眼花缭乱，创新的设计也不输传统样式。某些主题频繁重现，其中一部分样式用于线脚，参见P119、P121、P125和P127。其他一些几何形式则专门与古典设计相联系。

钥匙纹样式也称作希腊键纹、回纹或回纹波形装饰，其中线条在连续的直角处从一侧至另一侧（*a*），产生对称的连锁样式，进而发展为交替的T形结构（*b*），在端点处颠倒（*c*），尺寸减半后沿一个方向继续下去，如（*d*）和（*k*），然后再次转换方向（*e*）。还可以中断线条，插入其他样式（*f*）。可以将设计视为曲折线条、曲径、一系列连锁形状或钥匙纹，沿线条分配图案并且只采用对比颜色进行表达。希腊键纹也出现在其他文化中，考虑到它正方形的形状以及对反差效果的依赖，可以推测其起源于纺织品样式。

这种设计的含义不明，有观点认为其指连锁的指针。该样式进一步发展成万字饰（*g*），尽管其他文化中也包含这一样式，但是希腊装饰中的万字饰代表繁荣和古代风格的复兴，并成为该形式和独立形式中常见的装饰工具。将万字饰设计进行扩展（*h*）或使用它的三维版式（*i*）会产生迷宫的效果，此样式代表克里特和弥诺陶洛斯的传奇故事，象征冥界并且在中世纪指代地狱。古代的其他回纹饰（*j*和*l*）使人联想起迷宫。可以中断简单的曲径或将其改造进行转角，但是万字饰的翻转更为复杂（*m*和*n*）。

波纹样式（*q*）又称作希腊涡纹、波状涡纹、维特鲁威式涡纹和连续涡纹等，成为曲线版的定向钥匙样式（*d*），利用附加工具，如中心圆花饰或叶饰进行精心的设计（*r*）。该设计更多依赖连锁形状之间的对比，但可以展现出弯曲的线条（*s*）。弯角经翻转可以产生连续的涡纹方向（*o*）或颠倒涡纹方向（*p*）。

19. GEOMETRIC PATTERNS

KEYS AND WAVES

There is a great variety of geometric decoration in Classical design. In antiquity, shallow carving, mosaic, painting, and decorative plasterwork included a bewildering variety of geometric compositions, and invention was never far behind tradition. Certain themes recur and some of these are versions of the patterns used on mouldings, shown on pages 199, 121, 125, and 127. Some other geometric forms are specifically associated with Classical design.

The key pattern, also known as the Greek key, fret, or meander, is a series of developments of a line moving from side to side in a succession of right-angles (*a*). This produces a symmetrical interlocking pattern which can be developed as alternating T-shapes (*b*), can then be turned down at the ends (*c*), halved to go in one direction, (*d*) and (*k*), and then turned round again (*e*). The line can be broken and other patterns inserted (*f*). These designs can be seen as a wandering line, or meander, or a series of interlocking shapes, or keys, which can dispense with the line and be expressed only with contrasting colours. The Greek key is found in other cultures and, from its square form and reliance on contrast, probably originated as a textile pattern.

The meaning of this design is unknown, although ideas such as interlocking hands have been suggested. A further development of the pattern, however, creates a series of swastikas (*g*) which, although also shared by other cultures, signified prosperity and revival in Classical antiquity and was a common decorative device both in this form and in isolation. An expansion (*h*) or three-dimensional version (*i*) of the swastika design creates the impression of a maze or labyrinth, a pattern that represented Crete and the legend of the Minotaur, symbolising the underworld and, in the Middle-Ages, hell. Other fret designs from antiquity, (*j*) and (*l*), suggest a labyrinth. The simple meander can be broken or modified to turn a corner, but turning a swastika pattern can be more complicated, (*m*) and (*n*).

The wave pattern (*q*), also known as the Greek scroll, wave scroll, Vitruvian scroll, running dog, and running scroll, is a curved version of the directional key pattern (*d*) and is sometimes elaborated with additional devices (*r*), central rosettes or foliage. This design more frequently relies on the contrast between interlocking shapes but can be known as a meandering line (*s*). A corner can be turned to give a continuous direction (*o*) or to reverse the direction of the scroll (*p*).

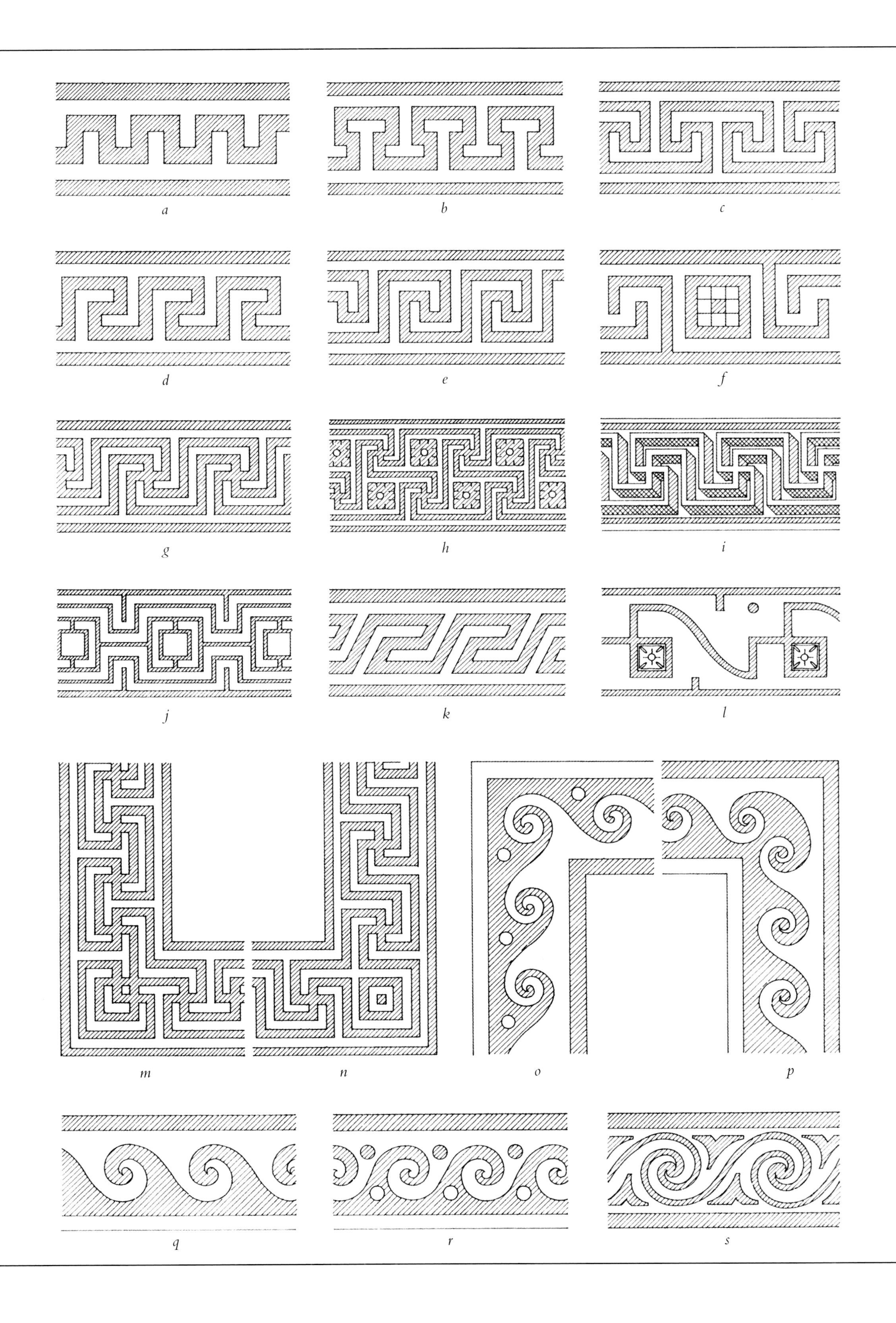

a
b
c
d
e
f
g
h
i
j
k
l
m
n
o
p
q
r
s

扭索饰、圆盘饰和对比特征

扭索饰又称交错样式，自古代出现以来经久不衰，是在古典装饰，尤其是罗马室内装饰中使用得最为广泛的一种交叉曲线设计和交错组合形式。交叉曲线样式被众多不同文化所接纳，并由北欧部落和伊斯兰艺术发展出高度复杂的形式，两者的灵感或许都得自罗马艺术。北欧部落传统和古典传统在罗马风格中融为一体。

最简单的扭索饰表现出连续的设计，其中两条缠绕在一起的带子或松（*b*）或紧（*a*）地将一排圆形转换到相反的方向。带子通常在两条平带的边界处向外或向内凸出，可以用绳子样式装饰或在顶部和底部翻转过来模仿平带条。有时将带子在一处系紧，以此减小中间插入的圆形的尺寸，将圆盘压至菱形可以产生不一样的效果。从而产生的中心空间常常形成凸面或者用圆花饰填充。交替布置中心圆形的尺寸或交错两个连续的扭索饰（*c*），可以使简单的连续边界设计更加精美。有两种方式可以用来包围两行圆盘以扩展该样式：一是盘绕和扭曲带状装饰使其变宽（*d*）；二是用交叉处的装饰细部设计创造出一系列的环形（*e*）。重复以上步骤可以进一步扩大到三行圆盘（*f*），甚至涵盖任何宽度。两行和三行圆盘可以丰富柱础上的圆盘线脚装饰，简易的拱墩有时带有扭索饰，这是唯一一种将该样式直接应用于柱式的传统方法。该设计常用来强化框架，诸如拱顶花格镶板的框架，或用作装饰带或束带层。

圆形装饰既属于几何形状又是自然形式。重叠的圆盘或隅石线脚（*h*）的表面可仿造长赌桌。相同的重叠样式形成一系列部分显露在外的圆花饰的基础（*i*）。许多圆花饰设计只是一般性地模仿真花，其几何结构与自然灵感一样显著，如（*g*）和（*j*）。

从古代开始，马赛克耐久的特征便保存了表面装饰中许多精致的设计。仅存的几处墙壁绘画装饰上的样式大多相似。这些发现揭示出人们十分迷恋几何组合样式和三维幻象，也因而制造出如此多样的设计（*k*～*p*）。当把如此装饰潜能与文艺复兴之后发展出的更加多样的样式放在一起时，很难限定哪些样式和特征可以称作古典主义，这种情形在古典设计的装饰细部设计之中十分常见。

GUILLOCHES, DISCS, AND CONTRASTS

The guilloche, or interlacing pattern, is a decorative device that has retained its popularity since antiquity. It is the most widely used version of a series of knot designs and interlacing compositions found in Classical decoration, particularly in Roman interiors. Knot patterns have been adopted by many different cultures and were taken to high levels of complexity by north European tribes and Islamic art. It is possible that both of these artistic developments had some stimulation from Roman art. The north European tribal tradition and the Classical tradition came together in the Romanesque.

The most simple form of guilloche is a running design formed of two intertwined bands which either turn tightly (*a*) or loosely (*b*) around a row of circles. These bands are often modelled to bulge outwards or inwards within a border of two flat strips and can be decorated with a rope pattern or turned over at the top and bottom to resemble a flat ribbon. The bands are sometimes brought tightly together to reduce the size of the intervening circles and a varied effect is created by flattening the discs to a lozenge shape. The resulting central space is often convex or filled with a rosette. The simple running border can be elaborated by alternating the size of the central circles or interlacing two running guilloches (*c*). The pattern is expanded by encompassing two rows of discs, either by winding and twisting the bands to a greater width (*d*) or by creating a series of loops (*e*), usually with a decorative detail at their intersection. A further extension to three rows of discs (*f*) could be repeated to occupy any width. Double and triple rows are used to enrich the torus moulding on column bases and simple arch imposts sometimes have guilloche decoration, but these are the only traditional direct application of this pattern to the Orders. The design is generally applied as an enrichment to the frame around features such as coffering in vaults or as a decorative band or string course.

Circular decorative shapes can be both geometrical and natural. Overlapping discs, or corn moulding (*h*), can have a modelled face that resembles a gaming counter. The same overlapping pattern forms the base for a series of partially revealed rosettes (*i*). Many rosette designs bear only a general resemblance to living flowers and the geometric construction of the form is as evident as its natural inspiration, (*g*) and (*j*).

The durable nature of mosaics has preserved from antiquity many elaborate schemes for surface decoration. Similar patterns often appear on the few surviving painted wall decorations. These finds reveal the simple and universal fascination with geometric composition and three-dimensional illusion that has created such a variety of designs. A few of these are illustrated, (*k*) to (*p*). When this range of decorative potential is seen alongside the even greater variety of patterns that have been developed since the Renaissance it becomes hard, as with much decorative detail in Classical design, to limit the patterns and features that might be called Classical.

a
b
c
d
e
f
g
h
i
j
k
l
m
n
o
p

20. 古典设计

对称与轴线

优秀的设计没有定式，但在古典传统内确实存在建筑布局和细部设计的一些组织方式，经发展、变化和精炼而成为古典建筑中出众的特征。

调整布局使其有序并且清晰。平面图通常在一条直线，即轴线的两侧对称，而这条直线或轴线从主入口开始笔直地穿过建筑。对称关系反映了大自然和人类身体的平衡性。双向对称比较少见，如帕拉迪奥1550年在威尼斯附近建造的圆厅别墅平面（*b*），单一轴线更为普遍，如1570年建造的科尔纳罗别墅（*a*）。调整的对象并不全是完全平衡的房间，平面图中的对称形式内部也可以包括不规则的样式，例如贝鲁奇1509年在罗马设计的法尔内西纳别墅（*h*）。当诸如位置和用途之类的条件允许进行统一的布局时，可以调整相连的对称房间和庭院周围，使平面图清晰有序，例如出自1世纪意大利中部庞贝城中的梅南德宅邸（*c*）内的入口门廊和周柱中庭。

轴线不仅是平面图的几何结构，也描绘出行人的通道和视野以及对建筑物的理解。希腊神庙用单轴（*e*）强调神的形象，而教堂则通常采用一系列次级轴线，如帕多瓦的安德烈・德拉・瓦莱（死于1577年）1532年建造的圣朱斯蒂娜教堂（*f*）。法尔内西纳别墅（*h*）的中轴线为不规则的房间布局而牺牲了整个建筑的对称形式。罗伯特・亚当1772年在伦敦设计威廉・韦恩庄园（*g*），主轴线在次轴线上改变方向，完成了狭窄的联排住宅平面图。

埃德温・勒琴斯有意在其众多房屋平面图上中断中轴线，并立即转动90° 至长长的走廊线上，如1908年在英国剑桥郡设计的米德尔菲尔德庄园（*i*）。尽管罗马设计常常基于严格的轴线布局，但也可以更加随意。别墅平面图和复杂的布局内包括单独的房间或空间群彼此独立，但由一系列次级轴线按不同的角度相连，如120年罗马附近哈德良别墅（*d*）中的浴场设计。

20. CLASSICAL DESIGN

SYMMETRY AND AXIS

There is no formula for good design. There are, however, certain ways of organising the layout and detail of buildings that have been developed, varied, and refined within the Classical traditional to become distinctive characteristics of Classical architecture.

Accommodation is organised into a layout that is orderly and easily understood. The plan is usually symmetrical on either side of a line, or axis, starting at the main entrance and running straight through the building. This symmetry reflects the balance of nature and the human form. Symmetry in two directions, such as the plan of Palladio's Villa Rotonda near Venice (*b*) of 1550, is rare and it is more often on a single axis, such as his Villa Cornaro (*a*) of 1570. The required accommodation does not always lend itself to equally balance rooms and the plan may have an irregular pattern within a symmetrical form, as does the Villa Farnesina in Rome (*h*), by Peruzzi in 1509. When circumstances, such as location or use, make a uniform arrangement impossible, the accommodation can be organised around linked symmetrical rooms and courts which give a clear order to the plan, as with the entrance atrium and peristyle court in the House of the Menander (*c*) from first-century-AD Pompeii in central Italy.

The axis is more than a geometric structure of the plan and describes the way someone would move through, view, and understand a building. Greek temples focused on the image of the god with the single axis (*e*) while churches, such as S. Giustina in Padua (*f*), by Andrea della Valle of Padua (*d.*1577) in 1532, often have a series of secondary axes. The central axis in the Villa Farnesina (*h*) gives the symmetry of the whole building to the irregular layout of rooms; a mayor axis changing direction on a secondary axis holds together the narrow town-house plan of William Wynn House in London (*g*) of 1772, by Robert Adam.

Edwin Lutyens deliberately interrupted the central axis in many of his house plans and, as in Middlefield House in Cambridgeshire in England (*i*) of 1908, immediately turned it through ninety degrees to the line of the long corridor hall. Roman planning, while often based on a rigid axial layout, could be more informal. Villa plans and complex arrangements such as the baths at Hadrian's Villa near Rome (*d*) of AD 120, have individual groups of rooms or spaces that are relatively independent of one another but linked at various angles by a series of secondary axes.

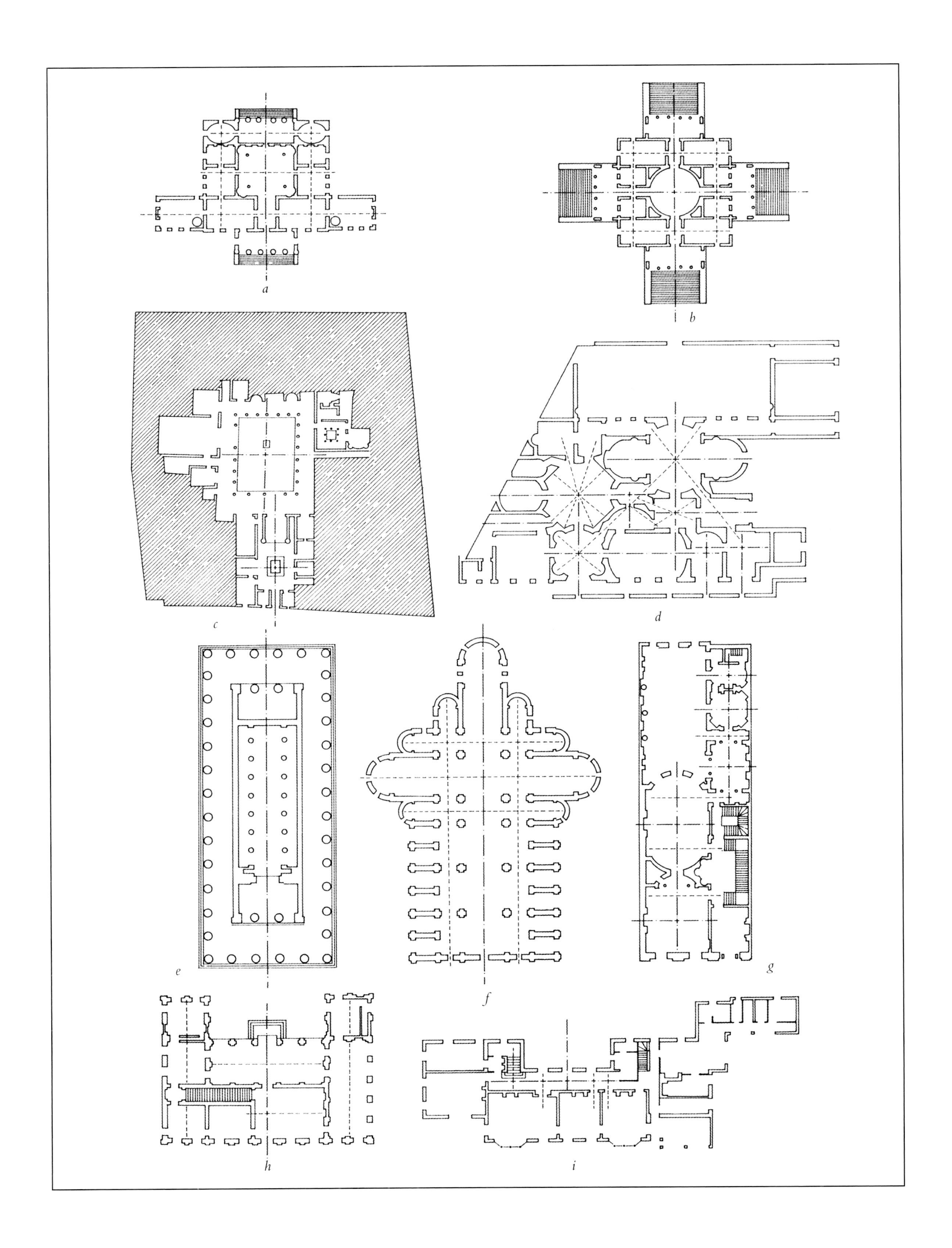
a
b
c
d
e
f
g
h
i

几何结构与空间

在设计封闭的区域，如房间或庭院时，应当解释空间内的几何结构，使人更加清楚地理解空间意义。墙壁、地面和天花板经装饰后可以强调空间的几何结构。建筑的结构或功能通常决定了几何框架，且当建筑形态简易或包括大房间时，室内空间可以包含在外部设计之中。

穹顶的几何结构常表现在其覆盖的空间之中。在约60年建于罗马的尼禄宫（*b*）内，两条走廊交会在简洁的穹顶下方。底下的鼓形座中有4个开口，用4个完全一样的矩形空间进行标记，此处走廊被支柱所遮蔽。矩形和圆形空间在地面与对角线和八边形结合在一起。维尼奥拉1554年在罗马建造的圣安德烈教堂（*c*）中，椭圆形穹顶覆盖简易的矩形空间，并在外部重复这些形状。矩形内墙上的椭圆拱形复制了上方的椭圆形状。结构和装饰中反映出4个椭圆圈的几何结构。入口支柱与每个大曲线的中心相连，对连接边墙支柱的地面进行装饰，标记出椭圆中大小圆圈交叉点处的线条。

几何结构变化后，一连串房间的特征也随之变化，建筑内的通行方式也变得多样化。勒杜1769年在巴黎设计的蒙莫朗西酒店（*f*）将主要的房间都设置在对角线上，最大化地利用了角落位置。圆形会客厅将轴线从道路转向占据内部角落的方形房间。居中的椭圆形房间没有外墙，向上穿过两层楼，由屋顶照明。索恩爵士1784年在英国萨福克设计的招标大厅（*d*）中包含相似的空间系列，但布置在房屋的中轴线上。外部的曲线与室内匹配，但是重复的曲线被矩形大厅所中断。室内装饰特征强化了房间之间的差异。

露天庭院使周围门窗的开口以及地平面成为一种框架，定义了封闭体积的几何结构。洛克·卢卡哥（1558—1597年其鼎盛时期）1564年在热那亚设计了市政宫中的庭院（*a*），在矩形平面图中，墙上的常规开口描述了平面图的比例。庭院位于入口门厅和大楼梯之间，形成一条通道，贯穿建筑中相连的开闭空间。庭院的设计凸显出楼梯前段更高的地平面的重要性。1547年，在罗马附近的卡普拉罗拉，维尼奥拉设计了法内斯宫（*e*）奇特的五边形平面图。圆形庭院的设计包括五边形的外部几何特征。在五边形墙壁角落和中心的前方，设置了10个支墩并额外添加了对柱。

GEOMETRY AND SPACE

The design of an enclosed area, such as a room or court, can make the space comprehensible by defining its geometry. Walls, floors, and ceilings can be decorated to emphasise the geometry of the space. The construction or function of a building usually introduces the geometry framework and, when a building has a simple form or large rooms, interior volumes can be a part of the external design.

The geometry of a dome is usually in the space it covers. Two corridors in one of Nero's palaces in Rome (*b*) of about AD 60 meet under a simple dome. The four openings in the circular drum beneath are marked with four identical rectangular spaces where the corridors are screened with columns. These rectangular and circular spaces are brought together with diagonals and an octagon on the floor pattern. Vignola's S. Andrea in Rome (*c*) of 1554 is a simple rectangular space covered with an oval dome and these shapes are repeated on the exterior. Elliptical arches in the internal walls of the rectangle reproduce the shape of the oval above. The geometry of the four circles of the oval is reflected in the structure and decoration. The entrance columns are in line with the centre of each large curve and the decoration in the floor linking the columns on the side-walls also marks the line of the intersection of the large and small circles of the oval.

Changing geometry can give a different identity to a sequence of rooms, adding variety to movement through a building. The Hotel de Montmorency in Paris (*f*), by Ledoux in 1769, makes maximum use of a corner position by arranging the major sequence of rooms on the diagonal. A circular salon turns the axis from the line of the street towards the square room which occupies the internal corner. The intermediate oval room has no external walls and rises through two floors to be lit from the roof. Tendering Hall in Suffolk, England (*d*), by Soane in 1784, has a similar series of spaces but placed along the central axis of the house. The curves on the exterior are matched on the interior, but their repetition is interrupted by a rectangular hall. Interior features reinforce the difference between the rooms.

Open courts are often designed so that the surrounding window- and door-openings and the floor levels act as a framework which defines the geometry of the enclosed volume. The court of the Palazzo Municipale in Genoa (*a*), by Rocco Lurago (*fl.*1558-97) in 1564, has a rectangular plan with proportions described by the regular openings in the walls. The court lies between an entrance vestibule and a grand staircase, creating a route through the building of linked open and closed spaces. The design of the court gives added significance to the higher ground level at the head of the stair. In 1547 at Caprarola near Rome, Vignola designed the Palazzo Farnese (*e*) in an unusually pentagonal plan. The court is circular, but the design includes features of the geometry of the outer pentagon. Ten piers with coupled columns lie in front of the angles and centres of the walls of the pentagon.

m
20
a
b
c
d
e
f
50
ft

垂直和水平等级

在大型建筑中，不同的部分需要具备各自的辨识特征，如建筑的入口。建筑的不同用途，或仅仅因为高度和长度单一乏味缺乏变化，便足以促使建筑师进行多样化的装饰设计，以实现区别不同要素的目的。调整外立面上细部设计装饰可以突出强调建筑中的任意部分，表明其重要的功用或改进某些结构。

如果建筑较长，可以沿其长度进行极为不同的设计，如甘登1781年在都柏林建造的海关大楼（*f*）。他改变了墙上的细部设计，以提供垂直界线以分隔中心和两端。作为结构中清晰的末端特征，即亭子，由中心的建筑要素组成，并且几乎自成建筑。这些亭子成功地终结了长长的外立面，但是由于中间有墙，因此也可以省略亭子，留下建筑的中心作为独立的设计。这一连串隔开的元素将设计向上延伸至山花和高耸的穹顶。伊尼戈・琼斯1635年在英国索尔斯伯里附近建造的威尔顿庄园（*e*）使用亭子作为支配性的垂直特征，勾勒出之间的低矮建筑，进而强调了精巧的中门设计。

意大利文艺复兴早期宫殿的设计师们建立了根据功能将建筑物分隔成清晰水平层次的传统，也确立了用多样化的细部设计来表达各层重要性或用途的传统。宫殿具有防御性作用，此外还要在有限的小块城市土地中提供用来服务和存储的房间，因此只能将更高级的房间安置在楼上，称作主楼层。在布拉曼特约1500年于罗马设计的托罗尼亚-纪劳宫（*d*）中，主楼层位于用粗琢工艺装饰的一楼上方，位置更高且细部设计更多，以此作为区分。主楼层发展成为宫殿设计中的一个重要特征，如米歇尔・圣米凯利（1484—1559年）1530年在意大利北部维罗纳设计的庞贝宫（*a*），在唯一可以看见的楼上部分使用完整装饰的多立克柱式，成为外立面上独一无二的装饰要素。如果二楼不是主楼层，那么更应当用相似的方法强调主房间所在的楼层。弗兰兹・古斯塔夫・乔基姆・佛尔斯曼（1795—1878年）1828年在德国汉堡设计的杰尼斯庄园（*c*）详细地将各层的大小和细部设计分成等级，人们能很清楚地看出主要的房间位于最底下的一层。

在20世纪美国的高层建筑中，大多数楼层往往重复相同的用途，因此产生新的问题，也使古典建筑师开始重新考虑在装饰中突出垂直差异的特征。亨利・D. 维特菲尔德（1876—1949年）和贝弗利・S. 金（1875—1935年）1906年在纽约市设计工程师俱乐部建筑（*b*）的时候，应用了一条原则，后来被众多高耸的古典建筑所采纳。精致的细部设计集中在低矮的楼层中，这些楼层大多可以从街面上看到，而那些只能从远处才能看到的上部楼层则添加了强烈的装饰特征。

VERTICAL AND HORIZONTAL HIERARCHY

On large buildings different parts, such as entrance, need to be capable of being identified. Diverse uses, or just the monotony of height and length, can also lead the architect to seek means of differentiating elements of the design by varying the decoration. The modification of detail across a façade can give emphasis to any part of a building, either to indicate its functional importance or to improve the overall composition.

Long buildings, such as the Customs House in Dublin (*f*), by Gandon in 1781, can vary considerably along their length. The details on the walls change to provide vertical divisions separating the centre and ends. The distinct end-features, or pavilions, are made up of architectural elements from the centre and are almost buildings in their own right. These pavilions provide a satisfactory termination to the long façade but, with the intermediate walls, could be omitted to leave the centre of the building as an independent design. This sequence of separate elements builds the design up to the pediment and tall dome. Wilton House near Salisbury in England (*e*), by Inigo Jones in 1635, uses the pavilions as dominant vertical features which, by framing the lower building between, help to highlight the elaborate central door.

The designer of early Renaissance palaces in Italy established a tradition of separating buildings into distinct horizontal layers according to their function and varying the detail to express the relative importance or use of each floor. The defensive role of these palaces and the provision of service and storage rooms within limited urban plots put the state rooms on the upper floor. This was the noble floor, or the piano nobile. On the Palazzo Torlonia-Giraud in Rome (*d*), designed by Bramante in about 1500, it can be seen above the rusticated ground floor and is distinguished by its greater height and increased detail. The piano nobile was developed to become a major feature in the design of palaces, such as the Palazzo Pompei in Verona in northern Italy (*a*), by Michele Sanmicheli (1484-1559) in 1530, where the use of a fully decorated Doric Order on the only visible upper floor is the single significant element on the façade. When the first floor above the ground is not a piano nobile it is often more appropriate to give a similar significance to the level of the principal rooms. In the Jenisch House in Hamburg, Germany (*c*), designed in 1828, Franz Gustav Joachim Forsmann (1795-1878) carefully graded the size and detail of each floor, making its evident that the main rooms are on the lowest level.

In the United States in the twentieth century the new problem of tall buildings, often with repetitive uses on most levels, made Classical architects reconsider vertical differences in decoration. Henry D. Whitfield (1876-1949) and Beverly S. King (1875-1935), in designing the Engineer's Club in New York City (*b*) in 1906, used a principle that was adopted for many tall Classical buildings. Fine detail was concentrated on the lower floors most visible from the street while on the upper floors, seen from a distance, strong decorative features were added.

a
b
c
d
ft 70
m 20
e
f

强调与细部设计

由于古典设计有机会改变柱式的装饰级别以适应建筑的规模、重要性或成本，因此具有极大的灵活性，并且能够应用于单独的特征或细部设计之中。

伦敦的两处大约建于1750年的房屋（*f*）以在各简单功能要素处的细部设计装饰数量变化为特征。窗缘装饰向中心递增。在顶层，只有中心窗子带有檐口，而在下面的一层中，所有窗子都有檐口，中间三扇还用山花装饰。这三扇相同的窗户下方带有小栏杆柱，但只有中心一扇带有支座和托架，山花也相应变宽。在一楼，相同的拱形开口由于添加了粗琢结构而在中心处产生了变化。最终，带有山花的檐口完成了从两侧到中心的装饰分级。大多数细部设计集中在建筑的中心，一般情况都是入口，但并不总是分得如此细致。位于法国北部圣洛的法院大楼（*e*）1823年由亨利·冯·克里普特（1792—1858年）设计，除完整的科林斯中心门廊和门之外几乎不含其他装饰。

中心细部设计可以仅限于门自身。意大利文艺复兴早期的教堂通常采用与其罗马风格和哥特风格前身相同的简单结构，如约1450年建于佛罗伦萨的圣费利切教堂（*c*），但是在该建筑中，精致的古典门缘取代了哥特风格的门，精美的细部设计与后面朴素的墙壁形成鲜明的对比。罗姆博特·凯尔德门斯二世（约1460—1531年）和盖约特·德·鲍尔格兰特（约1500—1549年）于1520年在比利时梅赫伦设计的萨瓦酒店（1515—1517年）（*d*），建筑末端狭窄无门，装饰集中在典型的北欧大角度斜屋顶上的山墙中。较低的两层除在上部中心窗子中添加了山花外再无其他装饰，而在窗子上方紧贴窗子的地方，用支柱、山花、涡卷形和瓶的华丽组合进行装饰。

小型建筑往往不适合过多的装饰主题，但是它的质感或周边也说明其中可以包括某些精美的细部设计。克里斯蒂安·弗雷德里克·汉森（1756—1845年）1803年在德国汉堡附近为自己建造的小房屋（*a*）用粗琢、转角石和檐口进行了简单装饰。这些朴素的细部设计与主楼层中的三扇由爱奥尼亚壁柱、山花和小型扶手环绕起来的大窗子形成鲜明的对比。塞缪尔·布瑞克·帕克曼·特罗布里奇（1862—1925年）和古德休·利文斯通（1867—1951年）1901年在纽约市的一处马厩上方设计的艺术工作室（*b*），通过工作室中的大型中心窗子实现了相似的效果。设计属于巴洛克风格，工作室位于上层，致使半圆形山花支撑在绚丽的托架上，置于精心装饰的阳台的上方，全部都与下方宽敞的马厩大门形成鲜明的对比。

EMPHASIS AND DETAIL

The opportunity to vary the level of decoration of the Orders to suit the size, significance, or cost of a building gives Classical design great flexibility and can be applied to individual features or details.

The design of two houses built in about 1750 in London (*f*) features variations in the amount of detail applied to each simple functional element. The decoration on the window-surrounds increases towards the centre. On the top floor only the centre window has a cornice, while on the floor below, all the windows have cornices with pediments on the centre three. These same three windows have small balustrades below, but only the centre window has pedestals and brackets, which increase the width of the pediment. On the ground floor identical arched openings are varied in the centre with the addition of rustication. A cornice with a pediment completes the grading of decoration towards the centre. Detail is most frequently concentrated on the centre of a building and, consequently, the entrance, but it is not always so carefully graded. The Courthouse at Saint-Lô in northern France (*e*), designed by Henry van Cleemputte (1792-1858) in 1823, has little decoration expect the full Corinthian central porch and door.

Central detail can be limited to the door itself. Early Italian Renaissance churches, like S. Felice in Florence (*c*), designed by an unknown architect in about 1450, were often the simple structures as their Romanesque and Gothic predecessors, but the rich Gothic doors were replaced with elaborate Classical door-surrounds, the fine detailing contrasting with the plain wall behind. The *Hôtel de Savoie*, Mechelen in Belgium (1515-17) (*d*), by Rombout II Keldermanns (*c.*1460-1531) and Guyot de Beaugrant (*c.*1500-49) in 1520 has narrow ends without doors and the decoration is concentrated on the gable of the typical, steep-sloping roof of northern Europe. The lower two floors have no decoration except a pediment added to the upper centre window, while immediately above this window there is an ornate composition of columns, pediments, scrolls, and vases.

Small buildings are often unsuitable for extensive decorative schemes, but their quality or surroundings may suggest the inclusion of some fine detail. The small house (*a*), built for himself in 1803 by Christian Frederick Hansen (1756-1845) near Hamburg in Germany, is simply decorated with rustication, quoins, and a cornice. There are no windows on the ground floor and the upper floor has a small plain window. The austerity of these details is contrasted with the three large windows on the principal floor, which are surrounded by Ionic pilasters, a pediment, and a small balustrade. An artist's studio over a stable in New York City (*b*), designed by Samuel Breck Parkman Trowbridge (1862–1925) and Goodhue Livingston (1867–1951) in 1901, achieves a similar effect with a large central studio window. The design is Baroque and the location of the studio on the upper floor gives the opportunity for a powerful half-round pediment supported on florid brackets over a richly detailed balcony, all of which contrast with the broad stable-door below.

a
b
c
d
e
f

延伸阅读

本书覆盖了诸多学术领域，并且在众多实例中，以一种前所未有的模式，将取材于不同资源的信息融合。目录中每种建筑元素的参考书目都十分庞大。因此，下面所列书目较有节制，且所推荐的延伸阅读仅针对部分章节进行拓展，这些章节虽在本书中篇幅有限，但在其他作品中涉猎较广。此外，还有一些针对历史上具有影响力的建筑作家作品的现代再版或译本的参考， 因为这些通常比史料摘要更能发人深省。

（建筑）通史

《建筑史》，班尼斯特・弗莱彻爵士著，丹・克鲁克香克编辑，第20版，1999年，新德里：CBS出版发行。

一部得到公认的，且内容极其广泛的参考书，其图解与描绘较之评述更有价值。

《西方建筑史》，大卫・沃特金著，第6版，2015年，伦敦：Laurence King Publishing出版发行。

可获取的最佳西方建筑通史。

柱式

《古典建筑：古典建筑语汇与基础知识简介（带有精选术语表）》，詹姆斯・史蒂文斯・科尔著，2003年，纽约：W.W. Norton & Co. Inc.出版发行。

关于古典建筑基本原理与各个方面的图解书籍。

《古典建筑柱式》，罗伯特・奇塔姆著，第2版，2005年，阿姆斯特丹；伦敦：Elsevier/Architecture Press出版发行。

极其详尽地描述理想柱式，所涉及范畴虽有限且教条，但很实用。

《建筑柱式：希腊、罗马及文艺复兴时期》，亚瑟・斯特拉顿著，1986年，Studio Editions出版发行。

1931年版的摹本，比罗伯特・奇塔姆的《古典建筑柱式》涉猎更加广泛，但例子选择较随意。

《建筑柱式之比较：希腊、罗马及文艺复兴时期》，查尔斯・皮埃尔・约瑟夫・诺曼德著，1946年，由R.A. 科丁利修订并重新编辑，伦敦：John Tiranti Co.出版发行。

FURTHER READING

This book covers a large field of study and in many cases draws together information from many sources in a format that has not been attempted before. A reading list that referred to each element in the contents would be prohibitively large. The list below is, consequently, of a limited nature and only recommends further reading to amplify sections of the book which have been limited by space but are covered more extensively in other works. There are also references to current reprints or translations of influential architectural authors from the past as these are often more revealing then historical summaries.

GENERAL HISTORY

FLETCHER, Banister Sir (1999) *A History of Architecture*; edited by Dan Cruickshank, 20th edition (New Delhi : CBS Publishers & Distributors).

A well-established and extremely comprehensive reference book, more valuable for illustrations and description than comment.

WATKIN, David (2015) *A History of Western Architecture*; 6th edition. (London : Laurence King Publishing).

The best general history of western architecture available.

THE ORDERS

CURL, James Stevens (2003) *Classical Architecture: an introduction to its vocabulary and essentials, with a select glossary of terms* (New York: W.W. Norton & Co. Inc.).

An illustrated book concerned with the fundamental principles and various aspects of Classical Architecture.

CHITHAM, Robert (2005) *The Classical Orders of Architecture*; 2nd edition (Amsterdam; London : Elsevier/Architectural Press).

A very detailed description of idealised Orders, limited in scope and rather doctrinaire, but useful.

STRATTON, Arthur (1986) *The Orders of Architecture : Greek, Roman, and Renaissance with selected examples of their application…* (London : Studio Editions).

A facsimile of the 1931 edition. More broad-based than Chitham but with arbitrary choice of examples.

NORMAND, Charles Pierre Joseph (1946) *Parallel of the Orders of Architecture : Greek, Roman, and Renaissance*, revised and re-edited by R.A. Cordingley (London : John Tiranti Co.).

不再印刷，却是不同版本柱式最实用的纲要。如果发现值得购买。

《工作中的古希腊建筑师：结构与设计问题》，J.J. 库尔顿著，第4版，1977年，牛津：Oxbow Books出版发行。

古希腊古典柱式之发展与演变的考古研究；引人入胜地阐释了试验与错误比理想主义更加重要。

No longer in print but the most useful compendium of different versions of the Orders. Worth purchasing if it can be found.

COULTON, J.J. (1977) *Ancient Greek architects at work: problems of structure and design*; 4th edition (Oxford : Oxbow Books).

An archeological study of how the Classical Orders developed and evolved in Ancient Greece; a fascinating exposition of how trial and error was more significant than idealism.

比例

《人文主义时代的建筑原则》，鲁道夫・维特科瓦著，1998年，伦敦：Academy Editions出版发行。

文艺复兴比例理论的经典之作。

《建筑中的比例理论》，P.H. 斯科菲尔德著，1958年，剑桥：Cambridge University Press出版发行。

关于比例理论及历史的综合论述，作者依据这些发展出自己的实证理论，这些理论协调了明显有矛盾冲突的对立体系，并且成为历史解读的关键。

PROPORTION

WITTKOWER, Rudolf (1998) *Architectural Principles in the Age of Humanism* (London : Academy Editions).

The classic work on Renaissance theories of proportion.

SCHOLFIELD, Peter Hugh (1958) *The Theory of Proportion in Architecture* (Cambridge : Cambridge University Press).

A comprehensive survey on the subject dealing with the history of the theory of proportion, and in doing so the author develops his own positive theory, which reconciles the apparent contradictions of rival systems and serves as a key to historical understanding.

细部设计与装饰

《装饰辞典》 菲莉帕・刘易斯和吉莉安・达利著，1990年，伦敦：Cameron Hollis与David & Charles 联合出版发行。

关于装饰主题及种类的广泛且透彻的概述。

《装饰手册》（修订版，译自德语版本），弗朗茨・塞尔斯・迈耶著， 1974年，伦敦：Duckworth出版发行。

1896年原版基础上的再版。广博且解析透彻，最实用的参考书，不可避免地带有19世纪后期的偏向性。

DETAIL AND DECORATION

LEWIS, Philippa & DARLEY, Gillian (1990) *Dictionary of Ornament* (London : Cameron Hollis, in association with David & Charles).

A wide-ranging and thorough survey of decorative themes and types.

MEYER, Franz Sales (1974) *Handbook of Ornament*; revised edition [translated from German] (London : Duckworth).

A reprint of the 1896 original. Extensive and analytical, a most useful reference book with an inevitable late-nineteenth-century bias.

原创作品

《维特鲁威：建筑十书》，维特鲁威・皮尼奥著，英格里德・D.罗兰译，1999年，剑桥：Cambridge University Press出版发行。

古代留存下来的唯一完整建筑专著。西方建筑史上称其为唯一重要著作，塑造了从文艺复兴时期到当代的人文主义建筑与建筑师形象。这一评论性的英文新版，表达了维特鲁威风格覆盖的范围，旨在重新塑造维特鲁威善于创造以及富于创新能力的思想家形象。

《十书中的建造艺术》，利昂・巴蒂斯塔・阿尔贝蒂

ORIGINAL WORKS

VITRUVIUS, Pollio (1999) *Ten Books on Architecture*; translated by Ingrid D. Rowland (Cambridge : Cambridge University Press).

The only full treatise on architecture and its related arts to survive from Classical antiquity. Claimed the single most important work of architectural history in the Western world having shaped humanist architecture and the image of the architect from the Renaissance to the present. This new, critical, English edition expresses the range of Vitruvius' style and aims to shape a new image of Vitruvius as an inventive and creative thinker.

ALBERTI, Leon Battista (1989) *On the Art of Building in Ten Books*; translated by Joseph Rykwert, Neil Leach, & Robert Tavernor, (Cambridge, Massachusetts ; London : The MIT Press).

著，里克沃特、尼尔·利奇与罗贝·塔费纳译，1989年，剑桥；马萨诸塞；伦敦：The MIT Press出版发行。

文艺复兴时期首部主要专著，约于1450年出版于意大利，其结构仿效维特鲁威作品。此版为优质新译本，无图解。

《塞巴斯蒂亚诺·塞里奥建筑五书》，塞巴斯蒂亚诺·塞里奥著，1996—2001年间出版，沃恩·哈特和彼得·希克斯译自意大利版本并评论，纽黑文；伦敦：Yale University Press出版发行。

塞巴斯蒂亚诺·塞利奥（1475—1554年）是16世纪最具影响力的建筑作家之一。本书整合了作者关于建筑原理的5本书，其中涉猎几何与透视法则，并且试图编纂设计法则。

《建筑四书》，安德烈·帕拉迪奥著，罗伯特·塔芙纳和理查德·斯科菲尔德译，2002年，剑桥；马萨诸塞；伦敦：The MIT Press出版发行。

1570年出版于意大利。新英文译本包括：帕拉迪奥的原版木刻印（在摹本的文本旁进行了复制）；术语表（在原文的文本中解释专业术语）；参考文献（对帕拉迪奥近期的研究）。

The first major Renaissance treatise, published in Italy about 1450, structured to resemble Vitruvius. This is an excellent new translation. Not illustrated.

SERLIO, Sebastiano (*c.*1996-2001) *Sebastiano Serlio on Architecture*; translated from Italian and commentary by Vaughan Hart and Peter Hicks (New Haven; London: Yale University Press).

Sebastiano Serlio (1475-1554) was one of the most influential architectural writers of the sixteenth century. This work brings together five volumes of his treatise on the principles of architecture, which deal with the rules of geometry and perspective, and attempt to codify the rules of design.

PALLADIO, Andrea (2002) *The Four Books of Architecture*; translated by Robert Tavernor & Richard Schofield (Cambridge, Massachusetts; London: MIT Press).

Published in 1570 in Italy. New English translation containing: Palladio's original woodcuts (reproduced in facsimile adjacent to the text); a glossary (that explains technical terms in their original context); and a bibliography (of recent Palladio research).

古典建筑的现代语汇

《新帕拉迪奥风格：21世纪建筑的现代化与持久性》，阿里礼萨·撒加驰和吕西安·斯泰尔著，2010年，伦敦：Artmedia出版发行。

案例展示了在21世纪开端，古典与民居传统如何在大量建筑公司、学者、学生、大学以及院校的实践下作为现代现象繁荣起来。

《论五书中的现代建筑》，乔治·索马里兹·史密斯著，2013年，牛津：Bardwell Press出版发行。

展示作者自己对托斯卡纳、多立克以及爱奥尼亚柱式的解读，带有乔治·索马里兹·史密斯在过去10年中设计并建造的不同建筑的图纸、照片。

《瑞典式优雅：被遗忘的现代风》，彼得·埃尔姆隆德和约翰·马特柳斯著，2015年，斯德哥尔摩：Axel & Margaret Ax:son Johnson基金会出版发行。

本选集讨论“瑞典式优雅”——主要是20世纪10年代晚期与20世纪20年代瑞典新古典建筑和设计。

A MODERN LANGUAGE OF CLASSICAL ARCHITECTURE

SAGHARCHI, Alireza & STEIL, Lucien (2010) *New Palladians : modernity and sustainability for 21st century architecture* (London : Artmedia).

Showcases how, at the outset of the 21st century, Classical and vernacular traditions are flourishing as modern phenomena practiced by a wide spectrum of architectural firms, scholars, students, universities, and institutions.

SAUMAREZ SMITH, George (2013) *A treatise on modern architecture in five books* (Oxford: Bardwell Press).

Presents the author's own versions of the Tuscan, Doric, and Ionic Orders, with drawings and photographs of various buildings that Saumarez Smith has designed and built over the last 10 years.

ELMLUND, Peter & MARTELIUS, Johan (eds), (2015) *Swedish Grace: The Forgotten Modern*, (Stockholm : Axel & Margaret Ax:son Johnson Foundation).

An anthology of works discussing 'Swedish Grace'—Swedish Neo-Classical architecture and design mainly of the late 1910s and the 1920s.

术语表

此表并非综合建筑术语表，其目的仅在于帮助读者理解文中不熟悉的词汇。专业词汇的定义与该词出现在同一页的，不列入本术语表。如需词条的其他信息也可查找索引。

冠板 柱头顶部平板，其形状与边缘轮廓依柱式种类不同而变化。

茛苕 虾膜花和刺老鼠簕。一种野生地中海植物，拥有宽大叶片与锯齿状叶缘。以固定形式见于科林斯柱头与其他古典装饰设计中。

山花雕像座 山花顶点与末端的雕像与底座总称。

神龛 以柱子、檐部及山花为框架的开口或凹陷。最初为圣祠，采用小型神庙正立面形式，并包含一座祭拜神像。

走廊 1. 巴西利卡中平行于中殿的空间，一排柱子或支墩将其与中殿分开，通常带有较低的天花板。

2. 同时也应用于表述任何狭窄的教堂过道。

装饰屋瓦 屋顶边缘的垂直支撑或装饰特征，以其原始形态支撑末端屋瓦。通常装饰有叶片、植物或人面纹案，有时见于屋脊处。

棕榈叶饰 一种规范化的植物设计，大概源于忍冬属植物，有连续的卷曲叶片与花瓣。

半圆形后殿 较大空间中的一个凹陷，通常为半圆形或多边形，带有半球形的穹顶。一般位于巴西利卡中殿或教堂末端，并带有圣像或神坛。

额枋/框缘 1. 额枋：檐部的下部或横梁。

2. 框缘：轮廓与之相同的门缘或窗框。

拱门饰 一种环绕于拱券正面下部的额枋。

中庭 罗马房屋的正房或大厅，传统意义上在屋顶正中有正方形或长方形开口。

GLOSSARY

This is not a comprehensive architectural glossary; it is only intended to assist the reader with unfamiliar words in the text. Specialised words that are only used on the same page as their definition are not entered in this glossary. Further information on many of the entries may also be found by referring to the index.

ABACUS — A flat slab on top of a *column capital*. The shape and edge profile vary according to the type of *capital*.

ACANTHUS — *Acanthus* mollis and *Acanthus* spinosus. A wild Mediterranean plant with large leaves with serrated edges. It is represented in a stylised form on the Corinthian *column capital* and other Classical decorative designs.

ACROTERION — (pl. acroteria) Both the sculpture and platform for the sculpture at the apex and ends of a *pediment*.

AEDICULE — An opening or recess framed with *columns*, an *entablature* and a *pediment*. Originally a shrine in the form of a miniature temple front containing a cult figure.

AISLE — **1.** A space parallel to the *nave* of a *basilica*, separated from the nave by a row of *columns* or *piers*, often with a lower roof.

2. Also used to describe any narrow passage in a church.

ANTEFIX — (pl. antefix) A vertical support or decorative feature on the edge of a roof which, in its original form, supported the last tile. They are often decorated with leaf or plant designs or faces and are occasionally found on the ridges of roofs.

ANTHEMION — A formalised plant design, perhaps derived from the honeysuckle, with a series of curling leaves or petals.

APSE — A recess, generally semicircular or polygonal, often with a half-domed ceiling, off a larger space. Usually found at the end of the *nave* of a *basilica* or a church and containing a cult image or alter.

ARCHITRAVE — **1.** The lower part, or beam, of the *entablature*.

2. A door-or window-surround with the same *profile*.

ARCHIVOLT — An *architrave* curved around the lower face of an arch.

ATRIUM — The principle room and hall of a Roman house, traditionally with a square or rectangular opening in the centre of the roof.

阁楼 檐部或山花之上的空间、墙壁或者楼层。

栏杆 较短的支柱或扶手，通常会有一处或多处隆起，并有其他细部设计。

矮护墙 成排的栏杆或低矮的护墙。

柱础 柱子、建筑特征或墙的底层支撑部分。

巴西利卡 一个带有大的矩形中央空间——中殿——的大厅，通常沿长边，偶尔沿短边，带有狭长的空间——走廊，由成排的柱子或支墩分隔。

开间
1. 成排柱子中，一个柱子中心线到另一柱子中心线的单元。
2. 也是墙长度上的分隔，常由柱子、支墩和壁柱定义，并且包含单独的门、窗或拱。

托架 凸出的支撑。

柱头 柱子顶部雕刻有装饰的部分。

女像柱 女性塑像作为支撑柱子。

天窗 巴西利卡中殿墙面上部，位于走廊屋顶的上方，通常带有通风窗。

花格镶板 天花板、拱顶或穹顶上不同形状的凹陷。

柱廊 成排的柱子。

柱子 细长的独立支撑物，切面通常为圆形，也可以是方形或多边形。不同种类的柱子是古典柱式的重要组成部分。

托臂 墙壁上的小型凸出物，或支撑托架。

檐口
1. 檐部上端，与神庙屋顶的屋檐一致。
2. 任何其他从屋檐演变而来的凸出装饰物。

挑檐滴水板 檐口上凸出的装饰线脚，带有扁平的垂直表面，通常位于波纹线脚之下。

交叉点 两个长方形空间相交形成的区域，通常具体表现为教堂的焦点，且通常覆以穹顶或塔楼。

波纹线脚 字面意义上为波浪。是一种轮廓，包括两条连接的凹形与凸形弧线，根据排列顺序称为正波纹线脚和反波纹线脚。

墩身/护墙板 又称方形柱脚。
1. 墩身：是柱子基座的主要部分，位于檐口及底座之间。
2. 护墙板：当柱式应用于墙壁之上时，也是在墙体上与基座相应部位一致的区域。

ATTIC The space, wall, or storey above and *entablature* or *pediment*.

BALUSTER A short post or railing, often with one or more swellings and other details.

BALUSTRADING Rows of *balusters* or low protective wall.

BASE The lower and supporting part of a column, feature, or wall.

BASILICA A hall with a large rectangular central space—the *nave*—and usually with narrow spaces along the long, and occasionally also the short, sides—the *aisles*—separated with rows of *columns* or *piers*.

BAY
1. One unit in a row of *columns* taken from the centre of one *column* to another.
2. Also a division in the length of a wall often defined with *columns*, *piers*, or *pilasters* and containing a single door, window, or arch.

BRACKET A projecting support.

CAPITAL A carved or decorated block at the top of a *column*.

CARYATID A female figure acting as a supporting *column*.

CLERESTORY The upper part of the wall of the *nave* of a *basilica* which stands above the lower roof of the *aisle* and usually contains clerestory windows.

COFFER A variously shaped recess in a ceiling, *vault* or dome.

COLONNADE A row of *columns*.

COLUMN A slender free-standing support, usually circular but can be square or polygonal in cross section. *Columns* of different types are one of the essential parts of a Classical *Order*.

CORBEL A small projection in a wall, or supporting *bracket*.

CORNICE
1. The upper part of an *entablature* which corresponds to the eaves of a temple roof.
2. Any other decorative projection that derives from the eaves.

CORONA The projecting *moulding* of the *cornice* with a flat vertical face, usually below a *cyma* moulding.

CROSSING The area formed by the crossing of two rectangular spaces, generally specific to the focal space of a church and often roofed with a dome or tower.

CYMA Literally a wave. A *profile* consisting of two joined curves alternately concave and convex and, according to their sequence, called cyma recta and cyma reversa.

DADO Also called a die.
1. The principal part of the *pedestal* of a *column* between the *cornice* and base.
2. Also the area of wall which corresponds to this part of the pedestal when an *Order* is applied to a wall.

齿状装饰 檐口上牙齿状的凸出装饰。

柱顶石 柱子顶部一块独立的檐部。

鼓形座 穹顶下圆柱形墙体。

突耳 门缘与窗框，即框缘上部两侧的凸出装饰。也称为肩。

钟形圆饰 柱子柱头上，位于冠板下的凸出装饰线脚。

檐部 古典柱式的最顶端，包含柱子支撑的所有水平部分。分成三部分：额枋、雕带和檐口。

柱子收曲分线 柱子上楔形或外凸楔形的部分，是古典柱式的特征。

扇形窗 门上方的窗户。

平缘 小的扁平凸出装饰线脚。

尖顶饰 屋顶上垂直凸出的装饰特征。

柱槽 垂直凹槽，通常带有凹面轮廓，最常见于柱子表面。

雕带 檐部的中间部分。

山墙 斜屋顶两侧斜坡中间的三角形墙。

穹棱 两拱顶交叉构成的线条。

扭索状装饰 模仿交织丝带的装饰纹案，通常为条状。

圆锥饰 小圆柱形木钉状装饰细部，位于三陇板之下，且在多立克柱式中置于飞檐托块的底面上。它们在文艺复兴后期成为更加常见的装饰特征。

拱墩 成块的石头或者其他材料，通常凸出且带有雕刻，拱于其上建起。

穹隆顶塔 小型垂直圆柱形或多边形设计，带有窗户，位于屋顶或穹顶上，且有独立的屋顶或穹顶。

板间平面 多立克式雕带中三陇板间的空间，有时带有装饰。

飞檐托饰 科林斯柱式和混合柱式中檐口上的小托架，置于挑檐滴水板之下。

模数 应用于柱式比例中的测量单位。本书中用柱子底部直径作为模数，分成10个部分（其他出版物中，可能使用半径作为模数，分成30份）。

装饰线脚 垂直或水平细部设计的轮廓或轮廓线。

DENTIL Tooth-like projection in a *cornice*.

DOSSERET An isolated piece of *entablature* above a *column*.

DRUM A cylindrical wall below a dome.

EARS Projections on the sides of the upper parts of door- and window-surrounds or *architraves*. Also known as shoulders or croisettes.

ECHINUS A projecting *moulding* on a *column capital* below the *abacus*.

ENTABLATURE The uppermost part of a Classical *Order* consisting of all the horizontal elements supported by the *column*. It is divided into three parts: the *architrave*, the *frieze*, and the *cornice*.

ENTASIS The taper or bulging taper of a *column*-characteristic of the Classical *Order*.

FANLIGHT A window over the top of a door.

FILLET A small flat projecting *moulding*.

FINIAL A vertical projecting feature at the top of a roof.

FLUTING Vertical grooves, or flutes, usually with a concave *profile* and most commonly found running up the face of *columns*.

FRIEZE The middle part of the *entablature*.

GABLE The triangular piece of wall between the two sloping sides of a pitched roof.

GROIN The line formed by the points of intersection of two *vaults*.

GUILLOCHE A pattern resembling interlaced ribbons, generally linear in form.

GUTTA (*pl.* guttae) Small cylindrical peg-like decorative detail below *triglyphs* and on the *soffit* of *mutules* in the Doric *Order*. They became a more general decorative feature in the later Renaissance.

IMPOST The block of stone or other material, usually projecting and often carved, from which an arch springs.

LANTERN A small vertical cylindrical or polygonal projection with windows in a roof, or dome with a roof, or dome of its own.

METOPE The space between the *triglyphs* in a Doric *frieze*, sometimes decorated.

MODILLION Small *bracket* in the *cornice* of the Corinthian and Composite *Orders* below the *corona*.

MODULE The unit of measurement used in proportioning the *Orders*. In this book the diameter of the lower part of the *column* is used as the module, divided into ten parts. (In other publications the module can be half the diameter and divided into thirty parts.)

MOULDING The *profile*, or contour, of a continuous vertical or horizontal detail.

飞檐托块 扁平凸出托架， 原本与其上的屋顶倾斜方向一致，位于多立克柱式的挑檐滴水板下，并且其底面上装饰有圆锥饰。

中殿 巴西利卡中的大厅，因此成为教堂公用的西侧区域。

眼洞窗 穹顶上的圆形窗户或者环形孔洞。

柱式 五种规范的带有基座、柱子和水平檐部的体系，形成了古典建筑的基础。所有部分之间有着根据传统确立的关系，每种柱式都有其特有的装饰主题。五种柱式为：托斯卡纳、多立克、爱奥尼亚、科林斯和混合柱式。

凸圆形线脚装饰 凸形装饰线脚带有与半个鸡卵相似的轮廓，有时刻有卵与箭头或飞镖相间的纹案。

亭台/亭子
1. 亭台：建筑中独特的部分，通常位于长正立面的一隅或末端。
2. 亭子：一座小型独立建筑物。

基座 柱子下，带有装饰的块状物或类似的用于支撑雕像的设计。

山花
1. 起源是神庙中较浅的三角形山墙，尤其是包括屋顶边缘或边沿的斜檐口。
2. 或者其他位置的相同设计，例如窗户或者神龛上。

穹隅 凹形拱肩，以某一角度起始于两墙之间的一点，并且逐渐变宽，用于支撑鼓形座或穹顶边缘，填充剩余空间，这一剩余空间是在将环形穹顶置于方形或多边形平面上而形成的。

主楼层 源自意大利语，宏伟或高贵的楼层。意大利宫殿中的主要楼层，包括大厅，并且通常是一楼上面的一层或地下室上面的那一层。拥有较高的天花板和大量的外部与内部装饰。这一术语还延伸至其他建筑物类型中。

支墩 用于支撑墙体的独立长方形部分。

壁柱 半圆形或扁平柱子的一部分，从墙体中凸出。当凸出为半圆形，或超过柱子的半径时，称为附墙柱。

MUTULE A flat projecting *bracket*, originally sloping with direction of the roof above, below the *corona* of the Doric *Order* and decorated with *guttae* on its *soffit*.

NAVE The large hall of a *basilica* and hence the public western part of a church.

OCULUS A round window or circular aperture in the top of a dome.

ORDER The five formal systems of *pedestals*, *columns*, and horizontal *entablatures* which form the basis of all Classical architecture. All the parts have a traditionally established relationship with one another and each *Order* has its specific decorative themes. The five *Orders* are: Tuscan, Doric, Ionic, Corinthian, and Composite.

OVOLO A convex *moulding* with a *profile* resembling half an egg, sometimes carved with a pattern of alternating eggs and arrowheads or darts.

PAVILION
1. A distinct part of a building, usually on a corner or at the end of a long façade.
2. Also a small isolated building.

PEDESTAL A decorative block below a *column*. A similar design for supporting statues.

PEDIMENT
1. Originally the shallow triangular *gable* of a temple, specifically including a sloping *cornice* at the verge, or edge, of the roof.
2. Also the same design dissociated from its original location, such as over a window or *aedicule*.

PENDENTIVE A concave *spandrel* starting at a point in an angle between two walls and becoming wider to support the rim of a dome or a *drum*, so filling the residual space created when a circular dome is located in a square or polygonal plan.

PIANO NOBILE From the Italian, great or noble floor. The principal floor of an Italian palace containing the state rooms and usually the first floor above the ground or basement floor. It has higher ceilings and enhanced exterior and interior decoration. The term has been extended to other building types.

PIER An isolated rectangular section of supporting wall.

PILASTER A semi-circular or flat partial *column* attached to and projecting from a wall. When a projection is semi-circular and exceeds half the width of a *column*, it is an engaged *column*.

柱基	标记墙底部的水平特征，或是一个上面带有柱式、雕像、瓮或其他独立物体的方块。
入口	原指柱廊，现在特指带有柱子的门廊。
轮廓	当直线切割细部设计或展示其剖面时装饰线脚或其他细部设计的形状。
鼓突	大的凸面凸起，或垫石状装饰线脚。通常为雕带。
外墙角	石块或假石块，通常为粗琢，位于建筑一角，通常每一块长短交替。
墙面凸出部分	檐部的一部分，沿着檐部向外，朝相似的独立柱子方向凸出。
拱肋	1. 一个拱或者薄结构，用于支撑拱顶或穹顶，并且从拱腹向下延伸。 2. 或者是创造出花格镶板的完全装饰性特征。
圆花饰	规范化的花朵装饰。
粗琢	粗糙表面的石头。接缝处全部或部分凹陷，或其他材料模仿相同细部设计。
凹弧边饰	较深的凹陷装饰线脚。
拱腹/底面	1. 任何凸出或者架高的装饰特征的底面，例如横梁。 2. 同样也是天花板的另一种说法。
拱肩	当拱设置于长方形开口内部时，其上的墙面区域。
起拱线	拱升起的水平线。
楼梯斜梁/腰线	1. 楼梯斜梁：楼梯台阶上的倾斜支撑。同时也是一座建筑物中的水平线，通常带有装饰并且凸出。 2. 腰线：一条凸出的石或砖线称为腰线。
垂花饰	又称垂花雕饰或花冠。一束花朵、水果或其他物体聚集在一起，由丝带捆扎并垂挂于两个支撑物之上。
柱脚圆盘线脚	大型凸出装饰线脚，常见于柱础。
横梁式结构	一种建筑体系，由垂直支柱或柱子以及水平过梁、横梁或檐部构成，具有希腊与某些罗马建筑特征。
格栅	格构式栅栏或栏杆，材质随意，大部分应用于古代。

PLINTH	A horizontal feature which marks the lower part of a wall or a block on which stands a colume, statue, urn, or other free-standing object.
PORTICO	Originally a *colonnade*, but now specifically a porch with *columns*.
PROFILE	The shape of a *moulding* or other detail shown as a section or straight cut through the detail.
PULVINATED	A large convex bulging, or cushion, *moulding*. Usually a *frieze*.
QUOIN	Block or simulated block of stone, often *rusticated*, on the corner of a building and usually in alternating long and short courses on each face.
RESSAUT	A section of *entablature* projecting outwards from a continuous entablature over a similarly isolated *column*.
RIB	**1.** An arch or thin structural member supporting a vault or dome and projecting down from the *soffit*. **2.** Also used as a wholly decorative feature and to create *coffering*.
ROSETTE	A formalised decorative representation of a flower.
RUSTICATION	Rough-faced stone, stone with all or some of the joints recessed, or the same detail simulated in other materials.
SCOTIA	A deep concave *moulding*.
SOFFIT	**1.** The underside of any projecting or overhead feature such as a beam. **2.** Also another word for a ceiling.
SPANDREL	The area of wall above an arch when it is set inside a rectangular opening.
SPRINGING	The level from which an arch rises.
STRING	**1.** The sloping supports for the steps of a stair. **2.** Also a horizontal line on a building, generally decorative and often projecting. A projecting course of stone or brick is a string course.
SWAG	Also known as a festoon or garland. A bundle of flowers, fruit, or other objects gathered together, draped from two supports and tied with ribbons.
TORUS	A large convex *moulding*, most often found at the base of *columns*.
TRABEATION	The system of vertical posts or *columns* and horizontal lintels, beams or *entablatures* characteristic of Greek and some Roman architecture.
TRANSENNA	A latticed fence or *balustrade*, in any material, most frequently used in antiquity.

三陇板	多立克雕带中垂直凹槽装饰细部，与板间平面交替出现。
山墙饰内三角面	山花内部的三角形空间或墙面，或起拱线之上，拱内的半圆形空间或墙面。
拱顶	弧形屋顶或天花板。
体量	建筑物、房间或庭院所包含的空间。
螺旋饰	旋涡或卷轴状装饰，与爱奥尼亚柱式柱头相关，并且带有卷轴状托架。
拱砌砖	楔形砖或石块，与楔形砖或石块一同从拱中心向外辐射，形成拱结构。

TRIGLYPH	A vertically grooved repetitive detail in a Doric *frieze*, alternating with *metopes*.
TYMPANUM	The triangular space or wall inside a *pediment*, or the semicircular space or wall inside an arch above the line of *springing*.
VAULT	A curved roof or ceiling.
VOLUME	The space contained by a building, room, or courtyard.
VOLUTE	A spiral or scrolled decoration associated with the *column capital* of the Ionic *Order* and with scrolled brackets.
VOUSSOIR	A wedge-shaped brick or stone that, with others, radiates from the centre of an arch and forms its structure.

索引 / INDEX

Words for which a definition can be found in the glossary are indicated by bold type.